*Springer Series in Molecular Biology*

*Series Editor: Alexander Rich*

# Yeast Genetics

## *Fundamental and Applied Aspects*


Edited by
J. F. T. Spencer, Dorothy M. Spencer, and
A. R. W. Smith

With Contributions by

E. A. Bevan, B. L. A. Carter, I. W. Dawes, I. H. Evans,
S. Fogel, R. F. Freeman, J. C. Game, C. Gjermansen,
S. Holmberg, J. R. Johnston, E. W. Jones,
M. C. Kielland-Brandt, C. P. Kurtzman,
J. G. Litske Petersen, K. Lusnak, S. A. Meyer,
D. J. Mitchell, R. K. Mortimer, T. Nilsson-Tillgren,
J. F. Peberdy, H. J. Phaff, J. R. Piggott, G. H. Rank,
H. P. Reader, A. J. Robertson, I. Russell, R. Snow,
G. G. Stewart, E. F. Walton, D. Wilkie


With 49 Figures


Springer-Verlag
New York Berlin Heidelberg Tokyo

J. F. T. Spencer
Thames Polytechnic
Wellington Street
London SE18 6PF
England

Dorothy M. Spencer
Goldsmiths' College
University of London
London SE14 6NW
England

A. R. W. Smith
Thames Polytechnic
Wellington Street
London SE18 6PF
England

*Series Editor*:
Alexander Rich
Department of Biology
Massachusetts Institute of Technology
Cambridge, Massachusetts 02238 U.S.A.

Sponsoring Editor: Philip Manor
Production: Richard Ruzycka

**Library of Congress Cataloging in Publication Data**
Main entry under title:

Yeast genetics.

(Springer series in molecular biology)
Bibliography: p.
1. Yeast fungi—Genetics. 2. Fungi—Genetics.
I. Spencer, J. F. T. II. Spencer, Dorothy M.
III. Smith, A. R. W. IV. Series. [DNLM: 1. Yeast—Genetics. QW 180.5.Y3]
QK617.5.Y44 1983 589.2′33 83-10545


Softcover reprint of the hardcover 1st edition 1983


Typeset by University Graphics, Inc., Atlantic Highlands, New Jersey.
Printed and bound by Halliday Lithograph, West Hanover, Massachusetts.

9 8 7 6 5 4 3 2 1

ISBN-13: 978-1-4612-5493-5 e-ISBN-13: 978-1-4612-5491-1
DOI: 10.1007/978-1-4612-5491-1

# Series Preface

During the past few decades we have witnessed an era of remarkable growth in the field of molecular biology. In 1950 very little was known of the chemical constitution of biological systems, the manner in which information was transmitted from one organism to another, or the extent to which the chemical basis of life is unified. The picture today is dramatically different. We have an almost bewildering variety of information detailing many different aspects of life at the molecular level. There great advances have brought with them some breath-taking insights into the molecular mechanisms used by nature for replicating, distributing and modifying biological information. We have learned a great deal about the chemical and physical nature of the macromolecular nucleic acids and proteins, and the manner in which carbohydrates, lipids and smaller molecules work together to provide the molecular setting of living systems. It might be said that these few decades have replaced a near vacuum of information with a very large surplus.

It is in the context of this flood of information that this series of monographs on molecular biology has been organized. The idea is to bring together in one place, between the covers of one book, a concise assessment of the state of the subject in a well-defined field. This will enable the reader to get a sense of historical perspective—what is known about the field today—and a description of the frontiers of research where our knowledge is increasing steadily. These monographs are designed to educate, perhaps to entertain, certainly to provide perspective on the growth and development of a field of science which has now come to occupy a central place in all biological phenomena.

The information in this series has value in several perspectives. It provides for a growth in our fundamental understanding of nature and the manner in which living processes utilize chemical materials to carry out a variety of activities. This information is also used in more applied areas. It promises to have a significant impact in the biomedical field where an understanding of disease processes at the molecular level may be the capstone which ultimately holds together the arch of clincial research and medical therapy. More recently in the field of biotechnology, there is another type of growth in which this science can be used with immense practical consequences and benefit in a variety of fields ranging from agriculture and chemical manufacture to the production of scarce biological compounds for a variety of application.

This field of science is young in years, but it has already become a mature science. These monographs are meant to clarify segments of this field for the readers.

Cambridge, Massachusetts

Alexander Rich
*Series Editor*

# Preface

## The Dissector's Guide to the Genetic Galaxy

The object of this book is to provide a review of current knowledge in yeast genetics in terms intelligible to the research worker in the field and to biologists and geneticists wishing to use yeasts as a research tool, and so desiring a comprehensive overview of the field along with a detailed list of relevant references. It should be equally useful to the genetic engineer and the yeast breeder who wish to have an up-to-date volume that can be used as an aid in developing new yeast strains for the brewing, wine making, distilling, or baking and food yeast industries or for yeasts that can be used for production of hormones and other proteins not normally produced by these organisms. With a few exceptions we have not attempted to describe techniques, and we have assumed a basic knowledge of genetics, and in particular fungal genetics, on the part of the reader.

The book is divided into four sections, the first and largest of which is mainly devoted to chapters concerned with the nuclear genome in yeast. The authors and editors assume a certain amount of fundamental knowledge of the life history and elementary genetic behavior of *Saccharomyces cerevisiae,* of its relation to other fungi (i.e., that it belongs to the group of hemi- or half-ascomycetes, that it produces spores in a naked ascus without benefit of ascocarp, that the sporulation process occurs in diploid cells or cells of higher ploidy, but that the organism can readily exist as haploid cells of opposite mating type (i.e., sex exists in yeasts), which arise from single spores, and that the yeast ascus can be dissected and the spores separated and cultivated as haploid clones, so that all of the products of a single meiotic event can be recovered and characterized). The fact that some yeast strains cheat, as it were, by carrying a gene for homothallism, which brings about diploidization of some or all single-spore clones isolated from these strains, makes genetic analysis more complex but does not prevent it, and the phenomenon may even be useful to the intelligent geneticist in unraveling some aspects of the genetic behavior of the organism. The reader may also recall that yeasts are somewhat uncooperative in failing to possess chromosomes that can be stained and made visible under the light microscope, so that they can be counted and their behavior observed during mitotic and meiotic cell division, so that yeast chromosomes have been defined in the first instance as linkage groups. However, even these mysteries are now yielding to the skill of the electron microscopist, and the behavior of the yeast chromosomes can be observed in part at the level of the DNA strands made visible.

The yeast cell division cycle has been a topic of absorbing interest, especially in recent years, since the pioneering work of Hartwell (1974), whose genetic analysis was so successful that at one time he gloomily claimed to be the discoverer of "34 genes and not one gene product among them" (Hartwell 1973, overheard). The subject has since been the topic of numerous reviews (Nurse 1981, for example), but progress in the field has shown no signs of slowing, and Carter *et al.* (Chapter 1) present a great deal of new and significant information.

Meiosis and sporulation constitute an alternative cycle in the life of yeasts, triggered in the case of *S. cerevisiae* by the diploid (or polyploid) cell being presented with special and somewhat adverse conditions. Much more is known now of the nature of mating in yeasts, its hormonal control, and the onset of meiosis, with subsequent spore formation, than was known as recently as 10 or 15 years ago. Dawes and his collaborators have been responsible for a significant part of the progress in this field, which he describes in the second chapter of this book.

While attempts have been made to find analogies between the yeast and the bacterial spore, one must draw a long bow to bridge the gap between the two. Unlike the armor-plated bacterial spore, which is a mechanism for survival, the yeast spore is slightly more resistant to heat than the vegetative cell, it is a little more resistant to drying, and it is more resistant to ether killing, a characteristic that Dawes and Hardie (1974) turned to account as a means of finding and isolating the few viable spores produced by some of the less cooperative industrial yeast strains. However, the yeast cell appears to view its spore less as a fortified shelter against adverse conditions than as the end result of a process for reshuffling its DNA, since recombination appears now to be an essential and normally indispensable step in meiosis and leads to the appearance of new combinations of characters in the resulting progeny.

Fogel and co-workers discuss this phenomenon in molecular detail in Chapter 3. They describe the fundamental nature of the recombination and gene conversion processes, as a unitary phenomenon underlying the genetic behavior of the yeast cell. Fogel scans the genetic galaxy in his study of the phenomenon of gene conversion and gives the reader an insight into the process and its significance that it would be presumptuous of an editor to summarize or subject to commentary. However, it is desirable to bring to the reader's attention the comments made by Fogel on the fundamental (i.e., vital) importance of the proper understanding of the theory and techniques involved in the phenomena of gene conversion and its study to any investigator who is intent on the practical applications of transformation, somatic hybridization, and similar techniques to the production of hormones, enzymes, and other products of considerable interest to man but of no particular interest to the yeast itself.

Two topics not mentioned are mitotic recombination and yeast cytology. This is not to minimize their importance, but simply to admit a failure in the persuasive powers of the senior editor. However, mitotic recombination proceeds by the same basic mechanism as meiotic recombination and gene conversion, that is, through the formation of Holliday structures and their resolution. As Fogel points out, the subject has been discussed in recent reviews (Kuntz and Haynes 1981, for instance) and to a considerable extent by Fogel himself.

The practical importance of genetic studies by mitotic recombination is more clearly visible when the investigator is dealing with those residents in the garbage can of yeast taxonomy, which do not possess a sex life known to man, the members of the genus *Candida.* Whelan *et al.* (1980, 1981a,b) and Sarachek *et al.* (1981) have used artificially induced mitotic recombination to demonstrate the existence of naturally occurring recessive mutations in many strains of the pathogenic species *Candida albicans.* Further studies on the subject may lead to a better understanding of the nature of the pathogenicity of the organism, which has long been a nuisance to mankind, and which, with some other such species, becomes a greater danger to mankind with the increasing use of immunosuppressive drugs. In addition, with the development of improved techniques for obtaining somatic hybrids between (perhaps) unrelated yeast species, the use of mitotic recombination to determine the genetic nature of these hybrids may assume increasing importance.

Yeast cytology is a similar case. No volunteer appeared to discuss this topic. At first glance, perhaps this is a serious omission, since the genetic behavior of the cell is governed to a considerable extent by its physical structure. On the other hand, the physical structure is governed by its genetic nature, and that is the subject covered in splendid detail by the authors of the chapters in this book.

The chapter by Fogel *et al.* is followed by one concerned with radiation sensitivity in yeast and repair of genetic damage by UV and x-irradiation. Game gives a remarkably lucid discussion of a complex process, which, of course, is based on the same fundamental foundation in the yeast DNA as the processes of recombination and gene conversion, described in detail by Fogel. The practical importance of this chapter to the geneticist making use of mutants for research or for use in breeding programs for industry is, of course, obvious.

What is a nice chapter like Kurtzman's doing in a book like this? The reason is both simple and profound. Yeast taxonomy and yeast genetics, after a long separation, are on converging paths, since they are based ultimately on the same material, the nuclear DNA. (So far, the nature of the mitochondrial DNA, while differing from species to species, especially in size, has not been used as a taxonomic criterion.) The criteria for yeast classification have progressed from the morphological, through the restricted physiological tests described by Lodder and Kreger-van Rij (1952) and the extended physiological tests suggested by Wickerham (1951), to increasingly sophisticated immunological tests and determination of the composition of the DNA, to the ultimate criterion of the similarity in the structures of the DNA itself, according to the degree of homology between the DNAs of different species. Fingerprinting methods such as determination of the spectrum of soluble proteins or of the NMR spectra of the individual cell wall mannans (Spencer and Gorin 1970) are valuable tools for the *identification* of yeast isolates, but the ultimate decision on the relatedness of different yeast species rests with the nature of the DNA itself, whether determined by hybridization methods, as used by the taxonomist, or by "classical" hybridization by the geneticist. Kurtzman *et al.* describe the methods used by the taxonomist, for the benefit of the geneticist. The genetic engineer may also find the information in this chapter valuable, as the search spreads wider for a source of new genes to introduce characters initially foreign to *Saccharomyces cerevisiae,* to give new combinations and characters to suit the needs of the biotechnologists.

With the chapter on proteinases (Elizabeth Jones, Chapter 6) the focus of the genetic spotlight shifts from the processes occurring in the yeast nucleus. There are numerous proteases, peptidases, and protease inhibitors in the cytoplasm of the yeast cell, and their function has been the subject of considerable speculation, but these have not produced anything approaching a surfeit of information. Dr. Jones describes these proteinases and inhibitors in detail and discusses their genetic control with the same care. She devotes a considerable part of her chapter to a searching discussion of the role of the proteinases, peptidases, and inhibitors in the functioning of the cell and in the process slays a few enchanting theories and lays them to rest.

Flocculation is a cell-surface phenomenon that is important in brewing, especially in some of the newer processes, and that is not completely understood. The genetic control of the phenomenon is better understood than the chemical or biochemical nature of the process itself, as can be seen from the contribution by Johnston and Reader (Chapter 7), who have themselves contributed greatly to the understanding of the nature of the FLO genes. However, though the chromosomal genes controlling flocculation are well mapped and understood, the contribution of the mitochondrial genes to control of the flocculation process is much less well understood. It is well known that *petite* mutants of flocculent strains often lose the ability to flocculate, though the reason for this is not known. It has also been observed (Spencer *et al.* 1980) that some flocculent strains may lose their ability to flocculate when they are converted to the chloramphenicol- or erythromycin-resistant mutants. When these strains were sporulated and single-spore clones were isolated, it was observed that those segregants that had lost their antibiotic resistance had generally again become flocculent. Study of the phenomenon was complicated by the fact that not all the antibiotic-resistant mutants lost a significant degree of flocculence, and of those that had, many failed to sporulate. Further study of this phenomenon may lead to a better understanding of the interaction of the nuclear and mitochondrial genomes, at least where phenomena at the cell membrane and cell surface are concerned.

If the nucleus and its DNA lie at the center of the yeast world, the cell membrane defines the boundaries of its system. Metabolites must cross the membrane on the way into the yeast cell, and cellular products and enzymes, for instance, cross it on the way out, a fact of great importance to the student of yeast physiology and metabolism and to the industrial producer of yeast and its metabolites. The nature of the membrane lipids is reasonably well known, but much remains to be learned of the structure and function of the protein inclusions, their role in material and energy transport, and the genetic control of their functions. For instance, the precise site of attachment of membrane-bound invertase has recently been investigated by means of binding of membrane fractions to microspheres, coated with antisera and concanavalin A (Maurer and Muhlethaler 1982). The recent discovery of yeast strains that, when converted to the petite form, fail to utilize sucrose, may throw further light on this problem and others associated with membrane-associated enzymes. Rank (Chapter 8) gives us a picture of the structure and composition of the yeast plasma membrane, and some insight into its functioning, a necessary foundation on which to build any investigation of the genetic systems related in any way to the role of the membrane systems within the yeast cell.

Freeman and Peberdy (Chapter 9) concern themselves with one of the practical aspects of the cell membrane and its relationship to genetic manipulation, the formation of somatic hybrids by protoplast fusion. As a means of improving brewing yeasts, the process has had a checkered history. At one time it was regarded with optimism as a "quick fix" technique for producing new and improved yeast strains, and gloom and discouragement set in when it was realized that it was no substitute for a thorough and careful definition of what constituted the perfect brewing yeast (or the perfect wine yeast—see Snow, concluding remarks, Chapter 14), and a simiiar thorough and careful study of the genetic control of the characteristics of such a yeast. As a result of the hasty attempt to circumvent the need for good genetic investigation, the hybrids obtained by protoplast fusion, like those obtained from classical genetic crosses, were generally inferior to the parent brewing strains. However, a better understanding of the genetic behavior of brewing yeasts, and of the factors that characterize a desirable brewing yeast, allow the development of improved brewing strains by this method.

In the meantime, protoplast fusion has developed into a valuable tool for the fundamental study of factors controlling hybridization between unrelated species and genera of microorganisms.

As Freeman and Peberdy point out, the potential of the techniques for obtaining improved hybrids by fusion of more distantly related species has scarcely been tapped. Somatic hybrids between different species of *Kluyveromyces* (Whittaker and Leach 1978), between *Saccharomyces cerevisiae* × *Saccharomycopsis lipolytica* (de van Broock *et al.* 1980), and between other yeast species have been obtained. Recently we have developed a method for obtaining interspecific and intergeneric somatic hybrids by protoplast fusion (Spencer and Spencer 1982), using a petite strain of *Saccharomyces diastaticus* as one parent. Hybrids with *Candida pseudotropicalis, Saccharomyces rosei, Saccharomyces montanus, Pichia membranaefaciens,* and *Hansenula capsulata* were obtained. The hybrids seemed to form readily, and respiratory-competent strains having characteristics from both parents were obtained. The sporulating cultures obtained by fusion of *H. capsulata* × *S. diastaticus*($p^-$) were particularly interesting, as they can be used for determination of the segregation of characters from the two parents, and if sporulating cultures of this nature can be obtained from crosses with other *Hansenula* strains, carrying auxotrophic markers, it should become possible to determine many of the factors controlling integration of the chromosomes of the two unrelated nuclei. The hybrids appeared to be quite stable on the appropriate media, the use of starch as a carbon source maintaining the hybrids in a condition permitting sporulation. Respiratory-competent yeast strains have been recovered from fusion of *petite* yeasts and plant cell protoplasts as well (Spencer and Spencer, unpublished data).

The relative ease with which the hybrids were formed, and their apparent stability on appropriate media, may indicate that suppression of the mitochondria in one parent is necessary for the formation of stable interspecific and intergeneric hybrids, obtained in this way. This view is confirmed by the results obtained by Weide *et al.* (1982), who showed that the presence of mitochondria from both parents was apparently detrimental to hybrids formed between mouse and human cells. Investigators using protoplast fusion to form interspecific and intergeneric yeast hybrids may find not only that stable hybrids may be more readily formed by the preceding techniques, but that the method may permit further elucidation of the basis for interaction between the nucleus and the mitochondria.

Wilkie (Chapter 10) and Evans (Chapter 11) describe the genetics and biochemistry of the mitochondrial genome. Thanks to the discovery of mitochondrial mutants to antibiotic resistance, which aided greatly in the genetic analysis of this genome, the nature of the genetic control of a number of respiratory processes that take place in the mitochondrion is much better understood. Thanks to a wide range of investigations using restriction enzymes, the chemical nature and sequence organization of the mitochondrial DNA is also well understood. Evans describes the structure and function of the mitochondrial genome in masterly detail and presents these details in a manner that should make them invaluable to any investigator studying the genetics and function of the yeast mitochondrion. However, the relationship between the mitochondrial genome and a number of phenomena at the cell surface is only beginning to be comprehended. Besides the phenomenon of flocculation, previously mentioned, occurring at the extreme exterior of the yeast cell, transport of various sugars into the yeast cell is affected by mutations in the mitochondrial DNA, and transport of proteins (invertase, for instance) out may be under the control of the mitochondrion, though not directly. The *petite* mutation and some mutations to antibiotic resistance affect flocculation considerably in some yeast strains and influence sugar transport into the cell. The *petite* mutation apparently determines permeation of the membrane by these sugars, since mitochondrial activity is required by the cell for this phenomenon. We have recently observed that the same effect as is produced by the *petite* mutation is sometimes produced by subjecting the cell to anaerobiosis, fermentation of sucrose, for instance, being abolished both by the *petite* mutation and by anaerobiosis. In other words, in this case, loss of the respiratory function brings about loss of the ability to metabolize this sugar, no matter what method is used to inhibit respiration. It must be kept in mind that although most other sugars are transported into the cytoplasm before being metabolized, sucrose is not, invertase being transported outward, so that transport of a much larger (protein) molecule is involved than in the case of galactose or maltose, for instance. It is interesting to note also that although invertase transport, in a few of our strains, is affected by the *petite* mutation (Spencer *et al.* unpublished data), transport of the protein toxin produced by killer strains is not. Protease export is also affected by mutation to antibiotic resistance in some strains (Spencer and Spencer 1979).

The killer phenomenon and its genetic control, discussed by Mitchell and Bevan (Chapter 12), once an interesting curiosity, is a characteristic of some yeast strains that is becoming of greater importance. Besides the fact that it is a phenomenon that is not limited to *S. cerevisiae* alone, it is a manifestation of the activity of a whole series of dsRNA viruslike particles (mycoviruses), found in filamentous fungi as well as in yeasts, and responsible for such phenomena as the determination of virulence in strains of *Endothia parasitica*, the chestnut blight fungus. The killer phenomenon is also of some economic importance, as witness the fact that some years ago a major London brewery was forced to close down its operation for a time because of a contamination by a killer yeast strain that replaced the normal brewing strain and imparted a foul flavor to the beer. At the time, the senior editor of this volume proposed to attempt to introduce the killer character into British brewing yeast strains, to "immunize" the production strains against further invasions (killer strains are not affected by the toxin, and the killer character itself has no effect on beer flavor) and to prevent future contamination by nonkiller "wild" *Saccharomyces* strains. This suggestion was received with all the enthusiasm that would be accorded an assault upon motherhood, and the project was dropped. The merits of the idea were recognized independently, of course, by the Japanese investigators, and production strains of sake yeasts and wine yeasts, protected by possession of the killer character, were soon in wide use in Japan. By now the brewers, too, have seen the light.

The last section of the book is devoted to the application to industrial processes of the principles enunciated by the authors of the preceding chapters. Kielland-Brandt *et al.* (Chapter 13) provide a superb description of the use of genetic techniques old and new for the improvement of brewing yeasts, and Snow (Chapter 14) gives an equally readable discussion of the use of these techniques for the solution of problems in the genetic improvement of wine yeasts.

Finally, Stewart and Russell give an excellent account of the genetic basis for development of yeasts capable of the utilization of pentoses, hexoses, and other sugars, and starches and dextrins in industrial processes, which should be required reading for all investigators in fermentation and biotechnology laboratories everywhere.

What of the future? Yeasts are becoming even more important to mankind than in earlier times, as from the traditional food and beverage industries; they are being adopted by the glamour girls of the modern biotechnological world, the gene manipulators. After a brief flirtation with the seductive bacterial forms represented by *E. coli,* these companies are finding that there are definite advantages to what may be a lasting union with the solid citizens of the microbial world, the yeasts, which have no murky and pathogenic past to crop up to plague the users with visitations from officials of regulatory bodies, concerned with ill effects from past undesirable associations with disease-producing microbial outlaws. Progress is now being made in transforming yeasts with plasmids carrying DNA sequences including, for instance, the centromere, which causes the incorporated plasmid to behave as a minichromosome and to be retained and replicated as a stable part of the normal yeast genome (Hsiao and Carbon 1981; Fitzgerald-Hayes *et al.* 1982), so that the introduced character segregates in crosses as do those genes that have been long-standing members of the yeast genome, resident in the yeast nucleus. Still to be solved is the problem of the export of the proteinaceous products of the introduced gene, but as we have previously noted, this problem is probably not impossible of solution, and may be solved by further study of those nuclear-mitochondrial interactions that relate to control of the cell membrane and cell surface.

It should not be overlooked that the methods of classical yeast genetics have many things yet to offer the breeder of industrial yeast strains. Emeis (1971), for instance, has successfully hybridized *S. diastaticus* with a brewing yeast strain and selected from the segregants a strain that metabolizes starch and dextrins but that lacks the foul flavor produced by normal *S. diastaticus* strains. Goodey and Tubb (1982) have likewise used classical genetic techniques to eliminate the gene controlling formation of 4-vinylguaiacol from the genome of a brewing yeast hybrid that metabolizes starch. In other words, the brewing yeast, in particular, once regarded as an impregnable, nonmating, nonsporulating, Victorian fortress, inaccessible to the geneticist, has been shown to mate and to produce enough viable spores to permit genetic analysis and breeding of new strains (Anderson and Martin 1975; Spencer and Spencer 1977). Dawes and Hardie (1974) have devised a method for separating viable spores from the horde of vegetative cells associated with them. It may be possible to achieve the goal reached by the plant breeders long ago of a stable of breeding strains carrying desirable genes in an infinity of groupings, so that the industrial user may define the characters he desires and select the appropriate breeding strains from the stock to form the perfect yeast for his specific purpose. Behind these breeding stocks will be the clone banks, containing the genes necessary for the breeding of yeasts for industrial production of enzymes, hormones, and many other pharmaceutically valuable products, previously only obtainable by laborious extraction from raw materials of plant and animal origin. The authors of the chapters in this book point the way to the achievement of these goals.

## References


Anderson, E., Martin, P. A. (1975). The sporulation and mating of brewing yeasts. J. Inst. Brew. 81:242–247.

Dawes, I. W., Hardie, I. D. (1974). Selective killing of vegetative cells in sporulating yeast cultures by exposure to diethyl ether. Mol. Gen. Genet. 131:281–289.

de van Broock, M. R., Sierra, M., de Figueroa, L. (1980). Intergeneric fusion of yeast protoplasts. In: Current Developments in Yeast Research, edited by G. G. Stewart and I. Russell. Proc. 5th Int. Yeast Symposium, July 20–25, London, Ont., Canada.

Emeis, C.-C. (1971). A new hybrid yeast for the fermentation of wort dextrins. Am. Soc. Brew. Chem. Proc., 58–62.

Fitzgerald-Hayes, Molly, Buhler, J. M., Cooper, T. G., Carbon, J. (1982). Isolation and sub-cloning analysis of functional centromere DNA (CEN11) from *Saccharomyces cerevisiae* chromosome XI. Mol. Cell Biol. 2:82–87.

Goodey, A. R., Tubb, R. S. (1982). Genetic and biochemical analysis of the ability of *Saccharomyces cerevisiae* to decarboxylate cinnamic acids. J. Gen. Microbiol. 128:2615–2620.

Hartwell, L. H. (1974). *Saccharomyces cerevisiae* cell cycle. Bacteriol. Rev. 38:164–198.

Hsiao, C.-L., Carbon, J. (1981). Characterization of a yeast replication origin (ars2) and construction of stable minichromosomes containing cloned yeast centromere DNA (CEN3). Gene 15:157–166.

Kunz, B. A., Haynes, R. H. (1981). Phenomenology and genetic control of mitotic recombination in yeast. Ann. Rev. Genet. 15:57–89.

Lodder, J., Kreger-van Rij, N. J. W. (1952). The Yeasts, a Taxonomic Study. Amsterdam: North-Holland.

Maurer, A., Muhlethaler, K. (1982). Isolation and localization of plasma membrane-bound invertase in yeast *(Saccharomyces cerevisiae)*. Eur. J. Cell Biol. 26:219–227.

Nurse, P. (1981). Genetic analysis of the cell cycle. In: Genetics as a Tool in Microbiology, edited by S. W. Glover and D. A. Hopwood, Soc. Gen. Microbiol. Symp. 31, Cambridge Univ. Press, pp. 291–315.

Sarachek, A., Rhoads, D. H., Schwartzhoff, R. H. (1981). Hybridization of *Candida albicans* through fusion of protoplasts. Arch. Microbiol. 129:1–8.

Spencer, J. F. T., Gorin, P. A. J. (1970). Proton magnetic resonance spectroscopy—an aid for identification and chemotaxonomy of yeasts. Adv. Appl. Microbiol. 13:25–89.

Spencer, J. F. T., Spencer, Dorothy M. (1977). Hybridization of nonsporulating and weakly sporulating strains of brewer's and distiller's yeasts. J. Inst. Brew. 83:287–289.

Spencer, J. F. T., Spencer, Dorothy M. (1979). Osmotic sensitivity and tolerance and proteinase production in a strain of *Saccharomyces*. Z. Naturforsch. 34c:131–135.

Spencer, J. F. T., Spencer, Dorothy M. (1981). The use of mitochondrial mutants in hybridization of industrial yeasts. III. Restoration of mitochondrial function in *petites* of industrial yeast strains by fusion with respiratory-competent protoplasts of other yeast species. Curr. Genet. 4:177–180.

Weide, L. G., Clark, M. A., Rupert, C. S., Shery, J. W. (1982). Detrimental effect of mitochondria on hybrid cell survival. Somatic Cell Genet. 8:15–21.

Whelan, W. L., Magee, P. T. (1981). Natural heterozygosity in *Candida albicans*. J. Bacteriol. 145:896–903.

Whelan, W. L., Partridge, R. M., Magee, P. T. (1980). Heterozygosity and segregation in *Candida albicans*. Mol. Gen. Genet. 180:107–113.

Whelan, W. L., Beneke, E. S., Regus, A. L., Soll, D. R. (1981). Segregation of 5-fluorocytosine-resistant varieties by *Candida albicans*. Antimicrobial Agents and Chemotherapy 19:1078–1081.

Whittaker, P. A., Leach, S. M. (1978). Interspecific hybrid products between the yeasts *Kluyveromyces lactis* and *Kluyveromyces fragilis* by protoplast fusion. FEMS Lett. 4:31–34.

Wickerham, L. J. (1951). Taxonomy of Yeasts. U.S. Dept. Agric. Bull. No. 1029, pp. 1–56.

J. F. T. Spencer
Dorothy M. Spencer
A. R. W. Smith

# Contents

## 1. Genetic Control of Cell Proliferation
*B. L. A. Carter, J. R. Piggott, E. F. Walton*



## 2. Genetic Control and Gene Expression During Meiosis and Sporulation in *Saccharomyces cerevisiae*
*Ian W. Dawes*



## 3. Meiotic Gene Conversion in Yeast: Molecular and Experimental Perspectives
*Seymour Fogel, Robert K. Mortimer, Karin Lusnak*

# Contributors

E. ALAN BEVAN School of Biological Sciences, Queen Mary College, University of London, London E1 4NS, England

B. L. A. CARTER Genetics Department, Trinity College, University of Dublin, Dublin, Ireland

IAN W. DAWES Department of Microbiology, University of Edinburgh, Edinburgh, Scotland

IVOR H. EVANS School of Biological Sciences, Queen Mary College, University of London, London E1 4NS, England

SEYMOUR FOGEL Department of Genetics, University of California, Berkeley, California 94720 U.S.A.

RICHARD F. FREEMAN Patscentre International, Cambridge Division, Melbourn, Herts SG8 6DP, England

J. C. GAME 360 Donner Laboratory, Lawrence Berkeley Laboratory, Berkeley, California 94720 U.S.A.

CLAES GJERMANSEN Department of Physiology, Carlsberg Laboratory, Gamle Carlsberg Vej 10, DK-2500 Copenhagen Valby, Denmark

STEEN HOLMBERG Department of Physiology, Carlsberg Laboratory, Gamle Carlsberg Vej 10, DK-2500 Copenhagen Valby, Denmark

J. R. JOHNSTON Yeast Technology Group, Department of Bioscience and Technology, University of Strathclyde, Glasgow G1 1XW, Scotland

ELIZABETH W. JONES Department of Biological Sciences, Carnegie-Mellon University, Pittsburgh, Pennsylvania 15213 U.S.A.

MORTEN C. KIELLAND-BRANDT Department of Physiology, Carlsberg Laboratory, Gamle Carlsberg Vej 10, DK-2500 Copenhagen Valby, Denmark

C. P. KURTZMAN Northern Regional Research Center, Agricultural Research Service, U.S. Department of Agriculture, Peoria, Illinois 61604 U.S.A.

JENS G. LITSKE PETERSEN Department of Physiology, Carlsberg Laboratory, Gamle Carlsberg Vej 10, DK-2500 Copenhagen Valby, Denmark

KARIN LUSNAK Department of Genetics, University of California, Berkeley, California 94720 U.S.A.

S. A. MEYER Department of Biology, Georgia State University, Atlanta, Georgia 30303 U.S.A.

DIANE J. MITCHELL School of Biological Sciences, Queen Mary College, University of London, London E1 4NS, England

ROBERT K. MORTIMER Department of Biophysics and Medical Physics, University of California, Berkeley, California 94720 U.S.A.

TORSTEN NILSSON-TILLGREN Institute of Genetics, University of Copenhagen, Øster Farimagsgade 2A, DK-1353 Copenhagen K, Denmark

JOHN F. PEBERDY Fungal Genetics and Biochemistry Laboratory, Department of Botany, University of Nottingham, Nottingham, England

H. J. PHAFF Department of Food Science and Technology, University of California, Davis, California 95616 U.S.A.

J. R. PIGGOTT Genetics Department, Trinity College, University of Dublin, Dublin, Ireland

G. H. RANK Department of Biology, University of Saskatchewan, Saskatoon, Saskatchewan, Canada S7N 0W0

H. P. READER Cockburn Smithes & Cia, Lda., 4401 Vila Nova de Gaia Codex, Portugal

A. J. ROBERTSON Department of Biology, University of Saskatchewan, Saskatoon, Saskatchewan, Canada S7N 0W0

I. RUSSELL Labatt Brewing Company Ltd., London, Ontario, Canada N6A 4M3

RICHARD SNOW Department of Genetics, University of California, Davis, California 95616 U.S.A.

G. G. STEWART Labatt Brewing Company Ltd., London, Ontario, Canada N6A 4M3

E. F. WALTON Genetics Department, Trinity College, University of Dublin, Dublin, Ireland

D. WILKIE Department of Botany and Microbiology, University College, London, London, England

# 1

# Genetic Control of Cell Proliferation

B. L. A. Carter,* J. R. Piggott,* and
E. F. Walton*

## Introduction

A number of reviews of the *Saccharomyces cerevisiae* cell cycle have been published recently (Carter 1978; Hartwell 1978; Simchen 1978; Carter 1981; Nurse 1981; Pringle 1981; Pringle and Hartwell 1981). It would be inappropriate to rework the material presented in these reviews and, although some overlap is inevitable, we have tried to present points which are not emphasized in them. Inevitably, a consequence of this is that some topics of interest are discussed only briefly. For instance, we have presented a very short account of results bearing on the temporal and functional coordination of stage specific events because this subject has been dealt with exhaustively in a number of the reviews we have mentioned. We have attempted to concentrate on those areas of cell cycle work which are relevant to the control of cell division and have tried to show that the secret of any successes we have in unraveling the complexities of cell division in yeast will be because the ease of genetic analysis in this organism will permit directed physiological, biochemical, and molecular approaches.

## Genes Involved in Cell Division

If we are to understand the molecular mechanisms involved in cell division, a vital first step is to identify the genes involved in the process. Such genes have been identified by isolating mutants that are altered or defective in cell division. Most of the mutants that have been isolated become blocked at a particular stage of the cell cycle. As mutants blocked in cell division are lethal, their isolation has been dependent on the ability to isolate conditional mutants. Hartwell and his colleagues (for review see Pringle and Hartwell 1981) have isolated a large number of temperature-sensitive (TS) cell division cycle *(cdc)*

*Genetics Department, Trinity College, University of Dublin, Dublin, Ireland.

mutants that grow and divide at 24°C, but arrest at a particular stage of the cell cycle after a shift to 37°C; the stage in the cycle at which arrest occurs is dependent on which cell cycle gene has been mutated. Complementation analysis has been used to determine how many genes are defined by the mutant collection; at present 50 *cdc* genes have been identified.

It is likely that many cell cycle genes remain to be identified. Rai and Carter (1980) have isolated nonsense mutants in a strain carrying a TS amber nonsense suppressor. Out of six nonsense mutants analysed only two had mutations in previously identified genes (i.e., *cdc4* and *cdc20*). The remaining four mutants complemented each other and all the known *cdc* mutants, *viz., cdc1–cdc40* that were available at the time. Thus it is likely that while the isolation of more TS *cdc* mutants may be unrewarding as a means of identifying additional cell cycle genes, alternative approaches such as the isolation of nonsense suppressible *cdc* mutants are promising.

It is probable that new *cdc* mutants can be obtained by selecting for the secondary characteristics associated with some cell cycle mutants. Reed (1980) isolated new *cdc28* alleles and identified three new start genes by selecting for the secondary characteristic of ability to mate at 37°C associated with the *cdc28* phenotype.

Although the isolation of mutants that are blocked in cell division has permitted the identification of genes that are involved in cell division, many of these genes will code for proteins that are vital for successful cell division but are not involved in the regulation of cell division. Most, if not all, of the genes involved in the propagation of DNA synthesis are likely to fall into this category.

Genes that are involved in the regulation of a particular process are usually identified by isolating mutants that are altered rather than defective in the phenotype associated with a process. Carter and Sudbery (1980) have used this approach to isolate mutants that produce a bud when the parent is approximately half the size of wild type cells and it is possible that these mutants are altered in the rate at which cells initiate the cell cycle. Another gene has been identified by a mutation that results in continued cell division in circumstances where wild type cells arrest in the G1 phase of the cell cycle (Sudbery *et al.* 1981). Wild-type yeast cells like those of most eukaryotes arrest in G1 as nutrients become depleted in batch culture; the mutant continues to proliferate. Eventually the medium becomes completely exhausted and cells stop dividing, but they do not show specific cell cycle stage arrest; they arrest at whatever stage of the cycle they have reached at the time of medium exhaustion.

In addition to genes that have a regulatory as well as a catalytic function at the same stage of the cycle, it is possible that some genes are involved in two distinct phases of the cycle; indeed we present evidence later that *cdc28* is such a gene. If the execution points for the two functions are close together in the cycle, the mutant will show homogeneous cell cycle arrest; the vast majority of cells will arrest at the terminal phenotype of the first execution point and only a few will arrest at the second (Fig. 1.1). The mutant will be identified as defective in a cdc gene but the role of the gene in two stages of the cycle will go unappreciated; the *cdc28* gene has been an example of this situation. If the execution points are widely spaced and the terminal phenotypes are morphologically distinct (Fig. 1.1), then the mutant will arrest at two stages of the cycle and quite possibly will not be recognised as a cell cycle mutant.

It is too simple to subdivide genes involved in cell division into those that have a function that moves a cell from, for instance, stage A to stage B and those that are involved in the regulation of this process. There are a number of multifunctional polypeptides in eukaryotes (see review by Kirschner and Bisswanger 1976) and it is likely that genes coding for multifunctional polypeptides are involved in cell division. It is quite possible that such genes have already been identified, albeit unwittingly. Some of the genes identified by TS cell cycle mutations may fall into this category, but only mutations in the A to B function have thus far been obtained. Cell cycle mutants were originally isolated because they blocked at a specific stage of the cell cycle at the restrictive temperature; those mutants that arrested at different stages of the cycle were discarded as being defective in some aspect of general metabolism. It is not difficult to imagine that some types of TS mutants that are involved in the regulation of cell division might exhibit heterogeneous cell cycle arrest at the restrictive temperature.

The use of revertant analysis as a means of investigating cell cycle genes has been curiously neglected. Extragenic suppressors (revertants) of cell cycle mutations should identify genes whose products interact with the product of the original *cdc* gene. In addition, genes involved in alternative pathways of cell division should be readily identified by this approach.

It is important to determine which *cdc* mutations can be externally suppressed and in these cases to elucidate the nature of the external suppression. We have evidence for external suppression of *cdc28* mutations (Piggott and Carter, unpublished observations).

## Gene Products Involved in Cell Division

A major challenge in cell cycle research is to isolate and identify the products of genes involved in cell division. Although progress in this area is likely to be slow, it is an avenue that must be pursued. It is reasonable to expect that the products of genes involved in DNA synsthesis will be the easiest to identify and biochemically characterize because we have considerable information concerning the enzymes involved in this process in other systems. Indeed, the products of two *S. cerevisiae* genes involved in this system have been identified: *cdc9* codes for DNA ligase (Johnston and Nasmyth 1978) and *cdc21* codes for thymidylate synthetase (Game 1976).

The only other cell cycle gene thus far biochemically identified is *cdc19*, which codes for pyruvate kinase (Kawasaki 1979). It is rather surprising that

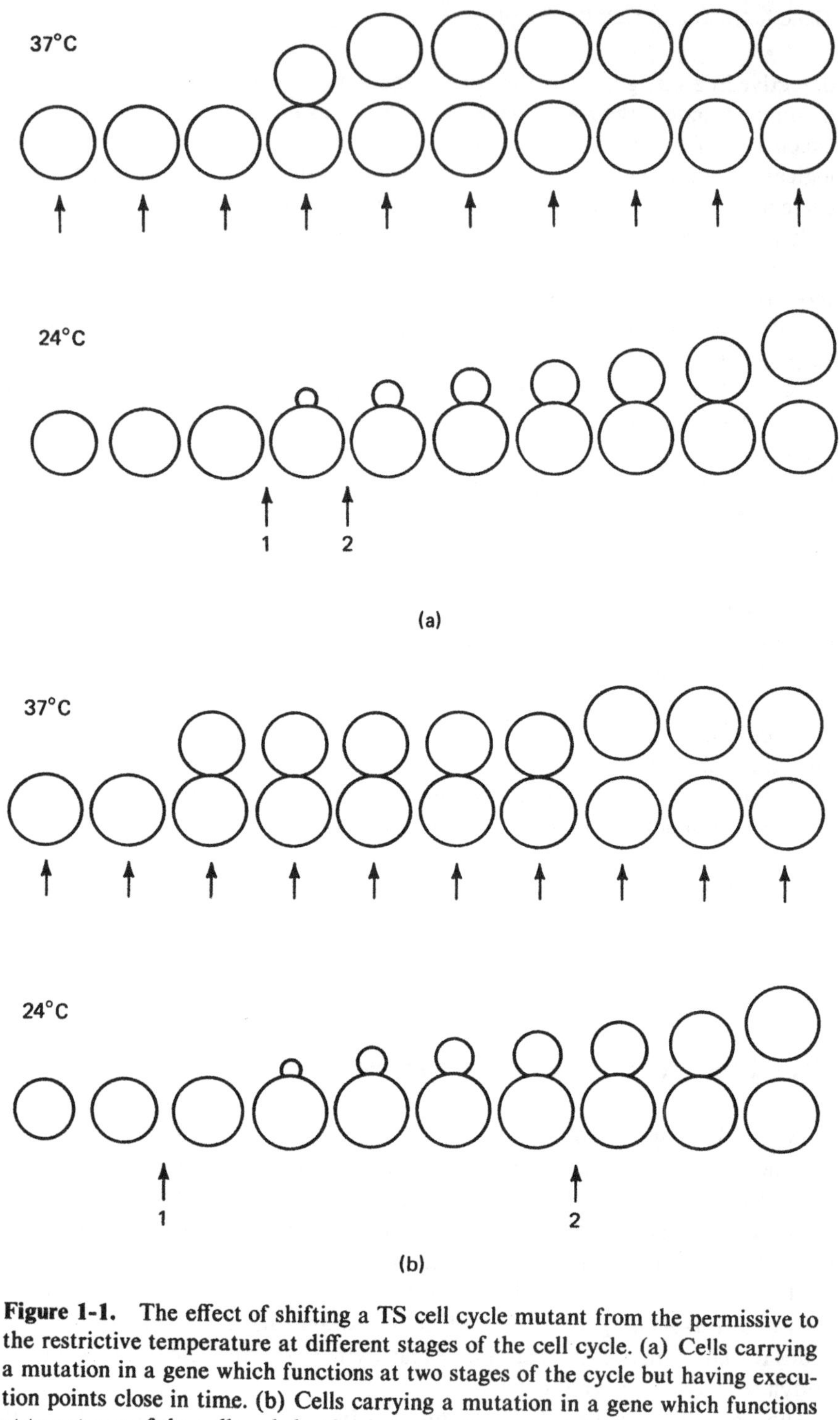


**Figure 1-1.** The effect of shifting a TS cell cycle mutant from the permissive to the restrictive temperature at different stages of the cell cycle. (a) Cells carrying a mutation in a gene which functions at two stages of the cycle but having execution points close in time. (b) Cells carrying a mutation in a gene which functions at two stages of the cell cycle but having execution points widely separated in time. The numbered arrows represent execution points.

a defect in this enzyme should lead to cells arresting in G1, but this may be related to nutrient deficiency effects discussed later.

Two general approaches have been adopted to identify the products of genes involved in cell division. Rai and Carter (1981) isolated nonsense mutations in cell cycle genes in strains carrying a TS amber nonsense suppressor, hoping that it might be possible to identify products by a comparison of the polypeptides made by wild type and mutant strains using two dimensional gel electrophoresis. The rationale is that the polypeptide pattern of a nonsense mutation in a cell cycle gene would lack a spot corresponding to the polypeptide of the wild type gene. It is quite possible that the products of some cell cycle genes will be identified using this approach, but there are potential difficulties. It is vital that the polypeptide can be resolved as a spot and has physicochemical properties distinct from other polypeptides that might otherwise obscure its identification. In addition, the products of some cell cycle genes, particularly those involved in regulation, might be present as only a few molecules per cell and may not be within the limits of detection.

A more direct approach has been adopted by Nasmyth and Reed (1980). Their approach depends on the isolation of the appropriate cell cycle gene, transcribing the DNA and translating the transcription products. Isolation of the *cdc28* gene was accomplished by complementation of a strain carrying a *cdc28* mutation by transformation with plasmid containing wild type yeast chromosomal DNA. Recently, Reed (personal communication) has identified the *cdc28* gene product. It is a polypeptide with a molecular weight of 27,000.

This approach will make it possible to identify other cell cycle gene products. It is, of course, one thing to physically identify a polypeptide, and quite another to identify its biochemical function. Nevertheless, it is difficult to see how we can understand the molecular mechanisms involved in cell division without isolating the gene products involved in the process.

## Temporal Organization of Cell Cycle Events

A small number of cell cycle events has been identified by biochemical and microscopic analyses of cells. A far larger number of events or functions that must be completed if cell division is to occur has been revealed by studies of cell cycle mutants. Although the order in which some of these events occur is probably unimportant, the temporal order of others is vital; to choose an obvious example, a cell must replicate its chromosomes before it segregates them at mitosis.

Hartwell and his colleagues have investigated in some detail how this temporal order is achieved, and their work has been reviewed a number of times and will not be dealt with at length here. The reader is referred to Hereford and Hartwell (1974); Hartwell (1976); Pringle (1978); Pringle and Hartwell (1981); Pringle (1981).

Although it is theoretically possible that a central timer might trigger cell cycle events at particular times, there is no evidence to support the existence of cell cycle clocks, although the behavior of *cdc4* mutants should be investigated further. There is, however, evidence that particular cell cycle events cannot occur without the prior occurrence of other events.

A functional sequence map describes the functional interrelations of cell cycle events. Such a map is shown in Fig. 1.2. Competion of the *cdc28* event

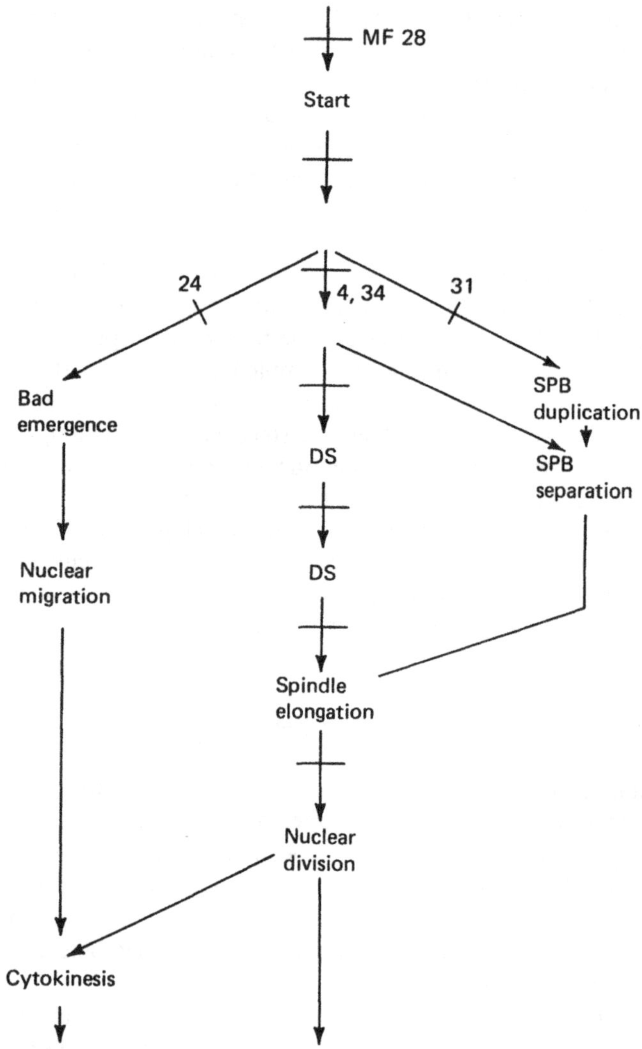


**Figure 1.2.** A functional sequence map. (Abbreviations: MF, mating factor step, cdc genes are represented by numbers; iDS, initiation of DNA synthesis; DS, propagation of DNA synthesis; SPB, spindle pole body (modified from Byers 1981).

is necessary for the initiation of three parallel pathways of mutually independent events. Each of these independent pathways can contain a number of events arranged in a dependent sequence. The pathways converge at cytokinesis, which is dependent on the prior completion of all the events in these pathways. At present we know nothing concerning the biochemical basis of any of these dependencies.

The last point in the cell cycle at which cells of **a** mating type are sensitive to G1 arrest by the mating pheromone $\alpha$ factor, the $\alpha$ factor step, coincides with the *cdc28* execution point. Reciprocal shift experiments (Hereford and Hartwell 1974) have shown that $\alpha$ factor arrested cells do not complete the *cdc28* gene function at 24 °C and *cdc28* cells arrested at 37 °C remain sensitive to $\alpha$ factor when returned to the permissive temperature. This has been interpreted to mean that the $\alpha$ factor step and the *cdc28* function are interdependent (Hereford and Hartwell 1974).

Among Hartwell's original *cdc* mutants, *cdc28* was the only mutant that mated at the restrictive temperature. As the biological function of $\alpha$ factor is to promote zygote formation, these results indicate that $\alpha$ factor and *cdc28*-induced G1 arrest lead to similar physiological conditions, a conclusion that is reinforced by the morphological similarity of cells exhibiting either type of arrest. Experiments in this laboratory, however, suggest that *cdc28* induced G1 arrest is not equivalent to the $\alpha$ factor induced response exhibited by **a** cells. Mutation to $\alpha$ factor resistance does not alter the morphological or cell cycle properties of *cdc28* mutants and arrest at the *cdc28* block does not relieve the sterility of $\alpha$ factor resistant mutants.

In view of this result the interdependence of the $\alpha$ factor step and the *cdc28* step reported by Hereford and Hartwell (1974) cannot be interpreted as implying that the *cdc28* gene product is a component of the $\alpha$ factor response system. Rather, since the completion of "start" is operationally defined by transition of *cdc28* and $\alpha$ factor steps, we must conclude that both kinds of arrest result in continued cell growth in the absence of "start" functions. Viewed in this light, the "interdependence" of the *cdc28* and $\alpha$ factor steps can be seen as the dependence of the "start" event on both *cdc28* function and the absence of $\alpha$ factor in the cell's environment, with *cdc28* and $\alpha$ factor each preventing the completion of "start" by independent mechanisms.

## The Start Concept

Cell division is a cyclic progress. Where does the cycle begin and end? The *cdc28* gene is involved in the earliest gene controlled event of all the *cdc* genes identified. Of itself this does not mean completing the *cdc28* event starts the cell cycle: it could be that it is merely the earliest functioning gene that has thus far been discovered. There is, however, some evidence that completion of the *cdc28* event marks the start of the cell cycle, and this evidence will be examined in some detail.

Completion of the *cdc28* event is necessary to set in train the three parallel pathways of dependent sequences described in Figure 1.2. Perhaps the most seductive reason for considering the *cdc28* event as start was the claim by Hartwell (1974) that cells that have completed the *cdc28* step are commited to cell division, whereas cells that have not completed this event can embark on alternative developmental pathways. Haploid cells can divide, mate, or enter stationary phase; diploid cells can divide, sporulate or enter stationary phase.

When proliferating, haploid or diploid cells deplete one of a number of nutrients in batch culture cells arrest at or prior to the *cdc28* event in G1. Any cell that completes the *cdc28* event as the medium is becoming depleted is able to complete cell division (Johnston *et al.* 1977).

Haploid cells have depleted a nutrient prior to the *cdc28* event can mate, but once this event has been completed they are committed to cell division. The evidence for this is derived from experimemts in which cell cycle mutants were arrested at the restrictive temperature and challenged to mate (Reid and Hartwell 1977). Only those cells arrested at the *cdc28* step could mate; mutant cells arrested at later gene controlled steps could not mate.

It cannot be said for sure that the point of commitment for cell division *vis-à-vis* stationary phase is the *cdc28* event; cells could be arrested at a prior but undiscovered event. The evidence for the *cdc28* event being a point of commitment for division *vis-à-vis* mating is more convincing, as cells arrested at the *cdc28* block could reasonably be expected to have completed hypothetical prior events. The situation is not absolutely clear, however, as Reid and Hartwell (1977) were unable to test whether cells arrested at a subsequent step, the *cdc4* event, could mate rather than divide, as the *cdc4* gene is required for mating as well as cell division.

Hirschberg and Simchen (1977) have challenged diploid cells arrested at various stages of the cycle using cell cycle mutants to sporulate. Somewhat unexpectedly, they found that cells arrested at the *cdc4* step, a gene-controlled event that occurs after the *cdc28* step, sporulated rather than proceeded to mitosis. Cells arrested at subsequent *cdc* steps could not sporulate when challenged, but completed cell division.

In conclusion, the evidence for the *cdc28* step being a point of commitment to cell division is not overwhelming; but the evidence does suggest that commitment does not occur before the *cdc28* step and that once cells have completed this step and the subsequent *cdc4* step, they are committed to mitotic cell division. Although the situation is not as simple as first envisaged (Hartwell 1974), the fact that cells cannot become committed to cell division until they have completed the *cdc28* event does support the view that passing this event starts the cycle.

Yeast cells can be grown at a variety of growth rates by altering media richness, by limiting a particular nutrient such as glucose, or by directly limiting protein synthesis. Whichever of these methods is employed to slow growth rate and lengthen the cell cycle, almost all the increased length can be accounted for by an increase in the time cells spend in the G1 interval prior to the *cdc28* step (von Meyenburg 1968; Hartwell and Unger 1977; Jagadish and Carter 1977, 1978; Slater *et al.* 1977; Carter and Jagadish 1978b; Tyson *et al.* 1979;

Lord and Wheals 1980). It is as if at slower and slower growth rates cells take longer and longer to complete the *cdc28* event, but once this event is completed cells proceed to cell division in a time that is relatively constant. Thus the rate of cell division in proliferating cells is dependent on the rate at which cells complete the cdc 28 event.

If all the evidence for the *cdc28* event is considered, some experiments suggest that start cannot be earlier than the *cdc28* event (commitment to mitosis *vis-à-vis* conjugation), whereas others indicate that it cannot be later (for instance, commitment to mitosis *vis-à-vis* stationary phase and the data from different growth rates).

The central role of the *cdc28* gene in cell division has recently been highlighted by the discovery that it is involved not just in the initiation of DNA synthesis, as previously thought, but also in nuclear division (Piggott *et al.* 1982). We have isolated a mutant that is defective in nuclear division but not in DNA synthesis. This mutant has a mutation at the *cdc28* locus. The fact that it failed to complement *cdc28-1* or *cdc28-6* alleles previously thought to carry defects in the initiation of DNA synthesis only, led us to investigate whether these alleles were also defective in nuclear division. At the time of writing, *cdc28-6* has been shown to cause a nuclear division defect and *cdc28-1* is under investigation. These results are discussed later in the section entitled "The Spindle Pole Body and Start."

Cells only complete start after a number of conditions have been fulfilled (see the following section). It is likely that a number of gene products are involved and three additional loci—*cdc36, cdc37,* and *cdc39*—have been identified that result in arrest at start (Reed 1980).

In the section entitled "Mating Hormones and Start" we present data indicating that the *cdc28* and *cdc37* gene products function in a very different way from *cdc36* and *cdc39* and that mutations in the latter two genes do not themselves result in cell cycle arrest at start.

## Conditions Necessary to Start the Cell Cycle

A number of conditions must be satisfied if a cell is to start the cell cycle. The experiments concerning the commitment described in the previous section show that haploid cells can mate rather than start the mitotic cycle and that diploids can sporulate rather than start the cycle. Therefore, one prerequisite for initiating the cycle is the absence of conditions favoring alternative developmental pathways.

### Nutrient Requirement

Cells will not start the cell cycle unless appropriate nutrients are present. As cells deplete their nutrients in batch culture, they stop dividing and arrest in

G1 phase. The depletion of a wide variety of nutrients results in G1 arrest; these include carbon, nitrogen, sulfate, phosphate, biotin, and potassium (Johnston *et al.* 1977). This might suggest that cells only start the cell cycle if they have accumulated sufficient resources to complete the cell cycle. That this is not strictly true can be deduced from the results of experiments in which glucose is abruptly withdrawn from proliferating cells; under such circumstances cells arrest at a variety of cell cycle stages (Pringle; unpublished observations). Pringle and Hartwell (1981) have suggested that since nutrients are necessary for macromolecular synthesis, failure to complete start may be a direct consequence of the cell's inability to satisfy the critical size requirement (see the following subsection) for start. This is not so; when cells approaching stationary phase arrest in G1 they continue to grow such that the mean cell size of the unbudded cells becomes greater than the critical size for bud initiation in exponential culture (Walton and Carter, unpublished observations). It is also apparent from comparisons of the behavior of wild-type cells and a mutant (*whi2*) that fails to show Gl arrest as it enters stationary phase that wild-type cells arrest in G1 when there is still sufficient glucose in the medium to support a further doubling in cell numbers (Walton and Carter, unpublished observations). At present we do not know how cells sense their nutrient supplies or how this information is used to control start.

An interesting illustration of the influence of nutrients upon start can be observed in a glucose limited chemostat. When haploid cells are grown at 25°C at mass doubling times faster than 4.5 hours, all of the cells in the population proliferate. At dilution rates equivalent to mass doubling times slower than 4.5 hours, a proportion of the population ceases to proliferate but arrests in G1 and adopts stationary phase characteristics (Bugeja, Piggott, and Carter, unpublished observations). The percentage of cells exhibiting G1 arrest increases at slower growth rates, but does not exceed 40% at any growth rate. In a chemostat all cells are in theory subject to identical environmental conditions and therefore the heterogeneity observed is rather surprising. It may be significant that cells with stationary phase characteristics do not appear in the chemostat when there is some residual glucose in the medium, but begin to appear at slow growth rates when the glucose concentration in the chemostat is negligible. Either cells have different capacities to enter stationary phase at slow growth rates or nutrient availability is not equal for all cells. The persistence of stationary phase cells implies a higher threshold of nutrient availability for stationary phase exit than entry.

### Size Requirement

There is some evidence that to pass start cells must grow to a crucial cell size. Yeast cells in a constant environment divide at the same cell size for generation after generation, and it is reasonable to assume that control mechanisms exist to coordinate growth with cell division. Johnston *et al.* (1977) proposed that such coordination could be achieved if the completion of some event in the DNA-division cycle were dependent on growth beyond a minimum size, provided that cells can always traverse the DNA-division cycle more rapidly than they can double their mass. They found that growth rather than progress through the DNA-division cycle was normally rate limiting for cell division. In addition, when small cells isolated from a nitrogen-starved culture were exposed to fresh medium, bud initiation was observed to occur when the parent cells reached the size characteristic of bud initiation in exponential culture. The size at which early cell cycle events were executed was determined by transferring *cdc* mutants to their restrictive temperature at intervals and examining the size at which *cdc* events were executed. It was observed that the *cdc4, cdc7, cdc25,* and *cdc28* mutants executed their gene controlled events when cells were approximately 30–35 $\mu m^3$. Hartwell and his colleagues concluded that an event in G1 at or before the event controlled by the *cdc28* gene product cannot be completed until a crucial cell size is reached.

As the interval from bud emergence to cell division is relatively constant over a range of growth rates, a predictable consequence of cells requiring to grow to a crucial size before start and bud initiation is that at slow growth rates the bud at cell division will be smaller than the parent cell and will subsequently have a longer interdivision time. This prediction is met when growth rate is slowed by direct limitation of protein synthesis (Hartwell and Unger 1977) or by limitation of carbon source in a chemostat (Jagadish *et al.* 1977; Carter 1978; Carter and Jagadish 1978a) or by alteration in media richness in batch culture (Jagadish *et al.* 1977; Carter 1978; Carter and Jagadish 1978a; Lord and Wheals 1980; Carter 1981). The cell cycles of yeast cells growing at slow and fast growth rates is shown in Figure 1.3. At fast growth rates the bud almost reaches the size of the parent at cell division and the parent and daughter (ex-bud) resulting from division have similar cycle time. At slow growth rates the bud at division is much smaller than the parent cell and takes a much longer time to produce a bud than the parent on the subsequent cell cycle. Thus at slow growth rates the cycle time of parents is shorter than the mass doubling time of the culture and that of daughters is longer.

The crucial cell size necessary for completion of a cell cycle event at or just before the *cdc28* event is dependent upon growth rate (Johnston *et al.* 1979; Lorincz and Carter 1979; Tyson *et al.* 1979; Lord and Wheals 1980). When cells are grown in a glucose-limited chemostat (Fig. 1.4) size at bud initiation in haploids and diploids varies with growth rate in the mass doubling time range 2.1–3.75 hours, but at mass doubling times slower than 3.75 hours the size at bud initiation remains constant; a minimum cell size is reached (Lorincz and Carter 1979). At all growth rates the ratio of diploid to haploid size at bud initiation is constant (1:17).

The best evidence that cells require to grow to a crucial size to start the cell cycle and produce a bud comes from experiments in which cells were shifted from poor to rich media and vice-versa (Lorincz and Carter 1979). Unbudded cells at various stages of the cycle in a synchronous culture (with a mass dou-

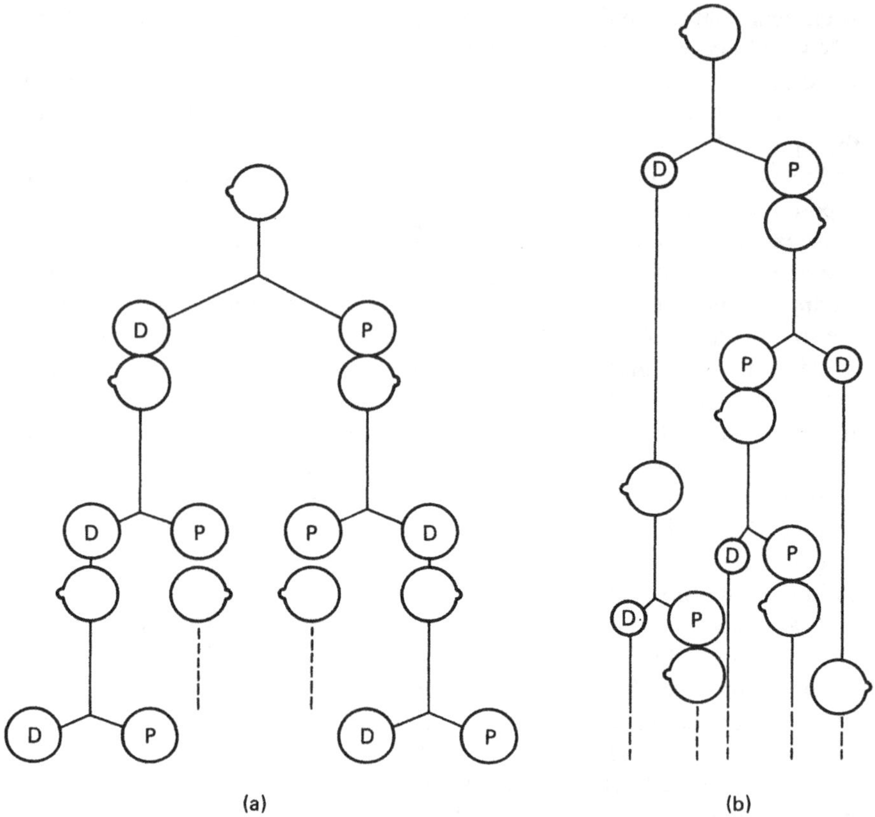


**Figure 1.3.** The cell cycle of cells growing at (a) fast growth rates and (b) slow growth rates. P, Parent cells; D. Daughter cells (Carter 1981).

bling time of 6.84 hours and a size at bud initiation of 26 $\mu m^3$) were shifted to a rich medium (supporting a mass doubling time of 2.4 hours and a size at bud initiation of 41.6 $\mu m^3$). It was observed that all cells, except those very close to the size characteristic of bud initiation on the poor medium, initiated a bud at the size characteristic of the rich medium. Only when cells were within 3 $\mu m^3$ of the size characteristic of bud initiation on poor media were they committed to producing a bud at that size even after a shift. All cells that were below the commitment size at the time of shift to rich media initiate their bud at the size of the rich media; no cells initiate a bud at a parent cell size intermediate between that of the two media.

When cells were shifted from rich to poor media, cells smaller than the size characteristic of bud initiation on the poor media grew until they reached that size and then produced a bud. Those that had exceeded this size at the time of shift produced a bud without the parent size increasing.

These results demonstrate clearly that it is size and not time that cells are measuring: they show that cells rapidly sense their new environment after a shift; they show that commitment to bud initiation at a particular cell size occurs at or very close to the size at which the *cdc28* event occurs; and they show that until they reach the commitment point any progress cells make toward bud initiation on poor media is irrelevant when they are shifted to rich medium.

If cells must grow to a crucial size before beginning a cell cycle, then because yeast divides by budding, once a cell has attained the crucial size it will in all subsequent cycles have fulfilled that size requirement. Thus if cell size itself were the only requirement to start a cycle, a parent cell would produce a bud without further growth. In fact, the size of parent cells at bud initiation increases with the genealogical age of the parent; it increases approximately 17% each generation (Hartwell and Unger 1977; Johnston *et al.* 1979; Lorincz and Carter 1979). This might be because we are measuring cell volume, but the cell is not. It may, for instance, monitor a specific protein whose synthesis is correlated with overall growth such that the parameter being monitored is correlated with cell volume imprecisely.

If cells must grow to a crucial size before starting the cycle, this provides an explanation of how cells coordinate growth and cell division such that cells divide at the same size generation after generation.

In an attempt to identify what genes are involved in the regulation of cell size at start, mutants have been sought that start the cell cycle and produce a bud at a smaller parent size than normal. Mutagenized cells were grown to

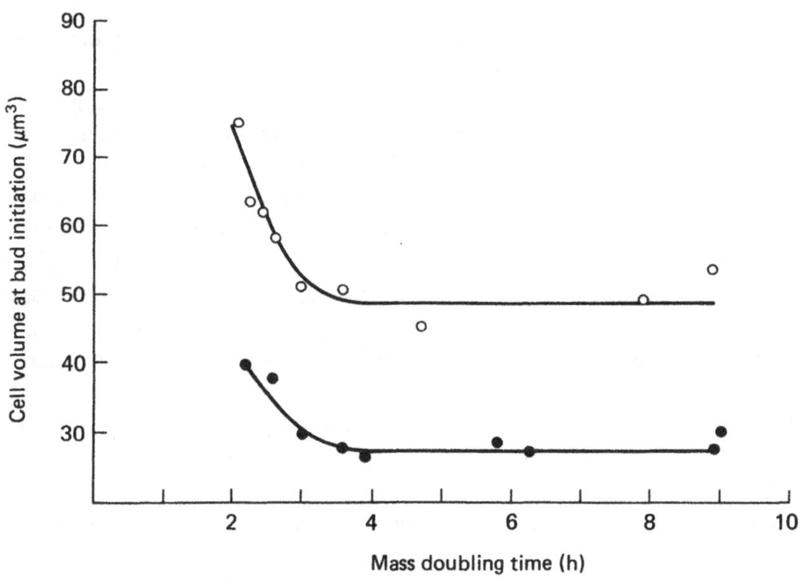


**Figure 1.4.** The size at bud initiation for haploids and diploids cultured at different growth rates in a glucose limited chemostat, zero bud scar (•) haploids and (○) diploids (Lorincz and Carter 1975).

stationary phase, small cells isolated (15 $\mu m^3$) and inoculated into rich medium in which wild-type cells are sensitive to $\alpha$-factor arrest until they reach 38 $\mu m^3$. The cells were allowed to grow synchronously and $\alpha$ factor was added for one generation time when the culture reached 30 $\mu m^3$. The rationale was that wild-type cells would show cell-cycle arrest but continue to grow, forming large, aberrant cells; small mutants that could produce a bud at, say, 25 $\mu m^3$ would, at the time of $\alpha$-factor addition, be past the $\alpha$-factor-sensitive stage of the cycle and would divide. Thus the size difference between wild type and mutant is exaggerated and small mutants can be isolated after the culture is separated according to size (Carter and Sudbery 1980).

Many of the mutants have proved genetically intractable, but five segregate the small phenotype in a 2:2 manner when crossed to normal cells, indicating that in these mutants the phenotype results from a single nuclear gene mutation. Genetic analysis indicates that all of the small mutants have a mutation in the same gene *(whi1)* that is co-dominant with the wild-type allele. The *whi1/whi1* diploid is the same size as a wild-type haploid.

The size at bud initiation of a strain carrying the *whi1* mutation has been determined at a variety of growth rates. The mutant behaves like the wild type in that at fast steady-state growth rates the size at bud initiation is dependent on growth rate, but at slower steady-state growth rates a minimum cell size is reached that is independent of growth rate.

The behavior of *whi1* mutants and wild-type cells differs markedly when cells are shifted from a poor medium to a rich medium. The percentage of budded cells in an exponential culture of *whi1* cells increases rapidly postshift, whereas it decreases after wild-type cells have been shifted (Wynne and Carter, unpublished observations).

The reason for this became apparent when small cells carrying the *whi1* mutation isolated from a culture growing in poor medium (ethanol medium supporting a mass doubling time of 75 hours) were inoculated into a rich medium (YEPD medium supporting a mass doubling time of 2.4 hours). The steady-state values for the size at bud initiation on these two media are 10.5 $\mu m^3$ for the poor medium and 19.3 $\mu m^3$ for the rich medium. Surprisingly, after a shift the mutant cells budded when the parent cell size reached the size characteristic of budding on the poor media (Fig. 1.5). This explains why in an exponential culture shifted to rich media the percentage of budded cells increases rapidly.

When homogeneously small wild-type cells isolated from poor medium were transferred to rich medium, they grew to the size characteristic of bud initiation when cells are grown for several generations on rich medium (Fig. 1.5). Bud initiation occurs at this size; the bud continues to increase in size, but the parent cell does not.

When *whi1* mutant cells are shifted from poor medium to rich medium, they bud at the size characteristic of poor medium, but the bud reaches the size of the parent cell and then both portions of the cell increase in volume such that at division the parent cell has reached the size that is characteristic of steady-state growth on rich medium (Fig. 1.5).

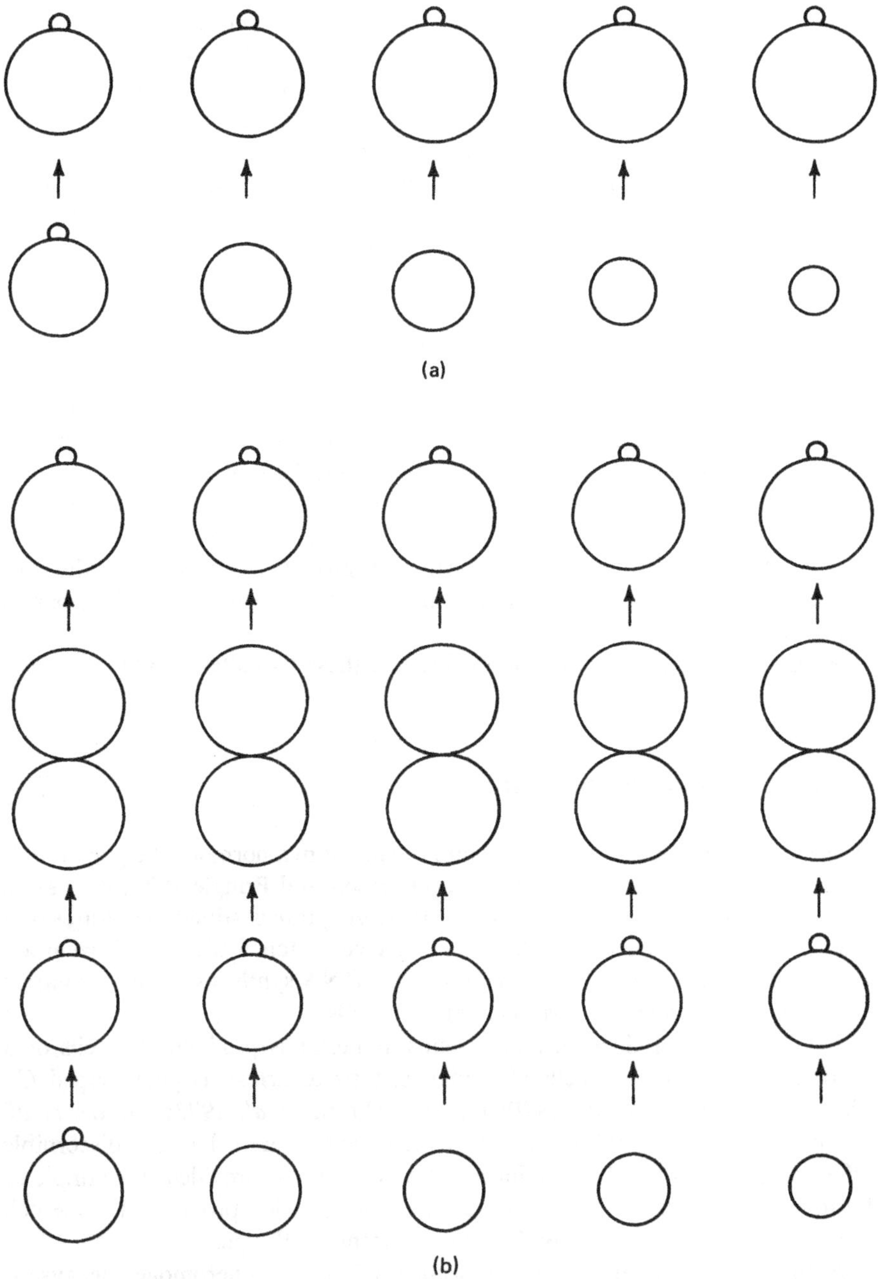


**Figure 1.5.** The behavior of wild-type and *whi1* mutants after a shift from poor media to rich media. (a) wild-type cells: (b) *whi1* cells.

Preliminary results suggest *whi1* cells must grow to a crucial size before they can complete a cell cycle event at or before nuclear division (Wynne and Carter, unpublished observations). Thus the *whi1* mutation reveals that, in addition to a size control over the initiation of DNA synthesis, there is a size control at the end of G2 that is not observed in wild-type cells at fast growth rates but may be revealed at slow growth rates. Interestingly, the fission yeast *Schizosaccharomyces pombe* has two size controls, one at the initiation of DNA synthesis and the other initiated by nuclear division, but in wild-type cells the latter control is expressed and the former is cryptic. It seems that both yeasts have two size controls, but in the budding yeast only the control over initiation of DNA synthesis is expressed at fast growth rates and in the fission yeast only the control over nuclear division is expressed.

An alternative interpretation of our results is that size at bud initiation occurs at a size independent of growth rate in *whi1* mutants, but that there is a similar interval between bud initiation and cell division in all media; in this period the bud reaches the size of the parent and then parent and bud grow to a larger cell size. At faster growth rates the size at cell division will be larger than at slower growth rates because cells will grow to a larger size in the time available. Thus size at bud initiation in the next cycle will depend on parent cell size at the previous division.

Experiments are in progress to distinguish these two alternatives.

## Mating Hormones and Start

Yeast cells of **a** mating type secrete into the medium a hormonelike polypeptide of eleven amino acids called **a** factor (Wilkinson and Pringle 1974; Betz *et al.* 1977; Betz and Duntze 1979). When this polypeptide is added to cultures of $\alpha$ mating type cells those cells which have not yet reached the cdc 28 event are prevented from doing so. They do not initiate DNA synthesis or bud formation but continue to grow and form pear-shaped cells.

A polypeptide of thirteen amino acids, $\alpha$ factor is produced by cells of $\alpha$ mating type and causes cells of **a** mating type to arrest as pear-shaped G1 phase cells (Duntze *et al.* 1970; Bucking-Throm *et al.* 1973; Duntze *et al.* 1973; Stotzler *et al.* 1976). Neither $\alpha$ factor nor **a** factor has any discernible effect on the cell type that produces them. Thus yeast provides an example of hormonelike control of cell proliferation; a polypeptide produced by one cell type prevents cell proliferation in cells of another cell type.

Alpha factor has been the better studied of the two pheromones because it is easier to purify. Mutants resistant to $\alpha$ factor have been obtained (Mackay and Manney 1974a, 1974b; Manney and Woods 1976; Hartwell 1980). These mutants have been analyzed genetically and 8 genes have been identified as being involved in the mechanism of $\alpha$ factor action. There must be functions in cells of **a** mating type which are responsible for the reception of the $\alpha$ factor signal distinct from functions in $\alpha$ cells that receive the signal from **a** factor.

Such functions can be identified as $\alpha$-factor-resistant **a** cells that are not **a**-factor resistant when the mutation is transferred to $\alpha$ cells. Thus cells of **a** mating type respond to $\alpha$ factor and cells of $\alpha$ mating type respond to **a** factor: the response is the same in both mating types; cell-cycle arrest at start. Functions involved in the transmission of the signals might be expected to be expressed in cells of both mating types. The foregoing describes a Y-shaped pathway for functions involved in hormone action with reception functions in the branches and transmission functions in the stalk.

Hartwell (1980) has identified 8 genes from genetic analysis of $\alpha$-factor-resistant mutants. Only one of these genes does not confer **a**-factor resistance when the mutation is transferred to $\alpha$ cells. It or a yet to be discovered gene with mutations exhibiting similar properties is a likely candidate for hormone receptor.

Little is known concerning the biochemical mode of action of either of the mating pheromones. A number of peptide hormones in mammalian cells exert their effect via cyclic AMP, and Liao and Thorner (1980) have examined whether an analagous situation pertains in yeast. They found that adenylate cyclase was membrane bound and that the enzyme activity was inhibited by $\alpha$ factor although inhibition never exceeded 60%. The authors reported that addition of cyclic AMP to cells of **a** mating type shortened the length of $\alpha$ factor arrest although they were unable to measure the cellular content of cyclic AMP.

Interestingly, *ste5* mutants, which at 34°C but not 23°C are resistant to $\alpha$ factor, possessed an adenylate cyclase which was insensitive to $\alpha$-factor inhibition when assayed at the higher temperature, but inhibited at the lower temperature. Representative strains carrying other *ste* gene mutations had normal adenylate cyclase activity. Mutations in *ste5* result in nonmating at the restrictive temperature and this phenotype occurs whether the cells are of the $\alpha$ or **a** mating type. In addition, Liao and Thorner (1980) observed that adenylate cyclase was present in membranes of $\alpha$ haploids and **a**/$\alpha$ diploids and in both cases was inhibited by $\alpha$ factor. In vivo $\alpha$ factor has no effect on $\alpha$ cells or **a**/$\alpha$ diploids. The authors suggest that the biochemical mechanism of $\alpha$ factor action involved its inhibition of the adenylate cyclase in the plasma membrane of **a** cells.

As $\alpha$ factor directly inhibits the adenylate cyclase and does so in preparations from both **a** and $\alpha$ cells, one must propose that *in vivo* $\alpha$ factor is in some way prevented from interacting with the adenylate cyclase of **a** cells. It is not easy to see how these data can be fitted into the Y-shaped pathway model of pheromone action.

Cells are sensitive to $\alpha$ factor for all of the unbudded phase of the cycle up until 15 minutes before bud initiation (Carter 1981), the $\alpha$ factor step. Thus cells can be prevented from proliferating at any time in the unbudded phase by $\alpha$ factor unless they have passed the commitment point for proliferation discussed earlier.

As mentioned previously, Hereford and Hartwell (1974) interpreted reciprocal shift data as showing that the *cdc28* event and the $\alpha$-factor step were interdependent and that $\alpha$ factor is involved in the same process as that blocked in *cdc28* mutants. We have suggested earlier that there is an alternative interpretation. Clearly, the relationship of start and $\alpha$ factor merits more attention. Reed (1980) identified three start genes in addition to *cdc28*. Strains carrying mutations in *cdc28, cdc36, cdc37*, or *cdc39* can all mate at 36°C (Reed 1980), whereas mutants blocked at other *cdc* steps at the restrictive temperature cannot mate (Reid and Hartwell 1977). Mutations in all four genes arrest as shmoos at 37°C; they are defective in the initiation of DNA synthesis but continue to grow.

The shmoo morphology is observed for *cdc28* or *cdc37* mutations in cells of the **a** or $\alpha$ mating type and in **a**/$\alpha$ diploids homozygous for *cdc28* or *cdc37*. By contrast, *cdc36* and *cdc39* mutants only exhibit the shmoo phenotype in haploid cells. Diploids (**a**/$\alpha$) homozygous for *cdc36* or *cdc39* arrest when shifted from 24°C to 38°C, but cells do not exhibit homogeneous cell cycle arrest; they appear to arrest at diverse stages of the cell cycle (Reed 1980). Intriguingly, *cdc28* and *cdc37* mutants also have karyogamy (*kar*) defects, whereas *cdc36* and *cdc39* do not.

The subdivision of these start mutants into two groups is further supported by our observations that $\alpha$-factor resistant mutants isolated in *cdc28* and *cdc37* behave differently from those studied in a *cdc36* or *cdc39* background (Connolly, Bugeja and Piggott, personal communication). Approximately 500 $\alpha$-factor resistant mutants were generated in strains carrying *cdc28, cdc36, cdc37*, and *cdc39* mutations, respectively. This number was chosen so that representatives of all the *ste* genes should be present as $\alpha$-factor-resistant mutants. All $\alpha$-factor-resistant mutants in *cdc28* or *cdc37* arrested as shmoos when shifted to 37°C in the absence of $\alpha$ factor; those in *cdc36* or *cdc39* arrested when shifted to 37°C in the absence of $\alpha$ factor, but not as shmoos. They appeared to be arrested at all stages of the cell cycle.

These results indicate that a block at any step in the pathway of $\alpha$-factor action results in a loss of the characteristic arrest phenotype of *cdc36* and *cdc39*. Thus mutations in these two genes do not themselves result in cell-cycle arrest at start, but may do so by activating the $\alpha$-factor arrest pathway. Thus *cdc36* and *cdc39* mutations result in a fundamentally different block to *cdc28* and *cdc37*. The nature of the defects in *cdc36* and *cdc39* is intriguing. Even in the absence of a functioning $\alpha$ factor, pathway mutations in these genes lead to a temperature-sensitive phenotype, albeit one that is characteristic of growth mutants rather than cell-cycle mutants. This suggests that lesions in these genes lead to metabolic defects other than specific cell cycle defects that may activate the $\alpha$-factor pathway.

## The Spindle Pole Body and Start

The cell cycle of yeast is conventionally divided into G1, S, G2, and M phases, with the phases occurring in the order stated. In fact, initiation of mitosis occurs before initiation of DNA synthesis insofar as the spindle pole body which is situated in the nuclear envelope and functions as a microtubule organizing center develops a small satellite structure late in G1. It then duplicates itself, and shortly thereafter initiation of DNA synthesis occurs.

Until recently it was thought that initiation of DNA synthesis was dependent on the prior duplication and separation of the spindle pole body. The fact that plaque duplication did not occur in *cdc28* mutants was used to support this notion (Byers and Goetsch 1974; Hartwell 1978). It is now clear that initiation of DNA synthesis can occur in the absence of spindle pole body duplication. Byers, quoted in Hartwell and Pringle (1981), observed that *cdc31* mutants can replicate their DNA, but spindle pole body duplication does not occur. Strains carrying *cdc31* mutations arrest as dumbbell-shaped cells with parent and bud of equal size. No nuclear division occurs: the nucleus occupies the isthmus between parent and bud. Interestingly, nuclear migration and elongation can occur in the absence of spindle pole body duplication.

Some time ago M. N. Jagadish and B. L. A. Carter (unpublished observations) measured the execution points of *cdc31* mutants grown at a variety of growth rates in a glucose-limited chemostat. We found that at mass doubling times ranging from 2.5 hours to 8 hours the execution point was always around 0.75 of the cycle previous to that at which cell cycle arrest occurred. Although leakiness cannot be ruled out as an explanation, it is possible that an event occurs at nuclear division in one cycle that is necessary for spindle pole duplication and nuclear division in the next. It is important to repeat these experiments with strains carrying other *cdc31* alleles. The association of *cdc31* with the spindle pole body and nuclear division is supported by the observation that at 30°C *cdc31* mutants diploidize (Byers 1981), since they duplicate their DNA without undergoing nuclear division.

If *cdc31* mutants can initiate DNA synthesis but do not duplicate their spindle pole body, what is the role of the *cdc28* gene? Pringle and Hartwell (1981) suggest that completion of the *cdc28* event results in an apparent doubling in size of the spindle pole body and commitment to division. They suggest that the *cdc28* product is a component of the spindle pole body and that the yeast cell measures its size at start by accumulating a sufficient amount of the subunits of the spindle pole body to complete its duplication. The argument depends on the behavior of mutants defective to karyogamy that were first identified by Conde and Fink (1976). Such mutations are dominant to the wild-type alleles in heterokaryons, but recessive in diploids. From this it is reasonable to suppose that the gene product is nucleus limited. Since nuclei fuse at their spindle pole bodies after conjugation (Byers and Goetsch 1975), it has been suggested that the gene products of karyogamy-defective mutants are structural components of the spindle pole body.

Dutcher (quoted in Pringle and Hartwell 1981) examined 40 *cdc* mutants to see if they had karyogamy defects. Two start mutants, *cdc28* and *cdc37*, as well as another mutant defective in the initiation of DNA synthesis, *cdc4*, were found to be karyogamy defective. Strains carrying mutations at *cdc28* or *cdc37* do not duplicate their spindle pole bodies at 36°C, and in *cdc4* mutants duplication occurs but the spindle bodies do not separate. The products of these three genes may be components of the spindle pole body.

Recently we have isolated *cdc28* **a** mutants that are not defective in the initiation of DNA synthesis at the restrictive temperature but are unable to undergo nuclear division (Piggott, Rai, and Carter, in press). The isolation of these mutants has led to the discovery that previously identified alleles of *cdc28* are defective in nuclear division as well as DNA synthesis.

The terminal phenotype of the new *cdc28* mutant allele is a budded cell in which parent and bud are approximately equal in volume. The nucleus remains within the parent portion of the cell; no nuclear migration is observed.

The fact that certain mutations at *cdc28* result in defects in both initiation of DNA synthesis and nuclear division, whereas others are defective in nuclear division only, suggests that different properties of the product are involved in the two processes. It is quite possible that *cdc28* mutants will be identified which are defective in the initiation of DNA synthesis only. Possibly such mutants exist already in Dutcher's collection. Reed, Ferguson, and Groppe (1981) quote Dutcher as having isolated *cdc28* mutants both with and without karyogamy defects. We would speculate that there is an association between karyogamy defects and nuclear division defects and that *cdc28* mutants with no karyogamy defects are defective only in the initiation of DNA synthesis. By the same reasoning we would suspect that *cdc37* but not the start mutants *cdc36* and *cdc39* would also have nuclear division defects.

What defect is likely to result in both karyogamy and nuclear division defects? It would perhaps be more accurate to define the nuclear division block as a nuclear migration defect where the nucleus remains within the parent portion of the cell. To answer this it is necessary to determine how the nucleus migrates and elongates and how, during conjugation, the spindle pole bodies of distinct nuclei meet and fuse.

It has been shown that in addition to intranuclear microtubules there are extranuclear tubules that emanate from the spindle pole body into the cytoplasm (Byers and Goetsch 1975). These extend into the bud. Nuclear migration and elongation could result from extension of microtubules between the duplicated spindle pole bodies or by a mechanism pulling the nucleus into the bud. The latter would require anchorage site(s) and presumably extranuclear microtubules connecting spindle pole bodies and anchorage sites.

It seems unlikely that the nuclear division defect in *cdc28* results from a mutation affecting microtubule formation because medial and late nuclear division are unaffected in *cdc28* mutants. It is possible that it derives from impaired construction of the nuclear cytoplasmic connections which permit coordinated nuclear movement relative to parent and bud portion of the cell. If the spindle pole body is impaired in its ability to make extranuclear connections in *cdc28* mutants, it could explain both nuclear division defects and the *kar* phenotype.

## Developmental Alternative to Mitosis

Cells can, as discussed earlier, undergo a number of developmental alternatives to mitosis, such as sporulation, conjugation, or entry into stationary phase. In addition, there is the vexing question of whether there is such a thing as a G0 phase. Lajtha (1963) suggested that cells arrested for a prolonged period in G1 leave the cell cycle and enter G0. It has proved to be immensely difficult to determine whether G0 really exists or is just a matter of semantics.

Recently we isolated a mutant that when shifted from exponential growth at 24°C to 37°C continued to proliferate, but when grown to stationary phase and shifted to fresh medium could resume proliferation at 24°C but not 37°C. It was TS for the ability to leave stationary phase. As the mutant was not TS when proliferating cells were shifted to the higher temperature, it is tempting to conclude that the mutant is not defective in any G1, S, G2, or M phase functions and therefore cells arrested in stationary phase leave the cycle and enter G0. The mutant would then be described as defective in the G0–G1 transition. An alternative interpretation would be that stationary phase cells are not qualitatively different from proliferating cells in G1, but that they may be quantitatively different with respect to certain products. The mutant may still make enough of the gene product to permit division when proliferating cells are shifted to 37°C, but not enough to permit product-depleted stationary phase cells to commence proliferation.

Studies of folded chromosomes isolated from yeast by gentle lysis led Pinon to conclude that stationary phase cells contain folded chromosomes that can be distinguished from those observed at different stages of the cell cycle in proliferating cells (Pinon and Salts 1977; Pinon 1978; Pinon 1979a; Pinon 1979b; Pinon and Pratt 1979). This suggests that stationary phase cells represent cells that have left the cell cycle and entered G0. Interestingly, $\alpha$-factor arrested cells have another distinctive form of folded chromosome (Pinon and Pratt 1979).

The control over the transition from the proliferating phase to the resting phase is of major importance if we are to understand the control of cell proliferation. A mutant, *whi2,* that is defective in this control has been isolated (Sudbery *et al.* 1980).

Wild-type cells stop cell proliferation, arrest in G1 phase prior to the *cdc28* event, and adopt stationary phase characteristics as the medium becomes depleted of glucose in batch culture. These characteristics include in excess of 95% unbudded cells in the population, accumulation of the carbohydrate reserves glycogen and trehalose (Chester 1963; Rothman-Denes and Cabib 1970; Patel and Ingledew 1975; Panek and Mattoon 1977; Gunja-Smith *et al.* 1977; Kuerzi and Fiechter 1977; Lille and Pringle 1980) increased heat resistance (Schenberg-Frascino and Moustacchi, 1972; Parry *et al.* 1976; Walton *et al.* 1979) and increased resistance to cell wall degrading enzymes (Deutsch and Parry 1974; Piedra and Herrera 1976). In addition, stationary phase cells appear bright under the phase contrast microscope, while proliferating cells appear dark, which may be due to the accumulation of carbohydrate reserves and increased mannan levels in the cell walls of stationary phase cells.

The *whi2* mutant phenotype is revealed in batch culture as the glucose in the medium becomes depleted and cells approach stationary phase. The mutant continues to divide and reaches a cell density approximately double that of wild-type cells.

In the final two doublings the mutant cells divide at a smaller size than wild-type cells. The mutant appears to have lost most of the properties of wild-type cells in stationary phase: (1) the cells appear dark under the phase contrast microscope: (2) they are sensitive to both heat shock and cell wall degrading enzymes; (3) the level of storage carbohydrates is reduced; and (4) an abnormal level of budded cells (24%) is present (Walton and Carter, unpublished observations). These properties are characteristic of cultures growing exponentially at a slow growth rate.

We can view *whi2* mutants cells as continuing to divide at glucose concentrations that result in wild-type cells arresting in G1. The mutant proliferates slowly, but eventually the medium becomes completely depleted of nutrients and cells stop at whatever stage in the cycle they have reached at this point. It is interesting to note that wild-type cells obviously stop proliferating when there is still some glucose in the medium, but at a critical level they arrest in G1 and utilize the remaining glucose for storage carbohydrate production. The mechanism whereby wild-type cells sense the declining glucose level is not known.

The *whi2* mutant may fail to exhibit G1 arrest in stationary phase either because it has lost the mechanism enabling it to do so or because it is defective in the sensing mechanism which indicates that conditions are appropriate for G1 arrest. The fact that when cells carrying the *whi2* mutation are starved of nitrogen rather than glucose they show the G1 arrest characteristic of wild-type indicates that the latter explanation is more likely. This hypothesis is compatible with some preliminary data which show that in exponential growth *whi2* cells divide at a faster growth rate than wild-type in media containing poorer carbon sources than glucose. We speculate that the mutant does not recognize the poorer nutritional status of the carbon source and uses the carbon source exclusively for growth, whereas wild-type cells divert some of the carbon source into reserve carbohydrates on anything less than the richest medium.

Mutation in the *whi2* gene causes cells to divide when they should not, mutations in the *spd1* gene were isolated because homozygous diploids sporulated when they should not (Dawes 1975). Strains carrying the *spd1* mutation sporulate after reaching stationary phase when glucose is the carbon source, but in complete medium containing glycerol as the carbon source they do not divide, but initiate sporulation immediately. Vezinhet *et al.* (1979) noted that haploid *spd1* mutants cannot proliferate in glycerol media; the lesion induced a particular form of starvation in which cells arrested in the G1 phase of the cycle at or prior to the *cdc28–1* execution point. The uptake of glycerol and the synthesis of the enzymes concerned with its entry into general metabolic pathways is not affected in the mutant. Vezinhet *et al.* (1979) concluded that the *spd* mutation affects a central metabolic function concerned with the utilization of some, but not all, nonfermentable carbon sources. It is possible that this mutant, like *whi2* mutants, is defective in the mechanism monitoring nutrient supply; the two mutations are not allelic.

Other interesting mutations that illustrate the importance of nutrient assessment to cells that have a choice of developmental alternatives are the *aas1* and the *tra3* mutations (Wolfner *et al.* 1975). Wild-type yeast cells derepress the biosynthetic enzymes for arginine, histidine, lysine, and tryptophan when the

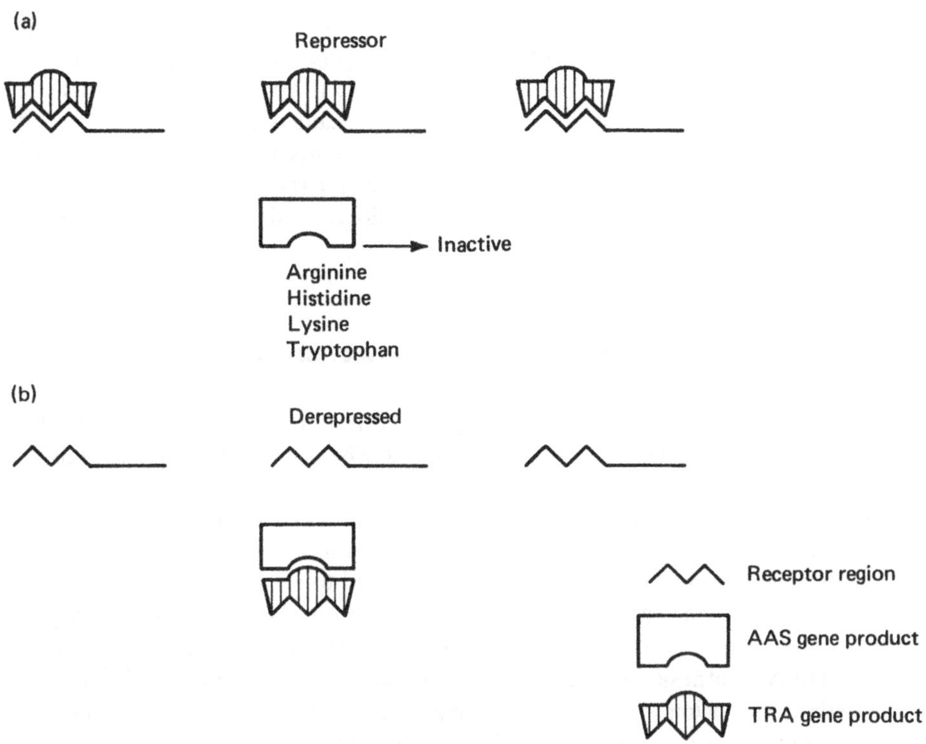


**Figure 1.6.** A model for the role of the *AAS1* and *TRA3* genes in the regulation of amino acid biosynthesis. Under *repression:* the *TRA3* gene product prevents expression of a number of structural genes. Repression could be at the level of transcription, translation, or processing of the messages. The *AAS1* gene product is inactivated in the presence of arginine, histidine, lysine and trytophan. The net effect is a low level of gene expression in the presence of all four amino acids. Under *derepression:* if cells are starved for one of the amino acids the *AAS1* product is now functional and inactivates the *TRA3* product. The net effect is to allow full derepression of the structural gene (Wolfner *et al.* 1979).

medium is depleted of any one of these amino acids. Two classes of mutant affected in the control of this process have been identified. One class, *aas*, is unable to derepress these enzymes and the other class, *tra*, is fully derepressed for the biosynthesis of these enzymes in all growth conditions. The *tra* mutation mimics nutrient depletion and it is interesting that at 36°C the mutant is temperature sensitive for growth. Strains carrying the *tra3* mutations arrest as single unbudded cells at the *cdc28/α* factor step when shifted to the restrictive temperature. If the *tra3* mutant cells arrest in G1 because they perceive themselves to be nutrient depleted, it would be interesting to determine whether strains carrying *aas* mutations (which in some respect behave as if they do not recognize nutrient depletion) arrest in G1 in stationary phase or, like *whi2*, continue to proliferate. The TRA and AAS genes could regulate the synthesis of as many as 30 or 40 genes (Wolfner *et al.* 1975) and a model compatible with the behavior of the mutants is shown in Figure 1.6. Wild-type yeast cells can as discussed earlier only start the cell cycle if appropriate nutrients are present and if any of a number of different nutrients are depleted cells arrest at "start." This implies that cells sense their nutrients and integrate this information at start. Wolfner *et al.* (1975) propose that the TRA genes are sensor genes whose product integrates amino acid biosynthesis into the cell cycle at start.

## *Saccharomyces Cerevisiae* V. *Schizosaccharomyces Pombe*

In a recent review Hartwell (1978) suggested that "the hypothesis of two sites for coordinating growth and division, one of which exists in some organisms *(S. cerevisiae)* and both of which exist in others *(Sch. pombe)*, will be sufficient to account for the various eukaryotic cell cycles." The two sites referred to are the initiation of DNA synthesis and nuclear division.

It is now apparent that the two yeasts referred to are more alike in the control of cell division than had been supposed. *S. cerevisiae*, like *Sch. pombe*, has two sites for coordinating growth and division. In *Sch. pombe* a size control over DNA synthesis is cryptic except under poor nutrient conditions and growth and division are usually coordinated at nuclear division. The converse is true of *S. cerevisiae* where coordination is achieved by a size control at the initiation of DNA synthesis and only at slow growth rates is a control at nuclear division exposed.

The finding that the *cdc28* gene plays a role in both the initiation of DNA synthesis and nuclear division is similar to the recent observation in *Sch. pombe* that the *cdc2* gene is involved in the initiation of DNA synthesis as well as nuclear division. An intriguing possibility is that the *cdc2* gene of *Sch. pombe* and the *cdc28* gene of *S. cerevisiae* fulfill like functions. It will be interesting to see whether plasmids containing the *cdc28* gene of *S. cerevisiae* (Reed, Ferguson and Groppe 1982) will complement *cdc2* mutants of *Sch. pombe*.

## References

Betz, R., Duntze, W. (1979). Purification and partial characterisation of *a* factor, a mating hormone produced by mating type a cells from *Saccharomyces cerevisiae.* Eur. J. Biochem 95: 469–475.

Betz, R., MacKay, V. L., Duntze, W. (1977). *a* factor from *Saccharomyces cerevisiae:* Partial characterisation of a mating hormone produced by cells of mating type a. J. Bacteriol. 132: 462–472.

Bucking-Throm, E., Duntze, W., Hartwell, L. H., Manney, T. R. (1973). Reversible arrest of haploid yeast cells at the initiation of DNA synthesis by a diffusible sex factor. Exp. Cell Res. 76: 99–110.

Byers, B. (1981). Multiple roles of the spindle pole bodies in the life cycle of *Saccharomyces cerevisiae.* In Alfred Benzon Symposium 16, Molecular Genetics of Yeast (D. von Weltstion, J. Friis, M. Kielland-Brandt A. Stenderup, eds.). Munksgaard, Copenhagen, 1981, pp. 119–131.

Byers, B., Goetsch, L. (1974). Duplication of spindle plaques and integration of the yeast cell cycle. Cold Spring Harbor Symp. Quant. Biol. 38: 123–131.

Byers, B., Goetsch, L. (1975). Behavior of spindles and spindle plaques in the cell cycle and conjugation of *Saccharomyces cerevisiae.* J. Bacteriol. 124: 511–523.

Carter, B. L. A. (1978). The yeast nucleus. Adv. Microbiol. 17: 243–301.

Carter, B. L. A. (1981). The control of cell division in *Saccharomyces cerevisiae.* In Society for Experimental Biology—Seminar Series 10—The Cell Cycle (P. C. L. John, ed.) pp. 99–119. Cambridge University Press, Cambridge.

Carter, B. L. A., Jagadish, M. N. (1978a). The relationship between cell size and cell division in the yeast *Saccharomyces cerevisiae.* Exp. Cell Res. 112: 15.

Carter, B. L. A., Jagadish, M. N. (1978b). Control of cell division in the yeast *Saccharomyces cerevisiae* cultured at different growth rates. Exp. Cell Res. 112: 373–383.

Carter, B. L. A., Sudbery, P. E. (1980). Small sized mutants of Saccharomyces cerevisiae, Genetics 96: 561–566.

Chester, V. E. (1963). The dissimilation of carbohydrate reserves of a strain of *Saccharomyces cerevisiae.* Biochem. J. 86: 153–160.

Conde, J., Fink, G. R. (1976). A mutant of *Saccharomyces cerevisiae* defective for nuclear fusion. Proc. Natl. Acad. Sci. USA 73: 3651–3655.

Dawes, I. W. (1975). Study of cell development using derepressed mutations. Nature 255: 707–708.

Deutsch, C. E., Parry. J. M. (1974). Sphaeroplast formation in yeast during the transition from exponential phase to stationary phase. J. Gen. Microb. 80: 259–268.

Duntze, W., MacKay, V. L. and Manney, T. R. (1970). *Saccharomyces cerevisiae:* a diffusible sex factor. Science (Washington, D.C.) 168: 1472–1473.

Duntze, W., Stotzler, D., Bucking-Throm, E., Kalbitzer S. (1973). Purification and partial characterisation of *α* factor, a mating type specific inhibitor of cell reproduction in *Saccharomyces cerevisiae.* Eur. J. Biochem 35: 357–365.

Game, J. C. (1976). Yeast cell cycle mutant *cdc21* is a temperature sensitive thymidylate auxotroph. Mol. Gen. Genet. 146: 313–315.

Gunja-Smith, Z., Patil, N. B., Smith, E. E. (1977). Two pools of glycogen in *Saccaromyces.* J. Bacteriol. 130: 818–825.

Hartwell, L. H. (1974). *Saccharomyces cerevisiae* cell cycle. Bacteriol. Rev. 38: 164–198.
Hartwell, L. H. (1976). Sequential function of gene products relative to DNA synthesis in the yeast cell cycle. J. Mol. Biol. 104: 803–817.
Hartwell, L. H. (1978). Cell division from a genetic perspective. J. Cell Biol. 77: 627–637.
Hartwell, L. H. (1980). Mutants of *Saccharomyces cerevisiae* unresponsive to cell division control by polypeptide mating hormone. J. Cell Biol. 85: 811.
Hartwell, L. H., Unger, M. W. (1977). Unequal division in *Saccharomyces cerevisiae* and its implications for the control of cell division. J. Cell Biol. 75: 422–435.
Hereford, L. M., Hartwell, L. H. (1974). Sequential gene function in the initiation of *Saccharomyces cerevisiae* DNA synthesis. J. Mol. Biol. 84: 445–461.
Hirschberg, J., Simchen, G. (1977). Commitment to the mitotic cell cycle in yeast in relation to meiosis. Exp. Cell Res. 105: 245–252.
Jagadish. M. N., Carter, B. L. A. (1977). Genetic control of cell division in yeast cultured at different growth rates. Nature 269: 145–147.
Jagadish, M. N., Carter, B. L. A. (1978). Effects of temperature and nutritional conditions on the mitotic cell cycle of *Saccharomyces cerevisiae.* J. Cell Sci. 31: 71–78.
Jagadish, M. N., Lorincz, A., Carter, B. L. A. (1977). Cell size and cell division in yeast cultured at different growth rates. FEMS Microbiol. Lett. 2: 235–237.
Johnston, L. H. and Nasmyth, K. A. (1978). *Saccharomyces cerevisiae* cell cycle mutant *cdc9* is defective in DNA ligase. Nature 274: 891–893.
Johnston, G. C., Pringle, J. R., Hartwell, L. H. (1977). Coordination of growth with cell division in the yeast *Saccharomyces cerevisiae.* Exp. Cell Res. 105: 79–98.
Johnston, G. C., Ehrhardt, C. W., Lorincz, A., Carter, B. L. A. (1979). Regulation of cell size in the yeast *Saccharomyces cerevisiae.* J. Bacteriol. 137: 1–5.
Kawasaki, G. (1979). Karyotypic instability and carbon source effects in cell cycle mutants of *Saccharomyces cerevisiae.* Ph.D. dissertation, Department of Genetics, University of Washington, Seattle.
Kirschner, K., Bisswanger, H. (1976). Multifunctional proteins. Annu. Rev. Biochem. 45: 143–166.
Kuenzi, M. T. and Fiechter, A. (1977). Regulation of carbohydrate composition of *Saccharomyces cerevisiae* under growth limitation. Arch. Mikrobiol. 84: 254–265.
Lajtha, L. G. (1963). On the concept of the cell cycle. J. Cell Comp. Physiol 62: Suppl. 1, pp. 143–144.
Liao, H., Thorner, J. (1980). Yeast mating pheromone $\alpha$ factor inhibits adenylate cyclase. Biochemistry 77: 1898–1902.
Lillie, S. H., Pringle, J. R. (1980). Reserve carbohydrate metabolism in *Saccharomyces cerevisiae:* responses to nutrient limitation. J. Bacteriol. 143: 1384–1394.
Lord, P. G. and Wheals, A. E. (1980). Asymmetrical division of *Saccharomyces cerevisiae.* J. Bacteriol. 142: 808–818.
Lorincz, A., Carter, B. L. A. (1979). Control of cell size at bud initiation in *Saccharomyces cerevisiae.* J. Gen Microbiol. 113: 287–295.
MacKay, V., Manney, T. R. (1974a). Mutations affecting sexual conjugation and related processes in *Saccharomyces cerevisiae.* I. Isolation and phenotypic characterisation of nonmating mutants. Genetics 76: 255–271.
MacKay, V., Manney, T. R. (1974b). Mutations affecting sexual conjugation and related processes in *Saccharomyces cerevisiae.* II. Genetic analysis of nonmating mutants. Genetics 76: 273–288.

Manney, T. R., Woods, V. (1976). Mutants of *Saccharomyces cerevisiae* resistant to the $\alpha$ mating type factor. Genetics 82: 639–644.

Nasmyth, K. A., Reed, S. I. (1980). Isolation of genes by complementation in yeast: Molecular cloning of a cell cycle gene. Proc. Natl. Acad. Sci. USA 77: 2119–2123.

Nurse, P. (1981). Genetic analysis of the cell cycle. Symp. Soc. Gen Microbiol., No. 31, "Genetics as a Tool in Microbiology", Eds. S. W. Glover & D. A. Hopwood, Cambridge Univ. Press, pp. 291–315.

Panek, A. D., Mattoon, J. R. (1977). Regulation of energy metabolism in *Saccharomyces cerevisiae.* Relationships between catabolite repression, trehalose synthesis, and mitochondrial development. Arch. Biochem. Biophys. 183: 306–316.

Parry, J. M., Davies, P. J., Evans, W. E. (1976). The effects of "cellage" upon lethal effects of physical and chemical mutagens in the yeast *Saccharomyces cerevisiae.* Mol. Gen. Genet. 146: 27–35.

Patel, G. B., Ingledew, M. W. M. (1975). The relationship of acid-soluble glycogen to yeast flocculation. Can J. Microbiol. 21: 1608–1613.

Piedra, E., Herrera, L. (1976). Selection of auxotrophic mutants in *Saccharomyces cerevisiae* by a snail enzyme digestion method. Folia Microbiol. 21: 337–340.

Piggott, J. R., Rai, R., Carter, B. L. A. (1982). A bifunctional gene product involved in two stages of the yeast cell cycle. Nature (in press).

Pinon, R. (1978). Folded chromosomes in non-cycling yeast cells. Evidence for a characteristic $G_0$ form. Chromosoma 67: 263–274.

Pinon, R. (1979a). A probe into nuclear events during the cell cycle of *Saccharomyces cerevisiae:* studies of folded chromosomes in cdc mutants which arrest in $G_1$ Chromosoma 70: 337–352.

Pinon, R. (1979b). Folded chromosomes in meiotic yeast cells: analysis of early meiotic events. J. Mol. Biol. 129: 433–447.

Pinon, R., Pratt, D. (1979). Folded chromosomes of mating-factor arrested yeast cells: comparison with $G_0$ arrest. Chromosoma 73: 117–129.

Pinon R., Salts, Y. (1977). Isolation of folded chromosomes from the yeast *Saccharomyces cerevisiae.* Proc. Natl. Acad. Sci. USA 74: 2850–2854.

Pringle, J. R. (1979). Proteolytic artifacts in biochemistry. In Limited Proteolysis in Microorganisms (G. N. Cohen, H. Holzer, eds.), pp. 191–196. U.S. Government Printing Office, Washington, D.C.

Pringle, J. R. (1981). Genetic approaches to study of the cell cycle. In: Cellular Dynamics: Mitosis/Cytokinesis (A. M. Zimmerman, A. Forer, eds.). Academic Press, New York, In press.

Pringle, J. R., Hartwell, L. H. (1981). *Saccharomyces cerevisiae* cell cycle. In: Molecular Biology of the yeast *Saccharomyces* (J. Broach, J. Strathern and E. Jones, eds.) in press. Cold Spring Harbor Laboratory, Cold Spring Harbor, New York.

Rai, R, Carter, B. L. A. (1981). The isolation of nonsense mutations in cell division cycle genes of the yeast *Saccharomyces cerevisiae.*

Reed, S. I. (1980a). The selection of *S. cerevisiae* mutants defective in the start event of cell division. Genetics 95: 561–577.

Reed, S. I. (1980b). The selection of amber mutations in genes required for completion of start, the controlling event of the cell division cycle of *S. cerevisiae.* Genetics 95: 579–588.

Reed, S. I., Ferguson, J., Groppe, J. C. (1981). A preliminary characterisation of the transcriptional and translational products of the yeast cell division cycle gene *cdc28.* In: Molecular and Cellular Biol. (in press).

Reid, B. J., Hartwell, L.H. (1977). Regulation of mating in the cell cycle of *Saccharomyces cerevisiae.* J. Cell Biol. 75: 355–365.

Rothman-Denes, L. B, Cabib, E. (1970). Two forms of yeast glycogen synthetase and their role in glycogen accumulation. Proc. Natl. Acad. Sci. USA 66: 967–974.

Schenberg-Frascino, A., Moustacchi, E. (1972). Lethal and mutagenic effects of elevated temperature on haploid yeast. I. Variations in sensitivity during the cell cycle. Mol. Gen. Genet. 115: 243–257.

Simchen, G. (1978). Cell cycle Mutants. Ann. Rev. Genet. 12: 161–191.

Slater, M. L., Sharrow, S. O., Gart, J. J. (1977). Cell cycle of *Saccharomyces cerevisiae* in populations growing at different rates. Proc. Natl. Acad. Sci. USA 74: 3850–3854.

Stozler, D., Kiltz, H. H., Duntze, W. (1976). Primary structure of $\alpha$ factor peptides from *Saccharomyces cerevisiae.* Eur. J. Biochem. 69: 397–400.

Sudbery, P. E., Goodey, A. R., Carter, B. L. A. (1981). Genes which control cell proliferation in the yeast *Saccharomyces cerevisiae.* Nature 288: 401–404.

Tyson, C. B., Lord, P. G., Wheals, A. E. (1979). Dependency of size of *Saccharomyces cerevisiae* cells on growth rate. J. Bacteriol. 138: 92–98.

Vezinhet, F., Kinnaird, J. H., Dawes, I. W. (1979). The physiology of mutants derepressed for sporulation in *Saccharomyces cerevisiae.* J. Gen. Microbiol. 115: 391–402.

von Meyenburg, H. K. (1968). Der Sprossungszyklus von *Saccharomyces cerevisiae.* Pathol. Microbiol 31: 117–127.

Walton, E. F. Carter, B. L. A., Pringle, J. R. (1979). An enrichment method for temperature-sensitive and auxotrophic mutants of yeast. Mol. Gen. Genet. 171: 111–114.

Wilkinson, L. E., Pringle, J. R. (1974). Transient G1 arrest of *S. cerevisiae* cells of mating type $\alpha$ by a factor produced by cells of mating type **a**. Exp. Cell Res. 89: 175–187.

Wolfner, M., Yap, D., Messenguy, F., Fink, G. R. (1975). Integration of amino acid biosynthesis into the cell wall of *Saccharomyces cerevisiae.* J. Mol. Biol. 96: 273–290.

# 2

# Genetic Control and Gene Expression During Meiosis and Sporulation in *Saccharomyces cerevisiae*

Ian W. Dawes*

## Introduction

Review articles frequently begin with an apologia to explain why the subject merits extensive study. Why for yeast meiosis and sporulation should this be necessary? The intelligent reader will doubtless appreciate that yeast are ideal subjects for experimental study, and that there is a fascination in trying to understand at a molecular level how an organism switches from one phase of its life cycle to another. He or she will be curious as to how meiosis, that process fundamental to those with any interest in sex (and variation), takes place and how it differs from mitosis. Some would see in the processes of yeast meiosis and concommitant spore formation an example of cellular differentiation and want to understand the mechanisms that control a cell developing from one form to another. Others may view the processes in the context of their possible involvement in the evolution of eukaryotes (Cavalier-Smith 1981), or from the more practical standpoint that with the recent insight into the regulation of the related phenomenon of homothallism in yeast (Klar *et al.* 1981; Nasmyth *et al.* 1981), sporulation may provide a system for studying in detail other mechanisms of eukaryotic regulation and that these might be exploited in practical ways. The overlap between the processes of cell division and mitosis, on the one hand, and spore formation and meiosis, on the other, may excite interest; failing all else there may remain to the geneticist curiosity concerning the tricks used to study the genetics of a process that only occurs in diploids or cells of higher ploidy.

A number of reviews cover some of the areas alluded to above. Earlier ones

*Department of Microbiology, University of Edinburgh, West Mains Road, Edinburgh, Scotland.

concerned with the physiology and biochemistry of sporulation include: Tingle *et al.* (1973); Haber and Halvorson (1972, 1975); Haber *et al.* (1975), and Fowell (1975). The genetics of yeast sporulation have been covered in some detail by Esposito and Esposito (1974, 1975). These latter articles, together with the reviews of meiosis in protists (Heywood and Magee 1976) and of the genetic control of meiosis (Baker *et al.* 1976), constitute a clear introduction to the more genetic aspects of the subject. A brief but lucid summary of progress in the genetics and molecular biology of sporulation by Esposito (1980) is also worth referring to if it can be obtained.

The present review, therefore, is concerned with briefly surveying the methods that have been found useful in the genetic analysis of sporulation, and reviewing those areas not covered in detail recently, especially the initiation of sporulation, gene involvement, and gene expression in sporulation of *Saccharomyces cerevisiae.*

## Outline of Sporulation in *Saccharomyces cerevisiae*

### The Life Cycle of Heterothallic Yeast

Figure 2.1 illustrates the life cycle of heterothallic strains of *Saccharomyces cerevisiae*. Both diploid and haploid organisms can exist stably undergoing cell division via mitosis and budding; when diploids that are heterozygous (**a**/$\alpha$) for the two alleles of the single gene determining mating type are starved under appropriate conditions they cease dividing and undergo meiosis. The four haploid nuclei produced within the mother cell (ascus) are packaged separately into ascospores. Sporulation is the term used for the integrated series of events from the completion of cell division through meiosis to the formation of mature ascospores. In normal four-spored asci two spores are of **a** mating type and two of $\alpha$ mating type. One-, two-, and three-spored asci are sometimes formed; in some strains these are common, especially when the growth medium for the cells prior to starvation was poor. In most such cases these result from a failure of some of the normal meiotic products to be packaged; however, in certain mutants (Grewal and Miller 1972; Klapholtz and Esposito 1980) two-spored asci may arise from defective meiosis.

If **a** and $\alpha$ spores are separated by dissection on germination they give rise to stable haploid strains capable of mitotic cell division. If **a** and $\alpha$ cells are mixed they undergo mating to produce a transient heterokaryon. Nuclear fusion occurs and the heterozygote undergoes budding to produce **a**/$\alpha$ diploids. This mating process has been studied as a morphogenetic process in its own right since it does illustrate some aspects of intercellular interaction that are not seen during sporulation (Sena *et al.* 1975).

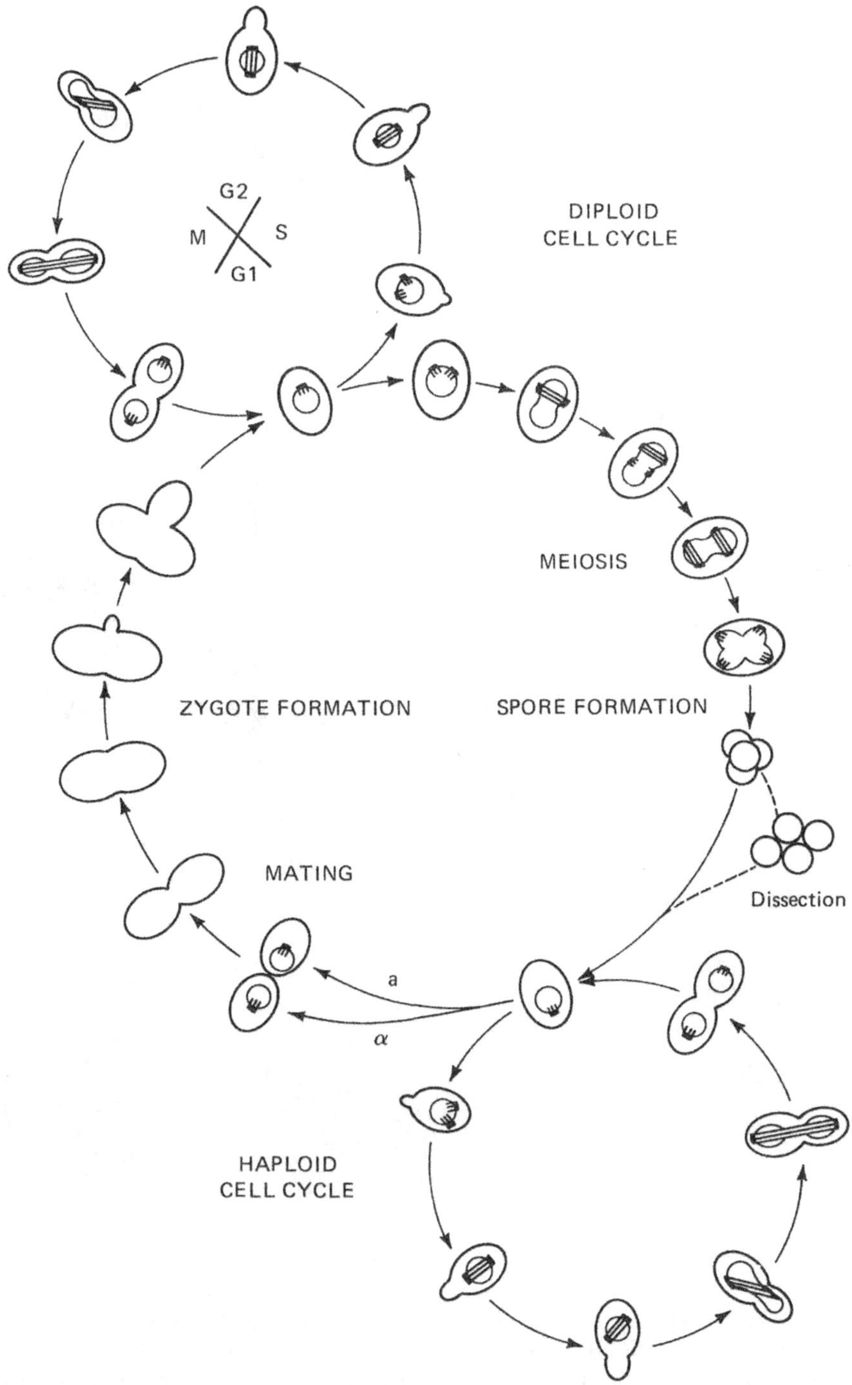


**Figure 2.1.** The life cycle of heterothallic strains of *Saccharomyces cerevisiae.* The nuclear configuration is indicated. G1, G2, S, and M refer to the phases of the cell division cycle.

## Sporulation Changes: Morphology

The morphological events of sporulation in wild-type *S. cerevisiae* have been analyzed in considerable detail by a number of groups (Engels and Croes 1968; Lynn and Magee 1970; Moens 1971; Moens and Rapport 1971; Guth *et al.*

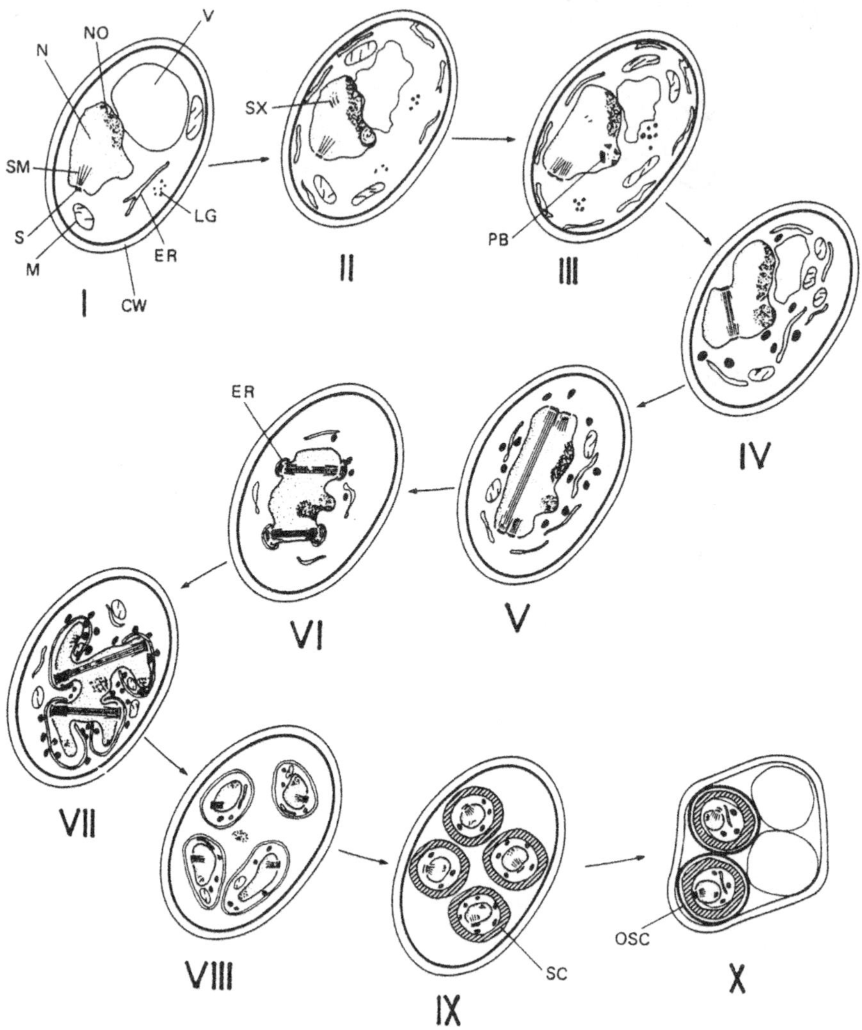


**Figure 2.2.** The cytology of meiosis and sporulation in *Saccharomyces cerevisiae*. Redrawn after Esposito and Esposito (1975) and Fowell (1975). The structures indicated are cell wall (CW), endoplasmic reticulum (ER), nucleus (N), nucleolus (NO), mitochondrion (M), spindle plaque (SP), spindle microtubule (SM), polycomplex body (PB), spore coat (SC), outer spore coat (OSC). (Reproduced with permission of the Society for General Microbiology from Dawes 1981.)

1972; Beckett *et al.* 1973) and in several asporogenous mutants by Moens *et al.* (1974). The changes are summarized in Figure 2.2, in which 10 stages of sporulation are indicated. This includes an additional stage to the nine previously proposed (Moens and Rapport 1971; Esposito and Esposito 1975), the tenth stage being added to indicate the deposition at a late stage of a thin electron-dense layer as an outer spore coat.

Stage I represents a vacuolate cell that has become starved with a nucleus containing a single spindle pole body associated with the nuclear membrane. In Stage II a "polycomplex body" appears in the nucleus, this structure containing multiple synaptonemal complex-like elements. During Stage III the spindle pole body divides. In Stage IV the first meiotic division proceeds with the two spindle pole bodies opposed on the nuclear membrane and connected by a spindle of microtubules; the synaptonemal complex elements of the polycomplex body disappear. During these early stages the vacuole decreases in size and may fragment and lipid granules begin to increase in size and number. Meiosis I ends during Stage V as the nucleus has become elongated by the spindle and the two opposed spindle pole bodies duplicate. In Stage VI each newly formed pair of spindle bodies migrates to oppose each other and spindles for the second meiotic division are formed. The endoplasmic reticulum begins to initiate a primordial prospore wall by forming a double membrane layer, beginning at the four spindle pole bodies. During Stage VII the spindles extend and the nucleus assumes a highly lobed form. The prospore wall membranes extend and begin to surround the lobes of the nucleus. This enclosure is completed in Stage VIII and the spindle pole body detaches from the prospore wall. In Stage IX the synthesis of spore coat material (glucan and mannan similar, but not identical in composition to that of the vegetative cell wall) occurs between the two membranes derived from the endoplasmic reticulum. Stage X has been added to highlight that an outer spore coat forms. This is thin, and electron dense; from electrophoretic evidence, it appears to be composed largely of protein (Briley *et al.* 1970) and is likely to contain the determinant of the spore surface antigen (Snider and Miller 1966).

The relatively thick spore wall layers contain electronegative polymers (mannan phosphate residues) and the yeast spore is more resistant than vegetative cells to solvents (Zakharov and Inge-Vechtomov 1964; Dawes and Hardie 1974) and to some extent also to heat (Put and Sand 1975), but it is still very sensitive when compared with the very resistant bacterial endospore.

## Physiological and Biochemical Changes

Some of the events that accompany sporulation are summarized in Figure 2.3 with an indication of their timing relative to the morphological changes. This is by no means complete since a fair amount of research has been done on biochemical changes such as the utilization of acetate (Croes 1967), induction of the glyoxylate cycle (Miyake *et al.* 1971), synthesis and degradation of var-

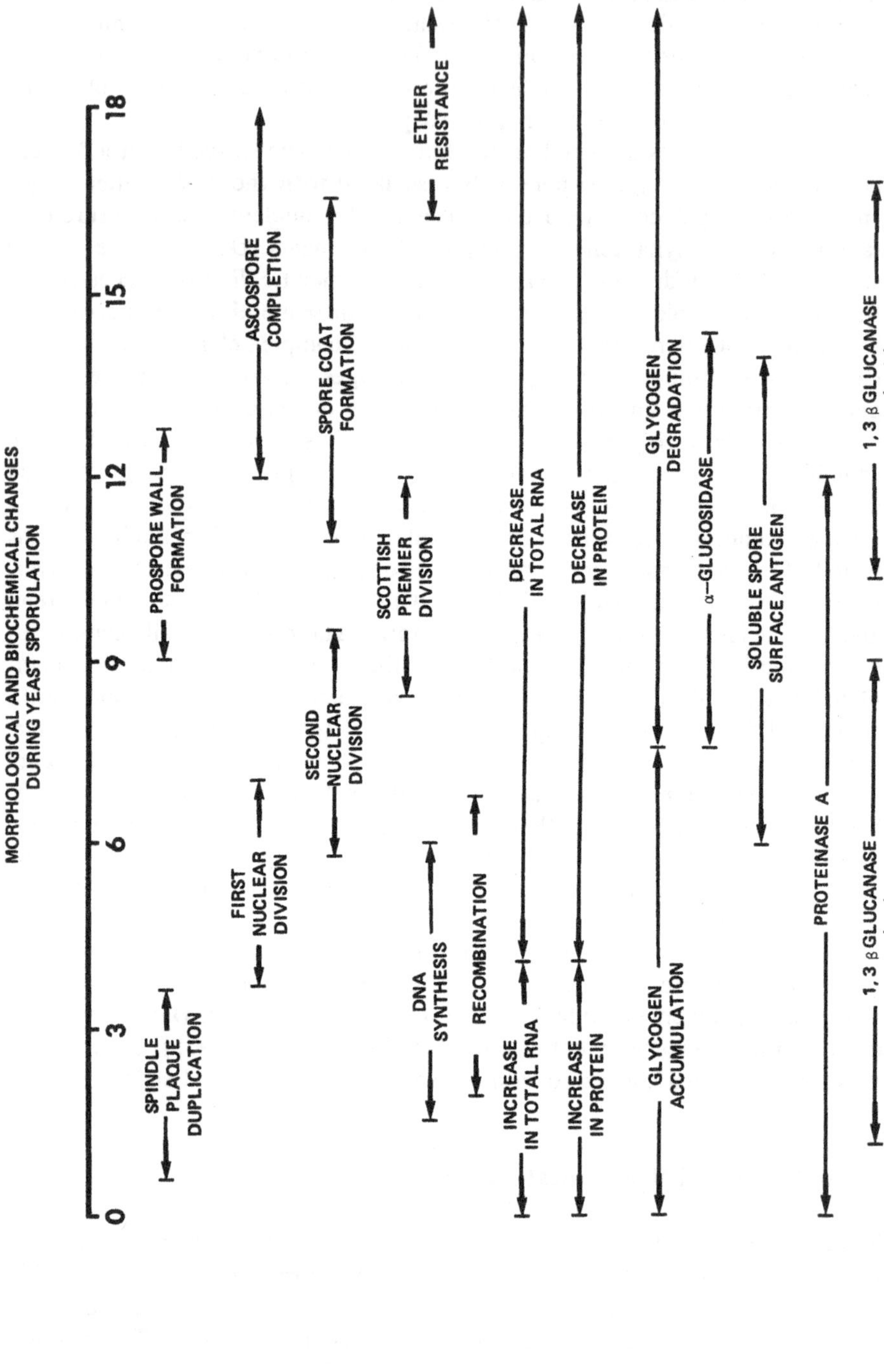
MORPHOLOGICAL AND BIOCHEMICAL CHANGES DURING YEAST SPORULATION
0
3
6
9
12
15
18
SPINDLE PLAQUE DUPLICATION
PROSPORE WALL FORMATION
FIRST NUCLEAR DIVISION
ASCOSPORE COMPLETION
SECOND NUCLEAR DIVISION
SPORE COAT FORMATION
DNA SYNTHESIS
SCOTTISH PREMIER DIVISION
RECOMBINATION
ETHER RESISTANCE
INCREASE IN TOTAL RNA
DECREASE IN TOTAL RNA
INCREASE IN PROTEIN
DECREASE IN PROTEIN
GLYCOGEN ACCUMULATION
GLYCOGEN DEGRADATION
α–GLUCOSIDASE
SOLUBLE SPORE SURFACE ANTIGEN
PROTEINASE A
1, 3 β GLUCANASE (exo)
1, 3 β GLUCANASE (endo)

ious classes of macromolecules such as DNA (Croes 1966; Esposito *et al.* 1969), RNA and protein (Croes 1967; Esposito *et al.* 1969; Mills 1972, 1980; Hopper *et al.* 1974; Curiale *et al.* 1976), and of storage polymers such as trehalose and glycogen (Kane and Roth 1974). A number of enzyme changes have also been studied (Betz and Weiser 1976; Matur and Berry 1978; Magee, 1975; Colunna and Magee 1978; del Rey *et al.* 1980).

In only a few cases has the relevance of these changes to sporulation been established. Sporulation is usually induced by a sudden shift of cells from a medium supporting rapid growth to one in which growth cannot occur (usually just 2% w/v potassium acetate), and under these conditions even nonsporulating strains (e.g., haploids, **a**/**a** and $\alpha/\alpha$ diploids) would undergo substantial shifts in metabolism. Some of these events may also be necessary to sporulation; others may be irrelevant. An obvious test of relevance is isolating a mutation that specifically affects the event in question (e.g., mutation in the structural gene of an enzyme) and testing its effect on sporulation. In only a few cases has this been done; for example, isocitrate lyase defective mutants do not sporulate and a functional glyoxylate cycle is, therefore, necessary for sporulation (Miclo and Guesdon 1980). This analysis has not been done in most cases and the relevance of events to sporulation is not clear. Another test of relevance is whether or not the event occurs in nonsporulating cells, such as haploids, placed under sporulation conditions. The quest for sporulation-specific processes has been considered important as a preliminary to analyzing regulation of gene action and gene expression during sporulation since such events are unlikely to be the result of starvation-induced changes that are unnecessary for sporulation. Again, few biochemical events have been characterized in adequate detail in this way: premeiotic DNA synthesis only occurs in **a**/$\alpha$ diploids, or in haploids disomic for chromosome III in **a**/$\alpha$ configuration, but not in haploids or **a**/**a** or $\alpha/\alpha$ diploids (Roth and Lusnak 1970), while the $\alpha$-glucosidase that appears relatively late in sporulation (Colunna and Magee 1978) is also specific to **a**/$\alpha$ cells. This enzyme is probably responsible for the degradation of glycogen, the only change in carbohydrate metabolism that was found by Kane and Roth (1974) to occur in **a**/$\alpha$ cells, but not **a**/**a** and $\alpha/\alpha$ diploids.

Another indirect method that can confirm the association of an event with sporulation is to test the effect of mutations conferring asporogeny *(spo)* on the occurrence of the event. This depends on the assumption that many sporulation events follow a dependent sequence in which later events depend on the successful completion of all prior events. This view has been shown to be only partly true for bacterial sporulation (Piggot and Coote 1976; see also Jenkinson *et al.* 1980) in which a much more detailed correlation between biochemistry

---

**Figure 2.3.** Timing of some morphological and biochemical changes during yeast sporulation. The times given are approximate only, some strains proceeding more rapidly through the program, others more slowly.

and morphological point of block of asporogenous mutants has been attempted, and it would be unwise to rule out a "late event" as being unimportant on the grounds that it took place in mutants blocked at a stage before that event would normally have occurred.

In comparing the timing of sporulation events, and relating them to each other, the commonly used changes include: premeiotic DNA synthesis (Croes 1966); nuclear configuration as seen by staining with Giemsa (Esposito *et al.* 1972) or fluorescent dyes (Williamson and Fennel 1975) and the appearance of phase-bright asci. The resistance of developing ascospores to diethyl ether can also be used as a late marker, occurring some time after asci can be readily distinguished, although care must be taken to standardize the time and temperature of the ether treatment since the extent of resistance to ether continues to develop for some time (Dawes and Hardie 1974 and unpublished observations).

Several events can also be detected provided suitably marked strains are used. For example, commitment to recombination at a particular locus (the time at which a cell appears as a recombinant after it has been plated on a selective growth medium) can be scored if the strain is heterozygous for two noncomplementing alleles of a single auxotrophic marker and commitment to meiotic segregation can be detected by screening for the formation of homozygosity from a heterozygous condition (Esposito and Esposito 1974a); for example, complementing *ade2-40* and *ade2-119* alleles in a heteroallelic diploid (white colonies) give rise to red *(ade2-40)* or pink *(ade2-119)* sectoring when cells in a culture have become committed to meiotic segregation (Dawes *et al.* 1980). The precise physiology of "commitment" as determined in this way is not known since it must represent the outcome of a number of processes of nutrient uptake and regulation that have yet to be studied in detail. Recently, however, Davidow and Byers (1980) have devised an elegant technique for detecting precisely when recombination at a particular locus has been completed by observing the formation of recombinant restriction fragments in a diploid that was heteroallelic for the pattern of restriction sites at a region around the *SUP4* locus. With this approach they were able to show recombinant molecules were present after about 6 h of sporulation at the approximate time of pachytene.

By density-gradient centrifugation it is possible to separate sporulating yeast populations into subfractions; at least eight distinct fractions can be obtained from cultures just beginning to produce asci visible by phase-contrast microscopy (Dawes *et al.* 1980). Cells in these fractions were found to differ in the stage they had reached in meiosis and sporulation, those more advanced in sporulation banding at lower density. Figure 2.4 illustrates this banding, and the heterogeneity of sporulating populations induced to sporulate by the commonly used procedure of Fast (1973). From this a number of important conclusions were drawn: (1) cells pass through a series of physiological states that affect their density and from the sharpness of the bands that can be obtained they undergo transition from one state to the next in a fairly rapid manner; (2)

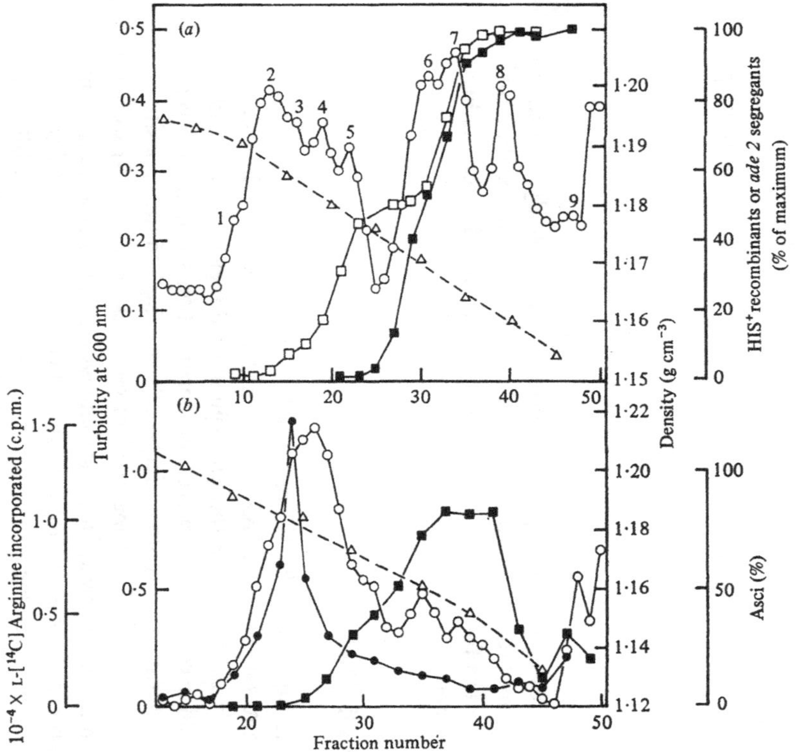


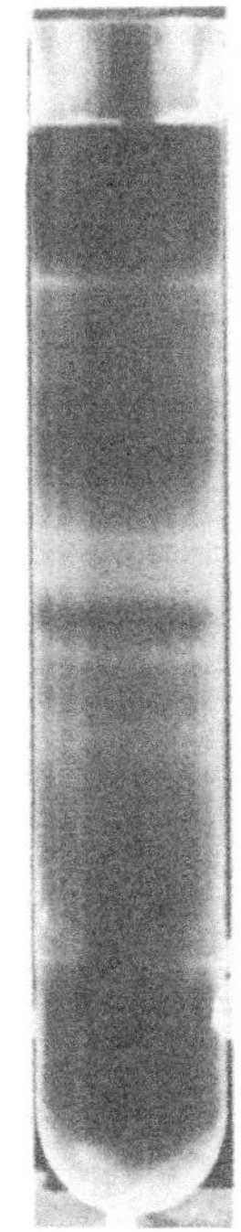

(c)

**Figure 2.4.** Heterogeneity of sporulating populations in cultures induced to sporulate by resuspension of vegetative cells in potassium acetate. (a) Separation of a 16.5-h sporulation culture on a linear Urografin gradient. Fractions were collected from the bottom of the gradient: ○, culture turbidity; □, commitment to recombination at the *HIS 4* locus; ■, commitment to meiosis; △, density of Urografin. Numbered peaks correspond to bands seen in the gradient. (b) Incorporation of labeled L-arginine into a 21-h sporulation culture. After pulse labeling with arginine for 10 min the culture was separated on a Urografin gradient. Fractions were collected and assayed for ●, arginine incorporation into trichloracetic acid-precipitable material; ■, percentage asci. (c) Cell banding seen for a 20-h sporulation culture after centrifugation through a linear Urografin density gradient. (Reproduced with permission from Dawes *et al.* 1981.)

the technique provides material for biochemical analysis that is considerably more homogeneous with respect to stage of sporulation than is commonly used, and it highlights the extensive heterogeneity of sporulating cultures derived from resuspension experiments; and (3) several labeled amino acids were incorporated into protein preferentially by cells uncommitted to recombination and meiosis and cultures in which a few asci were already appearing. Care therefore needs to be exercised in interpreting the results of pulse-labeling studies.

## Special Genetic Techniques

The basic techniques for genetic analysis in *Saccharomyces,* including methods for *in vitro* manipulation, have been described in detail elsewhere (Mortimer and Hawthorne 1966, 1969, 1973, 1975; Prescott 1975; Hicks *et al.* 1979; Beggs 1978) and a brief outline of these has been provided in an appendix. There are, however, several problems that arise in the genetic analysis of sporulation and these are discussed in more detail.

### Mutant Selection

Sporulation is expressed only in diploids or cells of higher ploidy, and the isolation of recessive mutations by straightforward methods would be difficult. A number of approaches for avoiding this problem have been devised, including the following methods.

#### Use of Homothallic Diploids

This method was introduced by Esposito and Esposito (1969, 1975). Homothallic yeast strains carrying the HO allele of the gene for homothallism exist as stable diploids in the **a**/$\alpha$ state and undergo normal meiosis and sporulation to produce haploid ascospores. However, on germination some of the progeny arising from each germinated spore undergo a "switching" to the opposite mating type (for a description of this mechanism see Hicks *et al.* 1979) so that **a** and $\alpha$ haploids that are otherwise isogenic mate within the microcolony formed on plating ascospores. This gives rise to a diploid colony that is homozygous at all loci except the mating-type locus. By treating ascospores with ultraviolet (UV) or x-radiation the Espositos obtained diploid mutants that were unable to sporulate because of the presence of homozygous recessive mutations; that is,

$$\underset{\text{diploid}}{\frac{\mathbf{a}\ HO}{\alpha\ HO}} \xrightarrow{\text{sporulation}} \underset{\text{spores}}{\begin{matrix}\mathbf{a}\ HO\\ \alpha\ HO\end{matrix}} \xrightarrow[\text{mutagenesis}]{\text{UV}} \underset{\text{spore}}{\mathbf{a}\ HO\ spo} \xrightarrow{\text{germination}} \begin{matrix}\mathbf{a}\ HO\ spo\\ \alpha\ HO\ spo\end{matrix} \xrightarrow{\text{mating}} \underset{\text{mutant diploid}}{\frac{\mathbf{a}\ HO\ spo}{\alpha\ HO\ spo}}$$

Ascospores are, however, relatively unaffected by the chemical mutagens ethyl methanesulfonate (EMS) or *N*-methyl-*N'*-nitro-*N*-nitrosoguanidine (NG), which are much more effective in inducing mutations than UV or x-radiation and which may provide a different spectrum of mutations. We have used a modification of the preceding method by treating vegetative populations of a homothallic diploid strain with the chemical mutagen, allowing the culture to recover for several generations in presporulation medium, then resuspending the cells in sporulation medium. Recessive mutations affecting sporulation were recovered by plating this sporulated culture after it had been subjected to harsh ether treatment (Dawes and Hardie 1974) to kill vegetative cells and most, but not all, ascospores.

## Strains Disomic for Chromosome III

Haploid strains that are disomic for chromosome III and heterozygous for mating type have been recognized, and when these are put in sporulation medium they exhibit some of the early sporulation events, including premeiotic DNA synthesis and enhanced levels of genetic recombination (Roth and Fogel 1971). By using **a**/$\alpha$ disomes that were heteroallelic at the *leu2* locus on chromosome III these authors have been able to isolate mutants that are defective in meiotic recombination.

## Normal Diploids; Mitotic Recombination

A special class of mutations that lead to asporogeny have been isolated, arising directly after mutagenesis of strains that are derepressed for sporulation (Dawes 1975). These mutations were not completely recessive, and both heterozygous and homozygous strains were obtained among revertants obtained by mutagen treatment of diploids. The frequency with which homozygotes were detected indicated that gene conversion or mitotic recombination events fairly often accompanied mutation—often enough for selection of asporogenous mutants by direct mutagen treatment of diploids. Dominant mutations have also been obtained by this route (Simchen *et al.* 1972).

## Selection of Specific Mutant Classes

The range of mutations affecting sporulation and knowledge of the physiology of sporulation has been extended by isolating mutations affecting a specific function and testing diploids homozygous for such mutations for their ability to sporulate. This rather haphazard approach is sometimes the result of a deliberate survey of the involvement of a biochemical function in sporulation; more often it is the result of an adventitious finding in the course of genetic analysis. If all newly isolated mutations had been tested for their effect on phases of the life cycle other than vegetative growth, the analysis of development might have been rather more rapid than it has been.

Examples of special interest include the following.

### *Mutations Affecting the Control Over Sporulation Exerted by the Mating-Type Locus.*

Hall and Hopper (1973) isolated these directly by mutagenesis of nonsporulating **a**/**a** and $\alpha/\alpha$ diploids and screening for sporulating mutants. Mutations affecting mating type have also been isolated (Strathern *et al.* 1981) and some of these have been shown to affect sporulation.

### *Some Radiation-Resistance (rad) Mutations*

These have also been shown to affect sporulation markedly and mutants carrying these have provided useful strains to probe the relationship between meiotic and mitotic recombination and DNA repair systems (Baker *et al.* 1976).

### *Cell Division Cycle (cdc) Mutations*

These have been isolated mainly by Hartwell and his colleagues in haploids (Hartwell 1974), have been studied in homozygous diploids by Simchen and his colleagues to analyze the dependence of sporulation on functions that are also essential for the control or maintenance of the mitotic cell cycle (Simchen, 1974; Shilo *et al.* 1978).

### *Mutations Affecting the Activity of Specific Enzymes*

These mutations have also been shown in several cases to affect sporulation directly. For example, the necessity for protein turnover or modification during sporulation has been demonstrated by the isolation of mutations affecting the structural gene for proteinase B (Zubenko *et al.* 1979); a mutation causing low activities of proteinase A also prevents sporulation, but it is probably not in the structural gene for this enzyme (Betz 1977). Certain mitochondrial mutations *(mit)* affecting known respiratory chain components do not prevent sporulation, showing that not all mitochondrial functions are essential to sporulation (Pratje *et al.* 1979). Some are essential, since respiratory-deficient *petite* strains are unable to sporulate (Ephrussi and Hottinguer 1951). Several other mutations affecting enzyme activities and also sporulation are listed in Table 2.1. This list is illustrative, but far from comprehensive. Unfortunately, although it does provide a clear test of the necessity of an event for sporulation this approach is

**Table 2.1.** Some Mutations Affecting Sporulation in *Saccharomyces*

| Mutation | | Reference |
|---|---|---|
| General sporulation mutations | | |
| Asporogeny | *spo1* to *spo13*; *SPO98* | Esposito and Esposito (1975) |
| (initiation) | *spo50, 51, 53*[a] | Dawes and Calvert (1981) |
| Derepressed | *spd1, 3, 4*[a] | Dawes (1975) |
| Mating control | *CSP1* | Hall and Hopper (1973) |
| Specific functions | | |
| Mating type | *mata, matα, rme1 sca* | Mackay and Manney (1974); Strathern *et al.* (1981); Kassir and Simchen (1976); Rine *et al.* (1981); Gerlach (1974) |
| Cell division cycle | *cdc* (DNA synthesis and nuclear division) | Simchen (1974) |
| | *cdc25, 28, 35; tra3*[a] | Shilo *et al.* (1978) |
| Radiation sensitivity | *rad6, 50, 52, 53, 55, 57* | Game and Mortimer (1974); Game and Cox (1971); Malone and Esposito (1981); Game *et al.* (1980); Prakash *et al.* (1980) |
| Meiotic recombination Gene conversion | *mei1, 2, 3; con1, 2, 3; cor1* to *4; rec 2, 3, 4* | Roth and Fogel (1971); Fogel and Roth (1974); Williamson and Fogel (1980) |
| Proteinase B | *prb1* | Zubenko *et al.* (1979) |
| Isocitratase | — | Miclo and Guesdon (1980) |
| Phosphoenolpyruvate carboxylase | — | Perea and Gancedo (1980) |
| Mitochondrial coded | *mit*⁻ (variable sporulation) | Pratje *et al.* (1979) |
| Suppressor genes | *SUP3*, etc. | Rothstein *et al.* (1976); Liebman *et al.* (1976) |

[a]Mutations affecting initiation of sporulation.

very tedious and time-consuming, and if undertaken in a deliberate way depends on a judicious, intuitive choice of biochemical function.

## Recognition of Sporulation Mutant Clones

Unlike a number of other sporulation systems that are commonly studied, there is no direct technique yet available for distinguishing a sporulating *Sacch. cere-*

*visiae* colony from a nonsporulating one; sporulated cultures are not pigmented, nor do they stain with $I_2$ vapor as for *Schizosaccharomyces pombe* (Bresch *et al.* 1968). For this reason, in selecting temperature-sensitive asporogenous mutants the Espositos used direct microscopic screening, a procedure that must have required considerable patience and effort. There are, however, a number of ways that have been useful in selecting ascospores or asci. One that is fairly general in application depends on the greater resistance of ascospores to diethyl ether compared with that of vegetative organisms (Dawes and Hardi 1974). The technique is suitable for selecting spores in vegetative populations even when spores are present at frequencies as low as 1 in $10^6$ cells; by using glass petri dishes and treating replicated colonies it was also possible to detect nonsporulating colonies on plates. The method has also been used for strain construction when spore progeny from a diploid cross need to be screened rapidly. We have found since publication of the original paper that in liquid suspension spores of some strains are killed too rapidly using the conditions described (shaking with an equal volume of ether at room temperature for 5 min); in such cases exposure to ether at 0°C for 1 min usually suffices. The resistance of ascospores to ether develops gradually, starting several hours after asci can be seen by microscopic examination, and at first the difference in resistance between vegetative cells (or sporulating cells) and spores is slight, increasing with maturation of the ascospores (Hardie and Dawes, unpublished).

Several other methods are available for detecting when sporulation (or at least recombination or meiotic segregation) has occurred. Diploids carrying recessive resistance markers (e.g., canavanine resistance, *can1;* or cycloheximide resistance, *cyh2*) in heterozygous state are sensitive to the antibiotic or antimetabolite; half of the haploid spores formed will carry the resistance mutation and for this reason canavanine resistance has been used widely in random spore analysis (Sherman and Roman 1963; Mortimer and Hawthorne 1969). Interestingly, whereas sporulation of homozygous cycloheximide-sensitive strains is inhibited by adding cycloheximide at any time up to the time of ascus formation, in strains that are heterozygous for *cyh1* sporulation becomes resistant to the drug some time before ascus formation, indicating that the altered ribosomal protein conferring resistance is synthesized during the later stages of sporulation and that there is some separate functioning of the haploid daughter genomes before they are completely packaged into ascospores (Emanuel and Magee 1981).

## Genetic Analysis of Asporogenous Mutations

A majority of the "classical" genetic manipulations used to characterize mutations in terms of complementation, dominance, epistasis, or linkage, or even just to ensure that a mutant is the result of a single gene mutation, depend on meiotic recombination and spore formation. Alternative strategies must, therefore, be adopted when handling asporogenous mutants. For example, Esposito *et al.* (1969) isolated temperature-sensitive asporogenous mutants so that genetic characterization by conventional techniques was possible at the permissive temperature.

When the mutation conferring asporogeny is not conditional (e.g., some regulatory elements are not likely to yield temperature-sensitive pheno types when altered by mutation), then there are still a few possibilities:

### *Mating to Haploids of Infrequent Spores*

Many mutations preventing sporulation are not absolute in their effect, and a very low proportion of viable spores are formed under sporulation conditions. By mixing populations of such diploids that have been exposed to sporulation conditions with a standard haploid strain carrying complementing auxotrophic markers, one can select for rare heterozygous diploids formed by the mating of a germinated spore to the haploid. This type of "spore mating" is the easiest way to carry out diploid construction when homothallic diploids are to be used as parents, for example,

$$\frac{\mathbf{a}\ HO\ arg\ 4}{\alpha\ HO\ arg\ 4} \longrightarrow \mathbf{a}\ HO\ arg4 \text{ spore} \times \alpha\ ade5 \text{ haploid } \alpha \longrightarrow \frac{\mathbf{a}\ HO\ arg4\ ADE5}{\alpha\ ho\ ARG4\ ade5}$$

$$\searrow \alpha\ HO\ arg4 \text{ spore}$$

$$\text{or even } \frac{\mathbf{a}\ HO\ ade5}{\alpha\ HO\ ade5} \longrightarrow \mathbf{a}\ HO\ ade5 \text{ spore} > \text{cross} \longrightarrow \frac{\alpha\ HO\ arg4\ ADE5}{\mathbf{a}\ HO\ ARG4\ ade5}$$

### *Tetraploid Genetics*

When a diploid mutant is completely asporogenous it is still possible to carry out genetic analysis if the mutation is recessive. By mitotic gene conversion (induced by low doses of UV) **a**/**a** and $\alpha/\alpha$ diploids can be formed from both mutant and wild-type strains. These can then be crossed to construct suitable tetraploids that will sporulate to generate four diploid ascospores. The assignment of sporulation genes to the diploid progeny can, however, be a nightmare since sporulation is also subject to mating-locus control and many tetrads will contain **a**/**a** and/or $\alpha/\alpha$ spores. Homothallism genes further complicate the issue since **a**/**a** and $\alpha/\alpha$ diploids in the presence of the *HO* gene will undergo mating-type switching.

## Gene Function and Gene Expression During Sporulation

Of the reasons for studying yeast sporulation, those that have been expressed most often are concerned with the process as an example of cell differentiation in a relatively simple eukaryote, and with an interest in understanding at a molecular level the important genetic events of meiosis and meiotic recombination. Research has, therefore, been directed largely toward identifying cytological, biochemical and genetic events that occur during sporulation and evaluating the extent to which such events are important to the process.

Some events may be merely a physiological response to the change in conditions that is needed to induce sporulation, and may play no part at all in meiosis or spore formation. Such changes would probably occur in nonsporulating cells in the same changing conditions. Other events of more interest are those that are *necessary* for sporulation to occur, these could also occur in nonsporulating cells or could be specific to sporulation. In bacterial sporulation it has been found that some events occurring during sporulation are not necessary for successful formation of a spore, but are necessary for the subsequent germination or outgrowth steps returning the spore to the vegetative state (Dion and Mandelstam 1980; Moir 1981).

Identifying and evaluating the changes involved is, however, only a small part of understanding the mechanism of meiosis and sporulation. It is the interaction between these events in time and space that is the essence of cell differentiation, and many questions of regulation, molecular architecture, and assembly of structures are as yet only partly answered. For example, how are the events controlled so that they occur to the right extent at the right time? Are there genes uniquely concerned with sporulation, when are they expressed, and how is their expression regulated? What switches the cell from the vegetative phase to sporulation? How are structures made from the components in the right arrangement at the right place?

More progress has been made in identifying events that occur during yeast sporulation than in determining their interaction, and despite considerable effort few essential sporulation events have been identified in biochemical as well as genetic terms. This is the result of a problem that is fundamental to, and hinders, all morphological research: It is relatively easy to identify genes that are essential to a morphological process by isolating mutants that are affected in their ability to complete it. To then determine the biochemical defect is very difficult, so that in few, if any, mutants isolated as being aspo-

**Table 2.2.** Requirements for Initiation of Meiosis and Spore Formation

1. Heterozygosity for mating-type alleles
   **a**/$\alpha$, **a**/**a**/$\alpha$, etc., sporulate whereas **a**, $\alpha$, **a**/**a**, $\alpha/\alpha$, **a**/**a**/**a**/, etc., do not.
2. Completion of mitotic cell division
   Sporulation is initiated only in G1 phase of the cell cycle.
3. Nutritional conditions
   a. Depletion or limitation of the carbon source, mitochondrial induction probably necessary.
   b. Depletion or limitation of the nitrogen source, especially $NH_4$.

rogenous *(spo)* has the mutant gene product been rigorously identified with a biochemical function. The alternate approach of identifying (or guessing) biochemical changes that may be important, then probing their involvement by isolating mutants defective in that biochemical function is, as was discussed earlier, tedious, slow, and to some extent dependent on serendipity. This gloomy view is, however, not warranted since many clues to the possible nature of genetic interactions involved in sporulation have been gleaned by studying *spo* mutants without knowledge of the biochemical lesion. Some progress toward analyzing gene involvement and gene expression (which are not necessarily the same thing) during sporulation is also emerging.

### Initiation of Sporulation: Mitosis and Cell Division versus Meiosis and Spore Formation

It is not *a priori* necessary that there should exist a single mechanism whereby vegetatively dividing cells are "switched" into a sporulation pathway: There may be several mechanisms, there may even be none (in the sense that no single event or series of events may differ in its regulation from that of many others). There are, however, several reasons to consider initiation as a separate phenomenon. It concentrates attention on the transition of the cell from mitosis to meiosis; it is fashionable to think in terms of switches and cells making "decisions." On more pragmatic grounds, there are mutations in a restricted number of genes that "derepress" sporulation, "switching" it on under conditions in which wild-type cells grow rather than sporulate, and from their existence it can be inferred that there is an initiation mechanism of some sort.

There are several conditions that must be met before a yeast cell initiates (or is at least committed to) sporulation, and these are summarized in Table 2.2. Of these, control from the mating locus is intrinsic, being a property determined by the genetic constitution of the cell and not subject to alteration once the diploid organism exists. On the other hand, nutritional control is extrinsic in that it reflects the cell's response to the status of the medium. Related to this response to external conditions is the physiological condition of the cells with respect to the stage they have reached in mitotic cell division, and these two aspects are considered together in the following.

### Mating-Type Gene Control

Meiosis and sporulation can only occur at significant levels in diploids or cells of higher ploidy in which at least both a and $\alpha$ alleles of the mating-type gene are expressed (Mortimer and Hawthorne 1969). The extent of mating locus control over sporulation is highlighted by the demonstration that **a**/**a** or $\alpha/\alpha$ diploids under "sporulation" conditions do not undergo premeiotic DNA synthesis or meiotic recombination, which are among the earliest events unique to sporulation. Moreover, by contrast **a**/$\alpha$ haploids disomic for chromosome III can initiate these events (Roth and Fogel 1971). The mating-locus control, therefore, appears to be a "master switch" that must be thrown before any sporulation events can take place, and this is often tacitly assumed to be the case since **a**/**a** and $\alpha/\alpha$ diploids are frequently used as "control strains" in the analysis of sporulation-specific events (see, e.g., Hopper *et al.* 1974). It is not, however, certain that this mating-type control does act from the outset in an "initiation" function; it may do so since other aspects of mating-type control, over cell division in haploids (**a**- and $\alpha$-mating factor activity: Hartwell 1974; Crandall *et al.* 1977) are expressed in the G1 phase of the cell cycle, from which point the cell division and sporulation pathways begin to diverge in diploids (see following sections). Moreover, it should be noted that a very low, but detectable, fraction of **a**/**a** and $\alpha/\alpha$ cells sporulate (Zakharov and Kozhira 1967).

The recent impressive advances in understanding homothallism (Hicks *et al.* 1977) and in the cloning and characterization of mating-type genes (Hicks *et al.* 1979; Nasmyth and Tatchell 1980; Klar *et al.* 1981; Nasmyth *et al.* 1981) have shown in normal haploid cells that there are three loci coding for mating-type information. One, 23 cM to the right of the centromere on chromosome III, is the site of active expression of mating-type information and carries either the *MATa* or the *MATα* allele. On the left arm of chromosome III is a silent copy of the $\alpha$ information (*HMLα*) and on the right arm, 60 cM distal to the *MAT* locus is a silent copy of **a** information (*HMRa*). The structure of these regions and transcription from them is illustrated in Figure 2.5. These aspects, together with the cassette model for homothallic conversion of mating type, the possible mechanism of this conversion, and mutations affecting homothallism and expression of mating-type information have been clearly and concisely reviewed by Hawthorne (1980).

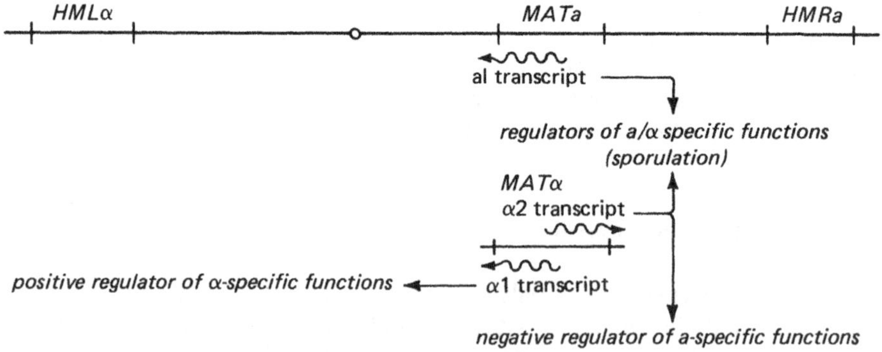


**Figure 2.5.** The arrangement, transcription, and control functions of mating-type information on chromosome III of *Saccharomyces cerevisiae*. *Mata* and Mat$\alpha$ are the alternative alleles of the mating type locus. *HMLα* and *HMRa* are silent copies of $\alpha$ and **a** information, respectively.

Here we are concerned with the salient features affecting the sexual processes of mating and sporulation. The results of detailed analysis of mutations conferring sterility *(ste)* on haploid **a** or $\alpha$ cells have indicated that *MATa* and *MAT*$\alpha$ genes do not code for all the cell functions that are determined by mating type, but control the expression of genes that are specific to cell type and are unlinked to the mating-type locus (Mackay and Manney 1974). Haploid cell-specific (**a** or $\alpha$) functions include: hormone production; mating-factor production; cell shape change (shmooing) in responses to mating factor from cell of opposite mating type; agglutination; participation in conjugation. Diploid cell-specific (**a**/$\alpha$) functions include sporulation and repression of haploid mating functions among others. These aspects have been reviewed extensively in Crandall *et al.* (1977). By studying the effects of mutations located in the mating-type loci Hicks *et al.* (1981) have identified two complementation groups, one in *MAT*$\alpha$ and one in *MATa*, and have shown the role of each of these in the control of **a**-specific, $\alpha$-specific and **a**/$\alpha$-specific functions:

*MAT*$\alpha$*1* is a positive regulator of $\alpha$-specific genes, and *mat*$\alpha$*1* mutants are sterile.

*MAT*$\alpha$*2* is a negative regulator of $\alpha$-specific genes (*mat*$\alpha$*2* mutants express both **a**-specific and $\alpha$-specific genes and are sterile) and is also a positive regulator of **a**/$\alpha$-specific functions since *mat*$\alpha$*2*/*MATa* diploids are unable to sporulate.

*MATa* has no role in the mating activity of **a** cells (this appears to be constitutive) but is required together with *MAT*$\alpha$*2* for the **a**/$\alpha$ phenotype since *mata*/

*MAT*$\alpha$ diploids mate as if they were $\alpha$ cells and do not sporulate.

With the cloning and sequencing of the various genes carrying **a** and $\alpha$ information (*MATa, HMRa, MAT*$\alpha$*, HML*$\alpha$; Hicks *et al.*, 1979; Nasmyth and Tatchell, 1980), the two groups have shown that two transcripts are produced from both *MATa* and *MAT*$\alpha$. The two from the *MAT*$\alpha$ locus ($\alpha$1 and $\alpha$2) have been identified with the two complementation groups within $\alpha$ by using *in vitro* mutagenesis of cloned *MAT*$\alpha$ genes (Nasmyth *et al.* 1981). Of the two transcripts from *MATa* only the a1 (ay of Klar *et al.* 1981) transcript is active. The DNA sequence of the a2 transcript shows no significant open reading frame. These interactions are summarized in Figure 2.5.

This research has entered an exciting phase and has stimulated a considerable amount of further work. It raises many important questions with respect to sporulation: (1) how do the products of *MATa*1 and *MAT*$\alpha$2 regulate the expression of other genes that are under **a**/$\alpha$ specific control? (2) How many genes are subject to **a**/$\alpha$ mating-type regulation? (3) When in the sequence of sporulation events do the *MATa1* and *MAT*$\alpha$*2* gene products need to act? (4) How does the mating-type control interact with the nutritional and cell division cycle controls operating on sporulation initiation?

One very interesting suggestion has been made by Hawthorne (1980). The *MATa1* transcript is curious in that it has an opal (UGA) codon at position

44 of a message that would otherwise code for a 147 amino acid protein. He has speculated that perhaps the 43 amino acid a-1 polypeptide is invovled in repression of transcription of the $\alpha$-1 message, while the 147 amino acid polypeptide, obtained by read-through of the opal codon, triggers sporulation. This read-through may be brought about by starvation conditions conducive for sporulation, since these might influence the a-1 product ratio by reducing the modification of trp-tRNA (the methylation of the first cytosine of the anticodon) which in turn would promote the ambiguous translation of the UGA codon in the a-1 message. This is indeed an elegant explanation, tying together the observations on mating type and on nutritional control of sporulation. We have tried to confirm this hypothesis by constructing diploids homozygous for the UGA-specific suppressor mutations *SUP76* and *SUP85* described by Hawthorne (1976) and testing to see if they were in any way derepressed for sporulation during vegetative growth. No derepression was observed but this does not rule out Hawthorne's suggestion since there are at least six amino acid substitutions that could be made to suppress a UGA codon and only one may be active.

## Nutritional and Cell Division Cycle Control

There has been much work on the external conditions that trigger sporulation (see Fowell 1975). From this it is clear that sporulation is triggered extensively by shifting vegetatively growing populations to media lacking in a rich (i.e., fermentable) carbon and energy source (Miller 1963; Vezinhet, 1970) and also lacking a readily metabolized nitrogen source such as $NH_4^+$ or glutamine (Miller 1963; Pinon 1977; Vezinhet *et al.* 1979). Derepression of mitochondrial activity is an essential prerequisite and sporulation is more synchronous and extensive if, before resuspension in sporulation media, populations are grown in media eliciting mitochondrial derepression (Fast 1973). Initiation of sporulation in a population of cells is not necessarily an all-or-none response to a shift in nutritional conditions since chemostat cultures of *S. cerevisiae* can be obtained in which a proportion of the cells formed by mitotic cell division undergo sporulation while the remainder continue dividing and maintaining the population (Calvert and Dawes, unpublished).

Very much related to this nutritional control is a dependence on the cell division cycle. Cells can switch to alternative development processes only if they are in the G1 phase of the cell cycle (i.e., between cytokinesis and the initiation of DNA synthesis). Thus Reid and Hartwell (1977) showed with haploids that switching from cell division to the processes of mating, and of entry into stationary phase, occurs during G1 before the step mediated by the product of the *cdc28* gene. The product of this gene is considered to act as a regulator of the start of the cell division cycle (Hartwell 1974). Once diploid cells have performed the function determined by the *cdc28* gene product, they are committed to mitosis and cannot undergo sporulation until they have completed the cell division cycle (Shilo *et al.* 1978).

Little is known of the molecular mechanisms that operate in the yeast cell to switch them into the various phases of the life cycle. However, there has been progress in determining how many genes are involved and the way these genes may interact. As far as sporulation is concerned, there are three main classes of mutants that have been studied: (1) mutants derepressed for sporulation *(spd)*, (2) asporogenous revertants *(spo)* of sporulation-derepressed mutants, and (3) cell division cycle *(cdc)* mutants that are also altered in their ability to initiate meiosis and spore formation.

Derepressed-sporulation mutants that sporulate extensively under conditions in which the wild type have not been isolated (Dawes 1975) and the physiology of one group affected in the nuclear gene designated *SPD1* has been studied in some detail (Vezinhet *et al.* 1979). Diploids homozygous for the *spd1* mutation sporulate much more extensively than the wild type after reaching stationary phase on media containing glucose as carbon source, and also unlike the wild type, initiate sporulation immediately when they are resuspended in a medium containing glycerol, pyruvate, acetate, or lactate as energy source. They are not, however, merely the result of a nuclear petite mutation since they grow (and do not sporulate) normally when ethanol is used as a carbon or energy source, and petites do not sporulate. It is interesting that *spd* mutants were initially isolated using protocols designated to select for mutants that were temperature sensitive in derepression for sporulation in a wide range of media. Despite varied attempts to select this later class of mutant, the medium-conditional *spd* mutant phenotype was until recently the only one obtained. The *spd* mutants are obtained at fairly low frequency using ether selection techniques to enrich for them. To date independent mutations with this phenotype have been obtained and these fall into four complementation groups. There are three unlinked loci that have been identified, defined by *spd1*, *spd3* and *spd4* mutations, of which mutations to *spd1* are by far the most commonly isolated (Dawes and Calvert, in press).

The physiology of the mutants carrying *spd1* has been studied in most detail. The phenotype is highly pleiotropic, not only with respect to growth and sporulation on different media, but also since the mutation leads to cell cycle arrest; in media in which the homozygous diploids sporulate *spd1* haploids arrest in the cell cycle as large budded or unbudded cells (i.e., in the G1 phase). By using double mutants carrying *spd1* (conferring medium-conditional growth) and *cdc28* mutations (temperature-sensitive growth) and exploiting the differences in conditions for each mutation it has been shown that the *spd1* mutation arrests cells in the cell cycle before or at the execution point of the *cdc28* gene (Vezinhet *et al.* 1979).

Repression of sporulation by $NH_4^+$ is also altered by the *spd1* mutation but this may not be a primary effect since the mutation is still expressed on carbon substrates on which $NH_4^+$ does not affect sporulation. Furthermore, $NH_4^+$ does not apparently inhibit the initiation of premeiotic DNA synthesis, but acts at two later stages of sporulation (Pinon 1977; Durieu-Trautmann and Delavier-Klutchko 1977).

From this complex phenotype Vezinhet *et al.* (1979) concluded that the *spd1* mutation of *S. cerevisiae* probably affects a central metabolic function concerned with the utilization of some, but not all nonfermentable carbon sources, that the lesion induces a particular form of starvation in which the cells are arrested in the G1 phase of the cell cycle under conditions that lead to relief of the repression of sporulation while the lesions could still reside in a function controlling sporulation. Although it is not necessarily the case that *spd1* mutants are affected directly in events that are unique to sporulation, they do, however, provide useful strains for studying those areas of metabolism and the molecular events associated with the initiation of sporulation. They also help in the study of switches controlling the "decision" of dividing cells to enter other phases of their life cycle.

One of the curious aspects of the *spd1* mutation is the frequency with which revertants rise either spontaneously or induced by mutagens, for example, at 4 $\times 10^{-4}$ spontaneous revertants per cell. These revertants are in most cases due to the occurrence of a second mutation that suppresses the *spd* mutations (Dawes 1975; Dawes and Calvert 1981). These second site events are very interesting mutations since they also confer asporogeny (spo) and they are discussed in more detail below. The frequence of reversion of *spd1* mutations is very high for a straightforward mutational event, and hence some other mechanism may be operative. There is indeed the possibility that transposition events are involved since this system resembles others in which transpositions of Tyl elements are involved; for example, insertion of Tyl elements some distance from structural genes can cause "derepression" of their expression (Struhl, 1980) and in some cases this is subject to control by the mating-type locus. In fact, the appearance of several polypeptides in *spd1* strains is apparently derepressed (Dawes *et al.* 1980).

## Asporogenous Mutations Affecting Initiation *(spo50, 52, 53)*

The assignment of a randomly selected *spo* mutation as one affecting initiation of sporulation is fraught with considerable difficulties due to the unidirectional pleiotropy associated with sporulation mutations and the lack of knowledge of the biochemical events concerned with initiation (and most subsequent sporulation events). However, by reverting derepressed mutants not for their sporulation character directly, but for growth under conditions in which the derepressed mutant hypersporulates, one is selecting for mutants that are affected in the initiation of sporulation, on the one hand, and cell division cycle initiation, on the other (Dawes 1975). This is a specific case of the more general "competitive growth selection" methods that have been used to select mutations affecting the initiation of sporulation in *Bacillus subtilis* (Piggot and Coote 1974). Figure 2.6 outlines the basis of the selection: If a population of cells is dividing with specific fission rate $\nu$ and is being initiated to sporulate at specific rate $\kappa$, then wild-type cells will have a reduced effective specific number growth rate (approximately $\nu - \kappa$). If a mutation occurs preventing the cells

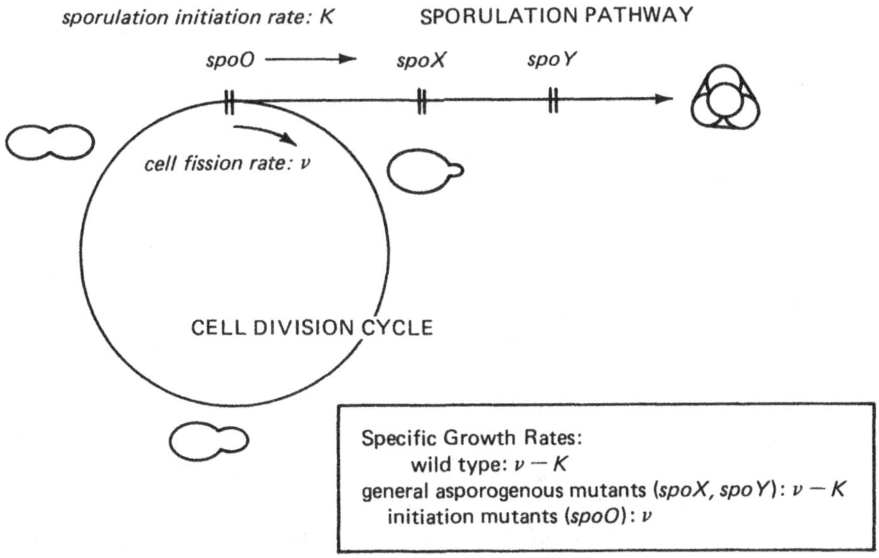


**Figure 2.6.** Selection of initiation mutants in populations of organisms that are partly growing and partly sporulating. Only mutations *(spo0)* that prevent the initiation of sporulation have a selective growth advantage over the wild-type cells.

from initiating sporulation, the cell will have an effective growth rate of $\nu$ and will outgrow the wild-type cell. Moreover it will also outgrow any later-blocked *spo* mutants since these also will have a reduced specific number growth rate ($\nu - \kappa$). Asporogenous mutants have thus been selected from chemostat cultures of wild-type yeast strains. These mutants have a very similar phenotype to those arising as revertants of homozygous *spd1* diploids (Calvert and Dawes, in press) and belong to the same complementation classes.

These *spo* mutants have an interesting, and complex, phenotype. They have a marked tendency to form pseudomycelial organisms, especially when approaching starvation, and homozygous diploids sporulate very poorly if at all. They lose viability rapidly when starved and appear to attempt to continue cell growth without DNA replication, giving rise to brown colonies after extended incubation on plates. They are also derepressed for the ability to reduce tetrazolium dyes (in this the wild-type cells are repressed by glucose, while *spd* mutants cannot reduce tetrazolium under any conditions) and are unable to produce the proteins whose synthesis is derepressed in the *spd1* mutants.

Each of these mutants is the result of a nuclear mutation at one of three unlinked loci *(spo50, spo51, spo53)*. The mutations show partial dominance since heterozygous diploids have a reduced ability to sporulate, and they are all epistatic to the *spd* mutations causing derepressed sporulation. Like the *spd* mutations they are unstable, homozygous diploids reverting to sporulating strains at a fairly high frequency.

## Cell Division Cycle Mutations Affecting Initiation of Sporulation

Simchen and his colleagues have extensively tested homozygous diploids constructed from available temperature-sensitive *cdc* mutations (Hartwell 1974) to determine the dependence of meiosis and sporulation on functions that are essential to mitosis and cell division (Simchen 1974; Shilo *et al.* 1978). Strains that were homozygous for the mutations *(cdc25, cdc28, cdc35,* and *tra3)* that affect the initiation of cell division (at the restrictive temperature these mutants are blocked at "start" of the mitotic cell cycle) were also affected in their ability to initiate sporulation. Mutations *cdc25* and *cdc35* lead to arrest of haploid cells as single unbudded cells in G1 and also derepress sporulation in acetate-containing media in a similar way to the *spd1* mutation. The other two, *cdc28* and *tra3,* led to very poor sporulation (for *cdc28* and 25°C as well as 34°C) and were, therefore, considered to affect *initiation* of sporulation adversely. No *spd* mutation or *spo50, 51* or *53* is linked to either *cdc25, 28,* or *35* (Calvert and Dawes, in preparation).

From these data Shilo *et al.* (1978) concluded that the initiation of meiosis includes some functions defined by the *cdc28* and *tra3* mutations that are shared with "start" of the mitotic cell cycle, as well as functions defined by *cdc25* and *cdc35* that are related to the choice between these two processes. Other interpretations are possible and from a number of recent results this view may need to be modified. We have tested two alleles of the *cdc28* mutation *(cdc28-1,* and *cdc28-15)* and have found (1) that the cell cycle arrest of *cdc28-1* is not expressed at 34°C (the highest temperature at which sporulation occurs in most strains), (2) that some strains homozygous for either of the alleles of *cdc28* can sporulate almost normally at 25°C and 34°C, and (3) that strains homozygous for both *spd1* and *cdc28* mutations sporulate well at temperatures up to 34°C. In searching for temperature-sensitive derepressed sporulation we have isolated a strain that is homozygous for a mutation that fails to complement *cdc25* mutations. This diploid is temperature sensitive for growth, growing at 25°C on either YEPG (yeast extract, peptone, glycerol) or galactose-glutamate minimal medium but it is partially derepressed for sporulation under these conditions. At 29°C the strain does not grow but sporulates extensively, whereas at 30°C the cells neither divide, nor sporulate, but lyse instead. Thus it appears that the product of the *cdc25* gene is needed in a fully functional form for cell division. When the activity of this product is impaired the cell survives but sporulation is triggered, whereas when the gene product is completely inactivated the mutant dies. This type of behavior is that which would be expected if the cells assess their nutritional status before committing themselves to a further division cycle by monitoring the concentrations of an essential metabolite (ATP or GTP or NADPH?) that must be generated in lower amounts than are needed for cell division for the cell to survive or sporulate.

Figure 2.7 summarizes a model of the possible involvement of the various mutations affecting the cell division cycle and sporulation. This has been sug-

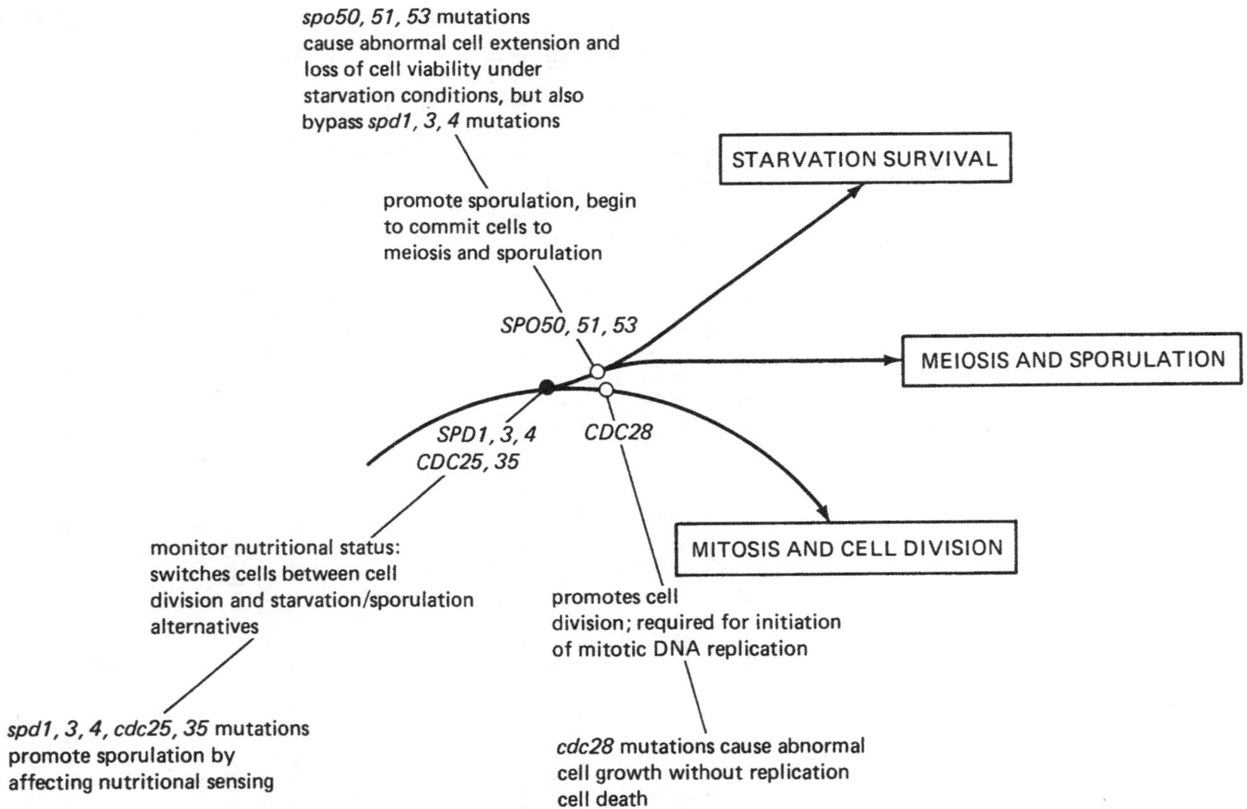


**Figure 2.7.** Model for the involvement of various gene products in the decision between cell division, sporulation, or adaptation to starvation conditions. The heavy lines represent the alternative courses of development for cells entering the G1 phase of the cell division cycle. The *CDC28* gene function is not given as being common to the sporulation and the cell division pathways for reasons discussed in the text, but this assignment remains tentative.

gested to account for the phenotypes of the various mutations concerned and for the interaction between them as can be observed by studying double mutants. Thus the products of the *SPD1, 3,* and *4* genes appear to act in the same way as those of *CDC25* and *CDC35* in determining whether the cell enters the cell division cycle or the sporulation sequence. The products of these genes presumably act in a positive way to promote cell division and hence act negatively on sporulation, since the derepression (of sporulation) and the inhibition (of cell division) mutations are recessive. The *spo50, 51,* and *53* mutations block sporulation very early after the *spd1* function in such a way that the cell bypasses the decision to sporulate and continues trying to grow and divide even under starvation conditions. The *cdc28* mutation has been placed on the cell division pathway in this figure on the basis that some strains carrying this mutation do sporulate. It certainly does not block sporulation once initiated in *spd1* strains.

## Gene Involvement in Subsequent Sporulation Events

Once initiated to sporulation the cells progress through the program of events associated with premeiotic DNA replication, recombination, meiosis, and spore formation. The extent of direct gene involvement in these events has not been clarified completely, although the work of Simchen (1974; also reported by Esposito 1980) with other *cdc* mutations than those discussed earlier has shown which cell cycle functions are also required during sporulation (essentially those required through DNA synthesis). He has also indicated that cells are committed to meiosis and to mitosis once spindle pole body separation has taken place. For the cell division cycle this occurs prior to DNA synthesis, whereas in meiosis this occurs after premeiotic DNA synthesis.

Genes that are uniquely concerned with sporulation have been (tentatively) identified by the isolation of asporogenous mutants, and the analysis of these by the Espositos has provided insights into the nature of the process and its control as well as a sound basis for more detailed molecular genetic study. These aspects have been reviewed extensively (Esposito and Esposito 1975; Baker *et al.* 1976) and are summarized only briefly here.

It has been estimated that about 50 genes are essential to sporulation and not to vegetative growth (Esposito *et al.* 1972). This estimate is subject to considerable uncertainty since it is based on a statistical analysis of the frequency of repeated isolation of mutations affecting the same gene, and to date considerably less than half this number of genes has been identified by mutation. The estimate is similar to that suggested for bacterial sporulation from a similar approach (Hranueli *et al.* 1974). An asporogenous mutation in general leads to arrest of development at a particular stage of the process and morphological and biochemical studies of *spo* mutants have therefore indicated the ordering of the correlation of, and to some extent the interaction between, crucial sporulation events. These aspects have been reviewed extensively by Esposito and

Esposito (1974b; 1975). Here examples of several conclusions they have drawn from their studies are given:

1. The ability of cells to complete the cell cycle in sporulation medium and premeiotic DNA synthesis are coordinately controlled.
2. Commitment to meiotic recombination does not commit cells to the reductional division of meiosis, and does not require duplication of the spindle bodies for the first division of meiosis.
3. The formation of the prospore wall does not require entry of nuclear material into the growing prospore, nor integrity of the nuclear membrane.
4. Ascospore maturation probably depends on the presence of some nuclear material in the closed prospore.

Using temperature-shift studies with three temperature-sensitive *spo* mutations, they were able to determine the period for which the products of several *spo* genes would normally be required to act, and they found that in two cases (for *spo2-1* and *spo3-1*) this preceded by some time the point at which the cells were blocked in their morphological changes (Esposito *et al.* 1970).

The value of mutant studies in analyzing the interrelationship of events occurring during meiosis and spore formation is clearly illustrated by recent work on the *spo13-1* mutation. This mutation eliminates the normal reductional division of meiosis I, but does not prevent sporulation that proceeds with an equational meiosis II division to form asci containing two diploid spores (Klapholtz and Esposito 1980). Malone and Esposito (1981) have used this property to study whether this mutation bypasses the meiotic defects (and therefore loss of spore viability) caused by various *rad* mutations (*rad6, rad50, rad52*) and have shown that only the effects of *rad50* could be bypassed, thereby unambiguously identifying this gene as a component of the meiotic recombination system. This approach has also been used with an *spo* mutation (*spo11*) that affects meiotic recombination and, as for *rad50 spo13-1* double mutants, the *spo11 spo13-1* double mutant produces viable spores. Thus not only does the *spo13* mutation show that certain events in meiosis can be dissociated, it can also be used to analyze the possible function of other genes affecting a specific meiotic function, that of recombination.

To date, while there is now considerable insight into the timing and ordering of sporulation events, and the interrelationships of genes affecting cell division, radiation repair, and sporulation, the precise biochemical lesion in any *spo* mutant has not been elucidated. This problem of assigning a biochemical defect to a morphological mutation is one faced (or often avoided) by geneticists attempting to determine the crucial molecular events in a developmental process, and it considerably hampers progress. Some clues to important biochemical functions that must participate in meiosis and spore formation have been gleaned by screening mutants with known biochemical defects, and some of these can be discerned by referring to Table 2.1. A number of biochemical events that may well be specific to the process, and that may therefore prove useful for detailed genetic analysis, have been identified. These include:

appearance of an $\alpha$-glucosidase activity thought to degrade glycogen during the later stages of sporulation (Colunna and Magee 1978), a specific $\beta$-glucanase (del-Rey *et al.* 1980), a surface protein on ascospores (Briley *et al.* 1970) that may be the spore-specific antigen (Snider and Miller 1966), and possibly a change in one of the three RNA polymerases (Magee 1975).

## Protein Changes and Gene Expression During Sporulation

Several attempts have been made to increase the range of known sporulation events by detecting polypeptide changes with the sensitivity of $^{35}$S-labeling and the high resolution of one- or two-dimensional gel electrophoresis. In those experiments using pulse labeling during sporulation with L-[$^{35}$S]methionine a number of polypeptide changes occurred during sporulation of **a**/$\alpha$ diploids, but these changes also took place in **a**/**a** and $\alpha/\alpha$ diploids and **a** and $\alpha$ haploids resuspended in sporulation medium (Hopper *et al.* 1974; Trew *et al.* 1979; Petersen *et al.* 1979; Kraig and Haber 1980). This failure to identify synthesis of proteins that are unique to **a**/$\alpha$ diploids should be interpreted with care, however, since sporulating yeast cells erect permeability barriers to amino acids and nucleic acid bases (Mills 1972). These increase during sporulation, and since not all cells in a population are usually induced to sporulate, the low concentrations of methionine added in pulse labels may be taken up preferentially by nonsporulating cells (Dawes *et al.* 1980).

Changes (in polypeptides) specific to **a**/$\alpha$ cells have been detected by labeling cells in $^{35}SO_4^{2-}$ during presporulation growth, then by resuspending them in sulfur-free sporulation medium (Wright and Dawes 1979). Of 400 polypeptides detected, 45 were altered markedly during sporulation, and of these 21 changes were specific to **a**/$\alpha$ cells. These changes occurred at different times during sporulation, and although most took place early there were nonetheless several later changes that were important since they defined the sequential nature of the process characterizing the later stages of sporulation (Wright *et al.* 1981). Clearly from the method of labeling used, these changes could have been due either to *modification* of existing polypeptides present during vegetative growth or to *de novo* synthesis of new gene products. It is not yet clear how many of these changes are due to modification, but a number of changes in phosphorylation of proteins have been detected in sporulating cells, including three polypeptides that were specific to **a**/$\alpha$ cells (Wright *et al.* 1981). Moreover, Ajam (1981) has shown that some (but by no means all) of the early **a**/$\alpha$-specific polypeptides appear even if cycloheximide is added at the transfer of cells to sporulation medium.

The results of two-dimensional gel electrophoresis studies are very important since they bring into question the extent to which a cellular differentiation process such as yeast sporulation depends on *de novo* expression of genes uniquely concerned with the process. It would be premature to suggest from the pulse-labeling studies that there are no changes in gene expression that are uniquely concerned with the process, or even to assume that all the **a**/$\alpha$-specific changes detected by prelabeling studies were the result of modifications, since (1) nonsporulating cells bias the results of pulse labeling (see discussion above), (2) the pulse-labeling experiments would not have detected changes in proteins containing few methionine residues, (3) some of the changes detected by prelabeling can be inhibited by the addition of cycloheximide during sporulation (Ajam 1981), and (4) at least one enzyme activity, $\alpha$-glucosidase, appears in an **a**/$\alpha$-specific way during sporulation (Colunna and Magee 1978).

Clearly much work needs to be done to resolve this issue, which is crucial to an understanding of molecular regulation during cell development. Since two-dimensional gel electrophoresis of cell extracts only detects 400 or so of the major polypeptides present in the cell, many important **a**/$\alpha$-specific changes may occur in minor protein species and not be detected by a straightforward approach. This would be expected for regulatory proteins (but what of the ones that are regulated?). One approach is, therefore, to fractionate cells to expand the sensitivity of the technique. Petersen and Piñon (1979) have begun this by investigating the proteins associated with folded chromosomes isolated by zone sedimentation in sucrose gradients, and have detected four proteins that are present in replicating chromosomes of meiotic cells, but not in chromosomes from resting phase cells. These proteins may yet prove to be meiosis specific. Another is to isolate particular classes of proteins, for example, by affinity chromatography, at different stages of sporulation. So far, proteins binding to AMP-sepharose or 2′,5′-ADP-sepharose have been studied without any sporulation-specific changes having been detected (Emsley and Dawes, unpublished).

Another approach to detecting *de novo* gene expression during yeast sporulation is to test for any changes in transcription patterns by *in vitro* translation of RNA species isolated from sporulating cells. In a preliminary study, Calvert and Dawes (1981) have detected **a**/$\alpha$-specific transcripts using the low resolution of one-dimensional gel electrophoresis, and more may be expected when the results of two-dimensional studies are available.

Finally, the full power of *in vitro* genetic analysis has yet to be brought to bear on this question of whether there are any genes uniquely read during meiosis and sporulation. By isolating polyadenylated RNA species prepared from **a**/$\alpha$ and **a**/**a** cells under vegetative and sporulation conditions, and using them as a probe a preliminary attempt at cloning sporulation genes on phage $\lambda$ vectors has been initiated, and several interesting clones have been obtained. These contain sequences hybridizing to poly(A)-containing RNA from **a**/$\alpha$ cells under sporulation conditions, but not from vegetative **a**/$\alpha$ or nonsporulating **a**/**a** cells resuspended in sporulation medium (Smith and Dawes, unpublished observations). Undoubtedly there is considerable energy currently directed toward cloning "sporulation" genes by the route of transforming *spo* mutants, and the next decade should see an exciting surge forward in our understanding of gene involvement and gene expression during sporulation.

It would be a pity if this excitement were at the complete expense of other biochemical studies. If one takes a balanced view of the field at the moment it may be that modifications to existing polypeptides are very important, and although there probably are genes that are uniquely read during sporulation, these may be fewer in number than had initially been envisaged by many of us. Studying modifications may lack the glamour of more exciting research using more fashionable techniques, but it will probably yield valuable information on the control of cell development.

It is maybe not really surprising to entertain the view that there may be only a few genes read uniquely during sporulation. There is a marked similarity in many of the processes and structures found during sporulation on the one hand, and cell division on the other. For example, a fairly extensive analysis of replicon size and formation has not revealed any significant differences between mitotic and premeiotic DNA replication (Johnston *et al.* 1982). Simchen and his colleagues have shown from their studies with *cdc* mutations that a number of gene functions that are essential for mitotic cell division are also required during sporulation (Simchen 1974; Shilo *et al.* 1978). Moreover, apart from the spore surface antigen the spore coats are composed of glucan and mannan of similar (although not identical) composition to the vegetative cell wall (I. W. Sutherland, personal communication). Indeed, if as has been suggested for bacterial spore formation (Hitchins and Slepecky, 1969), the process of sporulation in yeast evolved by a modification of the normal cell division process it would not be surprising if there were not a considerable sharing of gene function between the two processes. It is, however, possible that the cell division cycle in yeasts, exemplified by the unicellular budding process, may have evolved from an existing sporulation process by a modification of the germination and outgrowth processes in a hyphal ancestor. The chemistry of the yeast vegetative cell wall does resemble more that of the spore wall of other Ascomycetes than of hyphal structures. These and other evolutionary implications of sporulation have been discussed in more detail elsewhere (Dawes 1981); the recent work on what amounts to a bypass mutation (e.g., *spo13*) for the reductional division of meiosis I, and its effects on obviating meiotic defects conferred by mutations in recombination functions (*rad50* and *spo11*) are interesting since they highlight the involvement of specific genes in those aspects of sporulation that are peculiar to meiotic rather than mitotic cells.

## References

Ajam, N. (1981). Regulation of sporulation in *Saccharomyces cerevisiae*. Ph.D. Thesis, University of Edinburgh.

Baker, B. S., Carpenter, A. T., Esposito, M. S., Esposito, R. E., Sandler, L. (1976). The genetic control of meiosis. Ann. Rev. Genet. 10:53–134.

Beckett, A., Illingworth, R. F., Rose, A. H. (1973). Ascospore wall development in *Saccharomyces cerevisiae*. J. Bacteriol. 113:1054–1057.

Beggs, J. D. (1978). Transformation of yeast by a replicating hybrid plasmid. Nature 275:104–109.

Betz, H. (1977). Protein degradation and sporulation in yeast. In: Cell Differentiation in Microorganisms, Plants and Animals, edited by L. Nover and K. Mothes. Amsterdam: North-Holland, pp. 243–255.

Betz, H., Weiser, U. (1976). Protein degradation and proteinases during yeast sporulation. Eur. J. Biochem. 62:65–74; protein degradation during yeast sporulation: enzyme and cytochrome patterns. Eur. J. Biochem. 70:385–395.

Bresch, C., Müller, G., Egel, R. (1968). Genes involved in meiosis and sporulation of a yeast. Mol. Gen. Genet. 102:301–306.

Briley, M. S., Illingworth, R. F., Rose, A. H., Fisher, D. J. (1970). Evidence for a surface protein layer on the *Saccharomyces cerevisiae* ascospore. J. Bacteriol. 104:588–589.

Cavalier-Smith, T. (1981). The origin and early evolution of the eukaryotic cell. In: Molecular and Cellular Aspects of Microbial Evolution, 32nd Symp. Soc. Gen. Microbiol., edited by M. J. Carlile, J. F. Collins, and B. E. B. Moseley. London: Cambridge University Press, pp. 33–84.

Colunna, N. J., Magee, P. T. (1978). Glycogenolytic enzymes in sporulating yeast. J. Bacteriol. 134:844–853.

Crandall, M., Egel, R., Mackay, V. L. (1977). Physiology of mating in three yeasts. Adv. Microbial Physiol. 15, 307–398.

Croes, A. F. (1966). Duplication of DNA during meiosis in Baker's yeast. Exp. Cell Res. 41:452–454.

Croes, A. F. (1967). Induction of meiosis in yeast. I. Timing of cytological and biochemical events; II. Metabolic factors leading to meiosis. Planta 76:209–226, 227–237.

Curiale, M. S., Petryna, M. M., Mills, D. (1976). Ribonucleic acid synthesized in meiotic cells of *Saccharomyces cerevisiae:* effect of culture medium pH. J. Bacteriol. 126:661–667.

Dawes, I. W. (1975). Study of cell development using derepressed mutations. Nature 255:707–708.

Dawes, I. W. (1981). Sporulation in evolution. In: Molecular and Cellular Aspects of Microbial Evolution, edited by M. J. Carlile, J. F. Collins, and B. E. B. Moseley. London: Cambridge University Press, pp. 85–130.

Dawes, I. W., Calvert, G. R. (1981). Mutants of *Saccharomyces cerevisiae* defective in the initiation of sporulation. Soc. Gen. Microbiol. Quart. 8:249.

Dawes, I. W., Calvert, G. R., Martin, T. A. (1980). Initiation of meiosis and sporulation in *Saccharomyces cerevisiae:* regulation of protein synthesis in mutants altered in initiation. Abstracts 10th Int. Conf. Yeast Genet. Mol. Biol., Louvain-la-Neuve, p. 152.

Dawes, I. W., Hardie, I. D. (1974). Selective killing of vegetative cells in sporulated yeast cultures by exposure to diethyl ether. Mol. Gen. Genet. 131:281–289.

Dawes, I. W., Wright, J. F., Vezinhet, F., Ajam, N. (1980). Separation on Urografin gradients of subpopulations from sporulating *Saccharomyces cerevisiae* cultures. J. Gen. Microbiol. 119:165–171.

del-Rey, F., Santos, T., Garcia-Acha, I., Nombela, C. (1980). Synthesis of $\beta$-glucanases during sporulation in *Saccharomyces cerevisiae:* formation of a new sporulation-specific 1,3-$\beta$-glucanase. J. Bacteriol. 143:621–627.

Dion, P., Mandelstam, J. (1980). Germination properties as marker events characterizing later stages of *Bacillus subtilis* spore formation. J. Bacteriol. 141:786–792.

Durieu-Trautmann, O., Delavier-Klutchko, C. (1977). Effect of ammonia and glutamine on macromolecular synthesis and breakdown during sporulation of *Saccharomyces cerevisiae.* Biochem. Biophys. Res. Commun. 79:438–442.

Emanuel, J. R., Magee, P. T. (1981). Timing of ribosome synthesis during ascosporogenesis of yeast cells: evidence for early function of haploid daughter genomes. J. Bacteriol. 145:1342–1350.

Engels, F., Croes, A. (1968). The synaptonemal complex in yeast. Chromosoma, Berlin 25, 104–106.

Ephrussi, B., Hottinguer, H. (1951). On an unstable cell state in yeast. Cold Spring Harbor Symp. Quant. Biol. 16, 75–86.

Esposito, M. S., Esposito, R. E. (1969). The genetic control of sporulation in *Saccharomyces.* I. The isolation of temperature-sensitive sporulation-deficient mutants. Genetics 61:78–89.

Esposito, M. S., Esposito, R. E. (1974a). Genes controlling meiosis and spore formation in yeast. Genetics 78:215–225.

Esposito, M. S., Esposito, R. E. (1975). Mutants of meiosis and ascospore formation. Meth. Cell Biology, 11:303–326.

Esposito, M. S., Esposito, R. E., Arnaud, M., Halvorson, H. O. (1969). Acetate utilization and macromolecular synthesis during sporulation of yeast. J. Bacteriol. 100:180–186.

Esposito, M. S., Esposito, R. E., Arnaud, M., Halvorson, H. O. (1970). Conditional mutants of meiosis in yeast. J. Bacteriol. 104:202–210.

Esposito, R. E. (1980). Sporulation. In: Yeast Genetics and Molecular Biology 1980, edited by A. Goffeau and J.-M. Wiame. Workshop Reports 10th Int. Conf., Louvain-la-Neuve, pp. 19–25.

Esposito, R. E., Esposito, M. S. (1974b). Genetic recombination and commitment to meiosis in *Saccharomyces.* Proc. Nat. Acad. Sci. USA 71:3172–3176.

Esposito, R. E., Frink, N., Bernstein, P., Esposito, M. S. (1972). The genetic control of sporulation in *Saccharomyces.* II. Dominance and complementation of mutants of meiosis and spore formation. Mol. Gen. Genet. 114:241–248.

Fast, D. (1973). Sporulation synchrony of *Saccharomyces cerevisiae* in various carbon sources. J. Bacteriol. 116:925–930.

Fogel, S., Roth, R. (1974). Mutations affecting meiotic gene conversion in yeast. Mol. Gen. Genet. 130:189–201.

Fowell, R. R. (1975). Ascospores of Yeasts. In: Spores VI, edited by P. Gerhardt, R. N. Costilow and H. L. Sadoff. Washington, D.C.: American Society for Microbiology, pp. 124–131.

Game. J. C., Cox, B. S. (1971). Allelism of mutants affecting sensitivity to radiation in yeast and a proposed nomenclature. Mutation Res. 24:328–331.

Game, J. C., Mortimer, R. K. (1974). A genetic study of X-ray sensitive mutants in yeast. Mutation Res. 24:281–292.

Game, J. C., Zamb, T. S., Braun, R., Resnick, M., Roth, R. (1980). The role of radiation *(rad)* genes in meiotic recombination in yeast. Genetics 94:51–68.

Gerlach, W. L. (1974). Sporulation in mating-type homozygotes of *Saccharomyces cerevisiae.* Heredity 32:241–249.

Grewal, N. S., Miller, J. J. (1972). Formation of asci with two diploid spores by diploid cells of *Saccharomyces.* Can. J. Microbiol. 18:1897–1905.

Guth, E., Hashimoto, T., Conti, S. (1972). Morphogenesis of ascospores in *Saccharomyces cerevisiae.* J. Bacteriol. 109:869–880.

Haber, J. E., Esposito, M. S., Magee, P. T., Esposito, R. E. (1975). Current trends in genetic and biochemical study of yeast sporulation. In: Spores VI, edited by P. Gerhardt, R. N. Costilow, and H. L. Sadoff. Washington, D.C.: Am. Soc. Microbiol., pp. 132–137.

Haber, J. E., Halvorson, H. O. (1972). Regulation of sporulation in yeast. In: Current Topics in Developmental Biology, Vol. 7, edited by A. A. Moscona and A. Munroy. New York: Academic Press, pp. 61–82.

Haber, J. E., Halvorson, H. O. (1975). Methods of sporulation and germination of yeasts. Meth. Cell Biol. 11:45–69.

Hall, B. D., Hopper, A. (1973). Control of sporulation by mating type genes in *Saccharomyces cerevisiae.* Colloq. Int. Cent. Rech. Sci. 227:141–144.

Hartwell, L. H. (1974). *Saccharomyces cerevisiae* cell cycle. Bacteriol. Rev. 38:164–198.

Hawthorne, D. C. (1976). UGA mutations and UGA suppressors in yeast. Biochimie 58:179–182.

Hawthorne, D. C. (1980). Mating in yeast (regulation of transcription). In: Yeast Genetics and Molecular Biology, 1980, Workship Reports 10th Int. Conf., Louvain-la-Meuve, edited by A. Goffeau and J. M. Wiame, pp. 55–61.

Heywood, P., Magee, P. T. (1976). Meiosis in protists: some structural and physiological aspects of meiosis in algae, fungi and protozoa. Bacteriol. Rev. 40:190–240.

Hicks, J., Strathern, J., Herskowitz, I. (1977). The cassette model of mating type interconversion. In: DNA Insertion Elements, Plasmids and Episomes, edited by A. Bukhari, J. Shapiro, and S. Adhya. Cold Spring Harbor, N.Y.: Cold Spring Harbor Laboratory, pp. 457–462.

Hicks, J. B., Hinnen, A. and Fink, G. R. (1979). Properties of yeast transformation. Cold Spring Harbor Symp. Quant. Biol. 43:1305–1313.

Hitchins, A. D., Slepecky, R. A. (1969). Bacterial sporulation as a modified procaryotic cell division. Nature 223:804–807.

Hopper, A. K., Magee, P. T., Welch, S. K., Friedman, M., Hall, B. D. (1974). J. Bacteriol. 119:619–628.

Hranueli, D., Piggot, P. J., Mandelstam, J. (1974). Statistical estimate of the total number of operons specific for *Bacillus subtilis* sporulation. J. Bacteriol. 119:684–690.

Jenkinson, H. F., Sawyer, W. D., Mandelstam, J. (1980). Synthesis and order of assembly of spore coat proteins in *Bacillus subtilis.* J. Gen. Microbiol. 123:1–16.

Johnston, L. H., Williamson, D. H., Johnson, A. L. (1982). An analysis of the mechanism of premeiotic DNA synthesis in the yeast *Saccharomyces cerevisiae.* Exp. Cell Res., in press.

Kane, S. M., Roth, R. (1974). Carbohydrate metabolism during ascospore development in yeast. J. Bacteriol. 118:8–14.

Kassir, Y., Simchen, G. (1976). Regulation of mating and meiosis in yeast by the mating-type region. Genetics 82:187–206.

Klapholtz, S., Esposito, R. E. (1980). Recombination and chromosome segregation during the single division meiosis in *spo12-1* and *spo13-1* diploids. Genetics 96:589–611.

Klar, A. J. S., Strathern, J. N., Broach, J. R., Hicks, J. B. (1981). Regulation of transcription in expressed and inexpressed mating type cassettes of yeast. Nature 289:239–244.

Kraig, E., Haber, J. E. (1980). Messenger ribonucleic acid and protein metabolism during sporulation of *Saccharomyces cerevisiae.* J. Bacteriol. 144:1098–1112.

Leibmann, S. W., Sherman, F., Stewart, J. (1976). Isolation and characterization of amber suppressors in yeast. Genetics 82:251–272.

Lynn, R. R., Magee, P. T. (1970). Development of the spore wall during ascospore formation in *Saccharomyces cerevisiae*. J. Cell Biol. 44:688–692.

Mackay, V., Manney, T. R. (1974). Mutations affecting sexual conjugation and related processes in *Saccharomyces cerevisiae*. I. Isolation and phenotype characterization of non-mating mutants. II. Genetic analysis of non-mating mutants. Genetics 76:255–271, 173–288.

Magee, P. T. (1975). Changes in DNA-dependent RNA polymerase in sporulating yeast. Mol. Biol. Rep. 1:275–281.

Malone, R. E., Esposito, R. E. (1981). Recombinationless meiosis in Saccharomyces cerevisiae. Mol. Cellular Biol. 1:891–901.

Matur, A., Berry, D. R. (1978). The use of step enzymes as markers during meiosis and ascospore formation in *Saccharomyces cerevisiae*. J. Gen Microbiol. 109:205–213.

Miclo, A., Guesdon, X. (1980). Isocitratase (thre-D5-isocitrate glyoxalatelyase:EC 4.1.3.1) activity during *Saccharomyces carlsbergensis* sporulation. Bull. Fe. Natl. Super. Agron. Ind. Aliment. 22:23–27.

Miller, J. J. (1963). The metabolism of yeast sporulation. V. Stimulation and inhibition of sporulation and growth by nitrogenous compounds. Can. J. Microbiol. 9:259–277.

Mills, D. (1972). Effect of pH on adenine and amino acid uptake during sporulation in *Saccharomyces cerevisiae*. J. Bacteriol. 112, 519–526.

Mills, D. (1980). Quantitative and qualitative analysis of polyadenylated RNA sequences during meiosis of *Saccharomyces cerevisiae*. Abst. 10th Int. Conf. Yeast Genet. Mol. Biol., Louvain-la-Neuve, p. 55.

Miyake, S. Sando, N., Sato, S. (1971). Biochemical changes in yeast during sporulation. II. Acetate metabolism. Development, Growth & Differentiation 12:285–295.

Moens, P. B. (1971). Fine structure of ascospore development in the yeast *Saccharomyces cerevisiae*. Can. J. Microbiol. 17:507–510.

Moens, P. B., Esposito, R. E., Esposito, M. S. (1974). Aberrant nuclear behavior at meiosis and anucleate spore formation by sporulation-deficient *(spo)* mutations of *Saccharomyces cerevisiae*. Exp. Cell Res. 83:166–174.

Moens, P. B., Rapport, E. (1971). Spindles, spindle plaques and meiosis in the yeast *Saccharomyces cerevisiae* (Hansen). J. Cell Biol. 50:344–361.

Moir, A. (1981). Germination of a spore coat-defective mutant of *Bacillus subtilis*. J. Bacteriol. 146:1106–1116.

Mortimer, R. K., Hawthorne, D. C. (1966). Genetic mapping in *Saccharomyces*. Genetics 53:165–173.

Mortimer, R. K., Hawthorne, D. C. (1969). Yeast genetics. In: The Yeasts, edited by A. H. Rose and J. S. Harrison. London: Academic Press, pp. 385–460.

Mortimer, R. K., Hawthorne, D. C. (1973). Genetic mapping in *Saccharomyces*. IV. Mapping of temperature-sensitive genes and use of disomic strains in localizing genes. Genetics. 74:35–54.

Mortimer, R. K., Hawthorne, D. C. (1975). Genetic mapping in yeast. Meth. Cell Biol. 11:221–233.

Nasmyth, K. A., Tatchell, K. (1980). The structure of transposable yeast mating type loci. Cell 19:753–764.

Nasmyth, K. A., Tatchell, K., Hall, B. D., Astell, C., Smith, M. (1981). A position effect in the control of transcription at yeast mating type loci. Nature 289:244–250.

Peroa, J., Gancedo, C. (1980). A phosphoenolpyruvate carboxykinase mutant of *Saccharomyces cerevisiae*. Abst. 10th Int. Conf. Yeast Genet. Mol. Biol., Louvain-la-Neuve. p. 66.

Petersen, J. G. L., Kielland-Brandt, M. C., Nilsson-Tillgren, T. (1979). Protein patterns of yeast during sporulation. Carlsberg Res. Commun. 44:149–162.

Petersen, J. G. L., Pinon, R. (1979). *In vivo* labelling of proteins associated with folded chromosomes of yeast. Carlsberg Res. Commun. 44:395–402.

Piggot, P. J., Coote, J. G. (1976). Genetic aspects of bacterial endospore formation. Bacteriol. Rev. 40:908–962.

Pinon, R. (1977). Effect of ammonium ions on sporulation of *Saccharomyces cerevisiae*. Exp. Cell Res. 105:367–378.

Prakash, S., Prakash, L., Burke, W., Montelone, B. (1980). Effects of the *RAD 52* gene on recombination in *Saccharomyces cerevisiae*. Genetics 94:31–50.

Pratje, E., Shulz, R., Schnierer, S., Michaelis, G. (1979). Sporulation of mitochondrial respiratory deficient *mit*$^-$ mutants of *Saccharomyces cerevisiae*. Mol. Gen. Genet. 176:411–415.

Prescott, D. M. (ed.) (1975). Methods in Cell Biology, Vol. XI. New York: Academic Press.

Put, H. M. C., Sand, F. E. M. J. (1975). Some notes on the heat resistance of ascosporogenous yeasts. J. Appl. Bacteriol. 39(3):iii.

Reid, B. J., Hartwell, L. H. (1977). Regulation of mating in the cell cycle of *Saccharomyces cerevisiae*. J. Cell Biol. 75:335–365.

Rine, J., Sprague, G. F., Herskowitz, I. (1981). *rme1* Mutation of *Saccharomyces cerevisiae:* map position and bypass of mating locus control of sporulation. Mol. Cellular Biol. 1:958–960.

Roth, R., Fogel, S. (1971). A system selective for yeast mutants deficient in meiotic recombination. Mol. Gen. Genet. 112:295–305.

Roth, R., Lusnak, K. (1970). DNA synthesis during yeast sporulation: genetic control of an early developmental event. Science 168:493–494.

Rothstein, R. J., Esposito, R. E., Esposito, M. S. (1976). The effect of ochre suppression on meiosis and ascospore formation in *Saccharomyces*. Genetics 85:35–54.

Sena, E. P., Radin, D. N., Welch, J., Fogel, S. (1975). Synchronous mating in yeasts. Meth. Cell Biol. 11:71–88.

Sherman, F., Roman, H. (1963). Evidence for two kinds of allelic recombination in yeast. Genetics 48:255–261.

Shilo, V., Simchen, G., Shilo, B. (1978). Initiation of meiosis in cell cycle initiation mutants of *Saccharomyces cerevisiae*. Exp. Cell Res. 112:241–248.

Simchen, G. (1974). Are mitotic functions required in mitosis? Genetics 76:745–753.

Simchen, G., Pinon, R., Salts, V. (1972). Sporulation in *Saccharomyces cerevisiae:* premeiotic DNA synthesis, readiness and commitment. Exp. Cell Res. 75: 207–218.

Snider, I. I., Miller, J. J. (1966). A serological comparison of vegetative cell and ascus wall and the spore coat of *Saccharomyces cerevisiae*. Can. J. Microbiol. 12:485–488.

Strathern, J., Hicks, J., Herskowitz, I. (1981). Control of cell type in yeast by the mating type locus. The $\alpha$1-$\alpha$2 hypothesis. J. Mol. Biol. 147:357–372.

Struhl, K. (1980). Expression of cloned genes in yeast (regulation of transcription).

In: Workshop Reports, 10th Int. Conf. Yeast Genet. Mol. Biol. edited by A. Goffeau and J.-M. Wiame, Louvain-la-Neuve, pp. 51–54.

Tingle, M., Klar Singh, A. J., Henry, S. A., and Halvorson, H. O. (1973). Ascospore formation in yeast. In: Microbial Differentiation, 23rd Symp. Soc. Gen. Microbiol., edited by J. M. Ashworth and J. E. Smith, Cambridge: Cambridge University Press, pp. 209–243.

Trew, B. J., Friesen, J. D., Moens, P. B. (1979). Two dimensional protein patterns during growth and sporulation in *Saccharomyces cerevisiae*. J. Bacteriol. 138:60–69.

Vezinhet, F., Kinnaird, J. H., Dawes, I. W. (1979). The physiology of mutants derepressed for sporulation in *Saccharomyces cerevisiae*. J. Gen. Microbiol. 115:391–402.

Williamson, D. H., Fennel, D. J. (1975). The use of fluorescent DNA-binding agent for detecting and separating yeast mitochondrial DNA. Meth. Cell Biol. 12:335–351.

Williamson, M., Fogel, S. (1980). Correction defective *(cor)* mutants that alter the frequency and length of correction tracks in meiotic hybrid DNA. Abst. 10th Int. Conf. Yeast Genet. Mol. Biol., Louvain-la-Neuve, p. 44.

Wright, J. F., Dawes, I. W. (1979). Sporulation-specific protein changes in yeast. FEBS Lett. 104:183–186.

Wright, J. F., Ajam, N., Dawes, I. W. (1981). The nature and timing of sporulation-specific protein changes in *Saccharomyces cerevisiae*. Mol. Cellular Biol. 1:910–918.

Zakharov, I. A., Inge-Vechtomov, S. G. (1964). Ascospore isolation from yeast for genetic analysis without a micromanipulator. Issledovania po Genetike (English translation) 2:134–139.

Zakharov, I. A., Kozhina, T. N. (1967). The genetics of mating in yeast. Genetika 11:110–115.

Zubenko, G. S., Mitchell, A. P., Jones, E. W. (1979). Septum formation, cell division, and sporulation in mutants of yeast deficient in proteinase B. Proc. Natl. Acad. Sci. USA 76:2395–2399.

# 3

# Meiotic Gene Conversion in Yeast: Molecular and Experimental Perspectives[1]

Seymour Fogel*, Robert K. Mortimer**, and Karin Lusnak*

## Introduction

Overall, our understanding of the molecular events comprising gene conversion and genetic recombination is reasonably complete. Both phenomena can be accommodated within a unified theoretical framework based on the known structure and behavior of DNA molecules. Critical evidence drawn primarily from genetic investigations in fungi amenable to tetrad or octad analysis constitute the principal data bases. These are enhanced equally by structural analysis, at the electron microscope level, of recombination intermediates recovered from intact cells and the characterization of enzymes coded for by DNA sequences that are involved in bacterial and viral recombination.

Synoptic in design, this review is essentially didactic. Topical coverage is necessarily uneven and some topics are even omitted. Our aim is to present an overview and at the same time indicate the power and value of recombination as a tool for forging novel genetic reagents in both the academic laboratory and industry. A comprehensive review of the literature will not be attempted.

*Department of Genetics, University of California, Berkeley, California 94720 U.S.A.

**Department of Biophysics and Medical Physics, University of California, Berkeley, California 94720 U.S.A.

[1]This work was supported by a National Institutes of Health grant (GM 17317) to S. F. and by a grant to R. K. M. from the division of Biomedical and Environmental Research of the Department of Energy. S. F. acknowledges with gratitude the stimulating collegiality extended by the Stanford University Department of Biochemistry during a sabbatical leave. For conveying to him a working knowledge of recombinant DNA procedures and ongoing assistance, patient instruction, criticism, and gifts of DNA, S. F. is especially indebted to Professor R. W. Davis and his remarkable graduate students, postdoctoral fellows, and research associates. Special thanks are also extended to Professor J. L. Rossignol and his associates at the Université de Paris XI Sud, Orsay, France, for providing to S. F. a firsthand working introduction to the *Ascobolus* systems.

Excellent reviews that deal with various aspects are available. Included here are those by Clark (1974) on the regulation and control of bacterial recombination, by Hotchkiss (1974) on the comparative features of various molecular recombination models, by Stahl (1979a) on the function of special DNA sites in recombination and gene conversion in general (1969, 1978, 1979b). A vast quantity of data has been cogently organized in texts by Esser and Kunnen (1967), Whitehouse (1969), Grell (1974), Kushev (1974), Catcheside (1977), Fincham *et al.* (1979), Stahl (1979b). Reviews more specifically concerned with data drawn from fungi are those by Emerson (1969), Mortimer and Hawthorne (1973), Fogel and Mortimer (1971, 1971a,b, 1979), Radding (1973, 1979), Hastings (1975), Esposito and Esposito (1977), Pukkila (1977), Resnick (1979), Stahl (1979a), and Petes (1980). A recent overview of the yeast data is given by Fogel, Mortimer *et al.* (1981), and an overview for *Ascobolus* is given by Rossignol *et al.* (1979). A more complete key to the extensive literature can be obtained by consulting the Cold Spring Harbor Symposium (1978, Vol. II) and the recently published Cold Spring Harbor treatise on the Molecular Biology of *Saccharomyces* (Fogel *et al.* 1981). Also, provocative minireviews are given by Baltimore (1981) and Egel (1981). Mitotic recombination will not be considered in this paper. It is reviewed by Kunz and Haynes (1981) and by Esposito and Wagstaff (1981).

We shall approach the topic of meiotic gene conversion by (a) considering briefly a mechanistic overview of the phenomenon, (b) examining a specific molecular recombination model in some detail, (c) reviewing the salient properties and attributes of meiotic gene conversion based on genetic studies in fungi, (d) assessing newer knowledge derived from genetic analyses of recombinant DNA structures.

## A Mechanistic Overview

Detailed genetic studies in ascomycetous fungi, based mainly on the analysis of selected and unselected tetrads or octads, lead to the view that gene conversion is a signal of the molecular events underlying meiotic recombination. Gene conversions are associated with crossing-over between the markers flanking the converted gene. Moreover, the conversion-associated recombination alone is sufficient to account for the total length of the genetic map. More recently, our notions concerning gene conversion have been enlarged significantly by studies concerning the genetic behavior of structures generated by recombinant DNA strategies (Klein and Petes 1980). In addition to the classical interactions between allelic sequences occupying identical loci on homologous chromosomes, recombinationally interactive DNA segments may be nomadic (i.e., variously localized on nonhomologous chromosomes), distributed as nontandem duplications within a single chromosome arm or amplified as tandemly iterated gene sequences that may exhibit unequal crossing-over or sister chromatid exchange. In general, homologous DNA sequences interact to produce new recombinant forms regardless of where they might be localized within the nuclear genome (Scherer and Davis 1980).

Genetic recombinants are presumed to be the ultimate products of several enzymatically mediated reactions in which the DNA derived from two different parental molecules functions as the substrate. Nicking, unwinding, and reassociations of single DNA strands lead to the formation of a recombination intermediate, heteroduplex DNA (hDNA), that may undergo isomerization and be resolved as crossover products. In the meiotic process, with ordinary diploids, a failure to isomerize followed by resolution of the heteroduplex without crossing-over would result in haploid meiotic products that exhibited the parental array of flanking markers (see Figs. 3.2 and 3.3). However, if the hDNA included a heterozygous mutant site, a region of mispairing, nonpairing, or base-pair mismatch will occur, depending on the precise nature of the molecular difference between the wild-type and mutant alleles. Base pair substitutions, transitions, or transversions will yield single nucleotide mismatches, whereas single base additions or losses might yield subtle conformational changes. In contrast, deletions or insertions several hundred bases in length might be expected to generate single-strand loops within otherwise double-stranded DNA molecules. Even more complex configurations would be anticipated if each of the interacting parental alleles contained unique nonoverlapping regions that were not homologous to each other.

The various mismatches, nonpairings, and mispairings are thought to be sensed by an undefined cellular mechanism, possibly an endonucleolytic polymerase. These structures then become the substrates for an enzymatically mediated repair process that entails DNA excision-resynthesis over a chromosome segment subtending the mismatch. The net effect of the repair process is the nonreciprocal replacement of the information contained in one homologous (sequence) chromosome with a copy of the corresponding information contained in the other interacting homologous sequence. It should be noted that when meiosis is completed, the process leads to a change in the copy number of the two alleles (see Fig. 3.4). This is readily detected in organisms amenable to tetrad analysis, especially if the marker genes affect the cellular phenotype of the meiotic products as in *Sordaria* or *Ascobolus,* ascomycetes in which numerous mutants are known to alter the quantity of pigment, the pigment distribution pattern, or the shape of the haploid ascospores. Thus aberrant tetrad or octad segregation ratios are generated and these are known collectively as gene conversions. However, when heteroduplex repair fails to occur, this recombination intermediate is transmitted as such to the ascospore. On resolution via DNA replication, the heteroduplex intermediate is manifested as a pair of genetically dissimilar sister spores in octad forms and as a sectored ascosporal clone in tetrad forms. Asci displaying such spore patterns are designated as postmeiotic segregations *(pms),* since the hDNA generated at meiotic prophase is resolved and segregated as a consequence of the first DNA replication cycle required for the cell division that follows. Accordingly, mutations that affect the rates of hDNA formation might be expected to occur independently of those that alter the rates and patterns of hDNA correction and repair.

During the early decades of the chromosome theory of heredity it was widely held that meiotic recombination occurred only between genes but never within genes. In fact, the units of mutation, recombination, replication, and function were thought to be inseparable and coextensive within a single indivisible particle. Of course, such views are at best archaic and incorrect. Today, the study of intragenic or interallelic recombination is essentially the study of *gene conversion* (i.e., the nonreciprocal interchange of genetic information between homologous DNA sequences). We may approach the problem of intragenic meiotic recombination by posing several basic questions as follows:

1. What is the gene conversion phenomenon?
2. What methods are employed to study the conversion process?
3. What are the primary properties of gene conversion?
4. How can recombinant DNA strategies advance the analysis of intragenic recombination from the genetic to the molecular level.

## The Study of Gene Conversion

We begin by asking the operational question, "What is gene conversion?" In life forms adapted to tetrad analysis (e.g., where all the products of a single meiosis can be recovered and tested so that the allelic composition of each spore is unambiguously assigned), irregular segregations are sometimes observed. Identified as departures from the $4^+:4^-$ intraascal segregations expected for conventional heterozygous markers, such asci or spore clones display segregation ratios of $6^+:2^-$, $2^+:6^-$, $5^+:3^-$, $3^+:5^-$, $7^+:1^-$, $1^+:7^-$, aberrant $4^+:4^-$, and others. Because these ratios contradict the primary Mendelian segregation principle, they attracted the fungal geneticists' interest 50 years ago and more recently provoked the interests of molecular biologists interested in the organization and expression of DNA sequences (Baltimore 1981).

Since this report concerns meiotic gene conversion in yeast, some attention to the methods for scoring yeast tetrads is required. Yeast cultures sporulate readily when they are transferred to nitrogen-deficient media containing a nonfermentable carbon source such as potassium acetate. Ordinarily, two-, three-, and four-spored asci are encountered, but only the latter are taken for dissection with a micromanipulator device.[2] Unselected tetrads are randomly taken, dissected in a 6 × 6 or 12 × 12 mm grid on the surface of an inverted nutrient agar plate. After several days of incubation at 30°C, this plate is used as a master for replica-plating transfers to a series of petri dishes containing media diagnostic for each segregating marker. We designated this protocol as

[2]Routinely, we employ a large mechanical stage in combination with a spring-loaded micromanipulator fabricated by Lawrence Precision Instruments; 1416 West Winston Ave.; Hayward, California 94515.

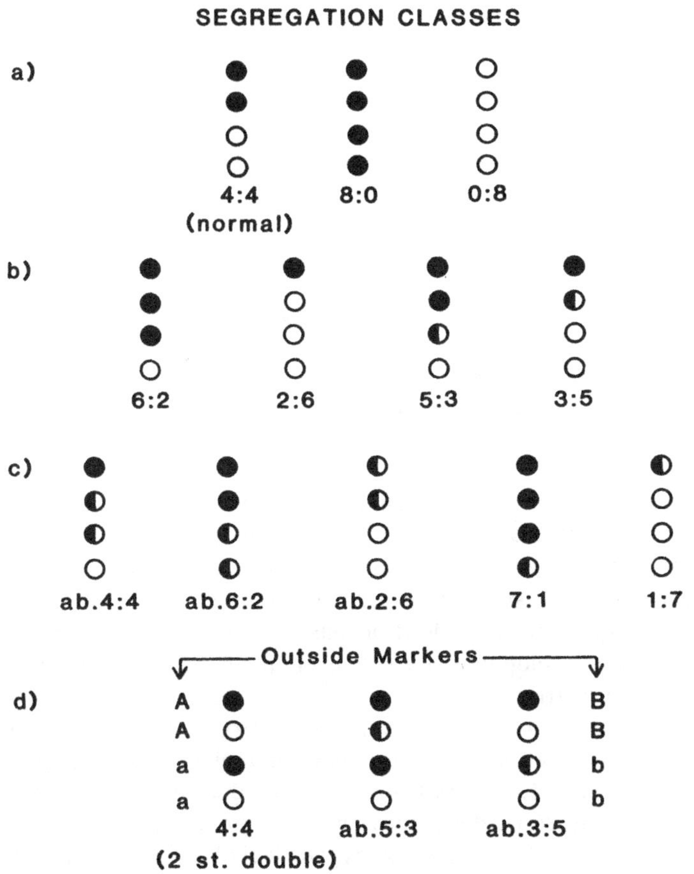


**Figure 3.1.** Diagram of various segregation classes of yeast tetrads. The ability to detect sectored ascosporal clones in effect converts tetrads into octads. (a, b, c) Normal segregation and 11 aberrant segregation classes. Classes shown in d are detected only if outside markers are available.

the *plate dissection method.*[3] Depending on their spatial separation, 10–20 tetrads can be accommodated on a single master. Most significantly, the procedure allows for the reliable, unambiguous detection of postmeiotic segregation *(pms)* as sectoring for the growth response on one or more plates. In effect, the plate dissection protocol converts biological tetrads into octads as follows: The spore germinates and if the first two buds are genotypically dissimilar with respect to a scorable marker, visibly sectored ascosporal clones will result provided that the original cells and their progeny grow in the plane of the agar substrate rather than intermix. For most yeast strains such contact inhibition actually prevails. Thus for a single segregating marker we can detect the 12 ascus patterns displayed in groups a, b, and c (Fig. 3.1). Only the first segre-

[3]Dr. Douglas Campbell contributed significantly to the plate holder design.

gation—a $4^+:4^-$ pattern—is normal; the others represent segregations that imply the loss or gain of informational copies by one chromosome at the expense of corresponding information contained in a homologous sequence.

The $8^+:0^-$ and $0^+:8^-$ classes will not be considered. Ample evidence indicates that they reflect clonally organized mitotic recombination events occurring prior to the onset of meiosis. Gene conversions are presumed to originate from transient heteroduplex hDNA—$6^+:2^-$ and $2^+:6^-$ ratios indicate enzymatic repair of the duplex, whereas $5^+:3^-$ and $3^+:5^-$ segregations imply persistent mismatches in the hDNA that failed to undergo repair and were subsequently resolved in the first postmeiotic replication cycle. Each sectored spore clone is scored as a *pms*. We deliberately emphasize that *pms* segregations constitute the primary genetic evidence for the existence of heteroduplex DNA whether it is generated reciprocally in a symmetrical manner on each of the interacting chromosomal sequences or is formed asymmetrically on only a single chromatid.

Reported first in mosses and fungi (see Fogel, Hurst, and Mortimer 1971), irregular or aberrant segregations and *pms* in yeast (Esposito 1971), were viewed as suspicious anomalies or artifacts. Indeed, a wide variety of conventional genetic mechanisms could easily be considered to generate virtually identical results. Thus, considerable effort has been given to the question of proper controls and the development of systems uniquely adapted to the critical study of irregular segregation.

Gene conversions are readily distinguished from other conventional genetic events when one monitors simultaneously the meiotic segregation behavior of many heterozygous sites—say, 15–20 distributed widely on the genome—some centromere linked, some distantly removed from centromeres, and others in tight linkage clusters flanking heterozygous loci of particular interest. Normal chromosome behavior among unselected asci is indicated by a $4^+:4^-$ segregation for each marker and exceptional conversion asci typically exhibit irregular segregation for only a single marker. Where a central marker flanked by two closely linked outside markers displays a gene conversion or postmeiotic segregation pattern and the proximal and distal markers show normal $4^+:4^-$ ratios, we assume that the conversional event is restricted to the DNA segment corresponding to the central marker and its immediate environs but does not extend into the realms of the outside markers. Frequently, however, the outside markers are recombined in such exceptional tetrads. Thus when we find nonreciprocal gene conversion or *pms* behavior for the central marker, the outside markers either are recombined reciprocally or retain the parental array with about equal likelihood. The obvious positive correlation between crossing-over and gene conversion implied a mechanistic linkage between them, and this soon led to the view that both processes reflected alternative aspects of a single basic event. On this basis, we proposed that gene conversion is a signal of the underlying recombination event in yeast (e.g., see Boon and Zinder 1971; Wagner and Radman 1975).

Now, given the major outlines of gene conversion, we may consider how it operates at the molecular level. Many molecular models have been proposed to account for both crossing-over and gene conversion as alternative aspects of a single underlying process. Comprehensive summaries and comparisons are given by Stahl (1969, 1978, 1979b), Pazewski (1970), Hotchkiss (1974), Catcheside (1977), Fogel *et al.* (1978, 1981) and in a series of critical studies by Rossignol (1979, 1980) and his collaborators. Each model conveys different predictions concerning the various segregation patterns depicted in Figure 3.1; space limitations prohibit their detailed consideration.

Here we present Holliday's original model (1964, 1974) (Fig. 3.2), one of several early molecular schemes that provided singular impetus to thinking about recombination in terms of the known behavior of nucleotide chains and DNA rather than the behavior of chromosomal elements visualized at the analytical level of the light microscope. It was among the first to account in molecular terms for the known behavior of heterozygous markers during the course of meiosis.

In its original version the Holliday model posits the formation of symmetrical hDNA regions on each of the two interacting nonsister chromatids (Fig. 3.2), and where the hDNA covers an included mutant site, such as a base substitution, two different base-pair mismatches are formed at the heterozygous site. The base-pair mismatches are correctable by two distinct enzymatic processes—excision of a single strand by an exonuclease and resynthesis or replacement by a polymerase along the remaining strand that serves as a template. Involved here is a modal length of the hDNA segment, not merely the mismatched base pairs themselves. Moreover, each heteroduplexed member of the symmetrical pair could undergo repair (or not) independently. Mutant sites in Figures 3.2 and 3.3 are indicated by the symbol m and the corresponding wild type by +. Accordingly, $6^+:2^m$ segregations result when mutant information is excised in each heteroduplex and replaced with a freshly synthesized copy of corresponding information contained in the remaining wild-type sequence. If both wild-type strands are excised and replacement synthesis takes place along the mutant template strand, $2^+:6^m$ segregations occur. Repair of only a single mismatch—one mismatch remaining uncorrected—generates a *pms,* i.e., $5^+:3^m$ or $3^+:5^m$ segregations, depending on whether the direction of correction was, respectively, toward wild-type or mutant. These relations are graphically displayed as Fig. 3.3, where only asymmetrical hDNA is considered. However, when both hDNA segments fail to undergo repair and are instead resolved by a single postmeiotic replication cycle, an aberrant $4^+:4^m$ ascus is produced.

In yeast, this distinctive category is distinguished by two sectored ascosporal clones accompanied by a spore clone of each parental type. The primary significance of these unique segregations was recognized first by Kitani *et al.* (1962), Kitani (1978), Kitani and Olive (1967), whose studies centered on the ascomycete *Sordaria fimicola.* Paquette (1978), Rossignol *et al.* (1977), Rossignol and Haedens (1980), and Kalogeropoulos and Rossignol (1980) working with *Ascobolus immersus,* have studied aberrant $4^+:4^m$ segregations exten-

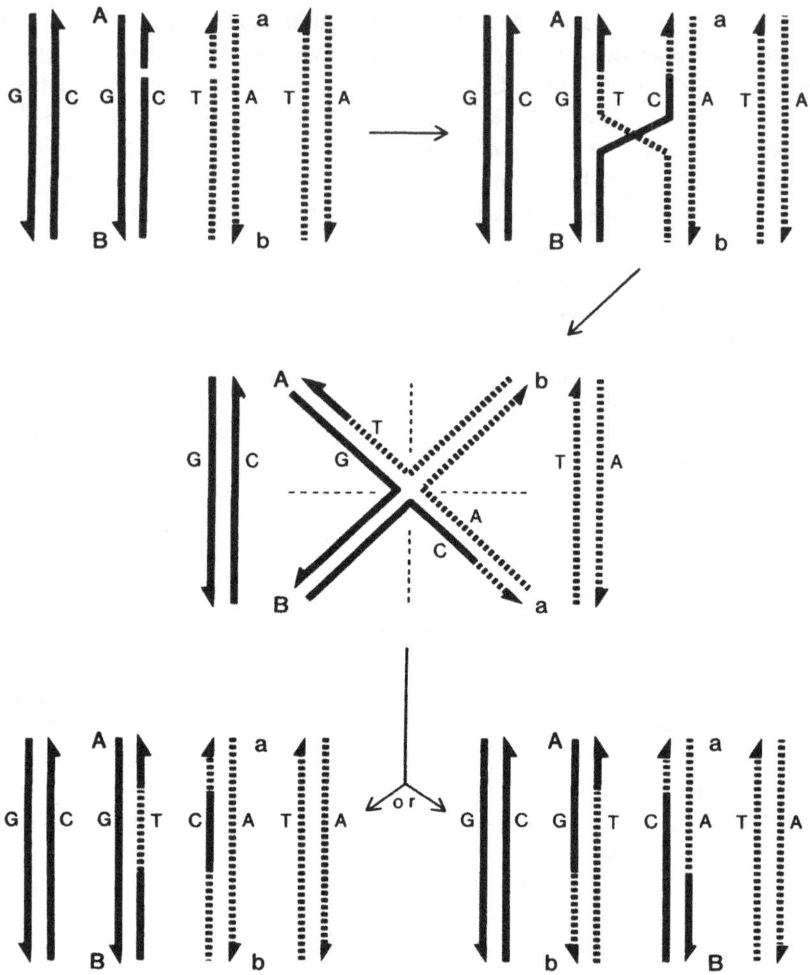


**Figure 3.2.** The Holliday model for recombination. At homologous special sites (recombinators), a bivalent enzyme nicks chains of the same polarity. Over part of their length, the nicked chains exchange partners and are ligated in their new positions. The resulting Holliday structure can be displayed as a *chi* form, which illustrates the structural equivalence of its two modes of resolution. A vertical cut yields patched chromatids, which are not crossovers for flanking markers. In this model, conversion results when a patch or splice falls on a marked site and one of the resulting heteroduplexes (+/m) enjoys mismatch correction, or both heteroduplexes are corrected to the same information content. In the former case, the resulting meiotic tetrad is a $5^+:3^-$ or a $5^-:3^+$; in the latter case, a $6^+:2^-$ or $2^+:6^-$ tetrad results. When correction fails on both chromatids, the resulting tetrad is an aberrant $4^+:4^-$.

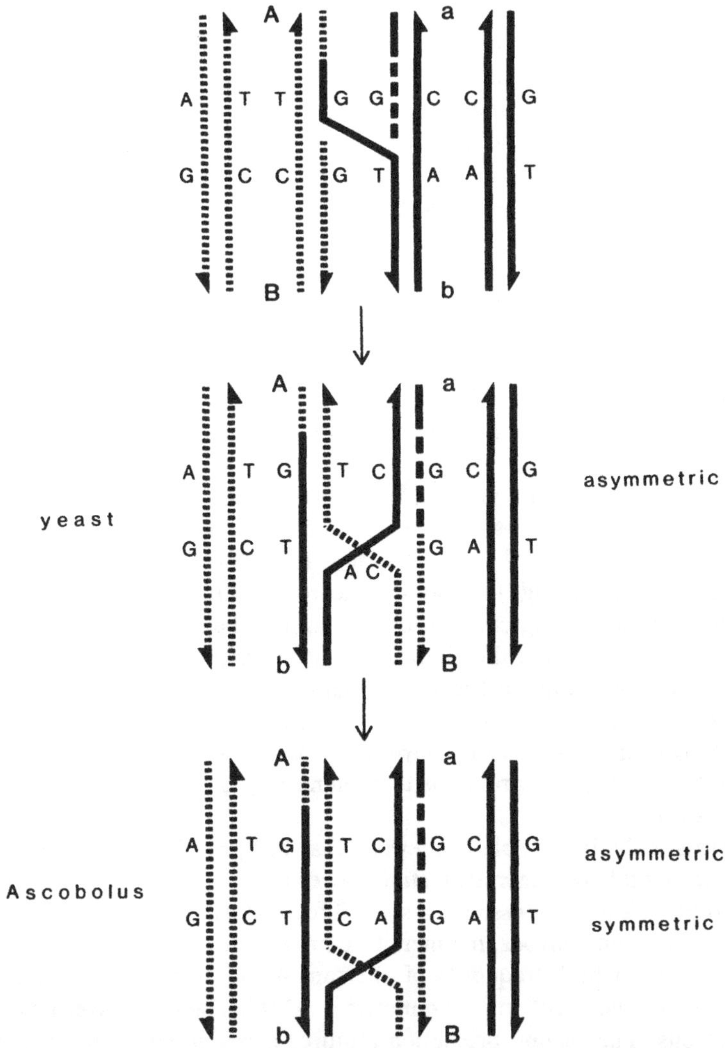


**Figure 3.3.** The Meselson-Radding model for recombination. DNA synthesis, initiated at a nick, displaces a chain, which then is taken up by the homologous chromatid. Coordinate chain extension and hydrolysis elongate the region of hybrid DNA formation. When the enzymes fall off, the structure can isomerize, and branch migration leads to hybrid DNA on both chromatids. This structure can then be visualized in a *chi* form as in Fig. 3.2; resolution can be to a partially asymmetric pair of either patches or splices.

sively. It may be noted that, given symmetrical hybrid DNA, corrections in opposite directions on each of the two chromatids would yield a normal $4^+:4^m$ or restoration to the parental state, and compelling evidence for such has been adduced by Rossignol's group. However, if outside flanking markers are present, these same events would appear as conventional two-strand double crossovers, as shown in Figure 3.1d. Clearly, these "two-strand double crossover" segregations arise as consequences of random repair in symmetrical hDNA rather than as classical multiple crossing-over, and in correction-efficient strains they should add substantively to the ranks of ordinary double crossovers involving the same two strands. Other segregation classes shown in Figure 3.1, including $7^+:1^m$, $1^+:7^m$, ab $6^+:2^m$, ab $2^+:6^m$, ab $5^+:3^m$, and ab $3^+:5^m$, are not readily accommodated within Holliday's original model. Lamb and Wickramaratne (1973), Lamb (1972, 1975), and Ghikas and Lamb (1977) suggest correctly that such categories are readily accounted for by the involvement of all four chromatids.

Central to the preceding molecular recombination model is a cross-stranded configuration (Fig. 3.2)—sometimes referred to as a half-chiasma or Holliday structure. Detectable at the electron microscope level, Holliday structures are produced by the nicking, unwinding, and reannealing of single DNA strands from the two interacting chromatids, which may be regarded as single duplex DNA molecules. Finally, the Holliday structure may be resolved by the occurrence of a double-stranded break in either of two possible planes indicated by dotted lines on the *chi* form shown as the middle diagram in Fig. 3.2. Thus heteroduplex structures can be resolved with or without outside marker exchange and they are prerequisite to gene conversion, pms, and associated crossing-over.

A heuristically invaluable framework that has provoked and guided numerous experimental and theoretical studies over the last 20 years, Holliday's symmetrical hDNA model possesses a special feature. It generates testable predictions in a clear, unequivocal manner. Two predictions are outstanding. The first requires a rather high frequency of aberrant $4^+:4^m$ segregations—a symptom diagnostic of noncorrection in symmetrical hDNA segments covering a heterozygous locus. The second prediction requires a much enhanced frequency of two-strand double crossovers among unselected nonconversional asci. These arise, as discussed earlier, from random corrections of paired symmetrical heteroduplexes flanked by closely linked markers. We first reported a conspicuous paucity of supportive data in both categories at the European Molecular Biology Organization's workshop on recombination held at Aviemore, Scotland in 1973. Our report concerning nearly 25,000 unselected tetrads and in which there was a scarcity of ab $4^+:4^m$ as well as two-strand double exchanges, prompted Meselson and Radding's (1975) formulation of a new recombination model, shown in Fig. 3.3. Constructed on the foundation of Holliday's symmetrical scheme, the new Aviemore (Meselson-Radding) model proposed an initial asymmetrical phase where hDNA formation occurred on only a single chromatid. The resultant recombination intermediate could isomerize, forming a typical chi-like Holliday structure that could be propelled by various forces to yield symmetrical paired hDNA tracks. Clearly, if the extent of the two phases could differ in different fungi, the seemingly conflicting observations in *Saccharomyces* and *Ascobolus* could be harmonized.

Yeast generates very little or quite possibly no symmetrical DNA, but *Ascobolus* typically initiates intragenic recombination asymmetrically and sooner or later passes over to the symmetrical mode. To date, mutants that conduce to symmetrical hDNA formation in yeast have not been found, despite critical attempts to identify them. To suppose that genes coding for this function are absent in the breeding stocks of yeast is plausible; however, a firm conclusion on the matter is premature. Taken collectively, but with a single exception, concerning the position of the associated exchange relative to flanking markers in *pms* tetrads, the vast quantities of data from both yeast and *Ascobolus* are readily accommodated by the Meselson-Radding revision of the Holliday model. Given this molecular, mechanistic overview we may now consider the principal methods used to study gene conversion and then inquire concerning the major features of the conversion process as revealed by such investigations.

## Methods for Studying Gene Conversion

Gene conversion may be investigated using three rather different protocols based on (1) selected recombinant or prototrophic spores, (2) selected tetrads containing a prototrophic spore, and (3) unselected tetrads. The first and most extensively utilized among these were the selective techniques that permit the detection, enumeration, and classification of "recombinant" wild-type spores or whole asci containing such prototrophic "recombinants." Typically, these are obtained by conventional dilution-plating experiments on selective media that utilized sporulated heteroallelic diploids $m1+/+m2$, where the input parental alleles are noncomplementing and of independent origin. Rapid and relatively easy to perform, such protocols soon led to the construction of remarkably detailed genetic fine structure maps that depict the spatial distribution of the various mutant markers in a single genetic locus (e.g., Esposito 1967, Korch and Snow 1973).

The maps are, of course, based on the assumption that recombination regularly occurs within a single unit of genetic function. In a sense, these maps stand as monumental contradictions to the cherished notion that the gene is a unit indivisible by recombination. But they yield little or no insight concerning the molecular details of the recombination process itself. However, extensive analyses of selected tetrads by Fogel and Hurst (1967) provided critical information on the mechanism underlying the production of such "prototrophic wild-type recombinants."

Heteroallelic crosses, in repulsion, of the general genotype

$$\frac{a \quad m1 \quad + \quad b}{+ \quad + \quad m2 \quad +}$$

were analyzed. These diploid hybrids carried closely linked outside markers along with numerous unlinked heterozygous sites that serve to monitor the regularity of meiotic behavior within each tetrad.

The protocol for recovering intact, four-spored tetrads containing a prototrophic recombinant is as follows: A heteroallelic diploid culture with genotype *m1+/+m2* and auxotrophic for a particular nutrilite such as histidine is sporulated. Washed asci are deposited on agar deficient for histidine, and after an appropriate incubation period on histidine-deficient agar, those asci with an actively budding spore are identified and selected for isolation and dissection.

About 1100 tetrads were identified, isolated, and analyzed. The allelic constitution for each and every spore was ascertained without any ambiguity by crossing all spore clones to strains carrying the input parental alleles, selecting the resultant diploid, and then assessing its recombination response rate. In some experiments, suppressibility or complementation are assessed, depending on the mutants used, since these are virtually "all-or-none" responses adapted to rapid, unequivocal scoring. As Figure. 3.4 illustrates, wild-type prototrophic recombinants in heteroallelic diploids arise as a consequence of $6^+:2^-$ segregations at either the proximal or distal allele or by reciprocal recombination between the mutant sites. The $6^+:2^-$ segregations for either allele represent conversion events that are nonreciprocal in character, and the double mutant expected as a consequence of a reciprocal, breakage–reunion crossover between mutant sites is not commonly encountered. Thus, for the heteroallelic diploid *m1+/+m2*, where *m1+* and *+m2* are the proximal and distal alleles, a $6^+:2^-$ conversion of the proximal mutant in the direction mutant to wild yields

| Normal segregations | Single-site conversions | | | | Symmetrical double-site conversions | | Reciprocal recombinants |
|---|---|---|---|---|---|---|---|
| (a) | (b)[a] | (c)[a] | (d)[b] | (c)[b] | (f)[c] | (g)[c] | (h) |
| *m1+* | *m1+* | *m1+* | *m1+* | *m1+* | *m1+* | *m1+* | *m1+* |
| *m1+* | *m1+* | *m1+* | *m1+* | *m1+* | *m1+* | *m1+* | *m1+* |
| *m1+* | *++* | *m1+* | *m1+* | *m1m2* | *+m2* | *m1+* | *m1m2* |
| *m1+* | *++* | *m1+* | *m1+* | *m1m2* | *+m2* | *m1+* | *m1m2* |
| *+m2* | *+m2* | *m1m2* | *++* | *+m2* | *+m2* | *m1+* | *++* |
| *+m2* | *+m2* | *m1m2* | *++* | *+m2* | *+m2* | *m1+* | *++* |
| *+m2* | *+m2* | *+m2* | *+m2* | *+m2* | *+m2* | *+m2* | *+m2* |
| *+m2* | *+m2* | *+m2* | *+m2* | *+m2* | *+m2* | *+m2* | *+m2* |

[a]Conversion of *+/m1*.
[b]Conversion of *+/m2*.
[c]Conversion of *+/m1* and *+/m2*.

**Figure 3.4.** Main types of asci recoverable from heteroallelic diploids.

the following tetrad: $++$, *m1*$+$, $+$*m2*, $+$*m2* and comparable $6^+:2^-$ conversions of the distal allele generate $++$, *m1*$+$, *m1*$+$, and $+$*m2*. Conversions in the opposite direction (i.e., wild to mutant) also occur, though they are not detected in experiments based on recovering prototrophic recombinants. Consider the proximal and distal conversions of the $2^+:6^-$ variety for the proximal and distal alleles. These are *m1*$+$, *m1*$+$, *m1m2*, $+$*m2*, and *m1*$+$, *m1m2*, *m2*$+$, *m2*$+$. Clearly, such asci contain double mutant spores, but prototrophic spores are absent; consequently, these events are not monitored by selective procedures or mere prototroph enumeration. In fact, they were first analyzed in detail through our studies on wholly unselected tetrads—a topic to be considered in a following section.

Nonreciprocal gene conversions represent about 90% of all the "wild-type" recombinant asci, and among these, proximal conversions were about six to seven times as frequent as distal conversions. Reciprocal events indicated by asci having the spore array $++$, *m1m2*, *m1*$+$, $+$*m2* comprise only 10% of the tetrads studied. Thus three major conclusions arise: (1) Nonreciprocal gene conversions occur more often than crossing-over between alleles. (2) Gene conversion is strongly polarized (i.e., marker alleles at one end of the locus are more prone to undergo conversion compared to markers at the other end of the gene). This property of gene conversion is often referred to as *polarity*. Analysis of gene conversion at several unrelated loci indicates that there are regular gradients of conversion frequency across genes. Polarity suggests that specific sites are involved in the origin or resolution of conversion events. (3) With regard to the behavior of the outside markers astride the converted locus, about half (42%) of the prototrophic spores among the gene conversions are recombined for the flanking markers, while the remainder (52%) retain the parental configuration. Clearly, a strong correlation exists between gene conversion and associated outside marker recombination. Moreover, because the prototrophic strand itself is nearly always involved in the outside marker exchange, molecular models were strongly constrained to accommodate these early findings. Finally, we may note a distinctive minority category among the prototrophic asci—those containing one spore each of the wild-type, double mutant, and each input parental mutant. These reciprocal recombinants between the intragenic sites almost invariably displayed outside marker recombination, and in virtually all instances, the prototrophic and double mutant-bearing strands are involved in this exchange. Not unexpectedly, chiasma interference in nearby marked intervals is pronounced within the group defined by reciprocal recombinants between the alleles. However, as Stadler (1959) first demonstrated with random spores in *Neurospora*, positive chiasma interference, in adjacent marked intervals, is not generated by gene conversions that retain their parental outside marker arrangements. Consequently, we are led to an inevitable conclusion—gene conversion is a basic signal of meiotic recombination, and it may occur about equally frequently with or without associated outside marker crossing-over. Hence considerable effort was directed toward formulating unified molecular schemes that forged a mechanistic linkage between these universal processes.

## Gene Conversion and Unselected Tetrads

Though analyses based on selective strategies with single strands or intact tetrads uncovered some of gene conversion's fundamental properties, the complete analysis of unselected tetrads disclosed frequent hitherto unknown events. Metaphorically, we may equate selective tactics with discovering the 20% above-water portion of an iceberg and unselective approaches with revealing the 80% submerged fraction.

To date, we have analyzed more than 25,000 unselected tetrads from two- and three-point heteroallelic crosses. Very simply, we are akin to astronomical observers in the sense that we are merely scanning the genetic heavens and taking inventory of the various meiotic events. Hybrids are sporulated, asci are dissected at random and the genotype of every spore is ascertained fully by a combination of replica-plating and scoring backcrosses to appropriate tester strains for recombination, suppressibility, or complementation responses. Because the rationale does not entail experimental treatments compared to controls, it could be construed as genetic voyeurism wherein the ultimate goal is to formulate a molecular model based on a complete inventory of the genetic events observed in single meioses.

Summarized in Figure 3.4 are the major varieties of asci recovered among unselected samples from heteroallelic diploids. Eight ascal categories are regularly found. However, omitted here are asci with one or more spore clones exhibiting postmeiotic segregation. Their crucial significance and importance will be discussed in a subsequent section. The segregations are displayed as octads—i.e., each of the four products arising from a single meiosis presumably undergoes a regular mitotic division, producing two sister cells (and their respective progeny) or two sister spores as in *Neurospora, Ascobolus,* or *Sordaria.* Displayed in Figure 3.4 are normal $4^+:4^-$ segregations expected for conventional Mendelian behavior of either allele—as in group (a). Single-site conversions for either the proximal or distal allele and directionally from mutant to wild or wild to mutant, as in groups (b, c, d, e), yield $6^+:2^-$ or $2^+:6^-$ segregations for each allele. Reciprocal recombinants between alleles are in group (h), and these possess all the features expected to follow an exchange between the heteroallelic sites. The novel categories, types (f) and (g), found only among unselected tetrads fall into two subsets that involve a $6^+:2^-$ segregation for one allele accompanied by a symmetrical $2^+:6^-$ segregation for the second heteroallele. Originally, we termed these *symmetrical double-site conversions.* But, following a suggestion by David Stadler, we designated them as *co-conversions.* We view co-conversion as a process of informational transfer (Fogel and Mortimer 1969), that is, the consequence of a single coordinated event leading to the nonreciprocal replacement of a segment of genetic information in one homolog with information *identical* to that contained in the corresponding section of a nonsister homolog. In fact, since the advent of recombinant DNA techniques, we may enlarge the preceding definition to include interactions between homologous DNA sequences regardless of whether they are distributed on homologous chromosomes, on nonhomologous chromosomes, within a single chromosome on the same or different arms, or on sister chromatids.

Is it conceivable that co-conversions arise not as single events affecting a given segment of DNA but as independent conversional events occurring at the two intragenic sites? In brief, we shall suppose autonomous activity for each mutant site. On this assumption, in addition to co-conversions, six additional ascal classes may be predicted to exist within unselected samples of tetrads, and these should occur with equal likelihood. The predictions are given in Figure. 3.5. How do the predictions match with our genetic accounting program? Some 6500 unselected tetrads from five isogenic hybrids were subjected to complete analysis. Ascus types (g) and (h), the co-conversion classes, are relatively common and amounted to several percent. However, representatives of the six remaining categories were conspicuously rare. In fact, types (a, b, c,) and (f) were never observed, and in consequence we concluded that co-conversion represented a single primary DNA event.

Considerations similar to the preceding led us to infer that gene conversions typically involve a sizable DNA segment rather than a narrowly restricted region such as a single base pair or a few base pairs. Alleles widely separated on a genetic fine structure map, for example, at opposite ends of a structural gene, convert independently of each other. However, as the distance between the mutant sites is diminished by choosing alleles in greater proximity, the co-conversion frequency rises at the expense of single-site conversions. Variable in length, the converted segment modally occupies several hundred nucleotides as estimated by x-ray mapping studies (Manney and Mortimer 1964; Fogel and Mortimer 1969). Our recent physical data on the genetic behavior of cloned DNA segments carrying several different genes are consistent with this estimate. Some relevant data may be considered.

Alleles *arg4-4* and *arg4-17* are internal markers lying at opposite ends of the structural gene specifying arginino-succinase. Of 697 unselected tetrads examined, 58 were exceptional but only three represented co-conversions. Sharply

| (a) | (b) | (c) | (d) | (e) | (f) | (g) | (h) |
|---|---|---|---|---|---|---|---|
| ++ | *m1m2* | ++ | *m1m2* | ++ | ++ | +*m2* | +*m2* |
| ++ | *m1m2* | ++ | *m1m2* | ++ | ++ | +*m2* | +*m2* |
| ++ | *m1m2* | ++ | *m1m2* | +*m2* | *m1*+ | *m1*+ | *m1*+ |
| ++ | *m1m2* | ++ | *m1m2* | +*m2* | *m1*+ | *m1*+ | *m1*+ |
| ++ | *m1m2* | *m1*+ | *m1*+ | +*m2* | *m1*+ | *m1*+ | +*m2* |
| ++ | *m1m2* | *m1*+ | *m1*+ | +*m2* | *m1*+ | *m1*+ | +*m2* |
| *m1m2* | ++ | +*m2* | +*m2* | *m1m2* | *m1m2* | *m1*+ | +*m2* |
| *m1m2* | ++ | +*m2* | +*m2* | *m1m2* | *m1m2* | *m1*+ | +*m2* |

**Figure 3.5.** Ascal types expected if heteroallelic sites in repulsion convert independently.

contrasted to these findings are the data for closely spaced mutants *arg4-16* and *arg4-17*. Among 3156 unselected tetrads, 334 aberrant segregations were found. Of these, only two were reciprocal recombinants, about one-quarter were single-site events, and the remaining 244 events, or 75%, were co-conversional. Similar data are readily available for many other loci widely distributed on the yeast genome. Also, the converted segment may extend across several closely linked genes. Co-conversion is doubtless ubiquitous. It has been reported in *Drosophila, Zea, Ascobolus, Sordaria, Schizosaccharomyces, Chlamydomonas, Sphaerocarpus,* and probably other forms subject to critical analysis.

The remaining sections summarize additional features of meiotic gene conversion, and these will be considered in the context of a molecular model developed by Meselson and Radding (1975). Their scheme requires the initial formation of asymmetric hybrid DNA on only a single chromatid, and like Holliday's original scheme, it also allows for the subsequent generation of symmetrical paired heteroduplex DNA regions. For the most part, the various molecular models do not consider the details relating to the repair of the heteroduplex DNA tracts.

## Mismatch Repair and Postmeiotic Segregation (*pms*)

More than 23,000 unselected tetrads, drawn from 20 related hybrids, were scored for their segregation patterns at all heterozygous loci (Fogel *et al.* 1981). Using the plate dissection replica-plating method, we scored *pms* and gene conversion at 30 different chromosomal sites. Tetrads are recorded as normal $4^+:4^-$ segregations or assigned to one of 11 other aberrant segregation classes listed in Table 3.1. Seven categories contain one or more sectored ascosporal clones.

Every site displays exceptional segregations, with frequencies ranging from 0.6% to 18%. The $8^+:0$ and $0^+:8^-$ classes are excluded. In yeast they undoubtedly represent mitotic events occurring prior to the meiotic divisions. However, in organisms such as *Ascobolus* or *Sordaria,* which do not possess extended diplophases in their life cycles, a meiotic origin is implicated. In fact, for these organisms Lamb and Wickramaratne (1973) emphasized that the occasional $8^+:0$ and $0^+:8^-$ asci occur as frequently as the $7^+:1^-$, $1^+:7^-$, ab $6^+:2^-$ and $2^+:6^-$ segregations. Such wider-ratio segregations, they presume, arise by the involvement of all four chromatids at a single site where two pairs of symmetrical hybrid DNA segments are generated coincidentally. In yeast we found only 11 wide-ratio segregations among the 221,151 segregations analyzed. The finding suggests either that symmetrical hybrid DNA is a rare occurrence in yeast or that they may be accounted for by positing the simultaneous formation of two independent asymmetrical hybrid DNA segments covering the same heterozygous marker. The conversion patterns of 80 *Ascobolus* mutants, representing 17 different single-cell ascosporal phenotypes, has been reported by Nicolas (1979). Conspicuous similarities and differences among various fungi are critically delineated in this comprehensive work and also in studies by Rossignol *et al.* (1979) and Rossignol and Haedens (1980).

The tabulated items (Table 3.1) are arrayed in order of their average conversion frequencies. And it may be noted that the percentage of *pms* among aberrant segregations is wholly uncorrelated with this arrangement. Accordingly, as we might have anticipated from purely enzymatic considerations, heteroduplex formation is a reaction sequence independent of heteroduplex correction or repair. Within this context, $6^+:2^-$ and $2^+:6^-$ segregations (gene conversions) are equated with correction-repair and $5^+:3^-$ and $3^+:5^-$ *(pms)* represent failure of repair within the hybrid DNA astride the mutant site. Clearly, some mutants, such as *ade8-18, arg4-16,* exhibit marked predilections toward *pms.* Thus, with a combined total of 1400 segregations of the $6^+:2^-$ and $2^+:6^-$ types, these markers also yielded 1053 *pms* tetrads or 43% pms/total aberrant asci. This behavior stands in marked contrast to the data for *met13,* where a single *pms* was found among some 10,500 segregations containing 933 gene conversions, or nearly 9%. Similarly, not a single *pms* is recorded among the 23,000 segregations for the yeast mating type locus, *mat1,* though nearly 200 ordinary gene conversions were identified. How can these diverse observations be rationalized within a simple, consistent molecular framework?

First, we consider the observed variations in the basic conversion frequency, which is defined as the frequency of all aberrant segregations at a locus; i.e., $6^+:2^- + 2^+:6^- + 5^+:3^- + 3^+:5^-, \ldots$ /total tetrads. Since aberrant segregations require the formation of hDNA, we take the basic conversion frequency to represent a minimal estimate of the probability that a given site will be covered by hDNA. However, this value doubtless underestimates the true value, because correctional repair or restorations to the parental state may occur—perhaps as often as conversions. On this same general notion, we might account for polarity of segregations across a gene locus by supposing that hDNA might be initiated at preferred sites and these are arrayed in a density distribution reflective of the observed polarity.

Second, we may consider the variations with respect to correctional repair, given that a site exists within a hDNA tract. Among the 221,751 segregations analyzed, only 4.5% were aberrant, and among these the median percentage of *pms*/ total events was 3.5.

Thus hDNA regions experience enzymatic repair for the most part on one strand or the other, with a slight bias, perhaps, in favor of correction to wild type over corrections to the mutant; that is, the overall summed conversion of $6^+:2^-$ and $2^+:6^-$ is 4521:3999. Only a decidedly minor fraction fails to repair. Resolved at the very first DNA replication cycle after meiosis has terminated, these noncorrections are detected operationally as sectored ascosporal clones.

It has been suggested that mutagenic origins and therefore DNA configuration primarily determine correction patterns. Rossignol and colleagues (1979), Yu-Sun *et al.* (1977), and Whitehouse (1963, 1969) propose that very small addition-deletion mutants generate hDNA arrangements that are conducive toward correction. Accordingly, they exhibit few *pms* events per $6^+:2^-$

**Table 3.1.** Distributions of Aberrant Segregations for 30 Heterozygous Sites

| Gene | Segre-gations | Hybrids[a] | 8:0 | 0:8 | 6:2 | 2:6 | 5:3 | 3:5 | ab4:4 | ab6:2 | ab2:6 | 7:1 | 1:7 | $\rho$ Events (%) | $\sigma p$ | $\chi^2$ Hetero-geneity[b] | pms/total (%) |
|---|---|---|---|---|---|---|---|---|---|---|---|---|---|---|---|---|---|
| *pet1* | 4,924 | 4 | 1 | 38 | 4 | 27 | 0 | 0 | 0 | 0 | 0 | 0 | 0 | 0.63 | 0.35 | * | 0 |
| *trp1* | 20,826 | 17 | 8 | 39 | 62 | 47 | 12 | 8 | 0 | 0 | 0 | 0 | 1 | 0.63 | 0.29 | | 16.2 |
| *mat1* | 23,135 | 20 | 0 | 8 | 93 | 104 | 0 | 0 | 0 | 0 | 0 | 0 | 0 | 0.85 | 0.34 | * | 0 |
| *ura3* | 2,315 | 3 | 0 | 0 | 10 | 9 | 2 | 1 | 0 | 0 | 0 | 0 | 0 | 0.95 | 0.84 | | 13.6 |
| *ade6* | 1,589 | 1 | 0 | 3 | 4 | 9 | 3 | 3 | 0 | 0 | 0 | 0 | 0 | 1.20 | — | — | 31.6 |
| *his5-2* | 2,315 | 3 | 2 | 2 | 14 | 10 | 4 | 1 | 0 | 0 | 0 | 0 | 0 | 1.26 | 0.17 | | 17.2 |
| *tyr1* | 8,391 | 7 | 7 | 6 | 59 | 44 | 5 | 1 | 0 | 0 | 0 | 0 | 0 | 1.30 | 0.41 | | 5.5 |
| *CUP1* | 18,016 | 15 | 20 | 16 | 123 | 124 | 6 | 3 | 0 | 0 | 0 | 0 | 0 | 1.42 | 0.52 | * | 3.5 |
| *gal2* | 2,416 | 2 | 1 | 3 | 36 | 17 | 0 | 0 | 0 | 0 | 0 | 0 | 0 | 1.78 | 0.09 | | 0 |
| *leu21* | 3,203 | 3 | 2 | 2 | 41 | 25 | 0 | 0 | 0 | 0 | 0 | 0 | 0 | 2.06 | 0.65 | | 0 |
| *trp5-48* | 2,315 | 3 | 3 | 1 | 23 | 31 | 1 | 0 | 0 | 0 | 0 | 0 | 0 | 2.38 | 0.56 | | 2 |
| *met1* | 1,589 | 1 | 1 | 1 | 18 | 17 | 18 | 6 | 0 | 0 | 0 | 0 | 0 | 3.72 | — | — | 40.7 |
| *met10* | 892 | 1 | 0 | 0 | 17 | 16 | 1 | 0 | 0 | 0 | 0 | 0 | 0 | 3.82 | — | — | 2.9 |
| *ura1* | 15,014 | 13 | 29 | 17 | 331 | 275 | 11 | 15 | 1 | 0 | 0 | 0 | 0 | 4.23 | 1.48 | ** | 4.3 |
| *ilv3* | 9,487 | 9 | 1 | 2 | 230 | 239 | 9 | 6 | 0 | 0 | 0 | 0 | 0 | 5.10 | 0.72 | | 3.1 |
| *lys1-1* | 2,315 | 3 | 3 | 1 | 51 | 78 | 3 | 7 | 0 | 0 | 0 | 0 | 0 | 5.76 | 0.20 | | 3.0 |

| | | | | | | | | | | | | | | | | | |
|---|---|---|---|---|---|---|---|---|---|---|---|---|---|---|---|---|---|
| *SUP6* | 892 | 1 | 0 | 0 | 22 | 33 | 0 | 0 | 0 | 0 | 0 | 0 | 0 | 6.16 | — | — | 0 |
| *thr1* | 21,220 | 18 | 13 | 15 | 691 | 594 | 14 | 22 | 0 | 0 | 0 | 0 | 0 | 6.23 | 1.15 | ** | 2.7 |
| *his4-4* | 12,533 | 11 | 11 | 12 | 411 | 303 | 39 | 41 | 1 | 0 | 0 | 0 | 1 | 6.36 | 1.16 | | 10.3 |
| *ade8-10* | 1,118 | 1 | 2 | 64 | 48 | 30 | 0 | 0 | 0 | 0 | 0 | 0 | 0 | 7.41 | — | — | 0 |
| *met13* | 10,454 | 10 | 151 | 15 | 459 | 474 | 1 | 0 | 0 | 0 | 0 | 0 | 0 | 9.07 | 2.15 | ** | 0.1 |
| *ade8-18* | 15,480 | 13 | 29 | 2221 | 351 | 239 | 375 | 294 | 10 | 1 | 3 | 1 | 2 | 9.72 | 0.89 | | 54.1 |
| *cdc14* | 892 | 1 | 1 | 0 | 48 | 39 | 1 | 1 | 0 | 0 | 0 | 0 | 0 | 9.99 | — | — | 2.2 |
| *ade7* | 1,028 | 1 | 3 | 159 | 47 | 27 | 11 | 8 | 0 | 0 | 0 | 0 | 0 | 10.7 | — | — | 20.4 |
| *his2* | 2,481 | 2 | 5 | 10 | 215 | 231 | 2 | 0 | 0 | 0 | 0 | 0 | 0 | 18.2 | 3.74 | ** | 0.5 |
| *arg4-4* | 1,188 | 1 | 0 | 0 | 5 | 13 | 0 | 1 | 0 | 0 | 0 | 0 | 0 | 1.6 | — | — | 5.2 |
| *arg4-3* | 2,405 | 2 | 0 | 0 | 28 | 23 | 5 | 3 | 0 | 0 | 0 | 0 | 0 | 2.45 | 0.26 | | 1.4 |
| *arg4-19* | 5,352 | 4 | 5 | 5 | 87 | 68 | 2 | 6 | 0 | 0 | 0 | 0 | 0 | 3.04 | 1.16 | ** | 4.9 |
| *arg16* | 14,490 | 11 | 8 | 11 | 508 | 302 | 95 | 289 | 4 | 0 | 1 | 1 | 0 | 8.29 | 1.58 | ** | 32.5 |
| *arg4-17* | 13,476 | 13 | 18 | 11 | 485 | 551 | 38 | 20 | 0 | 0 | 0 | 0 | 0 | 8.12 | 1.79 | ** | 5.3 |
| Total | | | | | | | | | | | | | | | | | |
| Percent | 221,751 | | 324 | 2662 | 4521 | 3999 | 657 | 740 | 16 | 1 | 4 | 3 | 3 | | | | |
| | | | | | 2.06 | 1.83 | 0.30 | 0.34 | 0.0073 | | | | | 4.54 | | | 3.5 (median) |

[a]Data pooled from segregations of hybrids G12, M141, M150, X3653, 5204, 5246, 5275, 5276, 5276, 5276-3, 5308, 5310, 5313, 5420, 5420-1, 5420-1-1.5462, 5463, 5475, and 5497.

[b]* and ** indicate significance at the 5% and 1% levels, respectively.

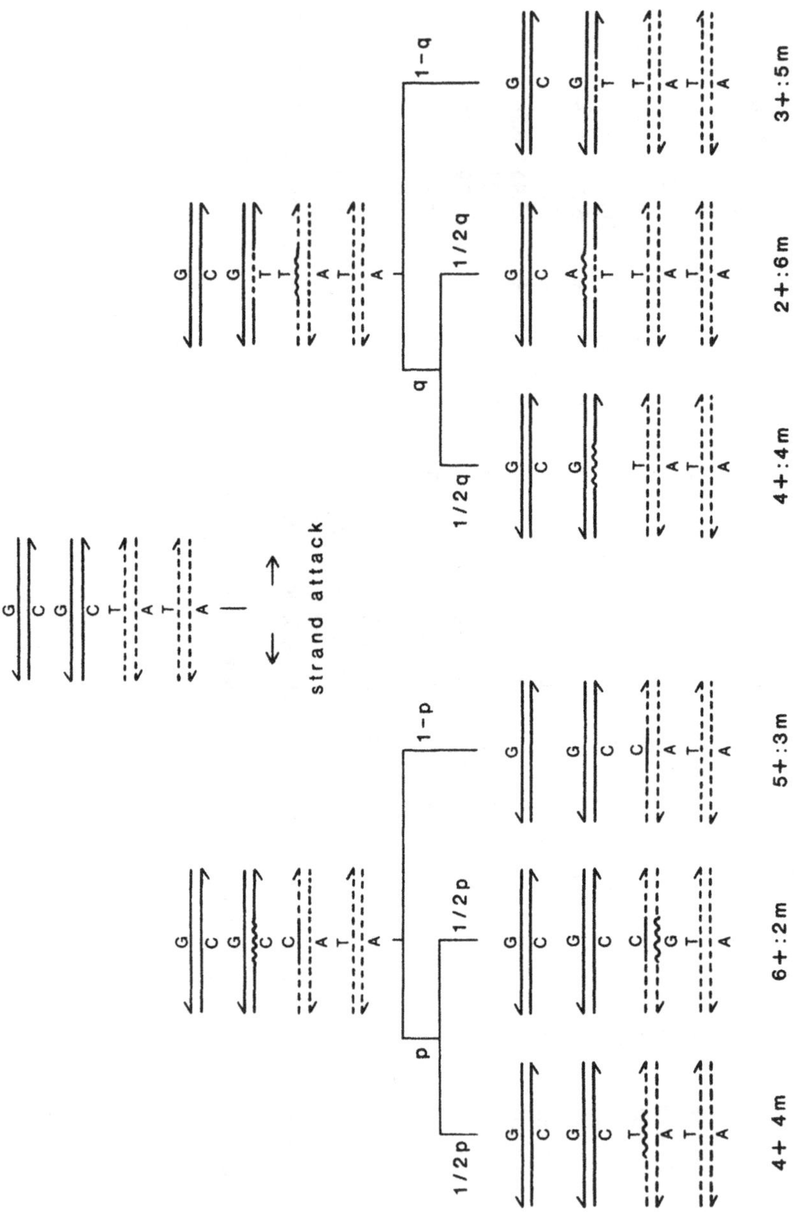
strand attack
p
1/2p
1/2p
1-p
q
1/2q
1/2q
1-q
4+:4m
6+:2m
5+:3m
4+:4m
2+:6m
3+:5m

or $2^+:6^-$, and within this general class, type A mutants yield significantly more $6^+:2^-$ than $2^+:6^-$ segregations. Type B mutants display the opposite pattern. In contrast, base substitution mutants may exhibit rather high proportions of *pms*/conversion. These are designated as type C and type D. In the former, $6^+:2^-$ patterns significantly exceed $2^+:6^-$ segregations, and the reverse pattern obtains for the latter. The correlation between mutagenic origin and conversion spectrum is readily verifiable in both *Sordaria* and *Ascobolus*. Indeed, the data are both extensive and impressive. However, similar effects are not known in yeast.

Taken collectively, the data arising from unselected yeast tetrads may be generalized by the term *parity*. This implies that $6^+:2^- = 2^+:6^-$ and $5^+:3^- = 3^+:5^-$ in many instances. At the molecular level (see Fig. 3.6), we suppose that strand attack is equally likely in either direction. A single DNA strand from a mutant chromatid might displace a corresponding strand of the same polarity in a chromatid bearing wild-type information at the locus or vice versa. Now, given the hDNA, formed on either parental chromatid, the recipient or the donor strand would be excised and replaced with the same probability. If strand attack were preferential or if correction favored one strand or the other, disparity would result. Some 12 statistically significant disparities can be identified within our yeast data. But overall the magnitude of the differences is small compared to the findings in *Ascobolus* or *Sordaria*. Thus the dissymmetry coefficients ($6^+:2^-/2^+:6^-$) for 30 yeast sites range from 0.38 to 2.12, whereas in *Ascobolus* and *Sordaria* they range from 0.01 to 100, and the largest disparities are found for addition–deletion mutants produced by ICR-170.

In our studies, all the mutants were induced with UV or EMS. At least 17 of the 30 mutants are identified as base substitutions on the criteria of complementation, suppressibility, or temperature sensitivity. To date, only a single frameshift mutant has been examined in our laboratory (M. Williamson, personal communication), i.e., *his4-51*, a mutant kindly provided by G. Fink. Among 978 unselected asci, we found 27 and 43 asci of the $6^+:2^-$ and $2^+:6^-$ types and one $3^+:5^-$ ascus, a finding consistent with parity and the general

**Figure 3.6.** A hypothetical molecular pathway to account for aberrant meiotic segregations in Yeast. Chromatids are represented as single DNA molecules composed of complementary antiparallel strands. At the top, synaptic pairing between replicated homologs is shown. This may occur just prior to or immediately after the round of meiotic DNA synthesis. Asymmetrical hybrid DNA covering the heterozygous mutant site (a base pair substitution) is shown to arise via single strand attack. Wild (GC) may attack on mutant (AT) as shown on the left or mutant (AT) may attack wild as shown on the right. Respectively, these events generate the base pair mismatches CA and GT. The probabilities of mismatched base pair corrections, via excision–resynthesis reactions, leading to gene conversions ($6^+:2m$ and $2^+:6m$) or restorations ($4^+:4m$) are ½ p and ½ q depending on the direction of attack and whether the donor or recipient strand is excised. Correction failures result in post meiotic segregations indicated by $5^+:3m$ or $3^+:5m$ ascal segregation ratios. These are signalled by sectored ascosporal colonies in the plate dissection method.

behavior reported for type A and type B mutants in *Ascobolus*. We do not wish to convey the notion that disparity may not occur in yeast. Simple, convincing evidence, in the absence of heterogeneity from hybrid to hybrid, is extraordinarily difficult to amass. Rossignol *et al.* (1979) correctly emphasize that small differences in the direction of correction and the variable occurrence of symmetrical hybrid DNA can readily account for parity departures.

## Fidelity of Conversion

Considerable significance attaches to the role of mismatched bases in hDNA. Holliday's original hybrid DNA model of gene conversion and recombination (1964) proposed that excision of the mismatched base pair itself, followed by random replacement, might lead to the production of new alleles. Admittedly, as it turned out at a later date, this feature was not crucial to the model. Three separate studies provide convincing evidence that gene conversion does not generate allelic diversity. Rather, it functions as a strictly conservative mechanism. Studies by Roman and Jacob (1958) and by Zimmerman (1968), as well as numerous early studies in *Neurospora,* established the approximate equivalence between a parental and converted allele. Subsequently, Fogel and Mortimer's (1968, 1970) studies used molecularly defined ochre and amber mutants to prove that the conversion-derived mutant alleles carried precisely the same alteration as the parental allele from which it was derived. Conversion neither creates nor destroys genetic information. Rather, with a fidelity operative down to the single nucleotide level, it leads to the nonreciprocal transfer of information from one DNA sequence to a corresponding homologous segment. A prudent revision of Holliday's hybrid DNA scheme to allow for single-strand excision and replacement neatly accommodated such seemingly contradictory findings.

## Conversion of Deletions and Insertions

Because conversion involving large deletions or insertions by the same mechanism proposed for single base substitutions would generate a hybrid DNA region with an extensive, unpaired single-strand loop, we might expect unusual conversion profiles marked by distinctive disparity, high-frequency *pms,* and ab $4^{+}:4^{-}$ or other features. Such deletion heterozygotes were studied first by Fink (1974), Fink and Styles (1974), and Lawrence *et al.* (1975). These studies and ours establish that deletions (1) convert in either direction with approximate parity (also Sherman, personal communication); (2) co-convert with other sites in the same or adjacent genes; (3) exhibit normal outside marker recombination. However, these studies were completed before the plate dissection method was developed.

Accordingly, we conducted a detailed analysis of two large heterozygous deletions studied earlier by Fink. From measurements on the gene product he estimated that *his4-Δ15* and *his4-Δ26* correspond to deletions of 400 and 300 bp, respectively. In a total of 364 conversions among 8239 tetrads analyzed, conversions in both directions were found. Also, modest twofold disparities were observed. Deletion Δ15 yielded an excess of $6^+:2^-$ conversions, whereas Δ26 generated more $2^+:6^-$ conversions. Thus the deletions convert with approximate parity at frequencies comparable to an included nonsense mutant—*his4-4*. But no *pms* asci were found among 364 conversions. This stands in marked contrast to the finding that 10% of the aberrant segregations at *his4-4* are *pms*. If the ratio of *pms* to total events is the same for the base substitutions and deletions, we would have expected to observe about 37 *pms* events. Thus, we may conclude, deletions do not exhibit *pms*.

Limited data with sizable insertions yield similar insights. Hinnen (unpublished) noted $6^+:2^-$ and $2^+:6^-$ conversions of a 10-kb insertion at LEU2. Similarly, Fogel (unpublished) observed two conversions among 360 tetrads of an 11-kb plasmid integrated at *ADE8*. Moreover, *pms* is conspicuously absent among the 200 aberrant segregations at *Mat1*. Here the two alleles differ by large nonhomologies of 600–750 bp, as shown by Hicks *et al.* (1977, 1979) and Nasmyth and Tatchell (1980). The extreme paucity of *pms* may be taken to imply that the "heteroduplex" spanning the deletion, or nonhomology, produced a distorted DNA configuration that is unfailingly detected by the cell's excision repair system and removed with high efficiency. Radding (1978) proposed a novel scheme for handling such results. Conversion of insertions and deletions is thought to be mediated by nonpairings, as against mispairings in the case of base substitution mutants.

An important generalized test of the notion that gene conversions (or restorations) and *pms*—respectively, corrections and noncorrections of underlying hDNA molecules—emerges from recent studies by Williamson and Fogel (1980). Recessive, correction-deficient *(cor)* mutants at four different loci were isolated. Pleiotropic in character, they display "hyper rec" phenotypes with respect to meiotic intragenic recombination and two orders of enhanced forward mitotic mutation rates to canavanine resistance.

In genetic tests with *cor* strains, we find map distances for several gene–gene and gene–centromere intervals unchanged compared to wild-type controls. But the ratios of gene conversions to *pms* $(6^+:2^- + 2^+:6^-)/(5^+:3^- + 3^+:5^-)$ are dramatically altered. If hDNA corrections yield conversion and *cor* mutants are impaired in correcting such molecular recombination intermediates, we should expect homozygous *cor* strains to display increased *pms* frequencies at the expense of a corresponding decline in gene conversions of the $6^+:2^-$ and $2/^+:6^-$ types. Data obtained for markers *arg4-16, thr1, lys1, met13,* and *his2* accord with the model's predictions and they are given as Table 3.2. Observations on *his4-4* and *ade8-18* will be considered elsewhere.

Strains doubly homozygous for two nonallelic mutations—*cor2 cor3* and *cor1 cor3*—may have additive effects. For most markers, the *pms* frequency in double mutant combinations is higher than that in either single mutant. Such

**Table 3.2.** Percentage Gene Conversion[a] and Postmeiotic Segregation[b] for Various Heterozygous Markers in Mutant *cor* Strains

| | Strain genotype | | | | | | | | | | | | | | |
|---|---|---|---|---|---|---|---|---|---|---|---|---|---|---|---|
| | +/+ | | | *cor1/cor1* | | | *cor2/cor2* | | | *cor3/cor3* | | | *cor2/cor2 cor3/cor3* | | |
| Marker | No. Tetrads | *gc* | *pms* | No. Tetrads | *gc* | *pms* | No. Tetrads | *gc* | *pms* | No. Tetrads | *gc* | *pms* | No. Tetrads | *gc* | *pms* |
| *his4-4* | 814 | 7.3 | 0 | 466 | 7.1 | 12.7 | 165 | 3.0 | 2.4 | 250 | 13.6 | 1.2 | 127 | 6.3 | 14.2 |
| *arg4-16* | 634 | 5.7 | 3.5 | 140 | 5.0 | 6.4 | — | | | — | | | — | | |
| *thr1* | 613 | 6.1 | 0.6 | 464 | 5.2 | 4.5 | 165 | 4.2 | 4.2 | 250 | 2.4 | 8.4 | 96 | 2.1 | 12.5 |
| *lys1-1* | 629 | 7.7 | 0.2 | 216 | 2.3 | 9.3 | 171 | 2.9 | 0.6 | 73 | 1.4 | 5.5 | 97 | 2.1 | 6.2 |
| *met13* | 894 | 9.9 | 0.1 | 277 | 7.6 | 1.8 | 171 | 6.4 | 4.1 | 257 | 7.4 | 12.1 | 103 | 3.8 | 16.5 |
| *Ade8-18* | 736 | 3.7 | 5.7 | 411 | 5.1 | 6.1 | 158 | 3.8 | 3.2 | 251 | 4.0 | 4.8 | 82 | 3.7 | 7.3 |
| *his2* | 892 | 12.9 | 0 | 474 | 9.3 | 11.4 | 165 | 8.5 | 1.2 | 175 | 12.6 | 5.1 | 173 | 7.5 | 11.0 |

[a]Gene conversion frequency ×100 denoted as *gc;* Gene conversion frequency = (6:2 + 2:6) segregations/Total segregations.
[b]*pms* frequency ×100 denoted as *pms; pms* frequency = Total *pms* spore clones/Total segregations.

results support a multiple reaction system for hDNA repair in meiosis—a system analogous to the multiple pathways involved in repairing radiation-damaged DNA.

We analyzed unselected tetrads from a *cor1* homozygote triply heteroallelic at *arg4*. Similarly, we also studied all four *cor* homozygotes in strains heteroallelic at *his4*. Though repair, and therefore conversion frequencies, are significantly reduced, some repair does take place. However, compared to wild-type strains, the repair tracts are notably shorter in the *cor* strains. Unique tetrad classes not found in *COR* wild-type strains are encountered—for example, tetrads with a spore clone segregating postmeiotically for the proximal alleles *arg4-3* and *arg4-16,* though the distal allele *arg4-36* segregates $2^+:6^-$. Such asci and other exceptionals are most simply explained by assuming heteroduplex formation on a single chromatid followed by limited correction that extends through the *arg4-36* mutant site but not into the second adjacent site *arg4-16*. Ordinarily *arg4-36* and *arg4-16* undergo co-conversion with a probability of 0.65 in wild-type strains. Additionally, single- and double-site *pms* events increase at the expense of potential co-conversions that would have resulted in restorations to the parental marker configurations at one of two sites included within the hDNA region (see Hastings *et al.* 1980). Shorter correction tracts alone are sufficient to account for the action of the various *cor* mutants. Clearly, modulating the correction tract length is under the control of at least four *cor* loci, in addition to genes such as *rec4,* which increase the length of the co-converted segment (Rodarte and Mortimer 1972).

Reduced heteroduplex correction over the entire genome in *cor* strains provides us with a novel opportunity to assess the occurrence of symmetrical hybrid DNA as proposed by Holliday's model or the later phase of the Meselson-Radding scheme, which allows for varying proportions of symmetrical and asymmetrical hDNA. If symmetrical hDNA occurred frequently in yeast, aberrant $4^+:4^-$ segregations should comprise 25% of all aberrant segregations when correction is reduced to 50%. This clear prediction is contradicted by our data. In *cor1/cor1* hybrids marked at *his2,* only 2 among 98 irregular segregations were ab $4^+:4^-$. Moreover, when data for all sites in all four *cor* mutants are considered, we find ab $4^+:4^-$ segregations are no more frequent than the ab $6^+:2^-$ or ab $2^+:6^-$ events. Presumably, these classes arise as a consequence of two simultaneous but independent events involving asymmetric heteroduplex DNA segments. Clearly, the genetic studies in yeast and at the W17 locus in Ascobolus (Stadler and Towe 1971) are readily interpreted as indicating that asymmetrical hDNA occurs much more frequently than symmetrical hDNA. However, Holliday (1974) and Nagylaki and Petes (1982) emphasize correctly that "the evidence is not completely unambiguous" and their models, along with others, allow for either or both interactions. We concur. In fact, symmetrical hDNA occurs regularly during mitotic gene conversion in yeast, as revealed by Esposito's (1978) critical studies involving long, marked segments of chromosome VII. However, we are not aware of any evidence supportive of the notion that symmetrical hDNA occurs with appreciable frequency during meiosis in yeast.

Undoubtedly, other mutants will play a major role in defining and distinguishing the functions required for initiating the formation and correction of hDNA distortions, mispairings, and nonpairings. Included here are mutants such as *pac1* (Davidow *et al.* 1979), *spo11* and *spo13* (Klapholz and Esposito 1980a,b), MIC1-4 (Maloney and Fogel 1980), *rad* mutants (Game *et al.* 1980), *mar1* (Klar *et al.* 1979), and *rem1* (Golin and Esposito 1977).

## Specific Molecular Mismatches in Conversion and *pms*

In Table 3.1 the 30 heterozygous sites display uncorrelated *pms* and gene conversion frequencies. The various heterozygosities probably represent a miscellaneous array of potential mismatches, and conceivably the potential specific mismatch constitutes a signal for hDNA formation and/or heteroduplex correction. To assess the role of specific mismatched base pairs, we constructed a diploid hybrid strain that carried an ochre allele from one parent opposite an amber allele contributed by the other parent—but in the same codon. Nonsense mutants of the ochre and amber types at *trp1, his4,* and *arg4* were derived by EMS mutagenesis from each progenitor nonsense allele following Hawthorne's (1969) isolation strategies. Our experimental diploid was heterozygous for ochre/amber at three different centromere-linked though independently assorting sites. During meiosis one of two possible base-pair mismatches could occur at each locus. Over all, we scored over 1100 unselected tetrads. These data, along with comparable controls for +/amber or +/ochre, are included for comparison in Table 3.3. Several conclusions may be drawn. First,

**Table 3.3.** Effect of Mismatch on Conversion

| Locus | Genotype[a] | Asci | Aberrant segregation 6:2 | 2:6 | 5:3 | 3:5 | other | Total events (%) | *pms*/ Events (%) |
|---|---|---|---|---|---|---|---|---|---|
| *arg4-17* | $+/o$ | 12,358 | 431 | 514 | 35 | 19 | 0 | 8.10 | 5.4 |
| | $a/o$ | 1,118 | 54 | 37 | 3 | 1 | 0 | 8.50 | 4.2 |
| *his4-4* | $+/o$ | 12,510 | 411 | 303 | 39 | 41 | 2 | 6.40 | 10.1 |
| | $a/o$ | 1,118 | 18 | 20 | 2 | 3 | 0 | 3.80 | 11.6 |
| *trp1-1* | $+/o$ | 17,061 | 48 | 37 | 10 | 7 | 0 | 0.59 | 16.7 |
| | $a/o$ | 1,118 | 1 | 3 | 0 | 0 | 0 | 0.36 | 0 |

[a] *a* is the amber allele; *o* is the ocher allele.

substituting a nonsense allele for a wild-type allele does not alter the basic conversion frequency. Essentially, this remains constant and characteristic for each locus. Second, the same potential mismatch is corrected with differing efficiency, depending on where it is located. Thus correction at *arg4* is more likely than at *his4* or *trp1*. Similar conclusions emerge from Nicolas (1979) study on various *Ascobolus* mutants. Clearly, our findings do not establish that base-pair mismatches play no role whatsoever in hDNA formation–correction reactions; rather, they strongly suggest that the mismatch influence is subordinate to the chromosomal location of the heterozygosity itself.

## The Effect of Adjacent Heterozygosity on *pms* Profiles

The genetic analysis of recombination is predicated on a generally accepted though unstated assumption. Generally, it is assumed that the mutant marker alleles used to study the recombination process do not themselves alter the process. For the most part, this appears to be warranted, though instances to the contrary have been noted and reported. Exceptions are subsumed under the term *marker effect*. For example, Gutz and Goldman (1971) and later Goldman (1974) reported unusual behavior for the specific allele *M26* in *Schizosaccharomyces pombe*. This allele converts more frequently than other mutant alleles located to either side on the genetic fine structure map. Furthermore, unlike other alleles that display approximate parity in their conversion spectra, *M26* converts almost exclusively in a single direction—from mutant to wild, and nearby markers are "pulled" into co-conversion. *M26* is a base substitution nonsense mutant of the amber variety. Clearly, the DNA alteration that led to its origin also produced a highly recombinogenic site.

The *chi* sites of phage lambda and the *cog* (Catcheside 1979) sites found in *Neurospora* share similar properties. They stimulate recombination markedly in their immediate vicinity, but the effect falls off with increasing distance and reaches a zero value at 20 kb. Sites such as *M26, cog,* and *chi* may produce their typical marker effects by serving as DNA binding zones for enzymes that mediate heteroduplex initiation and perhaps propagation. Given organized DNA sequences such as *M26*, *cog*, or *chi*, it follows, as emphasized by Stahl (1979b), that map distances and physical distance may be related by widely different factors in different regions of the genome (see Moore and Sherman 1974, 1975). The molecular cloning and detailed characterization of such sequences will doubtless be achieved within the next years and our knowledge concerning coarse and fine controls on recombination will expand explosively. Thus, as regards marker effects, our cherished assumptions have failed. They must be relaxed. Prudent scientists will necessarily take a strategic step backward, or forward, as the case might be, in the face of compelling evidence contrary to their accepted visions of reality. Marker effects call to mind the indeterminacy principle encountered in physical science.

Another kind of marker effect was uncovered and carefully documented by Leblon and Rossignol (1972, 1973) for the *b2* locus in *Ascobolus immersus*. At a given heterozygous site, they found a marked decrease in the *pms* : conversion ratio when an adjacent heterozygosity was present in the same gene. We studied the *arg4-16* allele of yeast (Fogel *et al.* 1979, 1981) from a similar viewpoint. When *arg4-16* is present as a single heterozygous site, it yields an average basic abberant segregation frequency of about 8.0%, and about 33% of the events are *pms*. When single heterozygosity is introduced close on either side, the basic conversion rate remains constant, but the proportion of conversions increases at the expense of the *pms* class, which in turn falls to about 10%, or a third of the control value. The effect of two adjacent heterozygosities is even more pronounced and fivefold shifts in *pms* are found. Observations of this sort, along with those stemming from analysis with *cor* mutants (Williamson and Fogel 1980), lend support to the general notion that gene conversions ($6^+:2^-$ and $2^+:6^-$) result from mismatch base-pair correction.

The high *pms* frequency observed when *arg4-16* or *ade8-18* are present as single heterozygous sites suggests that the corresponding potential base-pair mismatches and the adjacent hDNA segments are not efficient enzymatic binding sites for initiating the excision-resynthesis repair reaction sequence. Clearly, additional mismatches within the same heteroduplex tract constitute stronger signals and provoke more efficient repair.

## Conversion-Associated Outside Marker Exchange

Early experiments with *Neurospora*, reported by Case and Giles (1958, 1959, 1964) and by Stadler and Towe (1963), indicated that about 50% of gene conversions were associated with crossing-over between closely linked markers astride the converted region. Fogel and Hurst (1967) confirmed and extended this finding in their analysis of 1100 tetrads selected to contain a prototrophic recombinant at the *his1* locus. Later studies by Hurst, Fogel, and Mortimer (1972), based on unselected tetrads, yielded a similar conclusion. However,

Stadler (1973) and D. D. Perkins (personal communication) pointed out that recombination frequencies associated with gene conversion must be corrected for incidental exchanges within the flanking interval. The correction assumes that conversion-associated exchanges generate high or complete chiasma interference, as shown by Mortimer and Fogel (1974) and by Stadler (1973), but the remaining conversion events accumulate crossovers at the normal meiotic rate. We have reexamined our data (see Fogel *et al.* 1981, p. 318) for 12 hybrids segregating for chromosome VIII marker genes. Overall, about two-thirds of the conversions are associated with exchange and the values observed range from about 30% to 75%. When the appropriate correction removing incidental exchanges is applied, the average fraction of conversions associated with crossing-over is 35% and ranges between 18% and 66%. In terms of molecular models, this would imply that interacting molecules carrying a Holliday structure between them are more prone to resolution without associated crossing-over. Accordingly, resolution itself is probably not stochastic but is rather driven enzymatically.

Given that the average likelihood of conversion is 1.5% or 0.015, and the finding that the length of the converted DNA segment is about 1000 bp or nucleotides, we can organize an order of magnitude calculation to the effect that conversion-associated crossing-over alone can account for the total length of the genetic map in *Saccharomyces,* as follows:

Given: a. The average meiotic conversion frequency is 1.5%.
b. The total DNA-per genome is 15,000 kb.

Then: Total DNA converted per meiosis $= 1.5 \times 10^{-2} \times 1.5 \times 10^{7}$ bp $= 2.25 \times 10^{5}$ bp per meiosis.

Assume: Average length of converted segment is 1000 bp.

Then: Average number of conversions per meiosis $= (2.25 \times 10^{5}\ \text{bp})/10^{3}$ bp, or 225 conversions per meiosis.

If: 2 conversions = 1 crossover = 50 map units, then each meiosis generates 112.5 crossovers $\times$ 50 = 5625 map units or centimorgans.

This value compares favorably with the minimal value of 4600 cM obtained by Mortimer and Schild (1980).

Some of the assumptions inherent in the foregoing analysis are subject to a critical evaluation. This entails the examination of outside marker exchanges in a special aberrant segregation class—i.e., tetrads that exhibit *pms.* In such tetrads, exchanges are readily classifiable as associated or incidental. Associated exchanges are defined as those that necessarily involve the *pms* strand; incidental exchanges may or may not involve the hDNA bearing strand.

Incidental exchanges, as defined here, especially those that are clearly not associated with the *pms* strand, are uniquely identifiable (Fogel *et al.* 1979, 1981). A corrected value for the fraction of events associated with flanking marker exchange may be calculated by subtracting *twice the observed incidental-exchange frequency* from the frequency of associated exchanges found by direct examination. Thus for the chromosome VIII marker *thr1* (flanked by *arg4* and *cup1*), a total of 334 events were identified. Of these, 253, or 75.8%, displayed outside marker exchange. However, 65 exchanges were unequivocally incidental. When these observations are corrected for incidental exchanges, the corrected fraction of associated exchanges is 37%, a value consonant with those reported by DiCaprio and Hastings (1976) for conversion-associated exchange in the *SUP6* region of chromosome VI.

At this juncture, we may pose a related question: Does heteroduplex correction, via the presumed excision-resynthesis repair mechanism, significantly alter the probability that the underlying Holliday structure will experience resolution with or without outside marker exchange? We considered all conversion and *pms* segregations of the specific allele *arg4-16*, a complementing missense marker distinguished for a high basic conversion frequency and a high proportion of *pms*. In these hybrids 222 *pms* tetrads were identified and 115 (52%) displayed outside marker exchange compared to 480 gene conversions of which 57% were exchanged. Thus we may conclude that resolution of the recombinational intermediate Holliday structure occurs prior to correction-repair. Alternatively, repair may be viewed as a process independent of and without effect on the reactions leading to resolution.

## Position of the Associated Exchange

Contained in both the Holliday and Meselson-Radding molecular recombination models lies an implicit prediction concerning the precise location of the conversion associated crossover. As displayed in Figures 3.2 and 3.3, conversion-associated exchange mirrors resolution of the half-chiasma or single-strand exchange. The process requires nicks and rejoinings between the uncrossed strands at the terminus of the Holliday structure. In both schemes conversional polarity is easily accounted for by supposing that strand transfer initiates at a site adjacent to the high conversion region. Combined, these assumptions lead to the expectation that conversion (or *pms*)-associated crossovers should occur only on the lower conversion side of the polarity gradient. A critical analysis regarding this specific prediction can be achieved in tetrads where the central gene displays *pms* and the flanking markers are recombined. First, we can determine whether the exchange is incidental or associated. Second, and more significantly, we can determine the relative position of the exchange as proximal or distal to the *pms* site as displayed in Table 3.4. The relationships are illustrated for $3^+:5^-$ asci, but they apply with equivalent force to the $5^+:3^-$ category.

Among 5555 unselected, fully ascertained asci from three related hybrids we identified 173 *pms* at *arg4-16*. Of these 85 were recombined for their flanking markers. However, only in 73 was the exchange associated with the *pms* strand. Twenty exchanges could be assigned to the centromere proximal region (low end) while 44 fell into the distal region (high end). Nine asci had associated exchanges in both regions and five carried an incidental exchange along

**Table 3.4.** Position of Conversion—Associated Exchange[a]

| Not Exchanged | | | Exchanged | | | | | | | | |
|---|---|---|---|---|---|---|---|---|---|---|---|
| | | | Proximal | | | Distal | | | Not Associated | | |
| + | + | + | + | + | + | + | + | + | + | +/m | + |
| + | +/m | + | + | m | b | + | +/m | b | + | + | b |
| a | m | b | a | +/m | + | a | m | + | a | m | + |
| a | m | b | a | m | b | a | m | b | a | m | b |

[a]Method used to deduce the position of associated events in three-point crosses. The assumed parental genotypes are + + + and *amb*. The analysis applies to asci in which only the central marker exhibits pms. The presumption is that the sectored spore was originally ± and represents the exchange necessary to account for the observed segregation.

with an associated proximal or distal exchange. The observation that associated exchanges may occur both proximal and distal to the *pms arg4-16* site is contradictory to the idea that polarity indicates a relatively fixed initiation site near the high conversion end of the *arg4* locus with hDNA tracts extending into the gene for various lengths. Alternatively, we can imagine that hDNA is initiated and propagated in two different directions. P. Hastings suggested that the half-chiasma might be free to diffuse in either direction. However, reversed diffusion beyond the initiation site would generate a symmetrical hDNA tract. Accordingly, another restraint is required and the overall issue awaits clarification.

## Conversion of Adjacent Genes

In previous sections, the modal length of a hDNA gene conversion tract was estimated as about 1000 bp. There is no reason to suppose that this value is constant across the genome. Actually, data from at least three different experiments indicate that heteroduplex DNA may extend across the boundaries of adjacent genes. For example, in an experiment involving several comparable chromosome VI hybrids with the markers *cen6–his2–SUP6–met10,* and respective map intervals of 21, 6.5, 2.1 cM, respectively, some 2678 tetrads were analyzed. Total single-site conversions at *his2, SUP6,* and *met10* were 342, 290, and 12. Additionally, 40 coincident conversions were observed for the *his2–SUP6* pair and 17 for the *SUP6–met10* pair. If the sites in each pair experienced conversion independently, we would predict, on the basis of their respective single site frequencies, that 49 coincident events would be found for *his2–SUP6* and of these one fourth, or about 12, would exhibit a co-conversional pattern. The predictions accord with the observations, since only 8 "co-conversions" were found among the 40 coincident events observed. Clearly, *his2* and *SUP6*, are rarely if ever within the same meiotic hDNA tract. Instead, they convert as independent entities where hDNA spanning one locus does not extend into the other adjacent locus. A very different situation obtains for *SUP6–met10*. Here we would predict 3.8 coincident events, calculated on the basis of the respective single-site frequencies. However, 17 such events were encountered, and all were co-conversional. Thus, in this particular chromosomal region, we observe a marked discontinuity between *SUP6* and *met10* in conversion frequency, that is, 13% versus 1.1%. Yet more than half the conversions at *met10* extend proximally on the same strand to include *SUP6*. Hence closely adjacent loci may be included in a single heteroduplex. Consequently, the termini of genes do not stand as barriers to the propagation of hybrid *DNA*. Similar conclusions may be derived from our unpublished analyses of *thr3–his1*, *SUC1–MAL1*, and *trp1–gal3* and from Bassel and Mortimer's (1971) study of the cluster *gal7,10,1*. Insertions and deletions may, however, represent special cases, since large unpaired single-strand loops are expected rather than base-paired heteroduplexes. Conversions, in both directions (but no *pms*), have been observed, in our studies, with insertions amounting to 11 kb in length. Finally, it should be noted that hDNA extending over half a chromosome arm or more is commonplace in mitotic cells (Esposito 1978, 1981).

## Two-, Three-, and Four-Point Intragenic Crosses

Data for two-, three-, and four-point intragenic crosses are presented fully in our recent reviews (Fogel *et al.* 1979, 1981). Readers interested in the details of polarity, patterns of hDNA formation, and repair should consult these works as well as those of Rossignol *et al.* (1979, 1980), Lamb (1972, 1975), Sang and Whitehouse (1979), and Wickramaratne and Lamb (1978).

## Intrachromosomal Gene Conversion and Recombinant DNA Strategies

In recent years the definition of gene conversion as a nonreciprocal informational transfer (with or without associated outside marker exchange) was enlarged and extended because of the experimental potentialities inherent in recombinant DNA manipulations (e.g., see Klein and Petes 1981; Scherer and Davis 1980). Through the analysis of novel situations that arise as a consequence of plasmids with cloned genes integrating into homologous sequences at the corresponding locus or at other positions on nonhomologous chromosomes, insight into the interactions between comparable DNA segments has emerged rapidly. Petes (1980a,b) and Botstein and Davis (1982) reviewed the methodologies to identify and isolate specific genes, clone and characterize specific alleles, and reintroduce into yeast new alleles generated by *in vitro* manipulations of isolated DNA segments. Powerful and sophisticated, the collective strategies, though in a rapid state of flux, have already led to the emergence of an entirely new commercial enterprise—genetic engineering—a biotechnology concerned with the controlled production of critical substances such as insulin, interferon, growth hormones, vaccines, and enzymes. Briefly, we shall review here the impact of these developments on our knowledge of recombination in general.

Scherer and Davis (1980) introduced various *his3* alleles into several different positions on the genome and demonstrated that recombination or gene conversion between comparable DNA segments located on nonhomologous chromosomes could occur at predictable low frequencies. Similar results were achieved with *his4* (Fink, unpublished). In addition, Petes (1980a,b), Klein and Petes (1980), Szostak and Wu (1980), and Jackson and Fink (1981) reported on the occurrence of gene conversions between homologous nontandem duplication sequences that arise as a result of plasmids integrating into chromosomal sites at or near a corresponding genetic locus. This nonreciprocal transfer of information between the nontandem duplicated sequences has been termed *intrachromosomal gene conversion.* We may revise our view of recombination to include the process of hDNA production and repair so that it embraces interactions between homologous segments of DNA without regard to their genomic locations—the conversion frequency will, of course, depend on whether homologous chromosomes are involved, whether the interactive sequences comprise nontandem duplications or highly iterated genes. Related to the latter is sister chromatid exchange (SCE), a variety of recombination whose study was largely facilitated by recombinant DNA techniques (Petes 1980a,b; Szostack and Wu 1980). Some illustrative experimental work on intrachromosomal conversion is presented next.

## Conversion of Insertions[4] and Intrachromosomal Gene Conversion[5]

The observed conversional behavior of deletions poses special problems for any molecular model that purports to account for gene conversion and crossing-over. The very same considerations apply with equal force to the behavior of insertions. Accordingly, gene conversion analyses of insertions and deletions can provide valuable information concerning the topography of chromatid pairing at meiosis, structure of the synaptonemal complexes, and the frequency and lengths of pairings at any single site. Co-conversion estimates impinge on lengths of pairing segments, while single-site conversion data yield minimal estimates for frequencies of pairing in particular regions. These issues have

[4]Presented at the Herschel Roman Genetics Symposium, Seattle, 1980.

[5]These studies are a joint effort. They have been published elsewhere in greater detail (Fogel *et al.* 1982).

been addressed extensively by Wickramaratne and Lamb (1977). In ordinary laboratory yeast strains, deletions occur only rarely, but insertions and deletions can be generated readily by a coupling of *in vitro* recombinant DNA technologies with the yeast transformation system. The transformation procedure, the methods for identifying cloned genes, and the vector construction techniques were critically reviewed by Petes (1980b). This section presents some recent studies on the conversional behavior of relatively large insertion sequences 11 kb in length.

Primarily, the study aims to clone, analyze, and sequence the structural and regulatory sequences at *arg4-16, ade8-18,* and other selected high *pms* alleles for which a detailed background of recombinatorial and biochemical data is available. The study attempts to address at the DNA sequence level the long-term intriguing issues such as parity, fidelity, co-conversion, associated outside marker, postmeiotic segregation, position of the associated exchange, and polarity. In brief, we intend to exploit the resolving power of the recombinant DNA-transformation system to advance our understanding of recombination from the fine structure genetic level to the molecular level. Now it is possible to identify recombinant molecules containing a specified gene or gene segment. Isolated genes may be altered by *in vitro* or *in vivo* procedures and subsequently the effects of such alterations on gene expression may be assessed by returning the altered DNA segment to yeast cells via the transformation protocol. Sophisticated genetic studies may be bolstered powerfully by corresponding molecular studies.

A recipient diploid strain of genotype

$$\frac{\mathbf{a}\ \mathit{ade2}}{\alpha\ \mathit{ade2}}\quad \frac{\mathit{ade8\text{-}18}}{\mathit{ade8\text{-}18}}\quad \frac{\mathit{arg4\text{-}16}}{\mathit{arg4\text{-}16}}\quad \frac{\mathit{cup1}}{\mathit{cup1}}\quad \frac{\mathit{trp1\text{-}1}}{\mathit{trp1\text{-}1}}\quad \frac{\mathit{ura3\text{-}1}}{\mathit{ura3\text{-}1}}$$

was prepared by repeated crosses and isolations. Spheroplast transformability was assayed with a recombinant cloning vector, YRp17, known to carry the wild-type loci of *TRP1, URA3,* and *ARS1,* a yeast chromosomal origin of replication (Struhl *et al.* 1979; Stinchcomb *et al.* 1979). From a hybrid DNA pool prepared by ligating a BamHI restriction digest of DNA prepared from S288C into the YRp17 vector, also treated with the BamHI restriction enzyme, we isolated yeast transformants wild-type for *URA3* and *TRP1* functions. Among these, individuals displaying wild-type function with respect to either *arg4* or *ade8* were detected. Only studies concerning the *ade8* gene will be considered now. Those relating to *arg4* will be reported elsewhere. The cloning of *arg4* was first established by Clarke and Carbon (1978).

The presence of autonomously replicating recombinant or chimeric plasmids in the regenerated yeast protoplast recipients generates novel problems. Since the plasmid is presumed to carry random BamHI fragments, those including $ADE8^{+}$ might be expected to produce red yeast colonies, particularly in view of the fact that the mutant *ade8* block precedes the *ade2* block in the *de novo* purine biosynthesis pathway. However, this does not occur. Three factors operate in a contrary direction: (1) autonomously replicating plasmids are frequently lost during vegetative growth, even under selective conditions; (2) *URA*$^+$ *TRP*$^+$ cells excrete moderate quantities of nutrilites that partially satisfy the auxotrophy imposed by the mutants *ura3* and *trp1;* therefore, parental cells are carried forward; and (3), as demonstrated by Roman (1957, 1980), *ade2, ade8* white cells readily outcompete *ade2* red cells. Thus the presence of an *ADE8*$^+$ bearing segment within the YRp17 vector is not immediately apparent at the phenotypic level.

However, red sectors are frequently found among the *URA3*$^+$, *TRP1*$^+$ diploid transformants described earlier. When these are purified, sporulated, and subjected to tetrad analysis, we routinely find a segregation pattern of two red ascosporal clones that are simultaneously *TRP1*$^+$ and *URA3*$^+$ along with two white ascosporal clones that are auxotrophic for tryptophan and uracil. Complete linkage of *URA3*$^+$ *TRP1*$^+$ and *ADE8*$^+$ is expected if we assume that

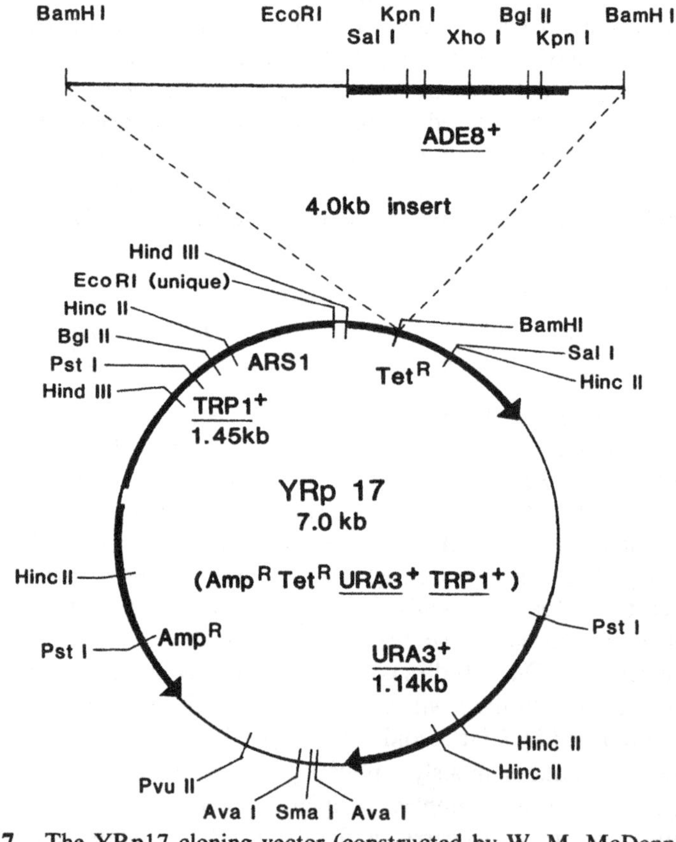


**Figure 3.7.** The YRp17 cloning vector (constructed by W. M. McDonnell and R. W. Davis, personal communication) is a 7.0-kb plasmid carrying pBR322 sequences and the wild-type sequences for the yeast genes *TRPI*$^+$ and *URA3*$^+$ in addition to an automomous replication sequence, *ars.* A 3.8-kb BamHI insert carrying *ADE8*$^+$ is shown.

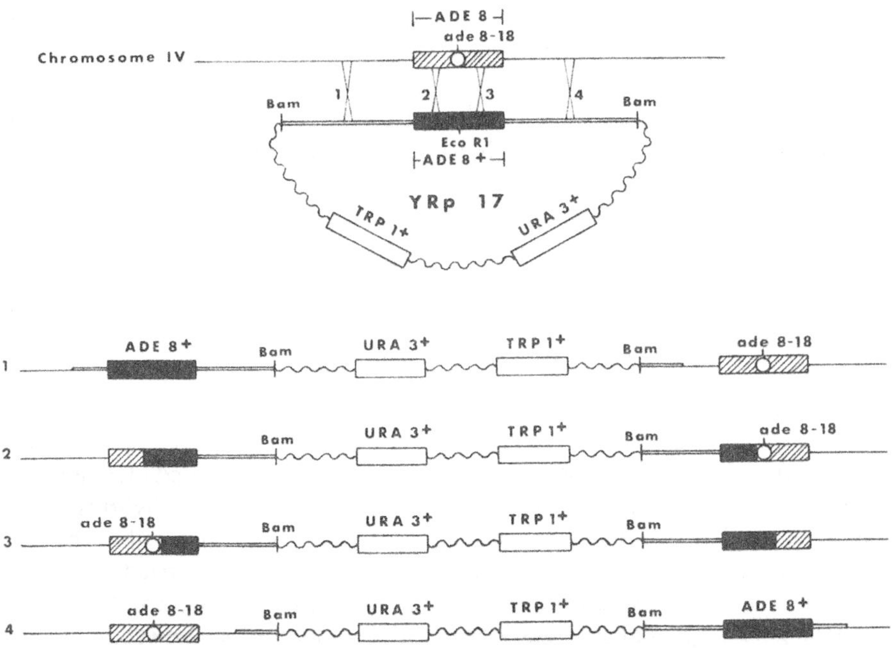


**Figure 3.8.** Integrations of the YRp insert-bearing cloning vehicle into the homologous chromosome IV region in the vicinity of *ade8-18*. (1–4) The consequence of heteroduplex formation with associated crossing-over.

integration of the plasmid into a chromosome site occurred by homologous recombination. Of 32 independent integration events, each signaled by the phenotypic white → red shift, a phenomenon designated here as the *Roman effect*, 28 correspond to the description given earlier. They are accounted for economically by assuming a single integration event involving only one homolog. The probable sequence of events leading to integration is depicted in Figure 3.8. The 7.0-kb cloning vector (McDonell and Stinchcomb, personal communication) carrying a 4.0-kb BamHI fragment is shown in Figure 3.7. Also, three additional red transformants were identified. Apparently homozygous for the inserted sequence, these might arise by two primary simultaneous but independent integration events. Alternatively, they might represent secondary mitotic gene conversions covering the entire integrated plasmid sequence or reciprocal mitotic recombinants (see Esposito 1981). Thus the question arises, "Are integrational events recombinogenic?"

Experimental designs to discriminate among these alternatives are readily devised. If the apparent homozygosity reflects primary rather than secondary events, a firm prediction emerges—that is, about half the events will carry insertions in opposite orientations. Thus, as in Figure 3.8, a diploid might carry strands 1 and 4 and these should display mostly meiotic segregations of 4 red:0 white. However, reciprocal recombination between the nontandem $ADE8^+$ and *ade8-18* sites would generate 3 red; 1 white ascosporal colonies. Among the former, two should carry duplications with both the wild and mutant *ade8-18* sequences and one spore should carry the wild-type sequence both proximally and distally. The white spore clone should carry two copies of the distinctive *ade8-18* allele in a nontandem duplication array. On the other hand, if the observed homozygosities arose secondarily, both homologs would display an identical orientation. In both cases, single site intrachromosomal conversions at either end are also detectable and equally predictable. Finally, a single red transformant displayed requirements for uracil and tryptophane. It probably arose secondarily as a Campbell-like excision event. Overall, these results are comparable to those obtained by Hinnen *et al.* (1978), Struhl *et al.* (1979), and Hicks *et al.* (1979).

Of the 32 red transformants, 19 were analyzed with respect to the genomic site at which integration occurred. Integration might have taken place at any one of three possible locations marked by the sequences immediately embracing *trp1*, *ura3*, or *ade8-18*. Tetrad analysis of the 19 resultant hybrids clearly established that in each of 16 instances integration occurred within the 4-kb segment carrying *ADE8*$^+$. Integrations at the *trp1* site occurred in 3 of 19 events. The observed frequencies are roughly proportional to the relative homologous lengths at *trp1* and *ade8* that are contained in the vector.

In the preceding experiment, 597 unselected tetrads were scored and classified via the plate dissection procedure. Thus in the nontandem duplication heterozygote, generated by a single integrational event, 36 aberrant segregation events were observed with respect to *ade8*. Of these, 21 were of the $6^+{:}2^-$ and $2^+{:}6^-$ variety (respectively, 18 and 3 in number), and 15 more exhibited *pms* in the $5^+{:}3^-$ and $3^+{:}5^-$ patterns (respectively, 8 and 7 in number). Overall, we observed a basic conversion frequency of 6% and a *pms*/total event ratio of about 42%. Comparable but conventional heterozygotes exhibit a basic conversion frequency of 10.1% and 63.5% *pms* at *ade8*, or almost twice the basic conversion value observed for the integration heterozygote with a lower *pms*/events ratio. A $\chi^2$ analysis shows that the basic conversion frequency and the *pms*/events are significantly different from control values at the 5% level. This result is expected if the mutant site in one homolog can pair equally well with the wild or mutant *ade8-18* sequence carried in the nontandem duplication. Detailed studies of this sort can shed considerable light on the molecular details in homologous chromatid pairing as well as heteroduplex formation and correction during meiosis. Clearly, other potentially valuable experimental designs are apparent for investigating homologous sequence interactions. And the interacting homologous sequences might be present in homologous or nonhomologous chromosomes or as repeated sequences in a single chromosome.

In this connection attention may be drawn to Klein and Petes' (1981) and Jackson and Fink's (1981) discovery of intrachromosomal gene conversion or to the Szostak and Wu (1980) report concerning sister chromatid exchange. The term "intrachromosomal recombination" or "conversion" implies either recombination between chromatids attached to the same centromere or sister-strand recombination, as well as recombination between repeated sequences within a single chromatid. In both instances it is highly probable that the underlying basic recombinational event entails the production of a heteroduplex DNA segment at the molecular level and a corresponding eventual gene conversion or *pms* at the cellular level.

As Hinnen *et al.* (1978) demonstrated, integration of the recombinant plasmid into the host genome generates a chromosome carrying a nontandem duplication of a yeast genomic segment separated by bacterial plasmid sequences of a known physical length. As shown in Figure 3.8, line 1, we have

$$_{\circ}\underline{trp1\text{-}1\text{–}ADE8^{+}\text{–}URA3^{+}\text{–}TRP1^{+}\text{–}ade8\text{-}18}$$

that is, duplications of both the *trp1* and *ade8* sequences (11 kb apart) where each duplication is represented by a specific mutant allele in addition to a wild-type sequence. From such haploid strains it is easy to derive isolates in which both copies are identical. Both may be mutant or both may be wild-type. Here, as in the case of mating-type switching or interconversions, there are good reasons to suppose that nonreciprocal informational transfers of a co-conversional character are involved. The observations reported here concur with the reports of Klein and Petes (1981) and Jackson and Fink (1981). Haber *et al.* (1980) have reviewed gene conversion at the mating-type loci.

A final experiment illustrates the enhanced resolving power provided by our recombinant DNA-transformation approach. Heteroallelic recombination may be pursued at the intrachromosomal level. To a haploid *ade8-18–URA3–TRP1–ade8-18* nontandem duplication isolate obtained by mitotic intrachromosomal nonreciprocal informational transfer, we crossed a strain bearing the ochre allele *ade8-10* provided kindly by M. E. Esposito. In addition to normal tetrads and single-site conversions to prototrophy (with or without *pms*), we could identify tetrads containing a spore clone with two different alleles on the same single chromatid. Such clones exhibit a very high papilla frequency on diagnostic media. The papilla represent single-site conversions at the heteroallelic sites and it must be recalled that the homologous interacting DNA sequences contain between them approximately 11 kb of yeast and bacterial plasmid DNA—a quantity that may be varied at will by *in vivo* or *in vitro* techniques. The possibilities for further analysis at the molecular and genetic levels are both rich and challenging.

At least four kinds of questions are being addressed with these structures:

1. Is the conversion rate sensitive to the size of the insertion? A systematic study of the response of conversion rate and degree of parity to the size of the insertion is possible by varying the size of the vector. Results from this experiment should reflect both the sensitivity of prerecombinational synapsis to the insertional disruption and the size restrictions on the formation and resolution of an actual DNA strand exchange intermediate.

2. How is gene conversion in the region of the insertion affected? The conversion rate for genes near the insertion reflects the sensitivity of conversion to a local pairing disruption. A series of different-sized insertions at the same posi-

tion in the genome could be used to determine whether pairing can be made rate-limiting for conversion.

3. Does the order of the duplication (i.e., *ade8-18–vector–ADE8 versus ADE8–vector–ade8-18;* Fig. 3.8) affect the conversion rate? If polarity reflects a site at which pairing and/or recombination is initiated, an insertion between the gene and that site will reduce the conversion rate. If, for example, conversion tracts initiate to the right of *ADE8* and propagate to the left, *ade8-18–vector–ADE8/ade8-18* diploids would show a higher conversion rate than *ADE-8–vector–ade8-18/ade8-18* diploids. If the initiation site were to the left of *ADE8* and propagated to the right, the second diploid would show the higher conversion rate.

4. Is intrachromosomal gene conversion sensitive to the size of the insertion? Fogel *et al.* (1982) generated a nontandem duplication of the *ADE8* locus that carries two different mutant alleles: *ade8-18–vector–ade8-10.* Haploid strains of this type exhibit high rates of Ade$^+$ papilla formation, indicative of an intrachromosomal conversion event. We may now monitor the sensitivity of papilla formation to changes in the insert DNA (size of insertion or screen for recombinogenic sites) and changes in the genetic background of the cell (various recombination-defective strains).

Combined, these approaches signal the beginning of a new understanding of recombination and gene conversion at the molecular level.

## References


Baltimore, D. (1981). Gene conversion: some implications for immunoglobulin genes. Cell 24:592–594.

Bassel, J., Mortimer, R. K. (1971). Genetic order of the galactose structural genes in *Saccharomyces cerevisiae.* J. Bacteriol. 108:179.

Boon, T., Zinder, N. D. (1971). A mechanism for genetic recombination generating one parent and one recombinant. Proc. Natl. Acad. Sci. USA 64:573.

Botstein, D., Davis, R. W. (1982). Principles and practice of recombinant DNA research with yeast. In: The Molecular Biology of the Yeast *Saccharomyces.* II. Metabolism and Gene Expression, edited by J. Strathern *et al.* Cold Spring Harbor Laboratory, Cold Spring Harbor, N.Y.: in press.

Case, M. E., Giles, N. H. (1958). Evidence from tetrad analysis for both normal and abnormal recombination between allelic mutants in *Neurospora crassa.* Proc. Natl. Acad. Sci. USA 44:378.

Case, M. E., Giles, N. H. (1959). Recombination mechanisms at the pan-2 locus in *Neurospora crassa.* Cold Spring Harbor Symp. Quant. Biol. 23:119.

Case, M. E., Giles, N. H. (1964). Allelic recombination in *Neurospora:* tetrad analysis of a three-point cross within the *pan-2* locus. Genetics 49:529.

Catcheside, D. G. (1975). Occurrence in wild strains of *Neurospora crassa* of genes controlling genetic recombination. Aust. J. Biol. Sci. 28:213.

Clark, A. J. (1974). Progress toward a metabolic interpretation of genetic recombination of *E. coli* and bacteriophage lambda. Genetics 78:259.

Clarke, L., Carbon, J. (1978). Functional expression of cloned yeast DNA in *Esche-*

*richia coli:* Specific complementation of arginino succinase lyase (arg H) mutations. J. Mol. Biol. 120:517.

Davidow, L., Goetsch, L., Byers, B. (1979). Genetic basis and consequences of a reversible thermal arrest of yeast meiosis at pachytene. Molecular Biology of Yeast (abstracts), August 1979. Cold Spring Harbor Laboratory, N.Y.

Di Caprio, L., Hastings, P. J. (1976). Gene conversion and intragenic recombination at the *SUP6* locus and the surrounding region in *Saccharomyces cerevisiae.* Genetics 811:697.

Egel, R. (1981). Intergenic conversion and reiterated genes. Nature 290:191–192.

Emerson, S. (1969). Linkage and recombination at the chromosome level. In: Genetic Organization, edited by E. W. Caspari, A. W. Ravin. New York: Academic Press, p. 267.

Esposito, M. S. (1967). X-ray and meiotic fine structure mapping of the adenine-8 locus in *Saccharomyces cerevisiae.* Genetics 58:507.

Esposito, M. S. (1971). Post-meiotic segregation in *Saccharomyces.* Mol. Gen. Genet. 111:297.

Esposito, M. S. (1978). Evidence that spontaneous mitotic recombination occurs at the two-strand stage. Proc. Nat. Acad. Sci. USA 75:4436–4440.

Esposito, M. S., Esposito, R. E. (1977). Gene conversion, paramutations and controlling elements: A treasury of exceptions. In: Cell Biology, a Comparative Treatise, edited by L. Goldstein, D. M. Prescott, Vol 1. New York: Academic Press, p. 59.

Esposito, M. S., Wagstaff, J. E. (1981). Mechanisms of mitotic recombination. In: The Molecular Biology of the Yeast *Saccharomyces,* edited by J. Strathern, E. Jones, J. Broach. Cold Spring Harbor, N.Y.: Cold Spring Harbor Laboratory, in press.

Esser, K., Kunnen, R. (1967). Genetics of Fungi, translated by E. Steiner. New York: Springer-Verlag.

Fincham, J. R. S., Day, P. R., Radford, A. (1979). Fungal Genetics, 4th ed. Berkeley: University of California Press.

Fink, G. R. (1974). Properties of gene conversion of deletions in *Saccharomyces cerevisiae.* In: Mechanisms in Recombination, edited by R. Grell. New York: Plenum Press, p. 287.

Fink, G. R., Styles, C. (1974). Gene conversion of deletions in the *his4* region of yeast. Genetics 77:31.

Fogel, S., Hurst, D. D. (1967). Meiotic gene conversion in yeast tetrads and the theory of recombination. Genetics 57:455.

Fogel, S., Mortimer, R. K. (1968). Meiotic gene conversion of nonsense mutations in yeast tetrads. Proc. Int. Cong. Genet. 1:6.

Fogel, S., Mortimer, R. K. (1969). Informational transfer in meiotic gene conversion. Proc. Natl. Acad. Sci. USA 62–96.

Fogel, S., Mortimer, R. K. (1970). Fidelity of gene conversion in yeast. Mol. Gen. Genet. 109:177.

Fogel, S., Mortimer, R. K. (1971). Recombination in yeast. Ann. Rev. Genet. 5:219.

Fogel S., Hurst, D. D., Mortimer, R. K. (1971). Gene conversion in unselected tetrads from multipoint crosses. Stadler Genet. Symp. 2:89.

Fogel, S., Mortimer, R. K., Lusnak, K., Taveres, F. (1979). Meiotic Gene Conversion: A Signal of the Basic Recombination Event in Yeast. Cold Spring Harbor Symp. Quant. Biol. 43:1325.

Fogel, S., Mortimer, R. K., Lusnak, K. (1981). Mechanisms of meiotic gene conversion, or "Wanderings on a foreign strand." In: Molecular Biology of the Yeast *Saccharomyces,* edited by J. N. Strathern, E. Jones, J. R. Broach. Cold Spring Harbor Laboratory, N.Y., pp. 289–339.

Fogel, S., Choi, T., Kilgore, D., Lusnak, K., Williamson, M. (1982). The molecular genetics of nontandum duplications at *ADE8* in yeast. Rec. Adv. Yeast Molec. Biol. 1:269–288.

Game, J. C., Lamb, T. J., Braun, R. J., Rernick, M. A., Roth, R. M. (1980). The role of radiation (rad) genes in meiotic precombination in yeast. Genetics 94:51–68.

Ghikas, A., Lamb, B. C. (1977). The detection, in unordered octads, of $6^+$:2m and $2^+$:6m ratios with postmeiotic segregation, and of aberrant 4:4s, and their use in corresponding-site interference studies. Genet. Res. 29:267.

Goldman, S. L. (1974). Studies on the mechanism of the induction of the site-specific recombination in the *ade-6* locus of *Schizosaccharomyces pombe.* Genetics 132:347.

Golin, J. E., Esposito, M. S. (1977). Evidence for joint genic control of spontaneous mutation and genetic recombination in *Saccharomyces.* Molec. Gen. Genet. 150:127–135.

Grell, R. F., ed. (1974). Mechanisms in Recombination. New York: Plenum Press.

Gutz, H. (1971). Site specific inductions of gene conversion in *Schizosaccharomyces* pombe. Genetics 69:317.

Haber, J. E., Rogers, D. T., McCusker, J. H. (1980). Homothallic conversions of yeast mating-type gene occur by intrachromosomal recombination. Cell 22:277–289.

Hastings, P. J. (1975). Some aspects of recombination in eukaryotic organisms. Annu. Rev. Genet. 9:129.

Hastings, P. J., Kalogeropoulos, A., Rossignol, J. L. (1980). Restoration to the parental genotype of mismatches formed in recombinant DNA heteroduplex. Curr. Genet. 2:169.

Hawthorne, D. C. (1969). Identification of nonsense codons in yeast. J. Mol. Biol. 43:71.

Hicks, J. B., Hinnen, A., Fink, G. R. (1979). Properties of yeast transformation. Cold Spring Harbor Symp. Quant. Biol. 43:1305.

Hicks, J. B., Strathern, J. N., Herskowitz, I. (1977). The cassette model of mating-type interconversion. In: DNA Insertion Elements, Plasmids, and Episomes, edited by A. Bukhari *et al.* Cold Spring Harbor, N.Y.: Cold Spring Harbor Laboratory.

Hicks, J. B., Strathern, J. N., Klar, A. J. S. (1979). Transposable mating-type genes in *Saccharomyces cerevisiae.* Nature 282:478.

Hinnen, A., Hicks, J., Fink, G. R. (1978). Transformation in yeast. Proc. Natl. Acad. Sci. USA 75:1929.

Holliday, R. (1964). A mechanism for gene conversion in fungi. Genet. Res. 5:282.

Holliday, R. (1974). Molecular aspects of genetic exchange and gene conversion. Genetics 78:273.

Hotchkiss, R. D. (1974). Models of genetic recombination. Ann. Rev. Microbiol. 28:445.

Hurst, D. D., Fogel, S., Mortimer, R. K. (1972). Conversion associated recombination in yeast. Proc. Natl. Acad. Sci. U.S.A. 69:101.

Jackson, J. A., Fink, G. R. (1981). Gene conversion between duplicated genetic elements in yeast. Nature 292:306.

Kalogeropoulos, A., Rossignol, J. L. (1980). Evidence for independent mismatch corrections along the same hybrid DNA tract during meiotic recombination in *Ascobolus.* Heredity 45:263.

Kitani, Y. (1978). Aberrant 4:4 segregation. Jpn. J. Genet. 53:301.

Kitani, Y., Olive, L. S. (1967). Genetics of *Sordaria fimicola*. VI. Gene conversion at the g locus in mutant wild-type crosses. Genetics 57:767.

Kitani, Y., Olive, L. S., El-Ani, A. S. (1962). Genetics of *Sordaria fimicola*. V. Aberrant segregation at the g locus. Amer. J. Bot. 49:697.

Klapholz, S., Esposito, R. E. (1980a). Isolation of *spo 12-1* and *spo 13-1* from a single meiotic division. Genetics 96:567.

Klapholz, S., Esposito, R. E. (1980b). Recombination and chromosome segregation during the single meiosis in *spo 12-1* and *spo 13-1* diploids. Genetics 96:589.

Klar, A. J. S., Fogel, S., MacLeod, K. (1979). *MAR-1*. A regulator of *HMa* and *HM* in *Saccharomyces cerevisiae*. Genetics 93:37–50.

Klein, H. L., Petes, T. D. (1980). Intrachromosomal gene conversion in yeast. Nature 289:144.

Korch, C. T., Snow, R. (1973). Allelic complementation in the first gene of histidine biosynthesis in *Saccharomyces cerevisiae*. Genetics 74:287.

Kunz, B. A., Haynes, R. H. (1981). Phenomenology and genetic control of mitotic recombination in yeast. Ann. Rev. Genet, edited by H. C. Roman. In press.

Kushev, V. (1974). Mechanisms of Genetic Recombination (translated by B. Haigh). New York: Plenum Press.

Lamb, B. C. (1972). 8:0, 0:8, 7:1 and 1:7 conversion ratios in octads from wild-type mutant crosses of *Ascobolus immersus*. Heredity 29:397.

Lamb, B. C., Wickramaratne, M. R. T. (1973). Corresponding site interference, synaptinemal complex structure, and 8+:0m and 7+:1m octads from wild-type mutant crosses of *Ascobolus immersus*. Genet. Res. 22:113.

Lawrence, C., Sherman, W., Jackson, M., Gilmore, R. A. (1975). Mapping and gene conversion studies with the structural gene for iso-1-cytochrome c in yeast. Genetics 81:615.

Leblon, G. (1972). Mechanism of gene conversion in *Ascobolus immersus*. I. Existence of a correlation between the origin of mutants induced by different mutagens and their conversion spectrum. Mol. Gen. Genet. 115:36.

Leblon, G., Rossignol, J. L. (1973). Mechanism of gene conversion in *Ascobolus* immersus. III. The interaction of heteroalleles in the conversion process. Mol. Gen. Genet. 122:165.

Maloney, D. H., Fogel, S. (1980). Mitotic recombination in yeast: isolation and characterization of mutants with enhanced spontaneous mitotic gene conversion rates. Genetics 94:825–839.

Manney, R. R., Mortimer, R. K. (1964). Allelic mapping in yeast by X-ray induced mitotic reversion. Science 143:581.

Meselson, M. S., Radding, C. M. (1975). A general model for genetic recombination. Proc. Natl. Acad. Sci. USA 72:358.

Moore, C., Sherman, F. (1974). Lack of correspondence between genetic and physical distance in the iso-1-cytochrome c gene in yeast. In: Mechanisms of Recombination, edited by R. Grell. New York: Plenum Press, p. 295.

Moore, C., Sherman, F. (1975). The role of DNA sequences in genetic recombination in the iso-1-cytochrome c gene in yeast. I. Discrepancies between physical distance and genetic distances by fine structure mapping procedures. Genetics 79:397.

Mortimer, R. K., Fogel S. (1974). Genetical interference and gene conversion. In: Mechanisms in Recombination, edited by R. Grell. New York: Plenum Press, p. 236.

Mortimer, R. K., Hawthorne, D. C. (1969). Yeast genetics. In: The Yeasts, Vol. I, edited by A. H. Rose, J. S. Harrison, New York: Academic Press, p. 385.

Mortimer, R. K., Hawthorne, D. C. (1973). Genetic mapping in *Saccharomyces.* IV. Mapping of temperature-sensitive genes and use of disomic strains in localizing genes and fragments. Genetics 74:33.

Mortimer, R. K., Schild, D. (1980). Genetic map of *Saccharomyces cerevisiae.* Microbiol. Rev. 44:519.

Nagylaki, T., Petes, T. D. (1982). Intrachromosomal gene conversion and the maintenance of sequence homogeneity among repeated genes. Genetics, in press.

Nasmyth, K. A., Tatchell, K. (1980). The structure of the transposable yeast mating type loci. Cell 19:753.

Nicolas, A. (1979). Variation of gene conversion and intragenic combination frequencies in the genome of *Ascobolus immersus.* Mol. Gen. Genet. 176:129.

Paquette, J. (1978). Detection of aberrant 4:4 asci in *Ascobolus immersus.* Can. J. Genet. Cytol. 20:9.

Paszewski, A. (1970). Gene conversion: Observations on the DNA hybrid models. Genet. Res. 11:55.

Petes, T. D. (1980a). Unequal meiotic recombination within tandem arrays of yeast ribosomal DNA genes. Cell 19:765.

Petes, T. D. (1980b). Molecular genetics of yeast. Ann. Rev. Biochem. 49:845–919.

Pukkila, P. (1977). Biochemical analysis of genetic recombination in eukaryotes. Heredity 39:193.

Radding, C. M. (1973). Molecular mechanisms in genetic recombination. Ann. Rev. Genet. 7:87.

Radding, C. M. (1979). The mechanism of conversion of deletions and insertions. Cold Spring Harbor Symp. Quant. Biol. 43:1315.

Resnick, M. A. (1979). The induction of molecular and genetic recombination in eukaryotic cells. Adv. Radiat. Biol. 8:175.

Rodarte-Ramon, M. S., Mortimer, R. K. (1972). Radiation induced recombination in *Saccharomyces:* isolation and genetic study of recombination deficient mutants. Radiat. Res. 49:133–147.

Roman, H. L. (1957). Studies of gene mutation in *Saccharomyces.* Cold Spring Harbor Symp. Quant. Biol. 21:175.

Roman, H. L. (1980). Recombination in diploid vegetative cells of *Saccharomyces cerevisiae.* Carlsberg Res. Commun. 45:1315

Roman, H. L., Jacob, F. (1958). A comparison of spontaneous and ultraviolet-induced allelic recombinations with reference to the recombination of outside markers. Cold Spring Harbor Symp. Quant. Biol. 23:155.

Rossignol, J. L., Haedens, V. (1980). Relationship between asymmetrical and symmetrical hybrid DNA formations during meiotic recombination. Curr. Genet. 1:185.

Rossignol, J. L., Paquette, N., Nicolas, A. (1979). Aberrant 4:4 asci, disparity in the direction of conversion, and the frequency of conversion in *Ascobolus immersus.* Cold Spring Harbor Symp. Quant. Biol. 43:1343.

Sang, J., Whitehouse, H. L. K. (1979). Genetic recombination at the *buff* spore colour locus in *Sordaria brevicollis.* Mol. Gen. Genet. 174:327.

Scherer, S., Davis, R. W. (1980). Recombination of dispersed repeated DNA sequences in yeast. Science 209:1380.

Stadler, D. R. (1973). The mechanism of intragenic recombination. Ann. Rev. Genet. 7:113.

Stadler, D. R., Towe, A. M. (1963). Recombination of allelic cysteine mutants in *Neurospora.* Genetics 48:1323.

Stadler, D. R., Towe, A. M. (1971). Evidence for meiotic recombination in *Ascobolus* involving only one member of tetrad. Genetics 68:401.

Stahl, F. W. (1969). One way to think about gene conversion. Genetics (Suppl.) 61:1.

Stahl, F. W. (1978). Summary. Cold Spring Harbor Symp. Quant. Biol. 43:1353.

Stahl, F. W. (1979a). Special sites in recombination. Ann. Rev. Genet. 13:7.

Stahl, F. W. (1979b). Genetic recombination. Thinking about it in phage and fungi. San Francisco: Freeman.

Stinchcomb, D. T., Struhl, K., Davis, R. W. (1979). Isolation and characterization of a yeast chromosomal replicator. Nature 282:39–43.

Struhl, K., Stinchcomb, D. T., Scherer, S., Davis, R. W. (1979). High frequency transformation of yeast: Autonomous replication of DNA molecules. Proc. Natl. Acad. Sci. USA 76:1035–1039.

Szostak, J. W., Wu, R. (1980). Unequal crossing over in the ribosomal DNA of *Saccharomyces cerevisiae.* Nature 84:426.

Wagner, R. E., Radman, M. (1975). A mechanism for initiation of genetic recombination. Proc. Natl. Acad. Sci. USA 72:3619.

Whitehouse, H. L. K. (1963). A theory of crossing over by means of hybrid deoxyribonucleic acid. Nature 199:1034.

Whitehouse, H. L. K. (1969). Towards an understanding of the mechanism of heredity, 2nd ed. New York: St. Martin's Press.

Wickramaratne, M. R. T., and Lamb, B. C. (1978). The estimation of conversion parameters and the control of conversion in *Ascobolus immersus.* Mol. Gen. Genet. 159:63.

Williamson, M., Fogel, S. (1980). Correction deficient *(cor)* mutants that alter the frequency and length of correction tracts in meiotic hybrid DNA. In: Tenth International Conference on Yeast Genetics and Molecular Biology, no. 135, Louvain-la-Neuve, Belgium: p. 44.

Yu-Sun, C. C., Wickramaratne, M. R. T., Whitehouse, H. L. K. (1977). Mutagen specificity in conversion pattern in *Sordaria brevicollis.* Genet. Res. 29:65.

Zimmerman, F. K. (1968). Enzyme studies on the products of mitotic gene conversion in *Saccharomyces cerevisiae.* Genetics 101:171.

# 4

# Radiation-Sensitive Mutants and Repair in Yeast[1]

J. C. Game*

## Introduction

*Saccharomyces cerevisiae* is an excellent organism for studying the genetics of DNA repair processes, since it has a versatile and well-characterized genetic system, and its unicellular nature is convenient for studies in radiation biology. Biochemical studies of DNA repair and related processes have proved more difficult, owing partly to difficulties with DNA-specific labeling. However, substantial progress is now being made in these areas also. Aspects of DNA repair and mutagenesis in yeast have recently been reviewed by several workers, e.g., Haynes and Kunz (1981), Lawrence (1982), and Lemontt (1980). In this chapter I shall briefly outline early studies in the radiation biology of *S. cerevisiae.* I shall then describe the major classes of radiation-sensitive mutants that have been isolated and discuss their contribution to understanding the genetics of repair in yeast. Finally, I shall describe some recent work indicating that genes that control repair of ionizing-radiation damage are also required for normal meiosis, as well as meiotic and mitotic recombination in yeast, and discuss the role of x-ray-sensitive mutants in current research in these areas.

## Radiation Biology of Yeast

Early work on the radiation biology of yeast demonstrated that the degree of resistance of cells to ionizing radiation depended both on the ploidy of the culture and on which stage of the mitotic cycle the cells were in when irradiated.

*360 Donner Laboratory, Lawrence Berkeley Laboratory, Berkeley, California 94720 U.S.A.
[1]I would like to thank Marsha Williamson for help in preparing the figures.

Lacassagne and Holweck (1930) observed that the onset of division in *Saccharomyces ellipsoideus* yeast cultures was accompanied by an increase in resistance to ionizing radiation, but the ploidy of the strain used is not clear. Latarjet and Epphrussi (1949) found that diploid *S. cerevisiae* strains showed a greater resistance to x-rays than haploids, and an even greater difference was observed by Zirkle and Tobias (1953), who found that the dose required to give 50% lethality in a diploid strain was more than 10 times that required for the same lethality in *a* and *α* haploids derived from it. In the same study, however, they also found that a small fraction (∼10%) of highly resistant haploid cells existed, so that the haploid survival curve exhibited a "tail" at high doses. In contrast, diploid strains exhibited a "shoulder" at low doses, with increasing slope at higher doses. These differently shaped survival curves for haploid and diploid strains (see Fig. 4.1) have subsequently been confirmed by many workers (e.g., Mortimer 1958; Laskowski 1960; Nakai and Matsumoto 1967; Game and Mortimer 1974), although recently Brunborg and Williamson (1978) have pointed out that resistance in some diploid strains is highly dependent on the stage of the cell cycle, and in these strains survival curves of G1 cells resemble those of haploids more than they do that of the diploids shown in Figure 4.1. Polyploid strains have also been studied by several workers (Mortimer 1958; Laskowski 1960). It has been found that higher levels of ploidy do not confer increased resistance, but on the contrary, resistance progressively decreases with increasing ploidy so that diploid strains are the most resistant, and hexaploid strains are the most sensitive, although more resistant than stationary phase haploids. This increased sensitivity of polyploids is probably a consequence of dominant lethal damage, which is induced at higher frequencies when more copies of the genome are present.

The occurrence of the resistant tail on haploid survival curves has been convincingly shown in several studies to be correlated with the cell division cycle (e.g., Beam *et al.* 1954), and it is now generally accepted that these resistant cells represent those in the G2 phase of the division cycle and that the resistance arises from repair systems that operate only when two homologous copies of DNA are present. Similarly, the high degree of resistance seen in many diploids can be attributed to repair, which could in principle operate either in G1 or in G2, since homologous DNA is present throughout the cell cycle in this case. Diploids with a sensitive G1, such as that described by Brunborg and Williamson (1978), probably represent those in which repair is primarily dependent on the presence of sister chromatids and is thus confined to the G2 phase. In contrast, those without a sensitive phase of the cycle are probably also able to do repair in G1, utilizing homologous chromosomes, and are thus resistant throughout the cycle (see Brunborg *et al.* 1980).

Evidence for this interpretation of ionizing radiation survival curves in yeast in terms of the cell division cycle and repair comes from work with synchronous cultures as well as many genetic and biochemical studies (e.g., Beam *et al.* 1954; Elkind and Sutton 1959; Game and Mortimer 1974; Perper 1975; Brunborg *et al.* 1980). A detailed discussion of x-ray-sensitive mutants and DNA

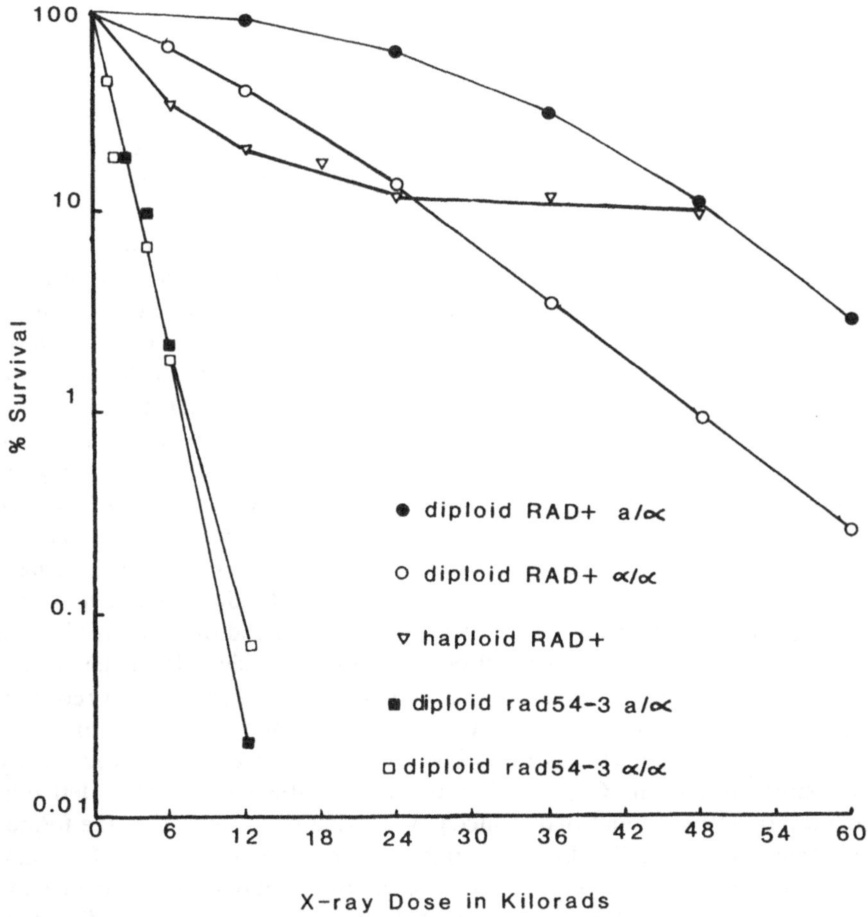


**Figure 4.1.** X-ray survival curves for log-phase cells of a haploid and two diploid wild-type *(RAD+)* yeast strains, and two diploid strains each homozygous for the *rad54-3* mutation. The effect of mating-type heterozygosity in wild-type strains and the absence of this effect in *rad54-3* strains can be seen by comparing the *MATa/MATα* diploid with the *MATα/MATα* diploid in each case. (Data of Game and Mortimer.)

repair is given later, but here it is appropriate to mention that Brunborg *et al.* (1980) have used highly synchronous cultures of strain NCYC 239 (a diploid strain with a sensitive G1 phase) to demonstrate biochemically that the ability to repair double-strand breaks in DNA is specifically associated with the radiation-resistant phase of the cell cycle, namely, G2. Genetically, it has been shown in several studies (e.g., Nakai and Matsumoto 1967; Game and Mortimer 1974) that mutants that abolish the resistant tail on haploid survival curves also abolish the shoulder seen in wild-type diploids with a resistant G1 phase, indicating that both types of resistance are at least largely under the same genetic control. This point is underscored by the existence of temperature-sensitive mutations such as *rad55-4* (Game and Mortimer 1974), in which the haploid tail and the diploid shoulder both become temperature dependent in a similar manner.

Studies of the effects of ultraviolet light on yeast have been at least as extensive as those involving ionizing radiation. In the case of UV light, survival curves for both haploid and diploid wild-type yeast strains show a pronounced shoulder at low doses, although diploids are usually somewhat more resistant than haploids. The tail that is seen on haploid survival curves after x-rays is absent from UV-survival curves, although cells in the G2 phase of the mitotic cycle are somewhat more resistant than G1 cells. This gives rise to a point of inflection in UV-survival curves of asynchronous populations, at the same survival level as the x-ray-resistant tail and arising from the same resistant G2 fraction (Elkind and Sutton 1959; Chanet *et al.* 1973). There is abundant evidence that the UV-resistant shoulder seen throughout the mitotic cycle in yeast survival curves arises from DNA repair activities, and such repair has been extensively studied, primarily through the use of UV-sensitive mutants (see later). The difference in shape between x-ray and UV-survival curves for haploid yeast probably arises from the fact that UV-light-induced damage is substantially repaired by the excision pathway (see later) and this operates in both the G1 and G2 phases of the cell cycle, producing a shoulder at low doses, whereas ionizing radiation damage is probably largely repaired by a recombinational pathway, which requires two copies of the DNA, leading to exponential killing of haploid G1 cells at low doses and a much greater relative sensivity compared to cells in G2. Another important difference between lethality induced by UV light and that induced by ionizing radiation is that the former is substantially reversible by visible light, whereas the latter is not. This photoreversibility has been studied in many organisms and depends on an enzyme (known as "photoreactivating enzyme") that splits dimerized pyrimidine residues *in situ* in DNA, in the presence of visible light (see Harm 1976 for a review). Such pyrimidine dimer residues comprise a major component of UV-light-induced damage and are also a substrate for more complex repair processes, discussed later, that do not require visible light. In the next sections I describe the genetic control of UV and x-ray repair processes and the properties of single and multiple mutants defective in one or more types of repair.

## Radiation-Sensitive Mutants

Following extensive work with radiation-sensitive mutants of bacteria (see Adler 1966 for a review) and other organisms (Harm 1963; Holliday 1965), the first *Saccharomyces cerevisiae* mutants sensitive to UV light and x-rays were reported by Nakai and Matsumoto (1967) and by Zakharov and Kozhina (1967). In a short time many other workers reported isolating numerous UV-light-sensitive mutants [notably Snow (1967), who described recessive mutants in six separate genes, and Cox and Parry (1968), who isolated 96 mutants that fell into 22 separate genes]. The usual method of isolation simply involved plating mutagenized cells on YEPD medium and replica-plating the resultant colonies to two additional plates. One of these was then exposed to a dose of UV light that left a sufficient number of surviving cells in the replica print of wild-type colonies to distinguish these from UV-sensitive mutants, where little or no survival occurred. A surprising feature of this work was the large number of genes that were found to mutate to give sensitivity to UV light, and this gave an early indication of the complexity of repair processes in yeast. Cox and Parry deliberately attempted to saturate the genome for mutations that conferred UV light sensitivity, hoping to identify all the yeast genes involved in UV repair, but although they identified 22 loci, statistical analysis of the number of independent mutants isolated per gene suggested the existence of at least several genes for which no mutant alleles had been found.

A standardized nomenclature for radiation-sensitive mutants was discussed at the yeast genetics meeting at Chalk River in Canada in 1970, and it was proposed that the general designation *RAD,* followed by a locus number, be used for genes that could mutate to confer radiation sensitivity. The system was to be used for currently known genes and for others that might subsequently be identified, and it was decided that locus numbers from 1 to 49 be set aside for genes whose mutant alleles primarily conferred sensitivity to UV light, whereas locus numbers from 50 upward were to be used for genes primarily controlling sensitivity to ionizing radiation. A series of allelism tests was performed by Game and Cox, which matched independently isolated mutants from different laboratories and provided standard locus and allele numbers for most of the known UV-sensitive mutants (Game and Cox 1971). These mutants comprised the genes *RAD1* through *RAD22.* Subsequently, Game and Mortimer (1974) published a genetic study of available x-ray-sensitive mutants, which designated eight genes, *RAD50* through *RAD57,* as separate from the UV-sensitive series. Genes such as *RAD6, RAD9,* and *RAD18,* whose mutants are substantially sensitive both to UV light and to ionizing radiation, are included among the UV-sensitive series. This implies a degree of arbitrariness, since many mutants highly sensitive to one form of radiation are at least marginally sensitive to the other, but the system proved workable and has gained acceptance. However, some radiation-sensitive mutants exist that have not yet been assigned *RAD* locus numbers, for example, the $r_1^s$ mutant described by Averbeck *et al.* (1970), and mutants that were isolated by Prakash and Prakash (1977) on the basis of sensitivity to the mutagen methyl methanesulfonate, some of which are also sensitive to radiation. The latter mutants define genes currently given *MMS* locus designations, except for those shown to be allelic with existing *rad* mutants (Prakash and Prakash 1977).

The absence of a simple method of specifically labeling DNA in yeast made biochemical studies of repair of UV-light-induced damage difficult. Early work largely focused on genetic techniques to study mutants and to provide information about UV-induced lethality, although recently some significant biochemical advances have been made (see later). Prior to this, genetic studies made considerable use of double and multiple mutants to work out a logically consistent model of repair processes in yeast, based on identifying "epistasis groups" (see the following discussion). In addition, UV-sensitive mutants were characterized with respect to other phenotypes, such as effects on induced and spontaneous mutation and recombination, as well as cross-sensitivity to ionizing radiation and chemicals that damage DNA. A third approach involved studying liquid-holding phenomena and photoreactivation in wild type and UV-sensitive yeast strains.

## Double-Mutant Interactions and Epistasis Groups

Information about the way different *RAD* genes control resistance to UV light can be obtained by constructing double-mutant strains and determining their UV sensitivity compared to the parental single mutants. If two different genes both confer sensitivity, when mutant, by the same mechanism—for example, by blocking different steps of a single DNA repair pathway—then in principle one may expect the double mutant to show a UV sensitivity no greater than that of the most sensitive single mutant. This follows from the fact that once the pathway has been blocked, additional mutations in subsequent steps should have no further effect on the cell. However, if each mutation blocks a different repair pathway (or confers sensitivity by other separate mechanisms), then one can expect an additive response in the double mutant, so that it is more sensitive than either of the parents. These two cases, epistasis and additivity, are illustrated in Figure 4.2, parts a and b, respectively. When double mutants were constructed and studied, many cases of epistasis were indeed found (e.g., Nakai and Matsumoto 1967; Game and Cox 1972), and one example, for *rad1* and *rad3,* is shown in Figure 4.3b. However, a surprising observation emerged from mutant combinations that did not show epistasis. It was found that in many cases the double mutant had a sensitivity much *greater* than that expected for the simple additivity predicted from Figure 4.2b (e.g., Nakai and Matsumoto 1967; Game and Cox 1973). This type of interaction is termed *synergistic,* and an example of it, involving *rad3* and *rad50,* is shown in Figure 4.3a. An interpretation of synergistic interactions is shown in Figure 4.2c, and is based on the idea that a more-than-additive interaction can occur if two separate repair pathways act on the same initial substrate. In this situation one can postulate that some damage is repaired by each pathway in the wild-type cell. If a block is now introduced into one or other pathway by either single mutant, it is likely that the remaining pathway will to some extent be able to compensate for this block by repairing more of the damage than it would have done if both pathways were open. However, in the double mutant, this compensatory effect will not operate, since neither pathway will be available to compensate for the defect in the other, and the full effect of both mutations will be expressed, leading to synergism. This model has been used to interpret

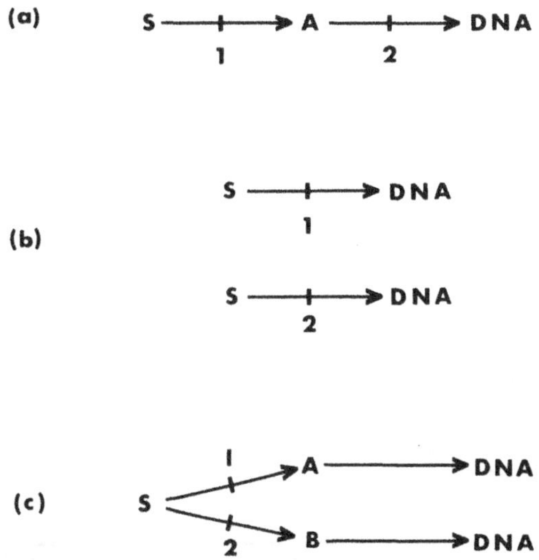


**Figure 4.2.** Repair pathway diagrams illustrating situations that are expected to give rise to three types of interactions of mutations in double mutant strains. S represents substrate (i.e., UV-damaged DNA), A and B represent intermediates in repair pathways, and DNA represents completed repair. The numbers represent mutations blocking the corresponding reaction steps. (a) *Epistasis* (Mutant 1 is *epistatic* to mutant 2, whereas mutant 2 is said to be *hypostatic* to mutant 1); (b) *additivity;* (c) *synergism.* See also Fig. 4.3 and text.

synergistic interactions that occur between UV-sensitive mutants in *S. cerevisiae* (Game 1971; Game and Cox 1973; Cox and Game 1974), and a similar suggestion was made independently by Fabre (1971) to account for synergism between *Schizosaccharomyces pombe* mutants.

A considerable number of double and multiple mutants have been studied in *S. cerevisiae* (see, for example, Nakai and Matsumoto 1967; Averbeck 1970; Game and Cox 1972, 1973; Brendel and Haynes 1973; Prakash and Prakash 1979), and although there are difficulties with certain detailed aspects of interpretation, some basic conclusions have emerged. Each *RAD* gene can be classified according to the way it interacts with other genes, and it has been found that there are three separate epistasis groups (Game 1971; Game and Cox 1973; Cox and Game 1974). Mutants within any one group show an epistatic interaction but interact synergistically (or sometimes additively) with mutants in other groups. Triple mutant strains involving all three groups in general show a further synergism, being even more sensitive than synergistic double

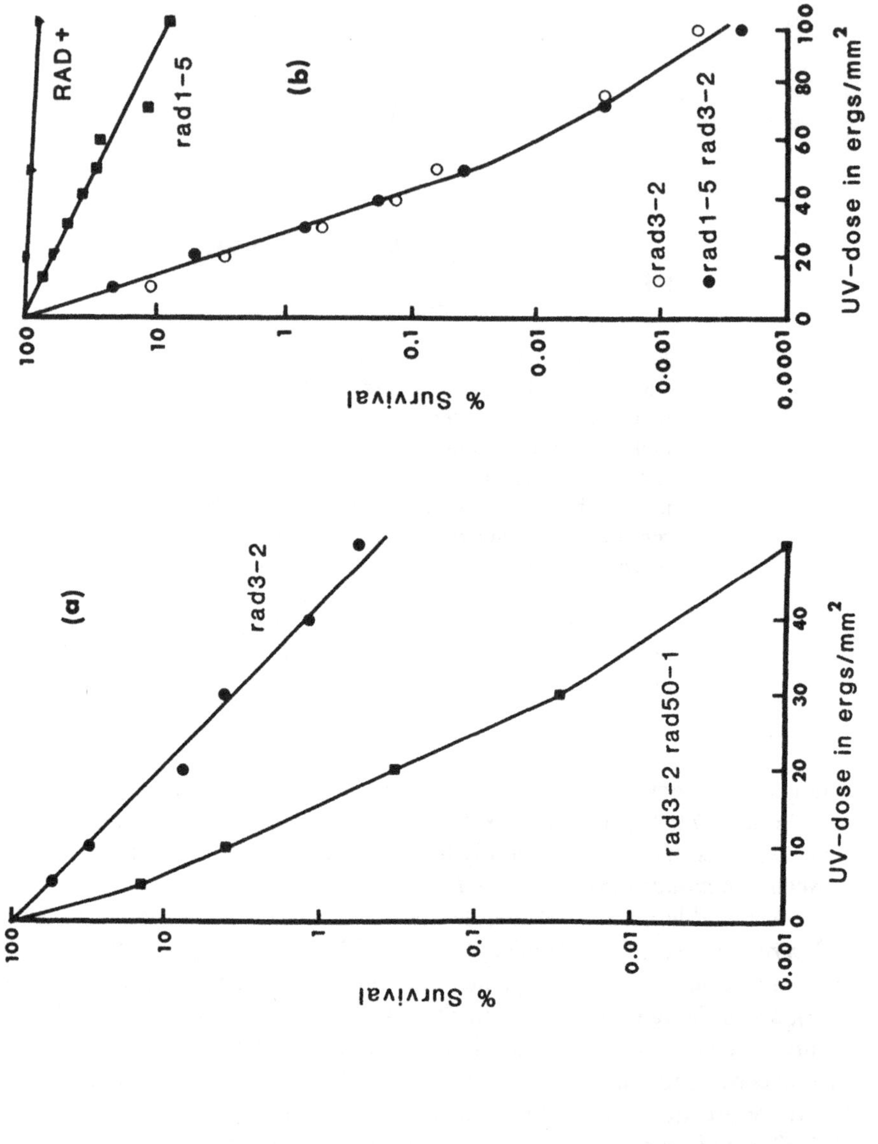
(a)
rad3-2
rad3-2 rad50-1
% Survival
UV-dose in ergs/mm2
(b)
RAD+
rad1-5
○rad3-2
●rad1-5 rad3-2
% Survival
UV-dose in ergs/mm2

mutants. Thus one can postulate three separate repair pathways, each acting on the same primary damage. This interpretation is supported by data from various workers (e.g., Lawrence and Christensen 1976) on the phenotypic characteristics of mutants (see later) in the three groups, and is outlined in Figure 4.4, which also shows some of the *RAD* loci believed to be involved in each of the three types of repair.

## Phenotypic Characteristics of Repair-Defective Mutants

Ultraviolet radiation, when applied to wild-type cells of yeast or other organisms, not only causes lethality but also induces mutation and recombination among the survivors. The way in which these processes are affected in mutants that are isolated on the basis of increased sensitivity to UV killing provides independent support for the existence of three types of repair as defined by the epistasis relationships outlined earlier. In addition, it provides important clues about the mechanisms of the three repair types and demonstrates the interrelationships of DNA repair, mutation, and recombination.

Ultraviolet radiation-induced mutation frequencies in UV-sensitive mutants have been extensively studied by Koval'tsova *et al.* (1970), Hunnable and Cox (1971), Lemontt (1971), Eckardt *et al.* (1975), Lawrence and Christensen (1976), and others. They found that mutants that are now known to be defective in the excision of pyrimidine dimers from DNA (see later) and that fall into the "excision repair" epistasis group (Fig. 4.4) consistently show UV-induced mutation rates per surviving cell that are higher than those of wild type after comparable doses of UV light. In contrast, mutants that fall into a second epistasis group ("error-prone repair," Fig. 4.4), show mutation rates after UV radiation that are lower than wild type at the same UV-light doses and in some cases (such as *rad6-1*) virtually abolish UV-induced mutagenesis.

An interpretation (Cox and Game 1974; Lawrence and Christensen 1976) of these results that has become accepted postulates that *RAD* genes in this epistasis group control a repair system that is error-prone and that UV-induced mutations arise (in wild type) as a consequence of errors in this system. Thus mutants that reduce or abolish this repair system also reduce or block UV-induced mutation. Mutants in the excision repair group, on the other hand, block an error-free repair system that in wild type does not lead to UV-induced mutations. The observation that excision repair mutants show more UV-

---

**Figure 4.3.** (a) An example of a synergistic interaction with respect to UV sensitivity between the *rad3-2* and *rad50-1* mutations. At the UV doses shown, neither wild-type nor *rad50-1* strains show any significant UV-induced lethality. (Data of Game and Cox.) (b) An example of an epistatic interaction with respect to UV sensitivity between the *rad1-5* and *rad3-2* mutations. The *rad1-5* mutation is "leaky," i.e., it only partially blocks the *RAD1*-mediated repair step, so that the order in which the gene products function cannot be inferred from these data. (Data of Game and Cox.)

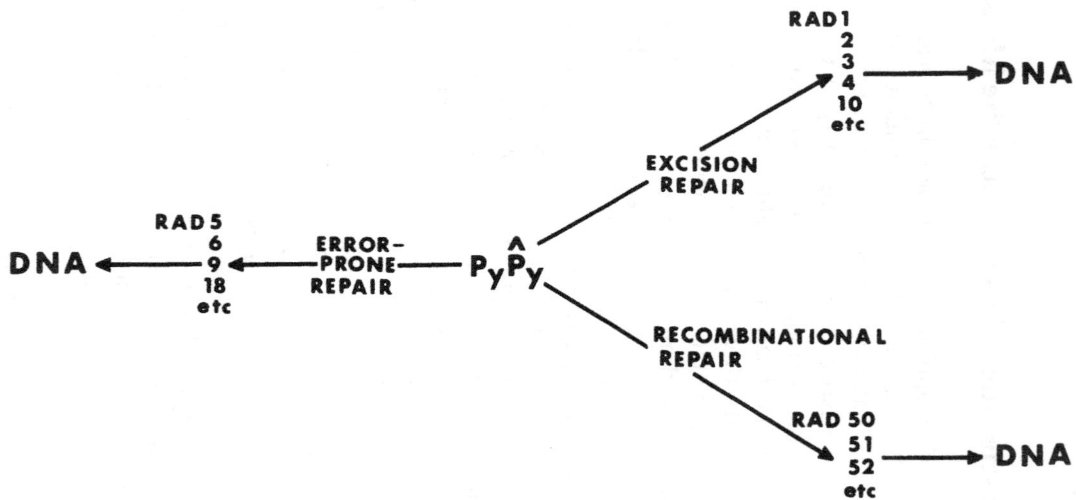


**Figure 4.4.** A schematic diagram for the three types of dark repair of UV-induced pyrimidine dimers in yeast, and some of the genes believed to be involved in each type. See text for discussion.

induced mutagenesis than wild type suggests that lesions that are repaired by this system in wild-type are channeled in its absence into the error-prone repair system, so that the proportion of UV-induced damage that is repaired in an error-prone manner is much higher in excision repair mutants than in wild type. This conclusion that alternate repair systems can operate on the same basic type of DNA damage is similar to that arrived at from a consideration of the synergistic interactions observed in double mutant survival curves and lends support to the interpretation of synergism described earlier. A complicating factor is the existence of the third UV repair system, mediated by genes in another epistasis group ("recombinational repair," Fig. 4.4). Lawrence and Christensen (1976) have found that mutants in this group show UV-induced mutation rates similar to those of wild type, suggesting that these genes do not mediate an error-prone system and (more surprisingly) that a block in this system does not channel a significant additional fraction of UV-induced damage into the error-prone system. This at first sight appears to conflict with double and multiple mutant survival data, which are interpreted to indicate that all three UV repair systems compete for the same basic UV-induced damage. However, the relatively slight increase in UV-induced killing seen in mutants in the third epistasis group indicates that this system represents a minor pathway for repair of UV-induced damage in wild type, so that the amount of additional damage channeled into the other systems by blocking it will be small. Furthermore, much of this damage will be channeled into the error-free excision repair pathway rather than the error-prone pathway, since the former represents the major repair system in wild type (see later), and in fact synergism with respect to UV-induced lethality between mutants in the error-prone and recombinational repair systems is most readily demonstrated in a background where excision repair is absent (Game and Cox 1973). Based on this reasoning, Lawrence and Christensen (1976) measured UV-induced mutation frequency in a *rad1 rad52* double mutant and found that it was nearly tenfold higher than it was in the strain carrying *rad1* alone, which in turn showed a frequency more than tenfold that of the wild-type strain. Thus *rad52,* which blocks UV-induced recombination (see Game *et al.* 1980; Prakash *et al.* 1980) and probably belongs to the "recombinational repair" epistasis group in Figure 4.4 (see the following discussion) does indeed appear to channel UV damage into an error-prone repair pathway, in the absence of excision repair.

In addition to the work on UV-induced killing and UV mutagenesis, measurements of UV-induced mitotic recombination rates in UV-sensitive mutants have also provided information about the three UV repair systems in yeast. Such studies usually involve diploid strains that are heteroallelic at one or more loci conferring auxotrophic requirements on the cell, so that frequencies of induced prototrophic recombinants can be determined by plating on nutrient-omission media that allow only such recombinants to grow. Hunnable and Cox (1971) showed that at least some of the excision-repair mutants in Figure 4.4 showed elevated levels of UV-induced intragenic recombination compared to wild type at the same UV doses. The increase was most marked in *rad2-5*, but was also evident in *rad1*, *rad3*, and *rad4* mutants. They also found some increase in *rad6-1* and *rad18-2*, which are mutants in the epistasis group that controls the error-prone system. Lemontt (1971) also found that *rad5-7* showed an increase in both intragenic and intergenic UV-induced recombination compared to wild type at the same doses, and Snow (1968) reported similar results for *rad5-1*, as well as finding elevated UV-induced recombination for several mutants now known to be defective in excision repair. Similarly, Kern and Zimmerman (1978) found that *rad3* and *rad6* mutants each showed elevated rates of spontaneous and UV-induced mitotic recombination. Thus it seems clear from these and other studies that abolition of either excision repair or error-prone repair leads to an increase in UV-induced recombination. This suggests that neither of these repair processes operate via a recombinational mechanism, but that the damage they normally repair can be channeled into a separate recombinational pathway when they are blocked.

Measurements of UV-induced mitotic intragenic and intergenic recombination in mutants in the third epistasis group (Fig. 4.4) indicate clearly that it is indeed this group of genes that controls the recombinational pathway of repair of UV-induced damage. For example, Morrison and Hastings (1979) found that *rad51-3* completely abolishes UV-induced mitotic prototroph formation in strains that are heteroallelic at the *his1* locus, and a similar abolition of UV-induced recombination has been reported for *rad52* (Prakash *et al.* 1980; Game *et al.* 1980). Both *rad51* (Game and Cox 1973) and *rad52* (Brendel and Haynes 1973) mutants interact synergistically with respect to UV-induced lethality with mutants in the excision and error-prone repair processes, although the interactions with excision mutants are much more pronounced in the case of *rad51*. Another mutant in the third epistasis group (Fig. 4.4), *rad50-1*, was reported by Hunnable and Cox (1971) to show a UV dose response for intragenic mitotic recombination that fell within the wild-type range, but subsequent findings (Game, unpublished observations) suggest that it shows less recombination than wild type, although some UV-induced recombination does occur.

In summary, we can say that *RAD* genes in the excision repair epistasis group control UV repair that is neither mutagenic nor recombinational; those in a second group control mutagenic repair that is not recombinational; whereas those in a third group control recombinational repair that is not mutagenic. The synergistic interactions with respect to sensitivity to UV-induced killing, as well as the elevation of UV-induced mutation or recombination when nonmutagenic or nonrecombinagenic pathways are blocked, all suggest that damage can be "channeled" from one pathway to another, and thus that the same primary lesions are operated on in wild type by all three repair processes. Hastings *et al.* (1976) have suggested that a similar channeling process occurs with respect to spontaneous lesions in DNA, to account for spontaneous mutation in wild type as well as in *rad* mutants and mutator *(mut)* mutants.

## Biochemical Studies on UV-Induced Damage and Repair

In addition to the genetic studies described earlier, UV-sensitive mutants have proved useful in studying the biochemistry of DNA repair in yeast, and in turn biochemical techniques have revealed information about the nature of UV-induced DNA damage and the molecular mechanisms for its repair.

Unrau *et al.* (1973) demonstrated the presence of pyrimidine dimers in UV-irradiated DNA from yeast by direct biochemical analysis and estimated that a dose of 1 $J/m^2$ applied to cells will generate about 240 dimers per haploid genome. From this it can be calculated (Cox and Game 1974) that a dose required to reduce survival in wild-type by 63% (an average of one "lethal hit" per cell) will generate about 27,000 dimers per haploid cell. Early studies on UV sensitivity in yeast argued from circumstantial evidence and by analogy with bacterial systems that such pyrimidine dimers were probably a major source of UV-induced lethality and that in wild-type cells they were a substrate for enzymatic repair processes that were defective in at least some UV-sensitive mutants. Some early evidence for the importance of pyrimidine dimers after UV treatment in yeast came from studies of photoreactivation. It was first shown by Saracheck (1954) that the lethal effects of UV radiation could be reversed to a substantial extent in yeast, as in bacteria, by treatment with visible light given immediately after the UV exposure. Such reversal is known as photoreactivation (PR), and extensive work by Harm (see Harm 1976 for a review) and others has shown it to be mediated by a single enzyme. This specifically binds to pyrimidine dimers in DNA and splits these *in situ* in the presence of visible light to restore DNA to its normal condition. In yeast, a mutant *(phr1-1)* was isolated by Resnick in 1969 that shows no photoreactivation and lacks photoreactivating enzyme activity. It does not affect the UV sensitivity of cells incubated in the dark after UV light and is thus not classified as a *rad* mutant.

The results of incubating UV-treated yeast cells in the dark in nongrowth conditions before plating, and the effects of visible light after such dark incubation, suggested that a major repair process that operates without light in yeast acts on the same UV-induced DNA damage as the photoreactivation system (Parry and Cox 1968). To summarize briefly, it was found that a gradual increase in survival occurs in wild-type yeast cells held in the dark for periods of as long as several days before plating, but that this recovery is absent in several of the most UV-sensitive mutants. If a dose of visible light is given immediately before plating, after various periods of dark holding, the increase in survival brought about by such visible light gradually declines in wild-type strains, concomitantly with the rise in survival that occurs in the dark. However, in those UV-sensitive mutants that abolish dark-holding recovery, the effect of visible light on viability is retained, so that cells can be photoreactivated even after several days in the dark in nongrowth conditions. Thus the major dark repair system in yeast appears to operate on potentially photoreactivable lesions, which have been identified from studies with photoreactivating enzyme as pyrimidine dimers (Setlow *et al.* 1965). The synergistic interactions between the three groups of UV-sensitive mutants described earlier also imply that the three dark repair systems act at least in part on a common substrate, and since pyrimidine dimers are inferred to be the major substrate for the excisionless mutants, it is likely from this reasoning that they are also a major substrate for the other two UV-repair systems, namely, error-prone and recombinational repair. Further support for this interpretation comes from observations (Cox and Game 1974) that mutant strains with blocks in all three dark repair pathways show exponential UV-survival curves, with a slope indicating that the UV dose required for an average of one lethal event per cell (i.e., 37% survival) is a little less than 0.01 $J/m^2$. This is consistent with estimates from the data of Unrau *et al.* (1973) of the dose required to give one to two dimers per haploid genome in yeast. It supports the interpretation (Cox and Game 1974) that pyrimidine dimers are the major source of UV-induced lethality in yeast, that one or two unrepaired dimers are lethal to a yeast cell, and that all dimer repair in wild-type yeast in the dark can be accounted for by the three identified repair systems. Excision repair appears to account for most dimer damage in wild type, since excision-defective mutants tend on the whole to be more sensitive than those in the error-prone group, which in turn are more sensitive than those defective in the recombinational repair of UV-induced damage (see Cox and Game 1974). An exception is the error-prone repair mutant *rad6-1,* which for unknown reasons is at least as sensitive as any mutants in excision repair (Cox and Parry 1968). Possibly, *RAD6* controls repair of other classes of UV-induced damage, in addition to the error-prone repair of pyrimidine dimers. Lawrence has argued from several lines of evidence (see Lawrence 1982 for a review) that *RAD6* controls some type of error-free repair as well as error-prone repair and has suggested that *RAD6*-dependent UV mutagenesis is functionally separable from *RAD6*-dependent resistance to lethality.

Recently biochemical studies have identified another UV-induced photoproduct that may occur at certain sites at about the same frequency as pyrimidine dimers in DNA from UV-irradiated human cells. Although the exact nature of the photoproduct is unknown, it results in lesions that lead to alkali-induced DNA strand breaks at positions of cytidine located 3′ to pyrimidine nucleosides (Lippke *et al.* 1981). Similar lesions may be responsible for alkali-induced strand breaks seen in bacterial DNA after UV irradiation (Haseltine *et al.* 1980), and Lippke *et al.* (1981) have suggested from the base changes seen in UV-induced mutations in the *lac i* gene of *Escherichia coli* (Coulondre *et al.* 1978) that such lesions may be a major source of UV-induced mutation. At the same time, however, other researchers have studied the fate of UV-induced pyrimidine dimers in wild-type and UV-sensitive yeast strains by biochemical methods, and there is little doubt that many mutants are defective in their excision repair. Unrau *et al.* (1971) showed, using chromatographic methods, that a *rad1-1* mutant strain failed to lose dimers from its DNA during incubation in growth medium in the dark, whereas such loss was observed in wild type. Resnick and Setlow (1972) demonstrated the loss of substrate for photoreactivating enzyme from the DNA of wild-type yeast incubated in the dark and the absence of such loss in a *rad2* mutant by assaying the amount of competition for yeast photoreactivating enzyme between crude extracts from UV-irradiated yeast cells and UV-irradiated *Haemophilus influenzae* transforming DNA. Competition was assayed by measuring the transforming ability of the *Haemophilus* DNA after standard photoreactivation treatments in the presence of the yeast extracts and PR enzyme. The loss of competition seen in extracts from wild-type log-phase cells during dark incubation was taken to indicate that cyclobutane dimers were being removed from DNA, or at least altered in some way that was dependent on the *RAD2* gene function.

More recently, dimer excision in yeast has been studied using UV-specific endonuclease preparations from T4 phage and *Micrococcus luteus* (Prakash 1975; Reynolds 1978). Some of these studies have used yeast strains that are permeable to deoxythymidine monophosphate (dTMP) (Jannsen *et al.* 1973; Wickner 1974; Fath *et al.* 1974; Little and Haynes 1979), so that DNA-specific labeling with [$^3$H]dTMP can be carried out. Other workers have used [$^3$H]uracil, which acts as a precursor for pyrimidines in DNA as well as labeling RNA. Such labeled cells are UV-light treated and are then converted to spheroplasts that are gently lysed to release their DNA. Such DNA can then be studied directly, using alkaline sucrose gradients to obtain single-stranded molecular weight profiles, or it can be treated with a UV-specific endonuclease prior to alkaline sucrose sedimentation. Such enzymes introduce single-strand nicks into DNA adjacent to pyrimidine dimers, so that the number of nicks introduced is a measure of the number of unexcised pyrimidine dimers in the DNA. Labeled RNA (derived from [$^3$H]uracil) is not a problem, since it is degraded by alkali to acid-soluble fragments (Reynolds 1978). Thus wild-type or *rad* mutant yeast strains can be incubated in the dark after UV-light treatment and monitored for their ability to make incisions in their own DNA, reflecting the initial step of excision repair, or for the loss of endonuclease-sensitive sites, reflecting the number of dimers remaining in their DNA after incubation. At least 12 different UV-sensitive mutants have now been studied with these techniques and the results may be summarized as follows: Dimer excision occurs in wild-type yeast nuclear DNA but not in mitochondrial DNA (Prakash 1975). Mutant alleles in *rad1, rad2, rad3, rad4,* and *rad10* completely block the initial incision step of excision repair (Prakash 1977a,b; Reynolds and Friedberg 1981), whereas *rad6, rad9* (Prakash, 1977a), and *rad18* (Reynolds and Friedberg 1981) mutants, which belong to a different epistasis group (see Fig. 4.4), are proficient in excision repair as monitored by loss of UV-endonuclease-sensitive sites. Mutants in four other genes— *rad7, rad14, rad16,* and *mms19*—are at least partially defective in loss of UV-endonuclease-sensitive sites (Prakash 1977b; Prakash and Prakash 1979; Reynolds and Freidberg 1981), and probably fall in the excision repair epistasis group, although interpretation of some of the survival data (Prakash and Prakash

1979) is difficult, owing to large differences in UV sensitivity of different mutants, so that epistasis is hard to distinguish from additivity. Additional evidence that it is specifically the incision step (rather than a subsequent stage of excision repair) that is blocked in *rad1, rad2, rad3, rad4,* and *rad10* (and partially blocked in *rad7* and *rad14*) mutants comes from work by Reynolds *et al.* (1981), who showed that cell-free extracts from such mutants are able to excise pyrimidine dimers from UV-irradiated DNA *in vitro,* provided this DNA has been treated with UV-endonuclease to introduce the initial incisions. Recently, Wilcox and Prakash (1981) have used the ligase-defective *cdc9-1* mutation (Johnston and Nasmyth 1978) in combination with *rad* mutants to increase their ability to detect DNA incision events taking place in these mutants after UV-light treatment. The absence of ligase activity prevents single-strand nicks that are introduced from being rejoined. They argued from such experiments that *rad14* differs from the other mutants studied and may be defective in a postincision step of excision repair, but data from R. Reynolds (Reynolds *et al.* 1981; R. Reynolds, personal communication) suggest that *rad14* is partially defective in the incision step. It is possible that the reduced amount of dimer excision may be a consequence of a reduced number of incisions made. Thus a surprisingly large number of genetic loci have been identified that are required for the primary incision step of excision repair. However, apart from *CDC9,* which is probably required (Johnston and Nasmyth 1978) for the final ligation step (and possibly *CDC8,* see the following discussion), genes whose products function in subsequent stages of excision repair have not yet been identified.

Biochemical studies relating to mutants from the other two UV epistasis groups (see earlier) are few and have not clearly revealed the molecular nature of error-prone or recombinational repair in yeast. However, a process referred to as "postreplication repair" has recently been studied in yeast by di Caprio and Cox (1981) and Prakash (1981). Starting with an excision-defective yeast strain lacking mitochondrial DNA, Prakash (1981) studied the size of newly synthesized nuclear DNA following various doses of UV light. She found that initially, the molecular weight of such DNA corresponds approximately to the average distance between pyrimidine dimers in the parental DNA strands. However, on further incubation in growth medium for periods up to 4 h, the molecular weight of the newly synthesized DNA increases until it corresponds in size to DNA seen from unirradiated cells. A similar phenomenon occurs in *E. coli* (Rupp and Howard-Flanders 1968) and is known as postreplication repair. Prakash (1981) found that in yeast the *rad6-1* mutation blocks this postreplicational rise in molecular weight in daughter strands after UV light and that the *rad18-2* and *rad52-1* mutations partially inhibit it. She suggested that more than one repair mechanism plays a role in postreplication repair and that both recombinational and nonrecombinational processes are involved, but the exact nature of these processes remains unknown.

Prakash *et al.* (1979) have also studied the role of another gene, *CDC8,* in repair. Mutants identified in the *CDC8* gene are temperature-sensitive lethals that block (at the restrictive temperature) both DNA synthesis and the cell division cycle (Hartwell 1971). Prakash *et al.* (1979) found that *cdc8* mutants are somewhat more sensitive than wild type to UV light at a temperature (25 °C) that permits growth and show a sharp reduction in UV-induced mutation frequency among survivors. They studied double-mutant combinations involving *cdc8* and several *rad* mutations and suggested that a *CDC8* gene function may be required by more than one repair system. More specific interpretations seem difficult to arrive at, since the *cdc8* mutant gene product must be at least partially functional in these experiments to permit cell division. However, it seems likely from their data that at least the error-prone repair system in yeast makes use of one or more gene products that are also required for normal DNA replication.

In summary, biochemical studies of UV repair in yeast have lagged behind those based on genetics. However, some progress is being made, and the advent of DNA-specific labeling with dTMP represents a significant advance that is helping molecular research on the many available UV-sensitive mutants.

## X-ray-Sensitive Mutants

In addition to the UV-light-sensitive mutants described earlier, many mutants that are sensitive to ionizing radiation have been described. There is considerable overlap in sensitivity between the two types of radiation, and indeed mutants that affect the recombinational repair pathway for UV-light damage are given *RAD* locus numbers from *RAD50* upward, since their most striking phenotype is their x-ray sensitivity. X-ray-sensitive mutants can be classified into several groups according to their phenotypes, although these groups are less clear-cut than is the case with UV-light-sensitive mutants. First, there are the x-ray-sensitive mutants that are also highly sensitive to UV light, and these block the error-prone UV-repair pathway. Such mutants appear to block x-ray mutagenesis (McKee and Lawrence 1979) as well as UV mutagenesis, and are not affected in mitotic recombination after x-rays (Saeki *et al.* 1981). This suggests that a common error-prone repair pathway exists for both UV-light-induced damage and x-ray damage, which does not involve recombination. This is borne out by the fact that mutants in this group, notably *rad6* and *rad18*, show a greater x-ray sensitivity than wild type, even in haploid cells in the G1 phase of the mitotic cycle. Hence the wild-type alleles of these loci must mediate a type of x-ray resistance that can function at least partially in the absence of homologous DNA. It is more difficult to explain the lack of resistance to x-rays in G2 cells in these mutants, since G2 repair of ionizing radiation damage is believed to involve recombination, yet *rad6* and *rad18* mutants, for example, have x-ray-induced recombination rates similar to those of wild type (Saeki *et al.* 1981).

A second well-defined group of x-ray-sensitive mutants, which can be called the *RAD52* group, consists of the three loci, *RAD51*, *RAD52*, and *RAD54*. Mutants in these three loci are very highly sensitive and have a striking radiation biology. The mutant *rad54-3* represents the clearest example of this: Its x-ray sensitivity is temperature dependent, and at the restrictive temperature (36 °C), survival curves both for haploids and homozygous diploids are exponential. The diploid survival curve, however, has twice the slope of the haploid curve, instead of being less sensitive as in wild-type and other classes of *rad* mutant (Game and Mortimer, unpublished data). This suggests not only the abolition of diploid and G2 haploid repair but that the lethality is due to dominant lesions, i.e., that an unrepaired lesion in only one copy of the genome in diploids will prevent cell division. Such dominant lethality has been confirmed by Ho and Mortimer (1973) for another mutant in the same group, *rad52-1*. They made use of the fact that haploid cells given a dose of x-rays within a certain range become unable to divide but are still able to mate with unirradiated cells of the opposite mating type. When this is done with wild-type *(RAD+)* cells, the zygotes formed in this way are viable, since the unirradiated nucleus is able to "rescue" the diploid and permit division. However, when both parents carry the *rad52-1* mutation, such zygotes are not rescued but show a survival at any given dose similar to that of the irradiated but unmated haploid control. Thus *rad52-1* strains are unable to repair dominant lethal damage. The *rad52-1* mutation confers somewhat less sensitivity than *rad54-3,* and the diploid has a slight shoulder, making it marginally more resistant than the haploid, but these differences are probably due to slight leakiness of the allele, since in other respects the phenotype resembles *rad54-3*. Other features of the phenotype of mutants in this group include both an almost complete abolition of mitotic recombination (Saeki *et al.* 1981) and a sharp increase, at least in the case of *rad51-3* (Morrison and Hastings 1979) and *rad52-1* (Mortimer *et al.* 1981), of spontaneous and radiation-induced chromosome loss, leading to monosomy in diploids. Mortimer *et al.* (1981) have suggested that such chromosome loss may arise because unrepaired double-strand breaks in DNA may lead to loss of telomeres, resulting in instability of the chromosomes concerned; or alternatively, recombinational repair may be affected so that unresolved exchanges between chromatids lead to aberrant chromosome segregations at mitosis. Some other x-ray-sensitive mutants, including *rad51-1, rad55-3,* and *rad57-1,* also show elevated frequencies of x-ray-induced chromosome loss, though probably to a lesser extent than *rad52-1* (D. Schild, personal communication).

Mutants in the *RAD52* group also show a sharp reduction, or (in the most sensitive mutants) a complete abolition, of the x-ray resistance normally associated with mating-type heterozygosity. For example, *MATα/MATa rad54-3/rad54-3* diploids at 36 °C show an x-ray survival curve identical to that for *MATα/MATα rad54-3/rad54-3* diploids (Game and Mortimer; see Fig. 4.1), whereas *MATα/MATα RAD* diploids (Mortimer 1958; Laskowski 1960) as well as those of mutants in other groups such as *rad50* (Game and Mortimer unpublished data), show a significantly greater sensitivity than *MATα/MATa* diploids. Hence the additional resistance conferred by mating-type heterozygosity is dependent on the *RAD52* group gene products, which is consistent with the hypothesis that the higher levels of mitotic recombination observed in wild-type *MATα/MATa* diploids compared with wild-type *MATα/MATα* or *MATa/MATa* diploids (Friis and Roman 1968) reflect a mating-type dependent recombinational repair function.

An interesting recent observation that also distinguishes the *RAD52* group from other yeast *rad* mutants concerns mating-type switching in homothallic strains. Malone and Esposito (1980) have reported that *rad52-1* homothallic strains are defective in mating-type switching, and similar observations obtain for *rad51* and *rad54* mutants (Game, unpublished data). Homothallic strains carrying those mutations are either inviable or in some genetic backgrounds are viable but fail to switch. Other x-ray-sensitive mutants, such as *rad50* and *rad57* alleles, are not defective in switching (Game, unpublished data), although a reduced frequency of switching has not been ruled out. The fact that those mutants that are defective in switching are specifically those with the strongest block in mitotic recombination suggests that normal switching utilizes enzymes involved in mitotic recombination.

Double mutant combinations within the *RAD52* group exhibit epistatic interactions (McKee and Lawrence, 1980) with respect to x-ray sensitivity, but are additive or synergistic in combination with mutants of the error-prone repair group, in particular *rad6* and *rad18* (Brendel and Haynes 1973; McKee and Lawrence 1980). With respect to UV-light sensitivity, *rad51-1* and *rad52-1* interact synergistically with *rad18* (Brendel and Haynes 1973; Cox and Game 1974) and *rad51-1* also interacts synergistically with at least some mutants in the excision repair group (Game and Cox 1973). There is no UV-light data for double mutants involving *rad54*. Brendel and Haynes (1973) found only a weak synergism between *rad52-1* and *rad1* and *rad2* mutants. This is surprising, since *rad52-1* blocks UV-induced mitotic recombination, and it therefore seems likely that it is involved in recombinational repair of UV-light-induced damage. Possibly the relative lack of synergism with *rad1* and *rad2* is explainable if both *rad52-1* and these excision-defective mutations block relatively late steps in their respective pathways, so that their wild-type gene products effectively do not compete for the same substrate but act on different biochemical intermediates, leading to an additive rather than synergistic interaction. Similar logic has been used to explain the lack of synergism between *rad2* and *rad50* (Cox and Game 1974). However, unless additional assumptions are made, such a model predicts that at least one of the single mutants should be much more sensitive to UV-light than the single mutants that block earlier steps in the pathway and do show synergistic interactions. This is not the case for *rad2* or *rad52*. This prediction follows from the expectation that all strains blocked in both pathways should be similar in sensitivity, so that additive interactions between mutants in late steps should lead to a sensitivity equivalent to that arising from synergistic interactions between early steps. The fact that *rad2 rad52* and *rad2 rad50* double mutants do not show this degree of sensitivity may be explained if various types of DNA damage in different intermediate stages of repair affect cells differently, but detailed pathway analysis may be of limited use here without more extensive biochemical data concerning the *RAD* gene products and their role in repair.

Some biochemical data about repair of x-ray-induced damage by the *RAD52* group mutants is already available. Ho (1975) and Resnick and Martin (1976) used neutral sucrose gradient techniques to demonstrate that *rad52-1* is defective in the repair of double-stranded DNA breaks induced by ionizing radiation. Subsequently, Budd and Mortimer (1982) showed that *rad54-3* mutant strains are defective in double-strand break repair at 36 °C, but not at 23 °C. Strains carrying the *rad54-3* allele are extremely x-ray sensitive at 36 °C (see Fig. 4.1), but only marginally sensitive at 23 °C (Game and Mortimer 1974; Ho and Mortimer 1975; Budd and Mortimer 1982) and are similarly temperature-conditional for spontaneous and UV-induced mitotic heteroallelic recombination (Game, unpublished data).

The remaining x-ray-sensitive mutants that have been given *RAD* locus numbers, but are probably not involved in error-prone repair, differ phenotypically from the *RAD52* group. They belong to the five loci *rad50, rad53, rad55, rad56,* and *rad57*. These are typically less sensitive to x-rays than *rad52* group mutants, and homozygous diploids are more resistant than the corresponding haploids. Double mutant combinations involving these mutants and *rad52-1* exhibit a sensitivity equivalent to that of the *rad52-1* single mutant (McKee and Lawrence 1980), but at least some double mutant combinations within these five loci (e.g., *rad50-1 rad57-1*) show a sensitivity greater than the corresponding single mutants, but no greater than that of *RAD52* group mutants (Saeki *et al.* 1981; Game and Mortimer, unpublished observations). Moreover, *rad50-1* diploids that are homozygous for mating type show a greater sensitivity than *MATa/MATα rad50-1* diploids. These interactions and phenotypic characteristics suggest that the *RAD52* group loci mediate common steps in all the repair functions mediated by these five loci but that the latter themselves mediate different subclasses of *RAD52*-dependent repair. Mitotic recombination data are consistent with this, as these mutants show UV- and x-ray-induced mitotic recombination rates lower than those of wild type, but higher than those of the *RAD52* group mutants (Saeki *et al.* 1981). Mutant alleles that confer temperature-sensitive x-ray sensitivity are known for at least three of these loci, and surprisingly in each case they are cold-sensitive rather than heat sensitive. Such cold-sensitive alleles have not been found among *RAD52* group mutants, which may again serve to distinguish between these two groups of loci.

Mutants in these five loci are only marginally sensitive to UV-light, and except for *rad50* there is little double-mutant data concerning their sensitivity in combinations with other UV-light-sensitive mutants. *Rad50-1* interacts synergistically with the excision-defective mutant *rad3-2* (but not with *rad2-1*, see earlier) and is in the same UV-epistasis group as *rad51-1*, since the *rad3-2 rad50-1 rad51-1* triple mutant has the same sensitivity as the *rad3-2 rad51-1* double mutant (Game and Cox 1973). However, it may (as seems to be the case for x-ray damage) only block part of the *RAD51*-dependent recombinational UV-repair system, since homozygous *rad50-1* diploids do show some UV-induced recombination (Hunnable and Cox 1971) and furthermore, *rad3-2 rad50-1* double mutants are less UV sensitive than *rad3-2 rad51-1* double mutant strains (Game and Cox 1973).

There is little biochemical data concerning the functions of the *RAD50* locus, or others of this type, although *rad50-1* does repair double-strand breaks (Budd and Mortimer, personal communication). This is to be expected if unrepaired double-strand breaks do indeed constitute dominant lethal events, since *rad50-1* strains resemble wild type in that diploids are less sensitive than haploids. Thus, although the *RAD50* locus is of considerable interest in meiotic studies (see the following discussion), the mechanism by which it and similar loci contribute to the repair of x-ray damage remains obscure.

## The Role of *RAD* Genes in Meiosis

Mutations in several of the *RAD* genes discussed earlier block the successful completion of meiosis, suggesting that the repair of radiation-damaged DNA has steps in common with aspects of normal DNA metabolism in meiosis. These *rad* alleles represent a class of meiotic mutants that are easy to isolate and monitor in haploid strains, using the phenotype of radiation sensitivity, but that are likely to be blocked in fundamental aspects of meiosis, since it is probable that they are involved in DNA metabolism rather than in some unrelated feature of sporulation in yeast. Hence they are of considerable interest in the study of meiosis and are the subject of current research in this area.

The *rad* mutants with a major defect in meiosis have the common property of conferring x-ray sensitivity on vegetative cells. Mutants belonging to the excision-repair UV-light epistasis group, which are only marginally sensitive to x-rays, appear to have a normal meiosis (Snow 1968; Dowling 1981), at least when the other UV-repair pathways are open. At least one mutant, *rad6-1*, in the error-prone UV-light repair pathway, is defective in meiosis, as are several mutants blocked in recombinational repair of UV-light damage, but all of these mutants are also highly sensitive to x-rays. Thus there appears to be a strong overlap between gene products needed for meiotic DNA metabolism and those needed for the repair of x-ray damage, although some x-ray-sensitive mutants, including many of those in the error-prone repair pathway, which are sensitive to UV light as well as x-rays, show apparently normal meiosis.

It is tempting to speculate that this apparent connection between x-ray resistance and normal meiotic DNA metabolism arises from a requirement for effective recombinational processes in both cases, and there is some evidence to support this view. The likelihood that x-ray repair processes operate via a recombinational mechanism has been discussed earlier, and recent evidence from meiotic studies suggests that at least several of the meiotic-lethal *rad* mutants are defective in some aspect of meiotic recombination. At a molecular level, it seems likely (see preceding discussion) that DNA strand breaks are an important source of x-ray-induced lethality, and it is also clear that DNA strands must be broken during the course of meiotic recombination. This may explain the requirement for common gene products, although it must be pointed out that meiotic recombination probably initially involves single-strand breaks, whereas it is specifically double-strand break repair that has been shown to be defective after x-rays in mutants such as *rad52-1*. Possibly it is the overall recombinational process that is defective in these mutants, so that any strand breaks requiring recombination for their resolution remain unrepaired. There is evidence that *rad52-1*, which fails to repair double strand breaks after x-rays, accumulates single-strand breaks (or gaps) during meiosis. These breaks have been detected in *rad52-1* by Resnick *et al.* (1981) using alkaline sucrose sedimentation techniques in a study of molecular weight changes at timed intervals after the start of meiosis in a relatively synchronous culture. The breaks, which do not accumulate in wild-type cells, occur in both parental and daughter DNA strands at approximately the same time as premeiotic DNA synthesis and are dependent on DNA synthesis, since they do not occur when this is blocked by hydroxyurea. They may represent the strand breaks that in wild-type would initiate recombination. If this interpretation is correct, then one can postulate that such breaks are not readily seen in wild-type strains because they are rapidly resolved, and too few are present at any one time for easy detection, whereas in *rad52-1* they accumulate because subsequent stages of recombination are blocked.

In addition to the preceding molecular study, *rad* mutants have been characterized in meiosis using two other techniques. These are (1) interrupted meiosis experiments and (2) studies of *rad* mutant behavior in a genetic background containing the *spo13-1* mutation, which enables strains to bypass the meiosis I division, to produce two diploid spores instead of four haploid ones (Klapholz and Esposito 1980a,b). These experiments are described in the following sections.

## Interrupted Meiosis Experiments

In yeast, meiotic recombination can be experimentally distinguished from meiotic division by interrupting the meiotic process at various times after its initiation in liquid culture and plating the cells on vegetative medium (Sherman and Roman 1963). When this is done it is found that diploid recombinants can be recovered among cells plated at early times, during and after premeiotic DNA synthesis but prior to commitment to meiotic centromere disjunction and haploidization. The frequency of such recombinants increases with time while cells are left in the sporulation medium, and each cell appears to undergo commitment to meiotic levels of recombination prior to reductional division. One can use these findings to ask whether a meiosis-defective *rad* mutant is defective in the formation of such diploid recombinants at early times in meiosis or whether the meiotic defect occurs in the process of haploidization itself. One can also ask whether such *rad* mutants are able to undergo premeiotic DNA synthesis, by monitoring DNA content per cell in the culture. So far, five *rad* mutants have been studied in this way. Game *et al.* (1980) examined *rad6-1, rad50-1, rad52-1,* and *rad57-1.* They found that the *rad6-1* mutant underwent a round of DNA synthesis in sporulation medium but failed to progress past this point in meiosis, confirming earlier findings (Cox and Parry 1968) that it is sporulation defective. No recombinants were formed among cells removed from sporulation medium and no major decline in viability of the cells in sporulation medium was seen, up to the time that meiosis would have been completed in wild type. It thus seems clear that *rad6-1* does not undergo commitment to meiotic recombination, although it is not clear whether it is defective in some aspect of recombination itself or whether it simply fails to enter this stage of the meiotic process because of a block in some prior step.

The other three *rad* mutants studied by Game *et al.* (1980) do sporulate but form inviable spores (see also Game and Mortimer 1974; Prakash *et al.* 1980). In interrupted meiosis experiments, *rad50-1* and *rad52-1* mutants fail to form the live diploid meiotic recombinants that can be seen in wild-type strains. They also fail to form viable haploids but do undergo a round of premeiotic DNA synthesis. Unlike *rad6-1,* they show a sharp drop in viability early in the meiotic process, so that each cell appears to undergo, or become committed to, a lethal event during the period when a wild-type cell would be undergoing recombination or commitment to meiotic division. These findings suggest that *rad50-1* and *rad52-1* are defective in some stage of meiotic recombination, and that the lethality observed is a consequence of this, since some diploid recombinants would be expected at early times if the primary defects were in a subsequent stage of meiosis. Evidence that appears to confirm this has come from experiments with these mutants in a *spo13-1* background (see later).

*Rad57-1* differs from *rad50-1* and *rad52-1* in that it can form some live diploid recombinants which are seen among cells plated at early times from meiotic cultures (Game *et al.* 1981). However, after an initial rise, the frequency of recombinants per live cell declines sharply among cells replated at later times, so that each recombinant clearly commits to a lethal event soon after committing to recombination, if left in sporulation medium. A consequence of this is that a rise in recombinants in *rad57-1* strains is only seen in relatively synchronous meiotic cultures. This was not seen earlier, in less synchronous cultures (Game *et al.* 1980) because at any one time there were few cells at the appropriate stage of meiosis, and the recombinants did not accumulate with time because of lethality. From these data it is hard to determine what is the nature of the meiotic defect in *rad57-1,* but despite the transient appearance of recombinants in interrupted meiosis experiments, data from *spo13-1 rad57-1* strains (see later) suggest that the *RAD57* gene does control some aspect of meiotic recombination. A similar transient appearance of recombinants in interrupted meiosis experiments has been seen in a *rad51-3* strain by Morrison and Hastings (1979).

## Experiments with *rad* Mutant Strains Containing the *spo13-1* Allele

In 1980, Klapholz and Esposito (1980a,b) described two genes in yeast whose naturally occurring recessive alleles, named *spo12-1* and *spo13-1,* each confer on cells the ability to undergo meiotic recombination and sporulation without undergoing meiotic division. Diploid strains homozygous for these alleles sporulate to form two diploid spores that represent products of a single predominantly equational meiosis II type of division, and the centromere disjunction normally seen at the first meiotic division does not usually occur. However, recombination occurs during the formation of these two-spored asci at or near normal meiotic frequencies.

The significance of these findings for studying *rad* mutants in meiosis lies in the fact that meiotic recombination can be studied in the absence of haploidization. The separation between these processes by bypassing the first meiotic division has an advantage over interrupted meiosis experiments in that the meiotic process can be allowed to go to completion in the former case, since haploidization does not occur at any stage. There is evidence from classical genetic and cytogenetic studies (see Baker *et al.* 1976 for a review) that normal meiotic centromere disjunction in most species is dependent on the occurrence of recombination. Thus a possible reason for meiotic lethality in *rad* mutants that are defective in meiotic recombination would be their likely inability to segregate chromosomes correctly at the first meiotic division. Using this reasoning, Malone and Esposito (1981) studied meiosis in *rad50-1* and *rad52-1* mutant diploids that were homozygous for *spo13-1,* to see whether bypassing the requirement for centromere disjunction would "rescue" the putative recombination-defective spores. They found that strains containing *rad50-1* are indeed "rescued" by *spo13-1.* Thus *rad50-1 spo13-1* diploid strains form two viable diploid meiotic spores; furthermore, this sporulation occurs without recombination, so that the spores contain heterozygous genetic markers in the same configurations as the original diploid. These findings confirm that *rad50-1* is indeed defective in meiotic recombination, as well as providing strong support for the idea that the first meiotic division is dependent on recombination for its successful completion. In contrast, Malone and Esposito found that *rad52-1* is not rescued by *spo13-1* and forms inviable two-spored asci in this background. However, *spo13-1* strains that contain both *rad50-1* and *rad52-1* resemble *spo13-1 rad50-1* strains in showing viable spores. This result can be interpreted to indicate that the *RAD50* and *RAD52* gene products function sequentially in meiosis and that *RAD50* mediates a step prior to that of *RAD52* in this dependent sequence. Since *rad50-1* blocks recombination, it seems likely that *RAD52* mediates a later step in meiotic recombination and that meiosis is lethal in *spo13-1 rad52-1* strains because they cannot resolve or repair structures that are formed as intermediates in the recombination process. Thus *rad50-1* "rescues" *rad52-1* in a *spo13-1* background by preventing the cell from entering a potentially lethal recombination process (Malone and Esposito 1981; R. Malone, personal communication). The absence of recombinants in interrupted meiosis experiments with *rad50-1* and *rad52-1*, as well as the single-strand breaks that occur in meiotic DNA in *rad52-1*, is consistent with this interpretation. More recently, Game and Esposito (unpublished data) have found that two other x-ray-sensitive mutants—*rad51-3* and *rad57-1*—resemble *rad52-1* in that they are not rescued by *spo13-1*. *Rad50-1 rad57-1* double mutants are rescued by *spo13-1*, however, and show no recombination. The viability seen here suggests that the lethality normally seen in *rad57-1* arises from some aspect of recombination, which is somewhat surprising in view of the transient appearance of recombinants in interrupted meiosis experiments. *Rad57-1* strains may be defective in some aspect of the control or regulation of recombination, or possibly are unable to repair DNA damage that may accumulate during the recombinational phase of meiosis. Clearly, much more work remains to be done both with genetic and molecular techniques in the study of meiosis and meiotic recombination, but it seems certain that radiation-sensitive mutants will continue to play an important part in this research.

## References

Adler, H. I. (1966). The genetic control of radiation sensitivity in microorganisms. Adv. Radiation Biol. 2:167–192.

Averbeck, D. (1970). Isolierung and Characterisierung von drei UV-sensiblen *Saccharomyces*-Mutanten. Dissertation, Freie Universitat Berlin.

Averbeck, D., Laskowski, W., Eckardt, F., Lehmann-Brauns, E. (1970). Four radiation-sensitive mutants of *Saccharomyces*. Mol. Gen. Genet. 107:117–127.

Baker, B., Carpenter, A. T. C., Esposito, M. S., Esposito, R. E., Sandler, L. (1976). The genetic control of meiosis. Ann. Rev. Genet. 10:53–134.

Beams, C. A., Mortimer, R. K., Wolfe, R. G., Tobias, C. A. (1954). The relation of radioresistance in budding in *Saccharomyces cerevisiae*. Arch. Biochem. Biophys. 49:110–122.

Brendel, M., Haynes, R. H. (1973). Interactions among genes controlling sensitivity to radiation and alkylation in yeast. Mol. Gen. Genet. 125:197–216.

Brunborg, G., Williamson, D. H. (1978). The relevance of the nuclear division cycle to radiosensitivity in yeast. Mol. Gen. Genet. 162:277–286.

Brunborg, G., Resnick, M. A., Williamson, D. H. (1980). Cell-cycle-specific repair of DNA double-strand breaks in *Saccharomyces cerevisiae*. Radiation Res. 82:547–548.

Budd, M., Mortimer, R. K. (1982). Repair of double-strand breaks in a temperature conditional radiation-sensitive mutant of *Saccharomyces cerevisiae*. Mutation Res. 103:19–26.

Chanet, R., Williamson, D. H., Moustacchi, E. (1973). Cyclic variations in killing and "petite" mutagenesis induced by ultraviolet light in synchronized yeast strains. Biochem. Biophys. Acta (Amst.) 324:290–299.

Coulondre, C., Miller, J. H., Farnbaught, P. J., Gilbert, W. (1978). Molecular basis of base substitution hotspots in *Escherichia coli*. Nature (London) 274:775–780.

Cox, B. S., Game, J. C. (1974). Repair systems in *Saccharomyces*. Mutation Res. 26:257–264.

Cox, B. S., Parry, J. M. (1968). The isolation, genetics, and survival characteristics of ultraviolet light-sensitive mutants in yeast. Mutation Res. 6:37–55.

Dowling, E. D. (1981). Meiotic recombination and sporulation in repair-deficient strains of yeast. Ph.D. Thesis, University of California, Berkeley, California.

di Caprio L., Cox, B. S. (1981). DNA synthesis in UV-irradiated yeast. Mutat. Res. 82:69–85.

Eckardt, F., Kowalski, S., Laskowski, W. (1975). The effect of three *rad* genes on UV-induced mutation rates in haploid and diploid *Saccharomyces* cells. Molec. Gen. Genetics 136:261–272.

Elkind, M. M., Sutton, H. (1959). The relationship between division and X-ray sensitivity, ultraviolet sensitivity, and photoreactivation in yeast. Radiat. Res. 10:283–295.

Fabre, F. (1971) A UV-supersensitive mutant in the yeast *Schizosaccharomyces pombe:* Evidence for two repair pathways. Mol. Gen. Genet. 110:134–143.

Fath, W. W., Brendel, M., Laskowski, W., Lehmann-Brauns, E. (1974). Economizing DNA-specific labelling by deoxythymidine-5-monophosphate in *Saccharomyces cerevisiae.* Mol. Gen. Genet. 132:335–345.

Friis, J., Roman, H. (1968). The effect of the mating type alleles on intragenic recombination in yeast. Genetics 59:33–36.

Game, J. C. (1971). A study of radiation-sensitive mutants in yeast. D. Phil. Thesis, University of Oxford.

Game, J. C., Cox, B. S. (1971). Allelism tests of mutants affecting sensitivity to radiation in yeast and a proposed nomenclature. Mutation Res. 12:328–331.

Game, J. C., Cox, B. S. (1972). Epistatic interactions between four *rad* loci in yeast. Mutation Res. 16:353–362.

Game, J. C., Cox, B. S. (1973). Synergistic interactions between *rad* mutations in yeast. Mutation Res. 20:35–44.

Game, J. C., Mortimer, R. K. (1974). A genetic study of X-ray sensitive mutants in yeast. Mutation Res. 24:281–292.

Game, J. C., Zamb, T. J., Braun, R. J., Resnick, M., Roth, R. M. (1980). The role of radiation *(rad)* genes in meiotic recombination in yeast. Genetics 94:51–68.

Game, J. C., Schild, D., Esposito, R. Easton (1981). The meiotic phenotype of the *rad 57-1* mutation. Abst. in: The Molecular Biology of Yeast, meeting abstracts, Cold Spring Harbor, N.Y.: Cold Spring Harbor Laboratory.

Harm, H. (1976). Repair of UV-irradiated biological systems: Photoreactivation. In: Photochemistry and Photobiology of Nucleic Acids, Vol. II, edited by S. Y. Wang. New York: Academic Press, pp. 219–263.

Harm, W. (1963). Mutants of phage $T_4$ with increased sensitivity to ultraviolet. Virology 19:66–71.

Hartwell, L. H. (1971). Genetic control of the cell division cycle in yeast II. Genes controlling DNA replication and its initiation. J. Mol. Biol. 59:183–194.

Haseltine, W. A., Gordon, L. K., Lindan, C. P., Grafstrom, R., Shaper, N. L., Grossman, L. (1980). Cleavage of pyrimidine dimers in specific DNA sequences by a pyrimidine dimer DNA-glycoslyase of *M. luteus,* Nature (London) 285:634–641.

Hastings, P. J., Quah, S. K., Von Borstel, R. C. (1976). Spontaneous mutation by mutagenic repair of spontaneous lesions in DNA. Nature 264:719–722.

Haynes, R. H., Kunz, B. A. (1981). DNA repair and mutagenesis in yeast. In: The Molecular Biology of the Yeast *Saccharomyces,* edited by J. Strathern, E. Jones, J. Broach. Cold Spring Harbor, N.Y.: Cold Spring Harbor Lab., p. 371.

Ho, K. S. Y., Mortimer, R. K. (1973). Induction of dominant lethality by X-rays in a radiosensitive strain of yeast. Mutation Res. 20:45–51.

Ho, K. S. Y. (1975). Induction of DNA double-strand breaks by X-rays in a radiosensitive strain of the yeast *Saccharomyces cerevisiae*. Mutation Res. 30:327–334.

Ho, K., Mortimer, R. K. (1975). Two mutations which confer temperature-sensitive radiation sensitivity in the yeast *Saccharomyces cervisiae*. Mutation Res. 33:157–164.

Holliday, R. (1965). Radiation sensitive mutants of *Ustilago maydis*. Mutation Res. 2:557–559.

Hunnable, E., Cox, B. S. (1971). The genetic control of dark repair in yeast. Mutation Res. 13:297–309.

Jannsen, S., Witte, I., Megnet, R. (1973). Mutants for the specific labelling of DNA in *Saccharomyces cerevisiae*. Biochem. Biophys. Acta 299:681–685.

Johnston, L. H., Nasmyth, K. A. (1978). *Saccharomyces cerevisiae* cell cycle mutant *cdc9* is defective in DNA ligase. Nature 274:891–893.

Kern, R., Zimmerman, F. K. (1978). The influence of defects in excision and error prone repair on spontaneous and induced mitotic recombination and mutation in *Saccharomyces cervisiae*. Molec. Gen. Genet. 161:81–88.

Klapholz, S., Esposito, R. (1980a). Isolation of *spo12-1* and *spo13-1* from a natural variant of yeast that undergoes a single meiotic division. Genetics 96:567–588.

Klapholz, S., Esposito, R. (1980b). Recombination and chromosome segregation during the single division meiosis in *spo12-1* and *spo13-1* diploids. Genetics 96:589–611.

Koval'tsova, S. V., Zheleznyakova, N. Y., Zakharov, I. A. (1970). Repair and mutagenesis communication I. Influence of mutations of radiosensitivity on UV-induced mutagenesis in *Saccharomyces cerevisiae*. Genetika 6:78–85.

Lacassagne, A., Holweck, F. 1930 Sur la radiosensibilité de la levure *Saccharomyces ellipsoideus*. Compt. rend de la Soc. de Biol., Paris, 104:1221–1223.

Laskowski, W. 1960 Inaktivierungsversuche mit homozygoten hefestammen verschiedenen ploidiegrades. Z. Naturforsch 15b:495–506.

Latarjet, Raymond, Ephrussi, Boris (1949). Courbes de survie de levures haploides et diploids soumises aux rayons X. Comptes Rendus, Académie des Sciences, Paris, 229:306–308.

Lawrence, C. W. (1982). Mutagenesis in *Saccharomyces cerevisiae,* Adv. Genet. 21:173–254.

Lawrence, C. W., Christensen, R. (1976). UV-mutagenesis in radiation-sensitive strains of yeast. Genetics 82:207–232.

Lemontt, J. F. (1971). Mutants of yeast defective in mutation induced by ultraviolet light. Genetics 68:21–33.

Lemontt, J. F. (1980). Genetic and physiological factors affecting repair and mutagenesis in yeast. In: DNA Repair and Mutagenesis in Eukaryotes, edited by F. J. de Serres *et al.* New York: Plenum Press, p. 85.

Lippke, J. A., Gordon, L. K., Brash, D. E., Haseltine, W. A. (1981). Distribution of UV light-induced damage in a defined sequence of human DNA: Detection of alkali-sensitive lesions at pyrimidine nucleoside-cytidine sequences. Proc. Natl. Acad. Sci. USA 78:3388–3392.

Little, J. G., Haynes, R. H. (1979). Isolation and characterization of yeast mutants auxotrophic for 2-deoxythymidine 5-monophosphate. Molec. Gen. Genet. 168:141–151.

Malone, R., Esposito, R. (1980). The *RAD52* gene is required for homothallic interconversion of mating types and spontaneous mitotic recombination. Proc. Natl. Acad. Sci. USA 77:503–507.

Malone, R. E., Esposito, R. (1981). Recombinationless meiosis in yeast. Mol. Cell. Biol. 1:891–901.

McKee, R. H., Lawrence, C. W. (1979). Genetic analysis of $\gamma$-ray mutagenesis in yeast I. Reversion in radiation-sensitive strains. Genetics 93:361–373.

McKee, R. H., Lawrence, C. W. (1980). Genetic analysis of $\gamma$-ray mutagenesis in yeast III. Double mutant strains. Mutation Res. 70:37–48.

Morrison, D. P., Hastings, P. J. (1979). Characterization of the mutator mutation *mut5-1*. Molec. Gen. Genet. 175:57–65.

Mortimer, R. K. (1958). Radiobiological and genetic studies on a polyploid series (Haploid to Hexaploid) of *Saccharomyces cerevisiae.* Radiation Res. 9:312–326.

Mortimer, R. K., Contopoulou, R., Schild, D. (1981). Mitotic chromosome loss in a radiation-sensitive strain of the yeast *Saccharomyces cerevisiae,* Proc. Natl. Acad. Sci. USA 78:5778–5782.

Nakai, S., Matsumoto, S. (1967). Two types of radiation-sensitive mutants in yeast. Mutation Res. 4:129–136.

Parry, J. M., Cox, B. S. (1968). The effects of dark holding and photoreactivation on ultraviolet light-induced mitotic recombination and survival in yeast. Genet. Res. Camb. 12:187–198.

Perper, T. (1975). Cyclic X-ray resistance to lethal and non-lethal damage in *Saccharomyces cerevisiae.* Radiation Res. 63:97–118.

Prakash, L. (1975). Repair of pyrimidine dimers in nuclear and mitochondrial DNA of yeast irradiated with low doses of ultraviolet light. J. Mol. Biol. 98:781–795.

Prakash, L. (1977a). Repair of pyrimidine dimers in radiation-sensitive mutants *rad3 rad4 rad6* and *rad9* of *Saccharomyces cerevisiae.* Mutat. Res. 45:13–20.

Prakash, L. (1977b). Defective thymine dimer excision in radiation-sensitive mutants *rad10* and *rad16* of *Saccharomyces cerevisiae.* Mol. Gen. Genet. 152:125–128.

Prakash, L. (1981). Characterization of postreplication repair in *Saccharomyces cerevisiae* and effects of *rad6, rad18, rev3* and *rad52* mutations. Mol. Gen. Genet. 184:471–478.

Prakash, L., Prakash, S. (1977). Isolation and characterization of MMS-sensitive mutants of *Saccharomyces cerevisiae.* Genetics 86:33–55.

Prakash, L., Prakash, S. (1979). Three additional genes involved in pyrimidine dimer removal in *Saccharomyces cerevisiae: RAD7, RAD14* and *MMS19.* Molec. Gen. Genet. 176:351–359.

Prakash, L., Hinkle, D., Prakash, S. (1979). Decreased UV mutagenesis in *cdc8,* a DNA replication mutant of *Saccharomyces cerevisiae.* Molec. Gen. Genet. 172:249–258.

Prakash, S., Prakash, L., Burke, W., Montelone, B. A. (1980). Effects of the *RAD52* gene on recombination in *Saccharomyces cerevisiae.* Genetics 94:31–50.

Resnick, M. A. (1969). A photoreactivationless mutant of *Saccharomyces cerevisiae.* Photochem. Photobiol. 9:307–312.

Resnick, M. A., Martin, P. (1976). The repair of double-stranded breaks in the nuclear DNA of *Saccharomyces cerevisiae* and its genetic control. Molec. Gen. Genet. 143:119–129.

Resnick, M. A., Setlow, J. K. (1972). Repair of pyrimidine dimer damage induced in yeast by ultraviolet light. J. Bacteriol. 109:979–986.

Resnick, M. A., Kasimos, J. N., Game, J. C., Braun, R. J., Roth, R. M. (1981). Changes in DNA during meiosis in a repair-deficient mutant *(rad52)* of yeast. Science 212:543–545.

Reynolds, R. J. (1978). Removal of pyrimidine dimers from *Saccharomyces cerevisiae* nuclear DNA udner non-growth conditions as detected by a sensitive enzymatic assay. Mutat. Res. 50:43–56.

Reynolds, R. J., Friedberg, E. C. (1981). Molecular mechanisms of pyrimidine dimer excision in *Saccharomyces cerevisiae:* Incision of ultraviolet-irradiated deoxyribonucleic acid in vivo. J. Bacteriol. 146:692–704.

Reynolds, R. J., Love, J. D., Friedberg, E. C. (1981). Molecular mechanisms of pyrimidine dimer excision on *Saccharomyces cerevisiae:* Excision of dimers in cell extracts. J. Bacteriol. 147:705–708.

Rupp, W. D., Howard-Flanders, P. (1968). Discontinuities in the DNA synthesized in an excision-defective strain of *Escherichia coli* following ultraviolet irradiation. J. Mol. Biol. 31, 291–304.

Saeki, T., Machida, I., Nakai, S. (1981). Genetic control of diploid recovery after $\gamma$-irradiation in the yeast *Saccharomyces cerevisiae.* Mutation Res. 73:251–265.

Saracheck, A. (1954). A comparative study of the retardation of budding and cellular inactivation by ultraviolet radiation in polyploid *Saccharomyces* with special reference to photoreactivation. Cytologia 19:77–85.

Setlow, J. K., Boling, M. E., Bollum, F. J. (1965). The chemical nature of photoreactivable lesions in DNA. Proc. Natl. Acad. Sci. USA 53:1430–1436.

Sherman, F., Roman, H. (1963). Evidence for two types of allelic recombination in yeast. Genetics 48:255–261.

Snow, R. (1967). Mutants of yeast sensitive to ultraviolet light. J. Bacteriol. 94:571–575.

Snow, R. (1968). Recombination in ultraviolet-sensitive strains of *Saccharomyces cerevisiae.* Mutation Res. 6:409–418.

Unrau, P., Wheatcroft, R., Cox, B. S. (1971). The excision of pyrimidine dimers from DNA of ultraviolet irradiated yeast. Molec. Gen. Genet. 354:359–362.

Unrau, P., Wheatcroft, R., Cox, B., Olive, T. (1973). The formation of pyrimidine dimers in the DNA of fungi and bacteria. Biochem. Biophys. Acta. 312:626–632.

Wickner, R. B. (1974). Mutants of *Saccharomyces cerevisiae* that incorporate deoxythimidine-5-monophosphate into deoxyribonucleic acid *in vivo.* J. Bacteriol. 117:252–260.

Wilcox, D. R., Prakash, Louise (1981). Incision and post incision steps of pyrimidine dimer removal in excision-defective mutants of *Saccharomyces cerevisiae.* J. Bacteriol. 148:618–623.

Zakharov, I. A., Kozhina, T. N. (1967). A yeast mutant hypersensitive to ultraviolet rays. Dokl. Acad. Nauk. SSSR, 176:1417.

Zirkle, Raymond E., Tobias, Cornelius A. (1953). Effects of ploidy and linear energy transfer on radiobiological survival curves. Arch. Biochem. and Biophys. 47:282–306.

# 5

# Nucleic Acid Relatedness Among Yeasts

C. P. Kurtzman,* H. J. Phaff,** and S. A. Meyer‡

## Introduction

The yeasts are a phylogenetically diverse group of fungi whose teleomorphs are distributed between the Ascomycotina and the Basidiomycotina. The grouping is one of convenience, bound together by the characteristics of predominantly unicellular growth resulting from either budding or fission and a sexual stage unenclosed in a fruiting body. The criteria accepted for use in speciation have, over time, paralleled many other areas of microbiology, first relying on the morphology of vegetative and sexual stages, followed later by definition of taxa through biochemical and genetic tests, and finally, owing to a partial failure of the foregoing to produce definitive results, a reliance on nucleic acid sequence relatedness.

The first widespread use of physiological tests in yeast taxonomy as an adjunct to the already accepted morphological criteria may be found in the monumental works of Stelling-Dekker (1931), Lodder (1934), and Diddens and Lodder (1942). Here reliance was placed on ability to ferment glucose, galactose, sucrose, maltose, lactose, and raffinose followed by growth on certain of these same sugars as well as ethanol, and the nitrogen sources ammonium sulfate, asparagine, urea, peptone and nitrate. Although helpful at first, the resolution afforded by this small number of compounds soon reached its limits, and further differentiation of taxa could only result from expanding the spec-

*Northern Regional Research Center, Agricultural Research Service, U.S. Department of Agriculture, Peoria, Illinois 61604 U.S.A.

**Department of Food Science and Technology, University of California, Davis, California 95616 U.S.A.

‡Department of Biology, Georgia State University, Atlanta, Georgia 30303 U.S.A.

trum of potentially assimilatable compounds. Bedford (1942) and especially Wickerham and Burton (1948) and Wickerham (1951) proposed the use of about 40 carbon compounds for differentiation of taxa, a convention now widely accepted in yeast systematics (Lodder 1970).

The realization that a single gene could confer on a yeast the ability to ferment or assimilate a sugar (Winge and Roberts 1949; Lindegren and Lindegren 1949) raised the possibility that species might be inadvertently and artifically established simply because of the functionality of one or a few genes. Verification of species relationships through hybridization studies is an ideal way in which to define taxa, but lack of fertility does not preclude conspecificity because only a few genes may affect the ability to mate (Fell *et al.* 1973; Hicks and Herskovitz 1976). The possibility of whole genome comparisons through DNA reassociation has been realized during the past 15 years, and it is now practical to define species as well as to assess present taxonomic methods using this molecular technique. Consequently, relationships determined from nucleic acid studies will be used in the ensuing discussion as a reference for evaluating many current taxonomic schemes and methodologies.

## Nucleic Acid Isolation and Purification

The methodology for isolation and purification of yeast DNA has been dealt with in detail elsewhere, and the following account will serve only as a brief review to direct attention to some of the more pertinent references. For many yeast strains, DNA isolation is relatively straightforward. Cell wall breakage may be effected by mechanical or lytic means. Mechanical methods have the advantage of being relatively rapid, but unless cell breakage is carefully monitored, considerable shearing of the DNA may result. The Braun cell homogenizer (Bronwill Scientific, Inc.) is one of the more frequently used mechanical devices and causes breakage by its shaking of a buffered cell suspension in the presence of 0.45-mm glass beads (Bak and Stenderup 1969; Price *et al.* 1978; Kurtzman *et al.* 1980a). An inexpensive variation of this technique involves placing the buffered cell and bead mixture into a test tube and "vortexing" the mixture (Lipinski *et al.* 1976; Van Etten and Freer 1978). Less frequently reported methods include grinding acetone-dried cells in a ball mill (Martini and Phaff 1973) and grinding cells in liquid nitrogen with a mortar and pestle (Storck and Alexopoulos 1970; Mendonça-Hagler *et al.* 1974). Sonication is seldom used for cell breakage, because it is very inefficient for yeasts and because of the danger of extreme DNA shearing.

Lytic methods offer the possibility of isolating DNA with a higher molecular weight than might be obtained by mechanical cell disruption. Meyer and Phaff (1969) and Bicknell and Douglas (1970) used a modification of the Smith and Halvorson (1967) technique in which cells undergo autolysis in the presence of saline EDTA, 2-mercaptoethanol, and sodium lauryl sulfate. The occasional failure of this method has prompted the widespread use of mechanical disruption. The enzyme preparations Glusulase and Zymolyase may be used to prepare osmotically fragile spheroplasts by cell wall digestion. Following this step, high-molecular-weight DNA is isolated from the lysed spheroplasts. Cryer *et al.* (1975) have given details of these procedures. Zymolyase generally promotes more rapid cell wall dissolution than Glusulase, but neither enzyme preparation is particularly effective with basidiomycetes. Brady (1981), however, reported the relatively rapid dissolution of *Rhodotorula* cell walls by an enzyme preparation from culture filtrates of *Penicillium lilacinum.* Another success in this area was achieved by Rhodes and Kwon-Chung (1982), who produced protoplasts from blastospores of *Cryptococcus neoformans* with the multienzyme preparation mutanase.

Following rupture or lysis of the cells, DNA isolation and purification may be carried out by the methods of Marmur (1961) and Bernardi *et al.* (1970) as detailed by Price *et al.* (1978). Basically, this consists of making the cell suspension 1 *M* in sodium perchlorate and in sarcosate and emulsifying with an equal volume of chloroform–isoamyl alcohol (24:1, vol/vol). After incubation and phase separation by centrifugation, DNA is precipitated from the aqueous phase with cold ethanol. The DNA is treated with $\alpha$-amylase, pancreatic RNase, and pronase. Following enzyme removal, the DNA is again ethanol precipitated, spooled, redissolved, and further treated with $\alpha$-amylase, pancreatic RNase, and $T_1$ RNase. Additional purification can include hydroxylapatite fractionation and banding in a cesium chloride gradient.

Some yeasts have active DNases that escape denaturation during the usual extraction procedures and cause sufficient degradation of the DNA to prevent spooling. Several methods have produced varying degrees of success in the deactivation of nucleases and merit testing if problems are encountered. Diethyl pyrocarbonate used at 0.5–1.0% at the time of cell breakage gave encouraging results in studies of *Lipomyces* spp. (Holzschu and Phaff, unpublished data) and *Sporidiobolus* (Holzschu *et al.* 1981). Ehrenberg *et al.* (1976) discussed the interaction of diethyl pyrocarbonate with nucleic acids and emphasized some of the precautions to be taken. Kurtzman and co-workers (unpublished data) found that the spermine–spermidine–sucrose buffer of Morris as cited by Timberlake (1978) much more effectively inactivated nucleases in *Aspergillus* spp. and *Neurospora* spp. than the sucrose buffer frequently used in yeast studies (Price *et al.* 1978). Heating the cells to 60°C for 20 min was also reasonably effective in deactivating nulceases in *Aspergillus* and *Neurospora* (Kurtzman *et al.*, unpublished data). Higher temperatures, if they do not result in denaturation of the DNA, may be even more effective.

Methods for assessing DNA purity include chemical analyses as well as spectrophotometric measurements and have been summarized by Price *et al.* (1978) and Johnson (1981).

Techniques for isolation of RNA have been presented by Gillespie and Spiegelman (1965), Bicknell and Douglas (1970), and Johnson (1981). These references also included procedures for RNA–DNA reassociation experiments. Although cellular RNA concentration is manyfold higher than that of DNA, the nearly ubiquitous occurrence of RNases and their marked stability make isolation of undegraded RNA more difficult than the isolation of high-molecular-weight DNA.

## DNA Base Composition

### Methodology

DNA base composition, expressed as molar percentages of guanine + cytosine (G + C), may be determined from thermal denaturation profiles (melts), buoyant density in cesium salt gradients generated by ultracentrifugation, chemical analysis, absorbance ratios, or high-pressure liquid chromatography of nucleotides or free bases (Johnson 1981). Most investigators use thermal denaturation (Marmur and Doty 1962), but Bak (1973) showed that results obtained by this method can be greatly affected by sample impurities and/or minor DNA species and must therefore be interpreted with caution. Cesium chloride buoyant density determinations (Schildkraut et al. 1962) generally show the greatest accuracy, since they are unbiased by the presence of contaminating RNA, mitochondrial DNA, and other impurities, such as carbohydrates and proteins.

### Taxonomic Uses of G + C Values

The taxonomic uses of G + C values are mainly exclusionary, because the 600 or so yeast species range in G + C content from approximately 30–70 mol % and overlap between unrelated species is inevitable. For example, *Debaryomyces hansensii* and *Debaryomyces marama* have a G + C content of 39.0 and 39.1 mol%, respectively, yet they show only 8.4% base sequence complementarity (Price *et al.* 1978). Consequently, since similar G + C content does not necessarily mean a similar base sequence in the DNA and thus that strains belong to the same species, we must ask how dissimilar G + C values need be to state categorically that two strains represent different species.

The answer to this question depends to some extent on the method used for determination of base composition. When buoyant density is used, the span of G + C values among strains of a species is about 1% (Price *et al.* 1978; Kurtzman *et al.* 1980a,b), which is approximately coincident with the accuracy of the method, i.e., a standard deviation of 0.00–0.50 mol %. Thermal denaturation shows greater variation, and strains of a species may have a G + C range of as much as 2% (Meyer *et al.* 1978). Thus strains showing a difference in G + C contents in excess of 1.0–1.5% by buoyant density or 2.0–2.5% by thermal denaturation may be expected to be different species.

Representative G + C values are given in Table 5.1. These values illustrate the interesting fact that the G + C content of ascomycetous yeasts is about

**Table 5.1.** Nuclear DNA Base Composition of Selected Yeast Taxa

| Species | Mol % G + C |
|---|---|
| Ascomycetes | |
| *Arthroascus javanensis* (Klöcker) von Arx | 32.5[a] |
| *Citeromyces matritensis* (Santa María) Santa María | 45.5[b] |
| *Debaryomyces hansenii* (Zopf) Lodder et Kreger-van Rij | 38.6[c] |
| *D. marama* di Menna | 39.1[c] |
| *D. vanriji* (van der Walt et Tscheuschner) Abadie, Pignal et Jacob | 33.2[c] |
| *Dekkera bruxellensis* van der Walt | 38[b] |
| *Hanseniaspora valbyensis* Klöcker | 30.0[d] |
| *H. occidentalis* Smith | 34.9[d] |
| *H. vineae* van der Walt et Tscheuschner | 40.2[d] |
| *Hansenula ciferrii* Lodder | 32.2[e] |
| *H. alni* Phaff, Miller et Miranda | 41.9[f] |
| *H. henricii* Wickerham | 49.6[g] |
| *Issatchenkia scutulata* (Phaff, Miller et Miranda) Kurtzman, Smiley et Johnson | 34.0[h] |
| *I. orientalis* Kudriavzev | 40.3[h] |
| *Kluyveromyces polysporus* van der Walt | 35.3[i] |
| *K. fragilis* (Jörgensen) van der Walt | 41.6[j] |
| *Lipomyces anomalus* Babjeva et Gorin | 43.8[k] |
| *L. lipofer* Lodder et Kreger-van Rij | 49.3[k] |
| *Lodderomyces elongisporus* van der Walt | 39.7[l] |
| *Metschnikowia reukaufii* Pitt et Miller | 42.2[l] |
| *M. pulcherrima* Pitt et Miller | 48.3[l] |
| *Pachysolen tannophilus* Boidin et Adzet | 43[b] |
| *Pichia kluyveri* Bedford ex Kudriavzev | 30[b] |
| *P. haplophila* Shifrine et Phaff | 39.5[a] |
| *P. rhodanensis* (Ramirez et Boidin) Phaff | 51.9[m] |
| *Saccharomyces cerevisiae* Hansen | 39.9[c] |
| *S. florentinus* (Castelli) Lodder et Kreger-van Rij | 42.4[c] |
| *Saccharomycopsis fibuligera* (Lindner) Klöcker | 40.5[b] |
| *S. capsularis* Schiönning | 43[b] |
| *Schizosaccharomyces pombe* Lindner | 42[n] |
| *S. octosporus* Beijerinck | 42[o] |
| *Schwanniomyces occidentalis* Klöcker | 35.3[c] |
| *Torulaspora delbrueckii* Lindner | 43.8[c] |
| *T. globosa* (Klöcker) van der Walt et Johannsen | 47.5[c] |
| *Wickerhamia fluorescens* (Soneda) Soneda | 38[b] |
| *Wingea robertsii* (van der Walt) van der Walt | 42.7[c] |
| *Yarrowia lipolytica* (Wickerham, Kurtzman et Herman) van der Walt et von Arx | 49.6[o] |
| *Zygosaccharomyces bisporus* Naganishi | 44.0[c] |
| Basidiomycetes | |
| *Filobasidiella neoformans* Kwon-Chung | 51.5[p] |
| *Rhodosporidium toruloides* Banno | 60[q] |
| *Sporidiobolus salmonicolor* Fell et Tallman | 63[o] |

**Table 5.1.** Nuclear DNA Base Composition of Selected Yeast Taxa (*Continued*)

| Species | Mol % G + C |
|---|---|
| Imperfects | |
| *Candida albicans* (Robin) Berkhout | 35.5[b] |
| *C. utilis* (Henneberg) Lodder et Kreger-van Rij | 45.8[r] |
| *C. zeylanoides* (Cast.) Langeron et Guerra | 57.6[s] |
| *Phaffia rhodozyma* Miller, Yoneyama et Soneda | 48.3[t] |
| *Rhodotorula minuta* (Saito) Harrison | 53[p] |
| *R. glutinis* (Fres.) Harrison | 67[b] |
| *Sterigmatomyces penicillatus* Rodrigues de Miranda | 49.5[u] |
| *S. halophilus* Fell | 56.2[u] |
| *Trichosporon pullulans* (Lindner) Diddens et Lodder | 57[b] |
| *T. cutaneum* (De Beurm, Gougerot et Vaucher) Ota | 62[b] |
| *Trigonopsis variabilis* Schachner | 47[b] |

[a]Nakase and Komagata (1968b).
[b]Nakase and Komagata (1968a) [the values reported by Nakase and Komagata (1968a,b), have been increased by 3% as suggested by Meyer and Phaff (1970)].
[c]Price *et al.* (1978).
[d]Meyer *et al.* (1978).
[e]Nakase and Komagata (1971a).
[f]Phaff *et al.* (1979).
[g]Fuson *et al.* (1979).
[h]Kurtzman *et al.* (1980b).
[i]Presley and Phaff, unpublished data.
[j]Bak and Stenderup (1969).
[k]Phaff and Holzchu, unpublished data.
[l]Meyer and Phaff (1969).
[m]Kurtzman *et al.* (1980a).
[n]Rost and Venner (1964).
[o]Storck (1966).
[p]Storck *et al.* (1969).
[q]Nakase and Komagata (1971b).
[r]Kurtzman *et al.* (1979).
[s]Stenderup and Bak (1968).
[t]Miller *et al.* (1976).
[u]Kurtzman *et al.*, unpublished data.

30–50%, whereas that of basidiomycetous yeasts is approximately 50–70%. Except for the narrow range of 48–52%, where some overlap occurs, the taxonomic class of imperfects can be reliably determined from their base composition. For species in the doubtful range, class determination may be made by examining cell wall structure under the transmission electron microscope (Kreger-van Rij and Veenhuis 1971) or through use of the diazonium blue B test (van der Walt and Hopsu-Havu 1976; Hagler and Ahearn 1981).

The range of G + C contents among species within a genus is quite often

10% or less as found in *Debaryomyces, Hanseniaspora, Issatchenkia, Kluyveromyces, Metschnikowia,* and several other genera (Table 5.1). By contrast, species in *Pichia* differ by as much as 22%, while the disparity in base composition among *Rhodotorula* spp. is 14%. It seems likely that *Pichia* is an amalgam of several genera, but this is not certain of *Rhodotorula.* It is also apparent from G + C values, as well as from other tests, that the genus *Candida* and its synonym *Torulopsis* (Yarrow and Meyer 1978) is comprised of anamorphs of both ascomycetes and basidiomycetes.

## DNA Relatedness

### Methodology

The methods for assessing DNA relatedness vary, but short of actual sequencing rely on measuring the extent and stability of renatured DNA strands from the test pair, i.e., the fidelity of complementary base pairing. Depending on the method, the DNA may or may not need to be labeled with radioisotopes. Because of technical difficulties, heavy isotope labels are seldom used. An interesting variation of the DNA–DNA hybridization technique was employed by Bak and Stenderup (1969). Radioactive complementary RNA was synthesized *in vitro* with each test DNA serving as a template, and the relatedness between taxa was then assessed from hybridization of single-stranded DNA with the synthesized RNA.

*In vivo* labeling of DNA is frequently done using $^{14}C$, $^{3}H$, or $^{32}P$. Each isotope has its particular advantages and disadvantages. The DNA may also be labeled *in vitro* with $^{125}I$ (Commerford 1971; Mendonça-Hagler and Phaff 1975; Holzschu *et al.* 1979) or by nick translation (Kelley *et al.* 1970; Maniatis et al. 1975). DNA labeled with $^{125}I$ *in vitro* must be highly purified, which requires the extra purification step of banding in CsCl in a preparative ultracentrifuge. Following labeling, DNA reassociation reactions may be carried out using the membrane method, which involves immobilizing single strands of one of the DNA species onto nitrocellulose filters and allowing sheared single strands of the second DNA to react with the immobilized DNA (Denhardt 1966). Alternatively, both DNA species may be allowed to react in free solution, and the extent of reassociation assessed by percent binding of the resulting duplexes to hydroxylapatite (Brenner 1973) or resistance to hydrolysis by $S_1$ nuclease (Crosa *et al.* 1973). Price *et al.* (1978) and Johnson (1981) provide references and detailed procedures for these methods.

In addition to following DNA reassociation with isotopically labeled preparations, the reactions may also be monitored spectrophotometrically by measuring the kinetics of duplex formation. This technique has been refined to the point where results are comparable to data obtained using isotopes (Subirana and Doty 1966; De Ley *et al.* 1970; Seidler and Mandel 1971; Kurtzman *et al.* 1980a). Relatedness may also be estimated from the thermal stability of the renatured heterologous DNA as compared with renatured homologous DNA. This may be done spectrophotometrically by monitoring thermal melts or with labeled DNA by thermal elution from hydroxylapatite columns (Ullman and McCarthy 1973). Brenner *et al.* (1972) equated one degree of difference in thermal stability with 1% mismatch in base pairs. Lower thermal stability usually correlates well with lower DNA homology.

A number of factors affect DNA renaturation reactions and so may have a marked effect on the results obtained. Highly purified DNA preparations are essential. Contaminating protein or RNA can result in falsely high or low relatedness values, depending on whether the contaminants act as single-stranded or double-stranded DNA in the assay system (Brenner 1973).

Several classes of DNA occur in eukaryotes and their presence affects estimates of relatedness between taxa, since it is the intent of most studies to measure genomic similarity from complementarity of single-copy nuclear DNA sequences. Christiansen *et al.* (1971) reported up to 15% rapidly renaturing sequences in the nuclear DNA (nDNA) of yeasts, which may be interpreted as analogous to the repeated sequences found in higher eukaryotes (Britten and Kohne 1968). However, the presence of rapidly renaturing, plasmidlike 2-$\mu$m circular DNA in *Saccharomyces cerevisiae* (Bak *et al.* 1972) with a buoyant density similar to nDNA suggests the proportion of repeated sequences in nDNA to be smaller than first thought. Similar circular DNAs have been reported from *Hansenula wingei* and *Torulopsis glabrata* (O'Connor *et al.* 1975, 1976). Linear DNA plasmids have been discovered in *Kluyveromyces lactis* (Gunge *et al.* 1981). These plasmids, which are lighter than the densities of both nuclear and mitochondrial DNA, appear to be associated with the killer character of this yeast. The work of Timberlake (1978) indicates the genome of the mold *Aspergillus nidulans* to consist of approximately 97–98% unique sequences and 2–3% repeated sequences. The repeated sequences appear to code for ribosomal RNA.

Nuclear DNA represents the preponderance of cellular DNA in yeasts and has a kinetic complexity of 6.5–14 $\times$ $10^9$ daltons or about two to five times that found in *Escherichia coli* (Christiansen *et al.* 1971; Petes *et al.* 1973). Between 5 and 20% of the total DNA in *Saccharomyces cerevisiae* is mitochondrial in origin. It is arranged in closed circles with a G + C content noticeably lower than the nDNA (Hartwell 1974). Sedimentation studies (Blamire *et al.* 1972) and measurements of contour length (Hollenberg *et al.* 1970; Petes *et al.* 1973) suggest a molecular mass of 46 $\times$ $10^6$ d. The mitochondrial DNA (mtDNA) contour length of several other yeasts has been reported to be 25–50% that of *S. cerevisiae* (O'Connor *et al.* 1975, 1976).

If comparisons of taxa are to be made from single-copy nDNA, mtDNA, plasmidlike circular DNA and repeat sequence nDNA must be removed. Circular plasmidlike DNA as well as circular mtDNA may be relatively easily separated from nDNA by centrifugation in cesium chloride in the presence of ethidium bromide. Williamson and Fennell (1975) emphasized that resolution is poor should mtDNA break into linear fragments during isolation, a not uncommon happening. However, they point out that linear mtDNA may be effectively resolved in cesium chloride gradients when reacted with 4′,6-diamidino-2-phenylindole (DAPI), a fluorescent dye that preferentially binds with AT-rich DNA. Repeated sequences may be removed from nDNA by partial reassociation of sheared DNA fragments followed by fractionation with hydroxylapatite. Since the rate of reassociation is affected by concentration, multiple copies duplex first and can be removed because of their ability to bind to hydroxylapatite (Britten and Kohne 1968). Price *et al.* (1978) used this method to remove repeated sequences from yeast DNA for taxonomic studies. The fractionation undoubtedly also removes mtDNA and plasmidlike DNA because both occur as multiple copies and their small molecular weight would allow rapid reassociation when in significant concentration. The same effect may be obtained in spectrophotometric determinations of relatedness if the time between melting and cooling to the reassociation temperature ($T_{m-25}$) is not too rapid. Under their conditions, Kurtzman *et al.* (1980a) found a cooling period of 12 min to allow duplexing of most repetitive sequences as well as of mitochondrial DNA. Relatedness values of strains common to the work of Price *et al.* (1978) and of Kurtzman *et al.* (1980a) were essentially identical, showing that similar results can be obtained with rather different methodologies.

For most studies, DNA fragment size and the ionic strength of the incubation medium may be maintained at any constant value within a fairly wide range without significantly affecting the results (Brenner 1973). However, increasing ionic strength increases the reaction rate. Seidler *et al.* (1975) and Kurtzman *et al.* (1980a) used relatively high concentrations of SSC (SSC = standard saline citrate, 150 m*M* NaCl–15 m*M* sodium citrate, pH 7.0) when assessing DNA relatedness spectrophotometrically to complete the determination within a few hours. Thus, when $C_0t$ values [$C_0t$ = initial nucleotide concentration in moles per liter × time in seconds (Britten and Kohne 1968)] are used as a measure of the extent of reassociation, the ionic strength of the reaction mixture must be known if values obtained in different studies are to be compared. For example, the incubation mixture used by Kurtzman *et al.* (1980a) allowed reassociation to proceed nearly five times faster than occurred in the buffer system employed by Price *et al.* (1978). As noted in the definition of $C_0t$, initial nucleotide concentration also affects the reaction rate; consequently, a solution containing 100 μg of DNA per milliliter would reassociate twice as fast as one containing 50 μg of DNA per milliliter.

Optimal DNA reassociation occurs at 25–30°C below the midpoint of the melting curve ($T_m$), and this is dependent on base composition, since GC base pairs have greater thermal stability than do AT pairs (Marmur and Doty 1962; Marmur *et al.* 1963). An example of the effect of incubation temperature on DNA relatedness may be found in Brenner's (1973) work with enteric bacteria (Table 5.2). Under the experimental conditions 60°C was the optimum reassociation temperature. Pairs that are highly related at 60°C also show high relatedness at the more stringent temperatures of 66–67°C and 75°C. Pairs

**Table 5.2.** Effect of Incubation Temperature on DNA Reassociation[a]

| Species | Relative binding (%) at | | |
|---|---|---|---|
| | 60°C | 66–67°C | 75°C |
| *Escherichia coli/E. coli* | 100 | 100 | 100 |
| *E. coli/Shigella boydii* | 89 | 87 | 85 |
| *E. coli/Salmonella typhimurium* | 45 | 28 | 11 |
| *E. coli/Enterobacter aerogenes* | 37 | 23 | 8 |
| *E. coli/Enterobacter hafniae* | 21 | 11 | 4 |
| *E. coli/Proteus mirabilis* | 6 | | 1 |

[a]Data from Brenner (1973).

showing only moderate relatedness at 60°C show even less binding at higher temperatures, reflecting the greater mismatch of base pairs present. These studies point out the need for proper reannealing temperatures and also show that higher incubation temperatures allow an estimate of stability that indicates the extent of base-pair mismatch. Temperatures below $T_{m-25}$ would show even greater reassociation for sparingly related pairs. Consequently, we must view DNA relatedness values obtained from whole genome comparisons as an estimate of overall base sequence similarity under defined conditions rather than as a certain percentage of the DNA that has precisely the same base sequence.

## Correlation with Genetic Hybridization

On the basis of extensive comparisons between species of enteric bacteria, Brenner (1973) proposed that strains showing 70% or greater DNA relatedness should be considered members of the same species. Johnson (1973) compared numerous anaerobic bacteria and arrived at essentially the same guidelines. Price *et al.* (1978) examined species in four yeast genera and proposed that strains having 80% or greater DNA relatedness were conspecific, an opinion expressed earlier by Martini and Phaff (1973) from their studies in *Kluyveromyces*.

The preceding guidelines concerning relatedness were based on the observation that strains of species characterized by traditional means shared 70–100% of their DNA base sequences. Species boundaries may be prescribed on the basis of a number of criteria, but ability to hybridize and produce fertile successive generations would seem to be of cardinal importance. Dobzhansky (1976) noted that among sexually reproducing and outbreeding organisms, species can be defined as Mendelian populations or arrays of populations that are reproductively isolated from other population arrays. Since conjugation in bacteria is infrequent and may not involve the entire genome, genetic studies could not form the basis for interpretation of DNA complementarity studies among these prokaryotes. Many yeasts, on the other hand, have known sexual stages and with such species hybridization offers a means for correlating results from DNA studies with actual biological phenomena.

Van der Walt and Johannsen (1979) and Johannsen (1980) approached this problem through hybridization studies in the essentially homothallic genus *Kluyveromyces* where they were able to compare their results with DNA relatedness data obtained earlier by Martini (1973) and Martini and Phaff (1973). Van der Walt and Johannsen mated auxotrophic yeast strains, attained by *N*-methyl-*N'*-nitro-*N*-nitrosoguanidine mutagenesis, on media that supported growth of prototrophic recombinant clones. Utilizing this method they observed from various combinations prototrophic colonies at a frequency of $10^{-2}$–$10^{-8}$. Taxa with high DNA relatedness showed a higher frequency of recombination, but in a few cases taxa having low DNA complementarity also gave a considerable number of recombinants. Lack of crosses through successive generations of progeny has made interpretation of these data difficult. For example, mixes of *K. thermotolerans* and *K. marxianus,* which have significantly different nuclear DNA base composition values, 46.2 mol % G + C and 41.2 mol % G + C, respectively, and essentially no DNA homology (Presley and Phaff, unpublished data; Martini and Phaff 1973; Martini *et al.* 1972), yielded recombinant clones. The $F_1$ generation from a cross of *K. marxianus* and *K. thermotolerans* appears to be most similar to *K. marxianus* on the basis of DNA composition, i.e., $F_1$ 40.6 mol % G + C; *K. marxianus* 41.2 mol% G + C; *K. thermotolerans* 46.2 mol % G + C (Martini *et al.* 1972; H. Presley, unpublished data). These data suggest that the $F_1$ generation isolated from mixes of *K. marxianus* and *K. thermotolerans* by Johannsen and van der Walt (1978) represents an aneuploid yeast strain. Additionally, other laboratories have not been able to confirm the fertility of the $F_1$ generation isolated from this cross (S. R. Snow, personal communication) or in a similar cross between *K. fragilis* and *K. dobzhanskii* (Douglas *et al.* 1969). Consequently, interspecific fusion seems to have occurred, but there appears to be insufficient genetic complementarity to permit analysis of meiosis products.

Kurtzman and co-workers (1980a,b) were able to correlate genetic relatedness and DNA complementarity by working with heterothallic strains, thereby circumventing problems inherent in the *Kluyveromyces* study where auxotrophic mutants of homothallic strains were employed. In addition, Kurtzman *et al.* (1980a,b) isolated progeny from crosses by micromanipulation. In the first of these studies, the newly described *Pichia amylophila* and *P. mississippiensis* were shown to mate well, but ascospores were not viable. The two species showed only about 25% DNA relatedness, thus correlating lack of fertility with low DNA complementarity. However, the relationship between *Issatchenkia scutulata* var. *scutulata* and *I. scutulata* var. *exigua* was somewhat different. These two taxa also showed only about 25% DNA relatedness, but genetic crosses gave 3–6% viable ascospores (Table 5.3). Furthermore, crosses between these $F_1$ progeny gave 17% viability among $F_2$ progeny. Reciprocal crosses among the $F_1$ progeny were fertile, as were backcrosses to the parents.

**Table 5.3.** Ascospore Viability of Intra- and Intervarietal Crosses of *I. scutulata* var. *scutulata* NRRL Y-7663 and *I. scutulata* var. *exigua* NRRL Y-10,920, Taxa which Show Low DNA Relatedness[a,b]

| NRRL no. of parent[c] | No. of ascospores viable/total no. of ascospores isolated[d] | (%) |
|---|---|---|
| Y-7663 | 23/23 | (100) |
| Y-7663-2 × Y-7663-3 | 41/68 | (60) |
| Y-10,920 | 8/36 | (22) |
| Y-10,920-3 × Y-10,920-4 | 38/46 | (83) |
| Y-7663-2 × 10,920-3 | 5/85 | (6) |
| Y-7663-3 × Y-10,920-4 | 3/102 | (3) |
| (Y-7663-2 × Y-10,920-3)-1 × (Y-7663-3 × Y-10,920-4)-1 | 9/53 | (17) |

[a]Comparisons of DNA relatedness between two strains of each variety gave a range of 21–26% for intervarietal pairings and 92–100% for intravarietal pairings.
[b]Data from Kurtzman *et al.* (1980b).
[c]NRRL Y-7663 and NRRL Y-10,920 are sporogenous heterothallic diploids isolated from nature. The other strains are single-ascospore isolates derived from these diploids, as indicated. When grown alone, mating types showed neither conjugations nor ascospores.
[d]The frequency of conjugation and the number of apparently mature ascospores from intervarietal crosses were similar to the frequency and number from intravarietal crosses.

From this, it appears that all stocks had essentially homologous chromosomes and that progeny were neither amphidiploids nor aneuploids. Clearly then, the lower limit of DNA–DNA homology values suggesting species delimitation is not yet well defined, but base sequence divergence, as estimated from whole genome comparisons, may be as great as 75% before genetic exchange can no longer occur.

Data in Table 5.4 summarize the preceding findings and provide further comparisons of other taxa. In all of these studies, progressively less DNA relatedness parallels decreasing mating competence and fertility. Exceptions to this trend include gene changes that prevent mating between highly related strains and amphidiploidy and aneuploidy where moderate to high relatedness between strains might occur, but the disparate chromosome complements could cause infertility.

In addition to providing information on DNA divergence and speciation, the data in Table 5.4 demonstrate the limits of taxonomic resolution provided by DNA–DNA reassociation. Resolution appears to go no further than the detec-

**Table 5.4.** Correlation of Mating Reaction and DNA Complementarity Among Closely Related Heterothallic Ascomycetous and Basidiomycetous Yeasts

| Species | Mating reaction | Percent DNA relatedness |
|---|---|---|
| *Filobasidiella neoformans* × *Filobasidiella bacillispora* | Fair conjugation, 0–30% basidiospore viability ($F_1$ progeny; $F_2$ not determined) | 55–63[a] |
| *Issatchenkia scutulata* var. *scutulata* × *Issatchenkia scutulata* var. *exigua* | Good conjugation, ascospores viable: $F_1$ = 5%; $F_2$ = 17% | 21–26[b] |
| *Pichia amylophila* × *Pichia mississippiensis* | Good conjugation, ascospores not viable | 20–27[c] |
| *Hansenula bimundalis* × *Hansenula americana* | Poor conjugation, ascospores not produced | 21[d] |
| *Hansenula alni* × *Hansenula wingei* | Poor conjugation, ascospores not produced | 6[e] |
| *Issatchenkia orientalis* × *Issatchenkia occidentalis* | Infrequent conjugaiton, ascospores not produced | 3–8[b] |

[a]Aulakh *et al.* (1981).
[b]Kurtzman *et al.* (1980b).
[c]Kurtzman *et al.* (1980a).
[d]Kurtzman, unpublished data.
[e]Phaff *et al.* (1979); Fuson *et al.* (1979).

tion of sibling species. Beyond this, all species, regardless of extent of kinship, show essentially less than 10% complementarity. Perhaps this should not be surprising since it has already been pointed out that detectable DNA homology presupposes a relatively high degree of nucleotide complementarity (Wilson *et al.* 1977; Baumann and Baumann 1978).

## Comparisons of Taxa

Bak and Stenderup (1969) appear to have been the first to make extensive use of nucleic acid reassociation as a means for determining relatedness among yeasts. Their study included several phenotypically similar *Candida* species of medical interest. Specifically, their data showed sufficiently high base sequence complementarity to regard *C. albicans, C. stellatoidea,* and *C. claussenii* as conspecific (Table 5.5). *C. tropicalis* showed little relatedness to these three taxa.

Additional advances in the clarification of imperfect taxa through DNA studies were made by Meyer and Phaff (1972), who found high degrees of

**Table 5.5.** DNA Relatedness Among Medically Important Species of *Candida*[a,b]

| Species | Percent Relative Binding | | | |
|---|---|---|---|---|
| | *C. albicans* | *C. claussenii* | *C. stellatoidea* | *C. tropicalis* |
| *C. albicans* | 100 | 69 | 81 | 9 |
| *C. claussenii* | 56 | 100 | 79 | 10 |
| *C. stellatoidea* | 65 | 66 | 100 | 12 |
| *C. tropicalis* | 6 | 4 | 6 | 100 |

[a]Data from Bak and Stenderup (1969).
[b]Determined from the extent of complementarity between single-stranded nDNA and RNA synthesized from nDNA templates.

homology between *C. lusitaniae* and *C. obtusa* and between *C. salmonicola* and *C. sake;* by Meyer *et al.* (1975), who demonstrated *C. cloacae* and *C. subtropicalis* to be synonyms of *C. maltosa;* and by Kurtzman *et al.* (1980b), who showed *C. krusei* and *C. sorbosa* to represent, respectively, the imperfect states of *Issatchenkia orientalis* and *I. occidentalis.* Kurtzman *et al.* (1979) and Manachini (1979) demonstrated *C. utilis* to be the imperfect form of *Hansenula jadinii.* The work of Mendonça-Hagler and Phaff (1975) provided clear evidence that *Saccharomyces telluris, Torulopsis bovina, T. pintolopesii,* and *C. sloofii* are conspecific. This study also convincingly demonstrated that pseudomycelium formation provides no distinction between *Torulopsis* and *Candida.*

The genus *Kloeckera* has long been regarded as the imperfect counterpart of *Hanseniaspora,* but the small differences in sugar assimilation reactions prevented unambiguous assignment of species within the two genera and uncertain correlation between perfect and imperfect species. Meyer *et al.* (1978) established the presence of six species in *Hanseniaspora* and which of the members of *Kloeckera* corresponded to these perfect species (Fig. 5.1). Most of the *Hanseniaspora* species exhibited less than 20% DNA relatedness, but *H. osmophila* and *H. vineae* showed a mean relatedness of 47%. In view of more recent interpretations of DNA relatedness (Kurtzman *et al.* 1980b), this would suggest these two taxa to be sibling species or possibly varieties of the same species.

Further clarification of speciation in ascosporogenous yeasts resulted from the work of Price *et al.* (1978). Of particular interest were the rough-spored species of *Saccharomyces,* which are sometimes classified in *Torulaspora.* Nine of these species, which included the well-known *S. rosei* and *S. delbrueckii,* showed 98% or greater DNA relatedness and must be considered conspecific (Table 5.6). All are now classified as *S. delbrueckii.* Further evidence for the inadequacy of traditional taxonomic criteria was their finding that *S. microellipsodes* var. *osmophilus* was synonymous with *S. delbrueckii* but *S. microellipsodes* var. *microellipsodes* with low homology represented a separate species.

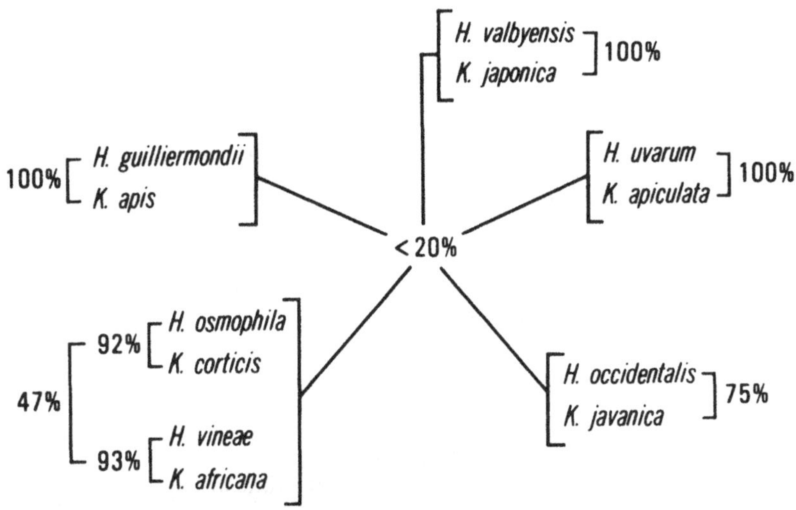


**Figure 5.1.** Schematic representation of DNA relatedness among species of *Hanseniaspora* and their imperfect counterparts in the genus *Kloeckera*. Of the ascosporogenous species, only *H. osmophila* and *H. vineae* show greater than 20% DNA base sequence complementarity. (Adapted from Meyer *et al.* 1978.)

**Table 5.6.** Extent of DNA Relatedness Between *Saccharomyces rosei* and Phenotypically Similar Species of *Saccharomyces*[a]

| Species | Percent DNA reassociation with *S. rosei* |
|---|---|
| *S. saitoanus* | 100.2 |
| *S. inconspicuus* | 99.9 |
| *D. nilssonii*[b] | 99.9 |
| *S. florenzani* | 99.6 |
| *S. fermentati* | 99.5 |
| *S. microellipsodes* var. *osmophilus* | 99.1 |
| *S. vafer* | 98.4 |
| *S. delbrueckii* | 98.0 |
| *S. pretoriensis* | 13.3 |
| *S. microellipsodes* var. *microellipsodes* | 10.7 |
| *S. kloeckerianus* | 10.5 |
| *S. cerevisiae* | 7.7 |
| *S. eupagycus* | 5.8 |
| *S. montanus* | 5.7 |
| *S. florentinus* | 5.5 |
| *S. bisporus* | 3.5 |

[a]Data from Price *et al.* (1978).
[b]Described as *Debaryomyces nilssonii*.

**Table 5.7.** Extent of DNA Relatedness Between *Debaryomyces polymorphus* and Phenotypically Similar Species with Roughened Ascospores[a]

| Species | Percent DNA reassociation with *D. polymorphus* |
|---|---|
| *D. cantarellii* | 96.4 |
| *D. phaffii* | 96.1 |
| *D. vanriji* | 23.9 |
| *D. formicarius* | 23.7 |
| *P. pseudopolymorpha*[b] | 21.0 |
| *D. castellii* | 19.8 |
| *D. hansenii* | 8.0 |
| *P. terricola*[b] | 1.9 |

[a]Data from Price *et al.* (1978).
[b]Assigned to the genius *Pichia* at the time of this study.

In an ultrastructural comparison of ascospores from species assigned to *Pichia* and *Debaryomyces,* Kurtzman and Smiley (1974) observed spores of *P. polymorpha* and *P. pseudopolymorpha* to be typical of those found in *Debaryomyces,* i.e., spheroidal and roughened. They suggested *P. polymorpha* and *D. cantarellii* to be conspecific because of this and other phenotypic similarities. This premise was confirmed by Price *et al.* (1978) (Table 5.7), and their DNA studies also showed *D. phaffii* to be a member of the taxon now designated *D. polymorphus* (Price and Phaff 1979).

## Macromolecules Other Than DNA as Indicators of Relatedness

As we have demonstrated earlier, complementarity studies involving whole genome DNA seem to resolve taxa no further than to the sibling-species level. Analysis of other macromolecules does, however, provide a means for determining more distant phylogenetic relationships. Some genes diverge quite slowly and that portion of genomic DNA coding for ribosomal RNA (rRNA) is highly conserved. Bicknell and Douglas (1970) appear to have been the first to examine distant relationships among the yeasts using rRNA–DNA homology. Their data indicate, as expected, that intrageneric relationships obtained by this method are not meaningful because of the high degree of conservation of the DNA sequences coding for rRNA (Table 5.8). However, because of this conservation, intergeneric distances can be resolved and it should be possible to construct an evolutionary hierarchy of the yeasts using rRNA–DNA homology. Additional rRNA–DNA complementarity studies among the yeasts were reported by Segal and Eylan (1974a,b, 1975). Comparative cataloging of oligonucleotides from $T_1$ ribonuclease digests of ribosomal RNA has provided information on phylogeny of the bacteria and might prove useful in defining relationships among the yeasts (Fox *et al.* 1977).

**Table 5.8.** Extent of DNA and Ribosomal RNA Relatedness Among Selected Yeasts[a]

| Species | Percent Relatedness with *Saccharomyces cerevisiae* | |
|---|---|---|
| | DNA | rRNA |
| *Saccharomyces cerevisiae* | 100 | 100 |
| *S. uvarum* | 40 | 100 |
| *S. kloeckerianus* | 20 | 97 |
| *S. rouxii* | 0 | 94 |
| *Kluyveromyces wickerhamii* | 0 | 87 |
| *Hansenula jadinii (Candida utilis)* | 0 | 77 |
| *Pichia pastoris* | 0 | 50 |

[a]Data from Bicknell and Douglas (1970).

Phylogenetic relationships of species within genera cannot, for the most part, be satisfactorily resolved using either DNA–DNA or rRNA–DNA homologies, because the first method is too specific and the second detects relatedness more distant than will usually be found within a genus. However, the DNA coding for certain enzymes such as glutamine synthetase and superoxide dismutase, although less conserved than that coding for rRNA, is still less divergent than is apparent from comparisons of whole genome DNA (Baumann and Baumann 1978; Baumann *et al.* 1980). By determining changes in the amino acid sequence of these enzymes through such techniques as quantitative microcomplement fixation, an estimate of intermediate relatedness may be found. Immunological studies of protein similarity are very sparse for yeasts, and the resultant information is highly dependent on the kinds of protein studied. Lachance and Phaff (1979) used exo-$\beta$-glucanases from species of *Kluyveromyces* but found that the enzyme was poorly conserved and immunological distances between most species were too great to be reliable for determining evolutionary relationships in that genus.

Electrophoresis of allozymes presents another means for estimating molecular diversity, and as with immunological studies, resolution depends on the extent of sequence conservation. The proportion of point mutations that are electrophoretically detectable is estimated at approximately 0.27 owing to the redundancy in the genetic code and the large proportion of amino acids that are electrically neutral (Baptist *et al.* 1971; Selander 1976; Holzschu 1981). The nearly universal occurrence of extensive protein polymorphism in natural populations has led some workers to believe that variation would prove physiologically important and therefore to be under selective control; others regard it as without phenotypic effect and thus selectively neutral (Selander 1976). Regardless, data derived by this technique do provide insight into evolutionary processes and taxonomy. Baptist and Kurtzman (1976) utilized comparative enzyme patterns to separate sexually active strains of *Cryptococcus laurentii* var. *laurentii* from nonreactive strains and from the varieties *magnus* and *flavescens*. Two new species, *C. magnus* and *C. heveanensis*, were proposed from the strains comprising the variety *magnus*. Holzschu (1981) recently studied the evolutionary relationships among some 400 strains of various cactophilic species of *Pichia*. His study of the banding patterns in starch gels of 14 metabolic enzymes allowed a determination of the genetic distances among the various yeast populations.

## Comparison of Relatedness from Nucleic Acid Studies with That Determined by Other Methodologies

In recent years it has become apparent that many of the criteria traditionally used for speciation of yeasts have their basis in only a very small portion of the genome. As is clear from genetic studies (Winge and Roberts 1949; Lindegren and Lindegren 1949; Kurtzman and Smiley 1976; Starmer *et al.* 1978), the assimilation and fermentation of carbon compounds, especially of di- and oligosaccharides, is frequently under the control of single structural genes. Consequently, carbon utilization tests should be regarded as a convenient laboratory means for separating taxa rather than as the foundation upon which they are defined. This is also true for the nitrate assimilation reaction. Ability to assimilate nitrate has been of cardinal importance in the definition of many species and genera since Stelling-Dekker (1931) first proposed its use over 50 years ago for the separation of *Hansenula* ($NO_3$+) and *Pichia* ($NO_3$−). Genetic studies show nitrate assimilation to be under the control of only a few genes (Marzluf 1981). Recently, Kurtzman *et al.* (unpublished data) found 68% DNA relatedness between *Hansenula mrakii* and *Pichia sargentensis*. They considered these two taxa to be conspecific and suggested that the species in the two genera be combined. Kurtzman *et al.* (unpublished data) also determined that the nitrate-positive *Sterigmatomyces halophilus* shared 100% DNA base sequence complementarity with the nitrate-negative *S. indicus*.

Nucleic acid studies have not proved entirely supportive of some conclusions drawn from numerical analysis, or from proton magnetic resonance spectra and serology of cell wall mannans. For example, using numerical analysis, Campbell (1973) proposed reduction of *Hansenula wingei*, *H. canadensis*, and both varieties of *H. bimundalis* to a single species. Fuson and co-workers (1979) showed from DNA sequence studies that *H. canadensis* and *H. wingei* are conspecific while this latter species and *H. bimundalis* show little relatedness. Kurtzman (unpublished data) demonstrated only 21% complementarity between the two varieties of *H. bimundalis*. This disparity of the two methods probably results from the relatively limited amount of phenotypic data available for numerical analysis, whereas DNA studies have the whole genome as a data base. This also appears to be the case for separations made using proton magnetic resonance, since Spencer and Gorin (1969) found the spectra of cell wall mannans from *H. beckii* and *H. canadensis* to be quite similar, but different from the pattern shared by *H. wingei* and the varieties of *H. bimundalis.* In further comparisons of this group, Tsuchiya *et al.* (1974) reported the cell surface antigens of *H. wingei, H. canadensis,* and *H. beckii* to be indistinguishable. The reason for this discrepancy is not at all clear, but Ballou (1974) has shown for *Saccharomyces* mannans that single gene changes can impart significant differences in immunological reaction. The difference between the two main PMR spectral types of mannan in *S. cerevisiae* was also found to be controlled by a single gene (Spencer *et al.* 1971).

Reliance on the presence or absence of pseudohyphae and true hyphae for taxonomic assignment has been questioned. Van Uden and Buckley (1970) as well as Yarrow and Meyer (1978) have pointed out that some taxa do not consistently form pseudohyphae. For primarily this reason, these latter authors transferred species of *Torulopsis* to *Candida,* the genus whose name has priority over *Torulopsis.* In 1979, Fuson *et al.* provided evidence that the presence or absence of true hyphal development is not necessarily a good trait for differentiation. This conclusion was based on the observation that the type strain of *Hansenula canadensis,* which is differentiated from *H. wingei* by its homothallism, ascospore number, and lack of true hyphae, has a higher degree of DNA homology (77.6%) with the type strain of *H. wingei* than do some strains identified as *H. wingei* (62.2%).

Differentiation of many ascosporogenous genera is based on ascospore shape and surface structure. An opportunity to test the soundness of this convention in the genera *Schwanniomyces, Saccharomyces, Debaryomyces,* and *Pichia* arose through a comparison of the DNA data of Price *et al.* (1978) with the scanning electron microscopy studies by Kurtzman and Kreger-van Rij (1976), Kurtzman and Smiley (1974, 1979), and Kurtzman *et al.* (1972, 1975). All four species of *Schwanniomyces* had high DNA relatedness (Price *et al.* 1978) and similar spore architecture (Kurtzman *et al.* 1972; Kreger-van Rij 1977). The least related, *S. persoonii* at 80% complementarity, showed fewer and less pronounced spore protuberances than the other three species, all four of which are now regarded as conspecific. Ascospore topography for many species in the *Torulaspora* section of *Saccharomyces* (van der Walt's [1970] group III *Saccharomyces* spp.) was essentially identical and also frequently indistinguishable from the majority of species in *Debaryomyces.* Although many species within their respective genera were found conspecific by DNA homology, others proved distinct despite spore similarity. The few species with recognizably different spores showed little DNA relatedness. Thus species with unlike spores can be expected to show little DNA complementarity, but no prediction can be made concerning species with similar spores. A few exceptions exist among other groups. Ascospores of *Pichia ohmeri* may be either spheroidal or hat-shaped, depending on mating type (Wickerham and Burton 1954; Fuson *et al.* 1980), and spore shape in *Saccharomycopsis lipolytica* is mating-type-dependent (Wickerham *et al.* 1969). Spore shapes among *Kluyveromyces* spp. with high degrees of DNA complementarity may be either spheroidal or kidney-shaped (Presley and Phaff, unpublished data).

Kreger-van Rij and Veenhuis (1973) demonstrated by transmission electron microscopy a basic difference in hyphal septa among ascomycetous yeasts. Three classes of septum ultrastructure were observed: a single central pore or closure line, plasmodesmata, and a dolipore configuration. All mycelial species of *Hansenula* and *Pichia* have the central pore, whereas plasmodesmata are common to all *Saccharomycopsis* species except *S. lipolytica.* Dolipores were found in species of *Ambrosiozyma.* Future comparisons of these species by molecular techniques that detect distant relatedness might show ultrastructural differences in septa to be of evolutionary significance.

An analysis of coenzyme Q or ubiquinone (Yamada and Kondo 1972, 1973; Yamada *et al.* 1973a,b,c, 1976a,b, 1977, 1980, 1981) has revealed that the number of isoprene units with side chains varies from 5 to 10 (Q-5 to Q-10) among yeast taxa. In many genera, all species have the same Q number, but heterogeneity exists in such genera as *Pichia, Hansenula,* and *Saccharomycopsis.* This may indicate that the atypical species are incorrectly assigned, and a comparison of these species by other molecular techniques could have interesting phylogenetic implications.

## Molecular Taxonomy as an Aid to Genetic Research

The discipline of systematics fulfulls our intellectual need to understand the relationships among organisms and has the practical advantage of providing a means for recognizing species of importance to many human activities. Numerous advances in basic and applied yeast genetics have stemmed from exploitation of the diversity found in natural isolates. The source of these isolates is frequently the microbiologist who is looking for additional strains to clarify taxonomic concepts. Once these strains have been properly identified, they are usually placed in culture collections for future reference and may in this way become available to the geneticist looking for new materials. Molecular techniques have served as powerful tools to ensure the most precise identification of strains for these potential genetic studies.

The fermentation industry has been highly dependent on strain improvement through breeding, and molecular taxonomy offers the potential for locating new stocks. An additional recent technique for strain improvement involves protoplast fusion, which may be used to add vigor to important strains or to combine useful properties not occurring together in nature (Ferenczy 1981). Protoplasts fused from closely related species are generally more stable than fusion products from distantly related taxa. Consequently, if the characteristics of interest may be obtained from similar species, a stable fusion is more likely. To date, molecular techniques have provided stronger guides for discerning which species are most closely related than have traditional morphological and physiological criteria.

## References


Aulakh, H. S., Straus, S. E., Kwon-Chung, K. J. (1981). Genetic relatedness of *Filobasidiella neoformans (Cryptococcus neoformans)* and *Filobasidiella bacillispora (Cryptococcus bacillisporus)* as determined by deoxyribonucleic acid base composition and sequence homology studies. Int. J. Syst. Bacteriol. 31:97–103.

Bak, A. L. (1973). DNA base composition in mycoplasma, bacteria and yeast. Curr. Top. Microbiol. Immunol. 61:89–149.

Bak, A. L., Stenderup, A. (1969). Deoxyribonucleic acid homology in yeasts. Genetic relatedness within the genus *Candida.* J. Gen. Microbiol. 59:21–30.

Bak, A. L., Christiansen, C., Christiansen, G. (1972). Circular, repetitive DNA in yeast. Biochim. Biophys. Acta 269:527–530.

Ballou, C. E. (1974). Some aspects of the structure, immunochemistry, and genetic control of yeast mannans. Adv. Enzymol. 40:239–270.

Baptist, J. N., Kurtzman, C. P. (1976). Comparative enzyme patterns in *Cryptococcus laurentii* and its taxonomic varieties. Mycologia 68:1195–1203.

Baptist, J. N., Shaw, C. R., Mandel, M. (1971). Comparative zone electrophoresis of enzymes of *Pseudomonas solanacearum* and *Pseudomonas cepacia.* J. Bacteriol. 108:799–803.

Baumann, L., Baumann, P. (1978). Studies of relationship among terrestrial *Pseudomonas, Alcaligenes,* and enterobacteria by an immunological comparison of glutamine synthetase. Arch. Microbiol. 119:25–30.

Baumann, L., Bang, S. S., Baumann, P. (1980). Study of relationship among species of *Vibrio, Photobacterium,* and terrestial enterobacteria by an immunological comparison of glutamine synthetase and superoxide dismutase. Curr. Microbiol. 4:133–138.

Bedford, C. L. (1942). A taxonomic study of the genus *Hansenula.* Mycologia 34:628–649.

Bernardi, G., Faures, M., Piperno, G., and Slonimski, P. P. (1970). Mitochondrial DNAs from respiratory-sufficient and cytoplasmic respiratory-deficient mutants of yeast. J. Mol. Biol. 48:23–43.

Bicknell, J. N., Douglas, H. C., (1970). Nucleic acid homologies among species of *Saccharomyces.* J. Bacteriol. 101:505–512.

Blamire, J., Cryer, D. R., Finkelstein, D. B., Marmur, J. (1972). Sedimentation properties of yeast nuclear and mitochondrial DNA. J. Mol. Biol. 67:11–24.

Brady, R. J. (1981). A technique for preparation of spheroplasts of *Rhodotorula.* Abst. Ann. Mtg., K65, p. 148. Am. Soc. Microbiol.

Brenner, D. J. (1973). Deoxyribonucleic acid reassociation in the taxonomy of enteric bacteria. Int. J. Syst. Bacteriol. 23:298–307.

Brenner, D. J., Steigerwalt, A. G., Fanning, G. R. (1972). Differentiation of *Enterobacter aerogenes* from klebsiellae by deoxyribonucleic acid reassociation. Int. J. Syst. Bacteriol. 22:193–200.

Britten, R. J., Kohne, D. E. (1968). Repeated sequences in DNA. Science 161:529–540.

Campbell, I. (1973). Numerical analysis of *Hansenula, Pichia* and related yeast genera. J. Gen. Microbiol. 77:427–441.

Christiansen, C., Bak, A. L., Stenderup, A., Christiansen, G. (1971). Repetitive DNA in yeast. Nature (London) New Biol. 231:176–177.

Commerford, S. L. (1971). Iodination of nucleic acids in vitro. Biochemistry 10:1993–1999.

Crosa, J. H., Brenner, D. J., Falkow, S. (1973). Use of a single-strand-specific nuclease for analysis of bacterial and plasmid deoxyribonucleic acid homo- and heteroduplexes. J. Bacteriol. 115:904–911.

Cryer, D. R., Eccleshall, R., Marmur, J. (1975). Isolation of yeast DNA, In Methods in Cell Biology, Vol. 12, Yeast Cells, edited by D. M. Prescott. New York: Academic Press, pp. 39–44.

De Ley, J., Cattoir, H., Reynaerts, A. (1970). The quantitative measurement of DNA hybridization from renaturation rates. Eur. J. Biochem. 12:133–142.

Denhardt, D. T. (1966). A membrane filter technique for the detection of complementary DNA. Biochem. Biophys. Res. Commun. 23:641–646.

Diddens, H. A., Lodder, J. (1942). Die anaskosporogenen Hefen, II Hälfte. Amsterdam: North-Holland.

Dobzhansky, T. (1976). Organismic and molecular aspects of species formation. In: Molecular Evolution, edited by F. J. Ayala. Sunderland, Mass. Sinauer Associates, pp. 95–105.

Douglas, H. C., Grindall, D. P., Talbott, H. (1969). Electrophoretic variants of phosphoglucomutase in *Saccharomyces* species. J. Bacteriol. 99:287–290.

Ehrenberg, L., Fedorcsak, I., Solymosy, F. (1976). Diethyl pyrocarbonate in nucleic acid research. Prog. Nucleic Acid Res. 16:189–262.

Fell, J. W., Hunter, I. L., Tallman, A. S. (1973). Marine basidiomycetous yeasts (*Rhodosporidium* spp. n.) with tetrapolar and multiple allelic bipolar mating systems. Can. J. Microbiol. 19:643–657.

Ferenczy, L. (1981). Microbial protoplast fusion. In: Genetics as a Tool in Microbiology, edited by S. W. Glover, D. A. Hopwood. Soc. for Gen. Microbiol. Symp. 31. Cambridge: Cambridge University Press, pp. 1–34.

Fox, G. E., Pechman, K. R., Woese, C. R. (1977). Comparative cataloging of 16S ribosomal ribonucleic acid: molecular approach to prokaryotic systematics. Int. J. Syst. Bacteriol. 27:42–57.

Fuson, G. B., Price, C. W., Phaff, H. J. (1979). Deoxyribonucleic acid sequence relatedness among some members of the yeast genus *Hansenula*. Int. J. Syst. Bacteriol. 29:64–69.

Fuson, G. B., Price, C. W., Phaff, H. J. (1980). Deoxyribonucleic acid base sequence relatedness among strains of *Pichia ohmeri* that produce dimorphic ascospores. Int. J. Syst. Bacteriol. 30:217–219.

Gillespie, D., Spiegelman, S. (1965). A quantitative assay for DNA-RNA hybrids with DNA immobilized on a membrane. J. Mol. Biol. 12:829–842.

Gunge, N., Tamuru, A., Ozawa, F., Sakaguchi, K. (1981). Isolation and characterization of linear deoxyribonucleic acid plasmids from *Kluyveromyces lactis* and the plasmid-associated killer character. J. Bacteriol. 145:382–390.

Hagler, A. N., Ahearn, D. G. (1981). Rapid diazonium blue B test to detect basidiomycetous yeasts. Int. J. Syst. Bacteriol. 31:204–208.

Hartwell, L. H. (1974). *Saccharomyces cerevisiae* cell cycle. Bacteriol. Rev. 38:164–198.

Hicks, J. B., Herskovitz, I. (1976). Interconversion of yeast mating types. I. Direct observations on the action of the homothallism (HO) gene. Genetics 83:245–258.

Hollenberg, C. P., Borst, P., van Brugger, E. F. J. (1970). Mitochondrial DNA. V. A 25-$\mu$ closed ciruclar duplex molecule in wild-type yeast mitochondria. Structure and kinetic complexity. Biochim. Biophys. Acta 209:1–15.

Holzschu, D. L. (1981). Molecular taxonomy and evolutionary relationships among cactophilic yeasts. Ph.D. Thesis. Univeristy of California, Davis.

Holzschu, D. L., Presley, H. L., Miranda, M., Phaff, H. J. (1979). Identification of *Candida lusitaniae* as an opportunistic yeast in humans. J. Clin. Microbiol. 10:202–205.

Holzschu, D. L., Tredick, J., Phaff, H. J. (1981). Validation of the yeast *Sporidiobolus ruinenii* based on its deoxyribonucleic acid relatedness to other species of the genus *Sporidiobolus*. Curr. Microbiol. 5:73–76.

Johannsen, E. (1980). Hybridization studies within the genus *Kluyveromyces* van der Walt emend. van der Walt. Antonie van Leeuwenhoek J. Microbiol. Serol. 46:177–189.

Johannsen, E., van der Walt, J. P. (1978). Interfertility as basis for the delimitation of *Kluyveromyces marxianus*. Arch. Microbiol. 118:45–48.

Johnson, J. L. (1973). Use of nucleic acid homologies in the taxonomy of anaerobic bacteria. Int. J. Syst. Bacteriol. 23:308–315.

Johnson, J. L. (1981). Genetic characterization. *In:* Manual of Methods for General Bacteriology, edited by P. Gerhardt. Washington, D.C.: American Society for Microbiology, pp. 450–472.

Kelly, R. B., Cozzrelli, N. R., Deutscher, M. P., Lehman, I. R., Kornberg, A. (1970). Enzymatic synthesis of deoxyribonucleic acids. J. Biol. Chem. 245:39–45.

Kreger-van Rij, N. J. W. (1977). Electron microscopy of sporulation in *Schwanniomyces alluvius*. Antonie van Leeuwenhoek J. Microbiol. Serol. 43:55–64.

Kreger-van Rij, N. J. W., Veenhuis, M. (1971). A comparative study of the cell wall structure of basidiomycetous and related yeasts. J. Gen. Microbiol. 68:87–95.

Kreger-van Rij, N. J. W., Veenhuis, M. (1973). Electron microscopy of septa in ascomycetous yeasts. Antonie van Leeuwenhoek J. Microbiol. Serol. 39:481–490.

Kurtzman, C. P., Kreger-van Rij, N. J. W. (1976). Ultrastructure of ascospores from *Debaryomyces melissophilus,* a new taxonomic combination. Mycologia 68:422–425.

Kurtzman, C. P., Smiley, M. J. (1974). A taxonomic re-evaluation of the round-spored species of *Pichia.* In: Proceedings of the Fourth International Symposium on Yeasts, Vienna, Austria, Part I, edited by H. Klaushofer, U. B. Sleytr. Vienna: Hochschülerschaft an der Hochschule für Bodenkultur, pp. 231–232.

Kurtzman, C. P., Smiley, M. J. (1976). Heterothallism in *Pichia kudriavzevii* and *Pichia terricola.* Antonie van Leeuwenhoek J. Microbiol. Serol. 42:355–363.

Kurtzman, C. P., Smiley, M. J. (1979). Taxonomy of *Pichia carsonii* and its synonyms *Pichia vini* and *P. vini* var. *melibiosi:* comparison by DNA reassociation. Mycologia 71:658–662.

Kurtzman, C. P., Smiley, M. J., Baker, F. L. (1972). Scanning electron microscopy of ascospores of *Schwanniomyces*. J. Bacteriol. 112:1380–1382.

Kurtzman, C. P., Smiley, M. J., Baker, F. L. (1975). Scanning electron microscopy of ascospores of *Debaryomyces* and *Saccharomyces*. Mycopathol. Mycol. Appl. 55:29–34.

Kurtzman, C. P., Johnson, C. J., Smiley, M. J. (1979). Determination of conspecificity of *Candida utilis* and *Hansenula jadinii* through DNA reassociation. Mycologia 71:844–847.

Kurtzman, C. P., Smiley, M. J., Johnson, C. J., Wickerham, L. J., Fuson, G. B. (1980a). Two new and closely related heterothallic species, *Pichia amylophilia* and *Pichia mississippiensis:* Characterization by hybridization and deoxyribonucleic acid reassociation. Int. J. Syst. Bacteriol. 30:208–216.

Kurtzman, C. P., Smiley, M. J., Johnson, C. J. (1980b). Emendation of the genus *Issatchenkia* Kudriavzev and comparison of species by deoxyribonucleic acid reassociation, mating reaction, and ascospore ultrastructure. Int. J. Syst. Bacteriol. 30:503–513.

Lachance, M. A., Phaff, H. J. (1979). Comparative study of molecular size and structure of exo-β-glucanases from *Kluyveromyces* and other yeast genera: evolutionary and taxonomic implications. Int. J. Syst. Bacteriol. 29:70–78.

Lindegren, C. C., Lindegren, G. (1949). Unusual gene-controlled combinations of carbohydrate fermentations in yeast hybrids. Proc. Natl. Acad. Sci. USA 35:23–27.

Lipinski, C., Ferro, A. J., Mills, D. (1976). Macromolecule synthesis in a mutant of *Saccharomyces cerevisiae* inhibited by *S*-adenoyslmethionine. Mol. Gen. Genet. 144:301–306.

Lodder, J. (1934). Die anaskosporogen Hefen, I Hälfte. Verhandel. Kon. Acad. Wetenschap. Amsterdam, Afd. Natuurkunde, Sect. II, 32:1.

Lodder, J. (1970). General classification of the yeasts. In: The Yeasts: A Taxonomic Study, edited by J. Lodder. Amsterdam: North-Holland, pp. 1–33.

Manachini, P. L. (1979). DNA sequence similarity, cell wall mannans, and physiological characteristics in some strains of *Candida utilis, Hansenula jadinii,* and *Hansenula petersonii.* Antonie van Leeuwenhoek J. Microbiol. Serol. 45:451–463.

Maniatis, T., Jeffrey, A., Kleid, D. G. (1975). Nucleotide sequence of the rightward operator of phage. Proc. Natl. Acad. Sci. USA 72:1184–1188.

Marmur, J. (1961). A procedure for the isolation of DNA from microorganisms. J. Mol. Biol. 3:208–218.

Marmur, J., Doty, P. (1962). Determination of the base composition of DNA from its thermal denaturation temperature. J. Mol. Biol. 5:109–118.

Marmur, J., Rownd, R., Schildkraut, C. L. (1963). Denaturation and renaturation of DNA. Progr. Nucl. Acid Res. Mol. Biol. 1:231–300.

Martini, A. (1973). Ibridazioni DNA/DNA tra specie di lieviti del genere *Kluyveromyces.* Ann. Fac. Agrar. Univ. Studi Perugia 28:1–15.

Martini, A., Phaff, H. J. (1973). The optical determination of DNA–DNA homologies in yeasts. Ann. Microbiol. 23:59–68.

Martini, A., Phaff, H. J., Douglass, S. A. (1972). Deoxyribonucleic acid base composition of species in the yeast genus *Kluyveromyces* van der Walt emend. v. d. Walt. J. Bacteriol. 111:481–487.

Marzluf, G. A. (1981). Regulation of nitrogen metabolism and gene expression in fungi. Microbiol. Rev. 45:437–461.

Mendonça-Hagler, L. C., Phaff, H. J. (1975). Deoxyribonucleic acid base composition and DNA/DNA hybrid formation in psychrophobic and related yeasts. Int. J. Syst. Bacteriol. 25:222–229.

Mendonça-Hagler, L. C., Travassos, L. R., Lloyd, K. O., Phaff, H. J., (1974). Deoxyribonucleic acid base composition and hybridization studies on the human pathogen *Sporothrix schenckii* and *Ceratocystis* species. Infect. Immunity 8:674–680.

Meyer, S. A., Phaff, H. J. (1969). Deoxyribonucleic acid base composition in yeasts. J. Bacteriol. 97:52–56.

Meyer, S. A., Phaff, H. J. (1970). Taxonomic significance of the DNA base composition in yeasts. Recent Trends in Yeast Research, Vol. 1 (Spectrum), edited by D. G. Ahearn. Atlanta: Georgia State University, pp. 1–29.

Meyer, S. A., Phaff, H. J. (1972). DNA base composition and DNA–DNA homology studies as tools in yeast systematics. In: Yeasts, Models in Science and Technics, edited by A. Kocková-Kratochvilová, E. Minarik. Prog. 1st Specialized Int. Symp. on Yeasts. Bratislava: Publ. House, Slovak Acad. Science, pp. 375–386.

Meyer, S. A., Anderson, K., Brown, R. E., Smith, M. Th., Yarrow, D., Mitchell, G., Ahearn, D. G. (1975). Physiological and DNA characterization of *Candida maltosa,* a hydrocarbon utilizing yeast. Arch. Microbiol. 104:225–231.

Meyer, S. A., Smith, M. T., Simione, Jr., F. P. (1978). Systematics of *Hanseniaspora* Zikes and *Kloeckera* Janke. Antonie van Leeuwenhoek J. Microbiol. Serol. 44:79–96.

Miller, M. W., Yoneyama, M., Soneda, M. (1976). *Phaffia,* a new yeast genus of *Deuteromycotina* (Blastomycetes). Int. J. Syst. Bacteriol. 26:286–291.

Nakase, T., Komagata, K. (1968a). Taxonomic significance of base composition of yeast DNA. J. Gen. Appl. Microbiol. 14:345–357.

Nakase, T., Komagata, K. (1968b). Taxonomic significance of base composition of yeast DNA. Yeast News Lett. 17:4–6.

Nakase, T., Komagata, K. (1971a). Further investigation on the DNA base composition of the genus *Hansenula.* J. Gen. Appl. Microbiol. 17:77–84.

Nakase, T., Komagata, K. (1971b). Significance of DNA base composition in the classification of yeast genera *Cryptococcus* and *Rhodotorula.* J. Gen. Appl. Microbiol. 17:121–130.

O'Connor, R. M., McArthur, C. R., Clark-Walker, G. D. (1975). Closed-circular DNA from mitochondrial-enriched fractions of four *petite*-negative yeasts. Eur. J. Biochem. 53:137–144.

O'Connor, R. M., McArthur, C. R., Clark-Walker, G. D. (1976). Respiratory-deficient mutants of *Torulopsis glabrata,* a yeast with circular mitochondrial deoxyribonucleic acid of 6 μm. J. Bacteriol. 126:959–968.

Petes, T. D., Beyers, B., Fangman, W. L. (1973). Size and structure of yeast chromosomal DNA. Proc. Natl. Acad. Sci. USA 70:3072–3076.

Phaff, H. J., Miller, M. W., Miranda, M. (1979). *Hansenula alni,* a new heterothallic species of yeast from exudates of alder trees. Int. J. Syst. Bacteriol. 29:60–63.

Price, C. W., Phaff, H. J. (1979). *Debaryomyces polymorphus* and *D. pseudopolymorphus,* new taxonomic combinations. Mycologia 71:444–445.

Price, C. W., Fuson, G. B., Phaff, H. J. (1978). Genome comparison in yeast systematics: delimitation of species within the genera *Schwanniomyces, Saccharomyces, Debaryomyces,* and *Pichia.* Microbiol. Rev. 42:161–193.

Rhodes, J. C., Kwon-Chung, K. J. (1982). A new efficient method for protoplast formation in *Cryptococcus neoformans.* Abst. Ann. Mtg. F61, p. 336. Am. Soc. Microbiol.

Rost, K., Venner, H. (1964). Untersuchungen an Nucleinsäuren. X. Isolierung und Untersuchung von Deoxyribonucleinsäure aus Hefen. Z. Physiol. Chem. 339:230–237.

Schildkraut, C. L., Marmur, J., Doty, P. (1962). Determination of the base composition of deoxyribonucleic acid from its buoyant density in CsCl. J. Mol. Biol. 4:430–433.

Segal, E., Eylan, E. (1974a). Genetic relatedness of *Candida albicans* to asporogenous and ascosporogenous yeasts as reflected by nucleic acid homologies. Microbios 9:25–33.

Segal, E., Eylan, E. (1974b). Nucleic acid homologies between *Candida albicans* and *Hansenula* species. Microbios 10:133–138.

Segal, E., Eylan, E. (1975). Nucleic acid homology studies among *Candida albicans, Syringospora albicans* and *Leucosporidium* species. Microbios 12:111–117.

Seidler, R. J., Mandel, M. (1971). Quantitative aspects of DNA renaturation: DNA base composition, state of chromosome replication, and polynucleotide homologies. J. Bacteriol 106: 608–614.

Seidler, R. J., Knittel, M. D., Brown, C. (1975). Potential pathogens in the environment. Cultural reactions and nucleic acid studies on *Klebsiella pneumoniae* from chemical and environmental sources. Appl. Microbiol. 29:819–825.

Selander, R. K. (1976). Genetic variation in natural populations. In: Molecular Evolution, edited by F. J. Ayala. Sunderland, Mass.: Sinauer Associates, pp. 21–46.

Smith, D., Halvorson, H. O. (1967). The isolation of DNA from yeast. In: Methods in Enzymology, Vol. 12, part A, edited by L. Grossman and K. Moldave. New York: Academic Press, pp. 538–541.

Spencer, J. F. T., Gorin, P. A. J. (1969). Systematics of the genera *Hansenula* and *Pichia:* proton magnetic resonance spectra of their mannans as an aid in classification. Can. J. Microbiol. 15:375–382.

Spencer, J. F. T., Gorin, P. A. J., Rank, G. H. (1971). The genetic control of the two types of mannan produced by *Saccharomyces cerevisiae.* Can. J. Microbiol. 17:1451–1454.

Starmer, W. T., Phaff, H. J., Miranda, M., Miller, M. W. (1978). *Pichia amethionina,* a new heterothallic yeast associated with the decaying stems of cereoid cacti. Int. J. Syst. Bacteriol. 28:433–441.

Stelling-Dekker, N. M. (1931). Die sporogenen Hefen. Verh. Kon. Akad. Wetenschap. Amsterdam, Afd. Natuurkunde. Sect. II 28:1.

Stenderup, A., Bak, A. L. (1968). Deoxyribonucleic acid base composition of some species within the genus *Candida.* J. Gen. Microbiol. 52:231–236.

Storck, R. (1966). Nucleotide composition of nucleic acids of fungi. II. Deoxyribonucleic acids. J. Bacteriol. 91:227–230.

Storck, R., Alexopoulos, C. J. (1970). Deoxyribonucleic acid of fungi. Bacteriol. Rev. 34:126–154.

Storck, R., Alexopoulos, A., Phaff, H. J. (1969). Nucleotide composition of deoxyribonucleic acid of some species of *Cryptococcus, Rhodotorula,* and *Sporobolomyces.* J. Bacteriol. 98:1069–1072.

Subirana, J. A., Doty, P. (1966). Kinetics of renaturation of renatured DNA. I. Spectrophotometric results. Biopolymers 4:171–187.

Timberlake, W. E. (1978). Low repetitive DNA content in *Aspergillus nidulans.* Science 202:973–975.

Tsuchiya, T., Fukazawa, Y., Taguchi, M., Nakase, T., Shimoda, T. (1974). Serological aspects on yeast classification. Mycopathol. Mycol. Appl. 53:77–91.

Ullman, J. S., McCarthy, B. J. (1973). The relationships between mismatched base pairs and the thermal stability of DNA duplexes. I. Effects of depurination and chain scission. Biochim. Biophys. Acta 294:405–415.

van der Walt, J. P. (1970). *Saccharomyces* Meyen emend. Reess. In: The Yeasts: A Taxonomic Study, edited by J. Lodder. Amsterdam: North-Holland, pp. 555–718.

van der Walt, J. P., Hopsu-Havu, V. K. (1976). A colour reaction for the differentiation of ascomycetous and hemibasidiomycetous yeasts. Antonie van Leeuwenhoek J. Microbiol. Serol. 42:157–163.

van der Walt, J. P., Johannsen, E. (1979). A comparison of interfertility and in vitro DNA-DNA reassociation as criteria for speciation in the genus *Kluyveromyces*. Antonie van Leeuwenhoek J. Microbiol. Serol. 45:281–291.

Van Etten, J. L., Freer, S. N. (1978). Simple procedure for disruption of fungal spores. Appl. Environ, Microbiol. 35:622–623.

van Uden, N., Buckley, H. (1970). *Candida* Berkhout. In: The Yeasts: A Taxonomic Study, edited by J. Lodder. Amsterdam: North-Holland, pp. 893–1087.

Wickerham, L. J. (1951). Taxonomy of yeasts. U.S. Dep. Agric. Tech. Bull. 1029:1–56.

Wickerham, L. J., Burton, K. A. (1948). Carbon assimilation tests for the classification of yeasts. J. Bacteriol. 56:363–371.

Wickerham, L. J., Burton, K. A. (1954). A clarification of the relationship of *Candida guilliermondii* to other yeasts by a study of their mating types. J. Bacteriol. 68:594–597.

Wickerham, L. J., Kurtzman, C. P., Herman, A. I. (1969). Sexuality in *Candida lipolytica*. In: Recent Trends in Yeast Research, Vol. 1, edited by D. G. Ahearn. Atlanta: Georgia State University, pp. 81–92.

Williamson, D. H., Fennell, D. J. (1975). The use of fluorescent DNA-binding agent for detecting and separating yeast mitrochondria DNA. In: Methods in Cell Biology, Vol. 12, Yeast Cells, edited by D. M. Prescott. New York: Academic Press, pp. 335–351.

Wilson, A. C., Carlson, S. S., White T. J. (1977). Biochemical evolution. Annu. Rev. Biochem. 46:573–639.

Winge, Ö., Roberts, C. (1949). Inheritance of enzymatic characters in yeast, and the phenomenon of long-term adaptation. Comp. Rend. Trav. Lab. Carlsberg 24:263–315.

Yamada, Y., Kondo, K. (1972). Taxonomic significance of the coenzyme Q system in yeasts and yeast-like fungi (2). In: Fermentation Technology Today, Proceedings of the IVth International Fermentation Symposium, edited by G. Terui, Osaka, Japan: Society of Fermentation Technology, pp. 781–784.

Yamada, Y., Kondo, K. (1973). Coenzyme Q system in the classification of the yeast genera *Rhodotorula* and *Cryptococcus*, and the yeast-like genera *Sporobolomyces* and *Rhodosporidium*. J. Gen. Appl. Microbiol. 19:59–77.

Yamada, Y., Arimoto, M., Kondo, K. (1973a). Coenzyme Q system in the classification of the ascosporogenous yeast genus *Schizosaccharyomyces* and yeast-like genus *Endomyces*. J. Gen. Appl. Microbiol. 19:353–358.

Yamada, Y., Kanematsu, Y., Ohashi, M., Kondo, K. (1973b). On the partly reduced coenzyme Q isolated from *"Rhodotorula lactosa"* IFO 1058 and its relation to the taxonomic position. Agric. Biol. Chem. 37:621–628.

Yamada, Y., Okada, T., Ueshima, O., Kondo, K. (1973c). Coenzyme Q system in the classification of the ascosporogenous yeast genera *Hansenula* and *Pichia*. J. Gen. Appl. Microbiol. 19:189–208.

Yamada, Y., Arimoto, M., Kondo, K. (1976a). Coenzyme Q system in the classification of apiculate yeasts in the genera *Nadsonia, Saccharomyces, Hanseniaspora, Kloeckera,* and *Wickerhamia.* J. Gen. Appl. Microbiol. 22:293–299.

Yamada, Y., Nojiri, M., Matsuyama, M., Kondo, K. (1976b). Coenzyme Q system in the classification of the ascosporogenous yeast genera *Debaryomyces, Saccharomyces, Kluyveromyces* and *Endomycopsis.* J. Gen. Appl. Microbiol. 22:325–339.

Yamada, Y., Arimoto, M., Kondo, K. (1977). Coenzyme Q system in the classification of some ascosporogenous yeast genera in the families Saccharomycetaceae and Spermophthoraceae. Antonie van Leeuwenhoek J. Microbiol. Serol. 43:65–71.

Yamada, Y., Takinami-Nakamura, H., Tahara, Y., Smith, M. Th. (1980). The coenzyme Q system in the classification of the ascosporogenous yeast genus *Dekkera* and the asporogenous yeast genus *Brettanomyces.* Antonie van Leeuwenhoek J. Microbiol. Serol. 46:595–599.

Yamada, Y., Sasaki, J., Hou, C. F., Tahara, Y., Miranda, M., Phaff, H. J. (1981). A new natural quinone, ubiquinone-5 isolated from an ascosporogenous yeast species *Nematospora coryli.* Agric. Biol. Chem. 45:2671–2673.

Yarrow, D., Meyer, S. A. (1978). Proposal for amendment of the diagnosis of the genus *Candida* Berkhout nom. cons. Int. J. Syst. Bacteriol. 28:611–615.

# 6

# Genetic Approaches to the Study of Protease Function and Proteolysis in *Saccharomyces cerevisiae*[1]

Elizabeth W. Jones*

## Introduction

In the yeast *Saccharomyces cerevisiae,* proteases have been implicated in septum formation (Ulane and Cabib 1976), sporulation (Chen and Miller 1968; Hopper *et al.* 1974; Klar and Halvorson 1975; Betz and Weiser 1976; Zubenko and Jones 1981), protein turnover (Lopez and Gancedo 1979; Hansen *et al.* 1977), catabolite inactivation (Hägele *et al.* 1978; Neeff *et al.* 1978; Funayama *et al.* 1980; Holzer 1976; Jušik *et al.* 1976; Molano and Gancedo 1974; Magni *et al.* 1977; Magni *et al.* 1978), carbon starvation-induced degradation of NADP-dependent glutamate dehydrogenase (Hemmings and Mazon 1979; Mazon and Hemmings 1979), nitrogen starvation-induced degradation of NAD-dependent glutamate dehydrogenase (Hemmings 1980), enzyme secretion (Perlman and Halvorson 1981), localization of mitochondrial enzymes (Neupert and Schatz 1981), processing of enzyme precursors (Hemmings *et al.* 1981; Hasilik and Tanner 1978a, Jones *et al.* 1981; Zubenko *et al.* 1981), production of the pheromone $\alpha$ factor (J. Kurjan and I. Herskowitz 1982; D. Julius, A. Brake, L. Blair, and J. Thorner, personal communication), destruction of $\alpha$ factor (Maness and Edelman 1978; Ciejek and Thorner 1979; Ciejek 1980) and degradation of missense proteins (Donahue and Henry 1981) and nonsense fragments (Bigelis and Burridge 1978; Bigelis and Fink 1981). One would guess that some of these proteolytic processes are likely to involve enzymes with restricted substrate specificities (destruction of $\alpha$ factor and, possibly, production of the pheromone), others may involve enzymes with group

*Department of Biological Sciences, Carnegie-Mellon University, Pittsburgh, Pennsylvania 15213 U.S.A.

[1]Work in the author's laboratory is supported by grants from the National Institutes of Health of the U.S. Public Health Service.

specificity (toward signal sequences for secreted enzymes or enzymes destined for the mitochondrion), whereas others are likely to involve enzymes with rather broad specificities (protein degradation during nitrogen starvation and/or sporulation may be examples). Both biochemical and genetic approaches are being employed to identify and characterize the proteolytic systems responsible for these events.

## Intracellular Proteases and Protease Inhibitors of *S. cerevisiae*

A number of proteases have been isolated from yeast and have been characterized in considerable detail. Others have been described to only a limited extent. A compilation of these enzymes appears in Table 6.1. Salient characteristics of a number of proteases are summarized in this section. Discussion of others will be found in the section on proteolytic processes.

### Endoproteinases

#### Proteinase A

Proteinase A [E.C. 3.4.23.6] is an endoproteinase that is active under acidic conditions (Hata *et al.* 1967a,b; Lenney and Dalbec 1967; Betz *et al* 1974; Saheki *et al.* 1974; Hinze *et al.* 1975; Magni *et al.* 1978; Muessdoerffer *et al.* 1980; Zubenko 1981). Estimates of the molecular size range from 41,000 to 60,000 d. Proteinase A is a glycoprotein (Hata *et al.* 1967b; Muessdoerffer *et al.* 1980) with 7.5–10% of the molecular weight of the glycoprotein being due to carbohydrate. Proteinase A is inhibited by pepstatin (Lenney 1975). The enzyme is located in the vacuole (Wiemken *et al.* 1979).

#### Proteinase B

Proteinase B (E.C. 3.4.22.9) is an endoproteinase with a pH optimum near neutrality (Lenney and Dalbec 1967; Hata *et al.* 1967a; Betz *et al.* 1974; Hinze *et al.* 1975; Sanada *et al.* 1975; Ulane and Cabib 1976; Looze *et al.* 1979; Fujishiro *et al.* 1980; Zubenko 1981; Kominami *et al.* 1981a,b). Estimates of the molecular size range from 31,000 to 44,000 d. Proteinase B is a glycoprotein (Ulane and Cabib 1976; Kominami 1981a); carbohydrate comprises 10% of the molecular weight. Proteinase B is inhibited by chymostatin and antipain as well as sulfhydryl reagents and serine reagents (Hata *et al.* 1967a; Lenney 1975; Fujishiro *et al.* 1980). It is thus classified as a seryl sulfhydryl protease. The enzyme is located in the vacuole (Wiemken *et al.* 1979).

## Carboxypeptidases

### Carboxypeptidase Y (Formerly Protease C)

Carboxypeptidase Y is a serine carboxypeptidase with a pH optimum near neutrality (Doi *et al.* 1967; Hata *et al.* 1967a,b; Hayashi *et al.* 1970, 1973; Kuhn *et al.* 1974, 1976). The enzyme has a molecular size around 60,000 d, of which about 10,000 d is carbohydrate (Hata *et al.* 1967b; Hasilik and Tanner 1978a,b, Aibara *et al.* 1971). Carboxypeptidase Y appears to contain four asparagine-linked oligosaccharides of the general formula (*N*-acetylglucosamine)$_2$-(mannose)$_{13}$ (Trimble and Maley 1977; Hasilik and Tanner 1978b). This "core" oligosaccharide is phosphorylated (Hashimoto *et al.* 1981). Carboxypeptidase Y is inhibited by serine and sulfhydryl reagents (Doi *et al.* 1967). The enzyme is located in the vacuole (Wiemken *et al.* 1979). It is likely that carboxypeptidase Y corresponds to carboxypeptidase $\beta$ of Felix and Brouillet (1966) and that carboxypeptidase S is carboxypeptidase $\alpha$ (see following discussion).

### Carboxypeptidase S

Carboxypeptidase S has not been purified. It has been detected by its ability to catalyze cleavage of Cbz-Gly-Leu (Wolf and Weiser 1977). It is inhibited by EDTA and can be reactivated by $Zn^+$ (Wolf and Weiser 1977). Its cellular location is unknown.

### Other Carboxypeptidases

Several additional carboxypeptidase activities capable of catalyzing cleavage of Cbz-blocked dipeptides have been detected in mutant strains lacking carboxypeptidases Y and S (Wolf and Ehmann 1981). They have not been characterized as yet.

## Aminopeptidases

At least four, and possibly more, aminopeptidases exist in yeast. Matile *et al.* (1971) detected four activities on starch gels, capable of catalyzing cleavage of leucyl-$\beta$-naphthylamide (leu$\beta$NA). Trumbly (1980) also detected four activities and generated four strains, each possessing only one of these activities. Masuda *et al.* (1975) described two activities and Röhm and collaborators have detected three species with aminopeptidase activity (Metz and Röhm 1976; Frey and Röhm 1978, 1979a). All of the data presented cannot be rec-

**Table 6.1.** Proteolytic Enzymes of *Saccharomyces cerevisiae*

| Enzyme | Characteristics | Molecular weight | Intracellular location | Inhibitors | References |
|---|---|---|---|---|---|
| Proteinase A | Acid endopeptidase (glycoprotein) | 41,000–60,000 | Vacuole | Pepstatin | See text |
| Proteinase B | Serine endopeptidase (glycoprotein) | 31,000–44,000 | Vacuole | PMSF; PCMB chymostatin; antipain | See text |
| Proteinase D | Sulfhydryl endopeptidase ($Me^{2+}$) | <100,000 | n.d. | pHMB, EDTA | D. Wolf, personal communication |
| Proteinase E | | 600,000 | n.d. | $Hg^{2+}$, $Zn^{2+}$ | D. Wolf, personal communication |
| Carboxypeptidase Y (protease C) | Serine exopeptidase (glycoprotein) | 61,000 | Vacuole | PMSF; PCMB | See text |
| Carboxypeptidase S | Exopeptidase ($Zn^{2+}$) | n.d. | n.d. | EDTA | See text |
| Aminopeptidase I (API) | Exopeptidase ($Zn^{2+}$) | 600,000 | Vacuole | EDTA | Metz and Röhm; 1976; Metz *et al.* 1977; Frey and Röhm 1978; Loffler and Röhm 1979 |
| Aminopeptidase II (APII) | Exopeptidase | 85,000 | Periplasm and intracellular | $Cu^{2+}$ | Frey and Röhm 1978 |

| | | | | | |
|---|---|---|---|---|---|
| Aminopeptidase III (APIII) | Exopeptidase | 30,000 | n.d. | | Frey and Röhm 1978 |
| Aminopeptidase III | Exopeptidase | Large | Vacuole | n.d. | Matile *et al.* 1971 |
| Aminopolypeptidase | | 670,000 | | | Johnson 1941 |
| Aminopeptidase I (API) | Exopeptidase ($Zn^{2+}$) | 200,000 | n.d. | EDTA | Masuda *et al.* 1975 |
| Aminopeptidase II (APII) | Exopeptidase | 34,000 | n.d. | EDTA | Masuda *et al.* 1975 |
| Leucine aminopeptidase I (LAPI) | Exopeptidase ($Me^{2+}$) | 85,000 | n.d. | EDTA | Trumbly 1980 |
| Leucine aminopeptidase II (LAPII) | Exopeptidase ($Me^{2+}$) | 85,000 | n.d. | EDTA | Trumbly 1980 |
| Leucine aminopeptidase III (LAPIII) | Exopeptidase | 94,000 | n.d. | None found | Trumbly 1980 |
| Leucine aminopeptidase IV (LAPIV) | Exopeptidase ($Zn^{2+}$) | 570,000 | n.d. | EDTA | Trumbly 1980 |
| Dipeptidylpeptidase IV | Cleaves X-Pro-Y or X-Ala-Y yielding X-Pro or X-Ala | n.d. | membrane | n.d. | Suarez-Rendueles *et al.* 1981; D. Julius, L. Blair, A. Brake, J. Thorner, personal communication |

onciled, but some equivalences can be made. Aminopeptidase I (using the nomenclature of Frey and Röhm 1978) is large (around 600,000 d) and is composed of 12 identical subunits. It contains 12.5% carbohydrate and is located in the vacuole. It requires $Zn^{2+}$ or $Co^{2+}$ and is inhibited by EDTA (Metz and Röhm 1976; Metz *et al.* 1977; Frey and Röhm 1978; Löffler and Röhm 1979). API corresponds to the aminopolypeptidase of Johnson (1941), to APIII of Matile *et al.* (1971), and to LAPIV of Trumbly (1980). Frey and Röhm (1978, 1979a) reported that aminopeptidase II is the only aminopeptidase to cleave lysyl *p*-nitroanilide (lysNA) and that about 40% of this activity is located in the periplasm with the rest being intracellular. The activity is reported to have a molecular size of 85,000 d. Trumbly (1980) reports two activities, LAPI and LAPII, capable of cleaving lysNA, and each of about 85,000 d. Possibly APII of Frey and Rohm is a mixture of two enzymes, only one of which is located extracellularly. Aminopeptidase III of Frey and Röhm (1980) is small (30,000 d). Whether APII of Masuda *et al.* (1975) (MW = 34,000) is the corresponding activity is uncertain, for it is reported not to cleave leuNA or leu$\beta$NA, unlike APIII of Frey and Röhm (1978). Additional species reported are LAPIII of Trumbly (1980) of 95,000 d and API of Masuda *et al.* (1975) of 200,000 d, which shares some characteristics with APII of Frey and Röhm (1978). Of all of the aminopeptidase species reported, only the large, 600,000-d species is located in the vacuole. Trumbly and Röhm's group worked with the isogenic strains, S288C and X2180. Hence the discrepancies do not originate in strain differences.

## Peptidases

Rose *et al.* (1979) have reported four peptidase activities to be present in *S. cerevisiae.* Trumbly (1980) reports that peptidases 1 and 2 cleave leuNA and that peptidase 1 of Rose *et al.* (1979) corresponds to the vacuolar aminopeptidase and peptidase 2 corresponds to LAPI.

## Other Proteolytic Activities

Achstetter *et al.* (1981) report a number of new proteolytic activities directed toward peptide substrates containing blocked amino groups. The activities are as yet uncharacterized but are likely to include endoproteinases as well as amidases. At least two of these activities, called proteinase M and proteinase P, are found in the membrane fraction.

Additional proteolytic activities have been detected, described, or inferred to exist in yeast. They will be discussed in the context of the proteolytic process in which they are implicated.

Of the preceding proteases and peptidases, proteinases A and B, carboxypeptidase Y, and the large aminopeptidase I are known to be located in the vacuole (Matile and Wiemken 1967; Lenney *et al.* 1974; Matern *et al.* 1974a; Frey and Röhm 1978; Wiemken *et al.* 1979). Carboxypeptidase Y, proteinase A, proteinase B, and possibly all vacuolar hydrolases in yeast are synthesized as inactive precursors (Hasilik and Tanner 1976a, 1978a,b; Jones *et al.* 1981; Hemmings *et al.* 1981; Zubenko *et al.* 1981, 1983; Mechler *et al.* 1982). In this characteristic, yeast vacuolar hydrolases are similar to some hydrolases present in the lysosomes of mammalian cells (Skudlarek and Swank 1979; Hasilik 1980; Hasilik and Neufeld 1980). The biosynthesis of carboxypeptidase Y requires at least two posttranslational modification events. The first carboxypeptidase Y precursor detected is approximately 65,000 d in size and is a glycoprotein that contains endoglycosidase H-sensitive core carboyhdrate moieties (Zubenko *et al.* 1981, 1983). Because no unglycosylated precursor has been revealed by pulse-chase studies (unpublished observation) and because the enzymes for dolichol-mediated glycosylation in yeast are located in the membrane fraction in which microsomal enzymes are located (Marriott and Tanner 1979), core glycosylation of this carboxypeptidase Y precursor may be a cotranslational process. The 65,000-d precursor is converted to a larger intermediate, about 67,000 d in size, by addition of sugars to the core carboyhdrate moieties. The smaller precursor accumulates in *sec* mutants which accumulate endoplasmic reticulum; the larger in *sec* mutants which accumulate Golgi membranes (Stevens *et al.* 1982). This larger intermediate then undergoes the scission of a carbohydrate-free amino terminal peptide to form the mature, active carboxypeptidase Y molecule, about 59,000 d in size (Zubenko *et al.* 1981, 1983). A 62,000-d precursor is converted to a 65,000-d intermediate, which is subsequently converted to the mature, active proteinase A molecule, a glycoprotein about 58,000 d in size. Although carbohydrate comprises about 10% of the proteinase A molecule by weight (Hata *et al.* 1967b; Muessdoerffer *et al.* 1980), neither of the proteinase A precursors appears to contain Endo H-sensitive carbohydrate moieties (Zubenko *et al.* 1981, 1983). For both carboxypeptidase Y and proteinase A the conversion of the smaller to the larger precursor and the conversion of the larger precursor to the mature, active enzyme occur with half-lives of about 5 min at 37°C (Zubenko *et al.* 1981, 1983).

## Intracellular Protease Inhibitors

### Proteinase A Inhibitors

Baker's yeast contains two heat-stable polypeptide inhibitors of proteinase A, called $I^A2$ and $I^A3$. The two inhibitors are specific for proteinase A and are the same size (7700 d), but differ in their isoelectric points (Saheki *et al.* 1974; Nunez de Castro and Holzer 1976; Muessdoerffer 1980). $I^A2$ and $I^A3$ are present in a 1:3 ratio in baker's yeast. Only $I^A2$ is present in *S. carlsbergensis;* only $I^A3$ is present in *S. cerevisiae* (Muessdoerffer 1980). The simplest explanation of the findings for baker's yeast is that there is a single gene encoding the inhibitor in each species, that the genes are alleles, and that the strains of baker's yeast examined were tetraploid and contained one copy of the *S. carlsbergensis* gene and three copies of the *S. cerevisiae* gene.

*In vitro* $I^A2$ and $I^A3$ are inactivated by proteinase B. Neither is inactivated by proteinase A or carboxypeptidase Y (Saheki *et al.* 1974).

### Proteinase B Inhibitors

Baker's yeast contains two heat-stable polypeptide inhibitors of proteinase B, called $I^B1$ and $I^B2$. The two inhibitors are specific for proteinase B and are the same size (8500 d), but differ in their isoelectric points (Betz *et al.* 1974; Bünning and Holzer 1977; Maier *et al.* 1979a). $I^B1$ and $I^B2$ are present in a 1:3 ratio in baker's yeast. Only $I^B1$ is found in *S. carlsbergensis;* only $I^B2$ is found in *S. cerevisiae* (Ulane and Cabib 1974; Betz 1975; Bünning and Holzer 1977). $I^B1$ and $I^B2$ differ by two amino acids (Maier *et al.* 1977b,c). The simplest explanation of the observations for baker's yeast parallels that given for the proteinase A inhibitors and is that there is a single gene encoding the inhibitor in each species, that the genes are alleles, and that the strains of baker's yeast examined were tetraploid and contained one copy of the *S. carlsbergensis* gene and three copies of the *S. cerevisiae* gene. Interestingly enough, in each case one *carlsbergensis* and three *cerevisiae* genes are required.

*In vitro* $I^B1$ and $I^B2$ are rapidly inactivated by proteinase A and slowly by proteinase B (Betz *et al.* 1974). Carboxypeptidase Y was not tested.

### Carboxypeptidase Y Inhibitor

Yeast contains a heat-labile polypeptide inhibitor of carboxypeptidase Y of about 24,000 d in size. The inhibitor is specific for carboxypeptidase Y, for it does not inhibit proteinases A and B or carboxypeptidase S. *In vitro* the inhibitor is inactivated rapidly by proteinases A and B. When complexed with carboxypeptidase Y, the inhibitor seems to be susceptible only to proteinase B (Matern *et al.* 1974b; 1979; Barth *et al.* 1978; Fischer and Holzer 1980).

The polypeptide inhibitors of proteinases A and B and carboxypeptidase Y are all located in the cytosol (Lenney *et al.* 1974; Matern *et al.* 1974; Hasilik *et al.* 1974).

## Genetics of Known Proteases and Protease Inhibitors

### Proteinase A

Several mutations that result in reduced levels of proteinase A activity have been identified (Jones 1977; Betz 1979; Jones *et al.* 1981; Zubenko 1981;

Mechler and Wolf 1981). The *pep4* and *pep7* mutations, which are pleiotropic (Jones 1977; Jones *et al.* 1981; Zubenko 1981), will be described later.

The *pra* mutation reported by Betz (1979) reduces the level of proteinase A by 80%, leaving levels of other proteinases unaffected. Betz concluded from his analysis that the mutation causing proteinase A deficiency did not reside in the structural gene. From a less than complete genetic analysis Betz concluded that a single recessive mutation imparted partial proteinase A deficiency and sporulation deficiency. The significance of this observation in unclear since the molecular basis of the phenotype is unknown.

Two recessive, allelic mutations that reduce proteinase A levels to 5–10% of wild-type levels, leaving proteinase B and carboxypeptidase Y levels unaffected, show dosage effects, as would be expected for mutations in the proteinase A structural gene (Jones *et al.* 1981; Zubenko 1981). As the known pleiotropic mutations *pep7-11* (corrected from *pep17-1*), *pep4-3,* and *pep4-5* also show dosage effects on levels of proteinase A (Jones *et al.* 1981; Zubenko 1981), an effect of gene dosage on enzyme levels does not provide an adequate criterion for identification of the structural gene. Hence it is uncertain whether these two mutations reside in the structural gene for proteinase A.

Mechler and Wolf (1981) have described mutations that lead to the absence of proteinase A activity, without affecting levels of proteinase B, carboxypeptidase Y, or $I^A$. Of all $Pra^-$ mutants reported, these are clearly the most deficient in proteinase A activity. Five of these mutations fail to complement and have been assigned the designation *pra1*. The mutations show gene dosage effects on levels of proteinase A activity and result in the absence of immunologically cross-reacting material. From these observations the authors infer that *PRA1* is the structural gene for proteinase A. As mentioned earlier, pleiotropic mutations can show gene dosage effects. Hence the assignment of *PRA1* as the structural gene for proteinase A must remain tentative.

## Proteinase B

Zubenko *et al.* (1979) have described four genes, *PRB1, PRB2, PRB3,* and *PRB4,* whose functions are required for the production of normal levels of proteinase B activity in yeast. A number of pleiotropic mutations also cause reduced levels of proteinase B (Jones 1977; Jones *et al.* 1981; Zubenko 1981).

*PRB1* is the structural gene for proteinase B. This assignment was made because mutations at the *PRB1* locus show a dosage effect on proteinase B activity, because two mutant *prb1* alleles result in the production of thermolabile proteinase B activities that are distinguishable from one another and from normal activity *in vitro,* and finally because extracts made from a tetraploid strain that carries three copies of a ts allele and one copy of a wild-type allele contain a thermolabile component of the total proteinase B activity in a proportion consistent with that expected based on gene dosage considerations. A total of 34 amber and 12 ochre mutations were found among the 126 *prb1* mutations isolated by these investigators (Zubenko *et al.* 1980). The *PRB1* locus maps 1.1 cM proximal to *CAN1* on the left arm of chromosome V (Zubenko *et al.* 1980). Mutations in *PRB1* leave levels of proteinase A and carboxypeptidase Y unaffected (Zubenko *et al.* 1979).

Wolf and Ehmann (1978a) have also reported isolation of mutants deficient in proteinase B activity. They possess normal levels of proteinase A, carboxypeptidase Y, and $I^B$. As no further genetic characterization of these mutants has been reported, the number and identity of the loci involved are unknown, as are the nature of the mutations.

## Carboxypeptidase Y

At least 18 genes are known whose functions are essential for the production of normal levels of carboxypeptidase Y activity (Wolf and Fink 1975; Jones 1977; Jones *et al.* 1981; Zubenko 1981; A. Mitchell, M. Hospodar, and E. Jones, unpublished observations). PRCI is thought to be the structural gene for carboxypeptidase Y since a *prcl* mutation causes production of an altered antigen (Wolf and Weiser 1977), the *prcl-229* mutant produces a nonsense fragment (Hemmings *et al.* 1981), and *prcl* alleles show gene dosage effects on levels of carboxypepidase Y (M. Saftner and E. Jones, unpublished observations). Mutations in the other 17 genes are pleiotropic and will be discussed later.

## Carboxypepidase S

Mutants lacking carboxypeptidase S activity have been isolated (Wolf and Ehmann 1981) as derivatives of *prcl* mutants no longer able to utilize N-CBz-Gly-Leu as a source of nitrogen. All four mutations isolated fell into one complementation group designated *cpsl.* Alleles at this locus show gene dosage effects on carboxypeptidase S activity. The authors infer that *CPSI* is the structural gene for carboxypepidase S. As indicated earlier, gene dosage effects are not compelling evidence for a structural gene assignment. *dutl* mutants also lack carboxypeptidase S activity (Zubenko 1981). Allelism tests with *cpsl* have not been performed.

## Aminopeptidases

Trumbly (1980) has isolated mutants with reduced aminopeptidase activity by successive rounds of mutagenesis, screening for reduced ability to cleave leu$\beta$NA. By four successive rounds of mutagenesis a strain was produced with a greatly reduced ability to hydrolyze leu$\beta$NA. Mutations in four loci—*LAP1, LAP2, LAP3,* and *LAP4*—were recovered, with the mutation in *LAP1* surfacing in the first round, in *LAP2* in the second round, and so on. By appropriate crosses Trumbly generated a family of strains, each possessing a wild-type allele for only one of these loci. Strains bearing only the wild-type allele for *LAP1* possess an aminopeptidase of 85,000 d. Strains bearing only the wild-type allele for *LAP2* also possess an aminopeptidase activity of 85,000 d, but this activity is distinguishable from that present in the *LAP1* strain. Whether the aminopeptidase activity present in the *LAP1* and/or *LAP2* strains is located in the periplasm was not determined and is of interest because of the results of Frey and Röhm (1979b). The *LAP3* strain has an aminopeptidase of 95,000 d. Strains bearing a wild-type allele only at the *LAP4* locus contain the 600,000 d, presumably vacuolar, aminopeptidase. Whether these loci are structural genes is unknown. The *LAP1* and *LAP4* loci are located on the left arm of chromosome XI between *MNN4* and *FAS1*. The *LAP3* locus maps near *PET2* on chromosome XIV (Trumbly 1980; Mortimer and Schild 1982).

## Pleiotropic Mutations Affecting Protease Levels

Many of the mutants isolated as deficient in carboxypeptidase Y activity proved to be deficient for proteinases A and B as well (Jones 1977). Mutations in six loci, *PEP3, PEP5, PEP6, PEP7, PEP12,* and *PEP14,* result in complex changes in phenotype. The phenotypes are summarized in Table 6.2. Mutations in these loci will genetically suppress alleles at the *CAN1* (arginine permease) locus. Six of 96 *can1* alleles are suppressible by mutations in these six *PEP* loci (unpublished observations). Preliminary evidence suggests that alleles at the *LYP1* (lysine permease) locus are also suppressed (R. Parker and E. Jones, unpublished observations). None of 160 *his4* alleles are suppressed (A. Myers and E. Jones, unpublished observations). Mutations in these loci cause the cells to be hypersensitive to amino acid and pyrimidine analogs (P. Szauter and E. Jones, unpublished observations), to be thermosensitive for growth (G. Zubenko and E. Jones, unpublished observations), to have reduced amino acid pools (G. Fabian and E. Jones, unpublished observations), to be unable to utilize glycerol as a carbon source (except for mutations in *PEP6*), to be unable to sporulate as homozygotes and prone to lyse in stationary phase (E. Jones and G. Zubenko, unpublished observations). These characteristics are summarized in Table 6.2.

Mutations in the *PEP4* gene are recessive and affect expression of at least five vacuolar hydrolases. The *pep4-3* mutation results in a 90–95% reduction in the levels of activity of proteinases A and B, carboxypeptidase Y, RNase, and nonspecific repressible alkaline phosphatase (Hemmings *et al.* 1981; Jones *et al.* 1981; Zubenko 1981; Jones *et al.* 1982). All of these enzymatic activities are located in the vacuole (Bauer and Sigarlakie 1975; Wiemken *et al.* 1979). The *PEP4* and *PHO9* genes appear to be identical (Kaneko *et al.* 1982). The action of the *PEP4* gene is not so general that it is required for the expression of all glycoproteins whose synthesis is dolichol mediated or of all enzymes that must traverse membranes before reaching their final cellular location (Jones *et al.* 1981; Zubenko 1981; Jones *et al.* 1982).

Mutations at the *PEP4* locus possess several peculiar genetic properties.

**Table 6.2.** Biological Effects of Some *pep* Mutations

| Mutation | Suppressor of *can1-11* | Sensitivity to analogs | Amino acid pools | Growth at 37°C | Use glycerol as carbon source | Sporulation competent | Lyse in stationary phase |
|---|---|---|---|---|---|---|---|
| None | | Normal | Normal | Yes | Yes | Yes | No |
| *prc1* | No | Normal | | Yes | Yes | Yes | No |
| *pep4* | No | Normal | Low | Yes | Yes | No | No |
| *pep3* | Yes | Hypersensitive | Low | No | No | No | Yes |
| *pep5* | Yes | Hyper | Low | No | No | No | Yes |
| *pep6* | Yes | Hyper | Low | No | Yes | No | |
| *pep7* | Yes | Hyper | Low | No | No | No | Yes |
| *pep12* | Yes | Hyper | Low | No | No | No | Yes |
| *pep14* | Yes | Hyper | Low | No | No | No | Yes |
| *pep1* | No | Normal | n.d. | Yes | Yes | Yes | No |
| *pep2* | No | Normal | n.d. | Yes | Yes | Yes | No |
| *pep8* | No | Normal | n.d. | Yes | Yes | Yes | No |
| *pep9* | No | Normal | n.d. | Yes | Yes | Yes | No |
| *pep10* | No | Normal | n.d. | Yes | Yes | Yes | No |
| *pep11* | No | Normal | n.d. | Yes | Yes | Yes | No |
| *pep13* | No | Normal | n.d. | Yes | Yes | Yes | No |

Diploids heterozygous for these mutations exhibit a dosage effect on the levels of activity of proteinases A and B, but not on the levels of activity of carboxypeptidase Y or nonspecific repressible alkaline phosphatase (Jones *et al.* 1981; Zubenko 1981; Kaneko *et al.* 1982; Jones *et al.* 1982). In addition, meiotic products from such a diploid show a pronounced lag in the expression of the *pep4* mutant phenotype, as judged by levels of proteinase B, carboxypeptidase Y, and nonspecific repressible alkaline phosphatase (Jones *et al.* 1981; Zubenko 1981; Zubenko *et al.* 1982). This phenotypic lag does not depend on any meiosis or sporulation-specific event, since *pep4* bearing haploid cytoductants formed from *PEP4/pep4* heterokaryons exhibit the lag. However, a functional proteinase B structural gene, *PRB1,* is required if phenotypic lag is to occur (Zubenko 1981; Zubenko *et al.* 1982).

## Proteinase A Inhibitor

The structural gene(s) for $I^A2$ and $I^A3$ have not been identified, although strains permitting this assignment may exist. As mentioned earlier, the simplest explanation of the finding that *S. carlsbergensis* contains only $I^A2$, *S. cerevisiae* contains $I^A3$, and baker's yeast contains both in a 1:3 ratio is that the genes encoding these two forms are allelic and bakers' yeast is a tetraploid hybrid. This postulate could be tested by crossing haploid *S. carlsbergenis* with haploid *S. cerevisiae* and seeing whether the two forms segregate from one another in meiosis 100% of the time as predicted by this hypothesis.

A mutation, *pai1-1,* has been isolated that has been inferred to alter the regulation, but not the structure, of the inhibitor of proteinase A (Beck *et al* 1980). Should the genes encoding $I^A2$ and $I^A3$ prove to be allelic, a direct test of this inference will be possible, for the *pai1-1* mutation should assort independently of the gene encoding $I^A3$ (assuming that such a regulatory gene will be unlinked to its regulated gene, as is usually the case in yeast).

The *pai1-1* mutation has no effect on growth at 23°C and results only in a reduced rate of growth of 30°C so long as the medium is buffered to pH 6.5. If the medium is unbuffered or the pH is reduced, the rate of growth and cell yield are reduced at 30°C. Utilization of ethanol in the later growth stages is impaired. Indeed, if ethanol is provided as carbon source at low pH, the *pai1-1* mutant cannot grow at all at 30°C. This effect on growth is correlated with inability of the cells to undergo the normal derepression of $I^A$ synthesis that occurs during late growth stages (Beck *et al.* 1980). Whether this phenotype results from the altered ratio of proteinase A/$I^A$ is as yet unknown.

## Proteinase B Inhibitor

The structural gene(s) for the proteinase B inhibitor has not been identified, although strains permitting this assignment may exist. As for the case of $I^A$, the simplest explanation of the finding that *S. carlsbergensis* contains $I^B1$ alone, *S. cerevisiae* contains $I^B2$ alone, and baker's yeast contains the two in a 1:3 ratio is that the genes encoding the two forms are allelic and baker's yeast is a tetraploid hybrid. As for the case of $I^A$, this postulate could be tested by crossing haploids of the two species and examining segregation patterns of the two forms.

Mutations in any of three genes, *BIN1, BIN2,* or *BIN3,* are recessive and result in a diminution of roughly 50% in the level of proteinase B inhibitor without affecting the levels of proteinase B (Zubenko 1981). Although mutations at any one of these loci result in the inability to utilize glycerol as a source of carbon at 37 °C, this phenotype is not the result of an excess of proteinase B activity or faulty regulation of proteinase B activity, for doubly mutant strains that bear a *bin* mutation and a *prb1* nonsense mutation exhibit an inability to use glycerol at 37 °C. The *BIN* loci are not closely linked to one another, to *PRB1,* or to a centromere (Zubenko 1981). A test of whether any one of these *BIN* loci encode the proteinase B inhibitor could be performed, if $I^B1$ and $I^B2$ prove to be encoded by allelic genes, by analysis of segregation of the *bin* mutations with respect to the inhibitor structural gene, as was suggested for the *pai1-1* mutation.

## Proteolytic Processes and the Role of Known Proteases in These Processes

### Specific Proteolytic Processes

#### Protein Maturation

Proteinases have been implicated in maturation of mitochondrially located polypeptides and secreted enzymes by removal of signal sequences, in removal of N-terminal methionines, in removal of N-terminal peptides from proenzymes, and in release of the α-factor tridecapeptide from a multivalent precursor. Each of these phenomena will be discussed.

Numbers of polypeptides located within the mitochondrion are synthesized as larger precursors that have additional sequences at the amino terminus (Maccechini *et al.* 1979a,b; Nelson and Schatz 1979; Lewin *et al.* 1980; Mihara and Blobel 1980, as examples). A protease that specifically processes yeast mitochondrial enzyme precursors to their mature sizes has been purified from yeast (Böhni *et al.* 1980; Neupert and Schatz 1981). It is inactive toward nonmitochondrial proteins. It is located in the mitochondrial matrix and is inhibited by EDTA. It does not correspond to any of the previously described proteases and, indeed, proteinases A and B degrade the precursors to small oligopeptides (Böhni *et al.* 1980). The existence of a signal peptidase within the intermembrane space of the mitochondrion is implied for enzymes, such as cytochrome c peroxidase, which are transported into this space.

A signal peptidase, possibly within the endoplasmic reticulum, is also implied for secreted enzymes such as invertase. The invertase polypeptide passes through the endoplasmic reticulum and the Golgi apparatus *en route* to secretion vesicles that deliver the enzyme to the membrane (Esmon *et al.* 1981; Novick *et al.* 1981). The initial translation product of the invertase structural gene contains a signal sequence (Perlman and Halvorson 1981).

As might be expected, mutations in the structural genes for the known proteinases A and B or carboxypeptidases Y and S do not lead to respiratory defects or inability to utilize sucrose.

Normally, the N-terminal methionine is removed from the iso-1-cytochrome c polypeptide (Stewart *et al.* 1971). Whether any of the known aminopeptidases is involved in this process is unknown.

Carboxypeptidase Y is synthesized via a larger precursor (Hasilik and Tanner 1976a, 1978a,b), which is inactive (Hemmings *et al.* 1981; Zubenko 1981; Zubenko *et al.* 1982). The final step in maturation of the enzyme involves scission of an N-terminal peptide from the enzyme (Hasilik and Tanner 1976a, 1978a; Hemmings *et al.* 1981). Because the same or a very similar cleavage can be catalyzed by proteinase B or trypsin *in vitro,* proteinase B was postulated to be involved in this maturation process (Hasilik and Tanner 1978a). Since mutants bearing nonsense mutations in the proteinase B structural gene, *PRB1,* have normal levels of carboxypeptidase Y (Zubenko *et al.* 1979) and the carboxypeptidase Y produced is the size of normal mature enzyme (Hemmings *et al.* 1981), proteinase B function is not required for maturation of carboxypeptidase Y enzyme. By the same line of reasoning proteinase A is unlikely to be involved in maturation of carboxypeptidase Y (Mechler and Wolf 1981).

Proteinase A is also synthesized via a larger inactive precursor (Zubenko *et al.* 1981, 1983; Mechler *et al.* 1982). As *prb1* nonsense mutations do not affect the levels of proteinase A, it is very unlikely that proteinase B is involved in maturation of this precursor (Zubenko *et al.* 1979).

Proteinase B is also synthesized via a larger precursor (Mechler *et al.* 1982). As *pra1* mutants retain proteinase B activity, it is unlikely that proteinase A is involved in this maturation (Mechler and Wolf 1981).

The function of the *PEP4* gene is required for the conversion of the larger of the two precursors of proteinase A and carboxypeptidase Y to the respective mature, active forms, for in *pep4* mutant strains the larger precursor forms accumulate (Hemmings *et al.* 1981; Jones *et al.* 1981; Zubenko 1981; Zubenko *et al.* 1981, 1983). Conversion of the smaller precursors of proteinase A and carboxypeptidase Y to the respective larger precursors does not depend on *PEP4* gene function (Zubenk *et al.* 1981, 1983). Since mutations in the *PEP4* gene impart a deficiency for at least five vacuolar enzyme activities, among them proteinase A, proteinase B, and carboxypeptidase Y, it is likely that many, if not all, vacuolar enzymes are synthesized as inactive precursors that mature to active forms by a common mechanism. Secreted enzymes such as invertase and acid phosphatase, which are localized in the periplasm (Sutton and Lampen 1962; McLellan and Lampen 1963), do not seem to share this common pathway (or portion of a pathway) since *pep4* mutations do not affect the secretion or activity of these enzymes (Jones *et al.* 1981; Zubenko 1981; Jones *et al.* 1982).

The maturation process is capable of distinguishing among the hydrolase precursors. The *pep4* mutation exhibits a dosage effect on the levels of activity of proteinases A and B, but not on the level of carboxypeptidase Y activity. As implied by the dosage effect, and revealed by pulse-chase experiments, the kinetics for maturation of proteinase A and carboxypeptidase Y are different in *pep4/PEP4* diploids, being slow and apparently rate limiting for proteinase A but wild type for carboxypeptidase Y (Zubenko *et al.* 1981, 1983).

It is, of course, possible that the *PEP4* gene encodes a protease or a component of a proteolytic system responsible for scission of the N-terminal peptide(s). Alternatively, mutations in the *PEP4* gene might prevent delivery of the precursors into some cellular compartment in which the proteolytic machinery resides. The former possibility seems more likely, since the carboxypeptidase Y precursor that accumulates in *pep4* mutants is found in vacuoles (Stevens *et al.* 1982). Further studies should clarify the nature of the proteolytic events and the proteases involved.

*S. cerevisiae* cells of $\alpha$ mating type produce a tridecapeptide pheromone, called $\alpha$ factor, which causes cells of **a** mating type to arrest in the $G_1$ phase of the cell cycle (Düntze *et al.* 1970; 1973; Stötzler and Düntze 1976; Bucking-Throm *et al.* 1973). Recently, a DNA fragment has been cloned whose nucleotide sequence encodes a polypeptide bearing four $\alpha$-factor tridecapeptide sequences separated by spacers (Kurjan and Herskowitz 1982). Noteworthy features of this polypeptide are summarized in Figure 6.1. This polypeptide shares features with multivalent precursors of hormones (see Steiner *et al.* 1980 for review), enkephalins (Stern *et al.* 1981), and promellitin (Kreil *et al.* 1980a,b). Release of $\alpha$ factor from this multivalent precursor could be effected by trypsinlike and chymotrypsinlike cleavages (between Arg and Glu and between Tyr and Lys, respectively), or by trypsinlike cleavages with trimming of the Lys-Arg C-terminal tails effected by carboxypeptidase activity. These cleavages could be followed by processive removal of X-Ala dipeptides from the N-terminal by an activity like dipeptidylpeptidase IV (Kenny *et al.* 1976). Processive removal of X-Pro or X-Ala dipeptides occurs in processing of promellitin (Kreil *et al.* 1980a,b).

The *ste13-1* mutation causes *MAT$\alpha$* cells to be sterile, but does not affect *MAT***a** cells (Sprague *et al.* 1981). *ste13* mutants do not produce $\alpha$ factor, but instead accumulate a mixture of polpeptides related to $\alpha$ factor. The predominant species are polypeptides four amino acids longer than normal, with the N-terminal extensions being Glu-Ala-Glu-Ala or Asp-Ala-Glu-Ala (D. Julius, L. Blair, A. Brake, and J. Thorner, personal communication). These $\alpha$-factor-like peptides appear to result from a reduced rate of processing by a dipeptidylpeptidase IV-like activity.

Yeast possesses a dipeptidylpeptidase IV-like activity (Suarez-Rendueles *et*

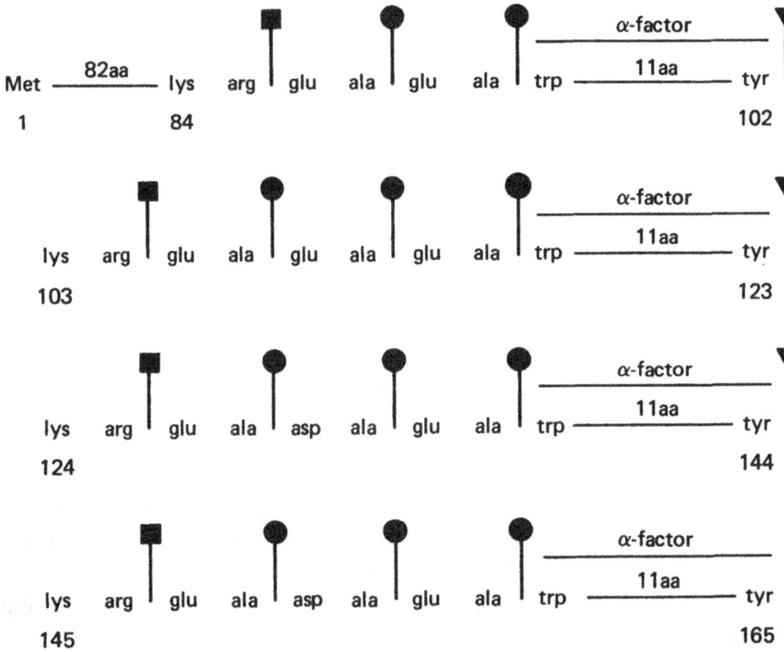


**Figure 6.1.** Elements of the amino acid sequence of the α-factor precursor as inferred from the nucleotide sequence (J. Kurjan and I. Herskowitz, 1982) and proposed processing sites. Numbers below the line indicate the number of the amino acid in the precursor with Met being the first residue. ▮, trypsinlike cleavage; ▼; chymotrypsinlike cleavage; ●, dipeptidylpeptidase IV-like processive cleavage.

*al.* 1981). This activity is present, but at reduced levels, in *ste13* mutants (A. Brake, D. Julius, and J. Thorner, personal communication). Interestingly enough, *ste13* mutants can be complemented by plasmids bearing either of two different yeast DNA sequences (G. Sprague and I. Herskowitz, personal communication) and each results in elevation of the level of the dipetidylpeptidase IV-like activity (A. Brake, D. Julius, and T. Thorner, personal communication). The simplest interpretation of these results is that there are two enzymes, each encoded by a separate gene and each gene capable of complementing the *ste13* mutation. The fact that *ste13* does not render *MAT***a** cells sterile raises interesting questions about the mechanism of production of **a**-factor by *MATa* cells.

*prb1* and *pep4* mutants produce normal levels of α factor (using the halo assay; unpublished observations) and mating of *pra1* mutants appears normal (Mechler and Wolf 1981). These results imply that, as expected, vacuolar proteinases are not involved in processing of the α-factor precursor.

Further studies of this processing system are certain to prove interesting.

Combined in one precursor molecule are features of structure that would appear to require processing machinery combining elements of several other processing pathways. As cells of **a** mating type also produce a pheromone, **a**-factor, comparative studies will be revealing. Analyses of regulation of these proteolytic processes should be particularly fruitful.

## Specific Activations

In yeast, the primary septum, which forms between mother and daughter cells during budding and which remains with the mother cell as a bud scar on cell separation, is composed of chitin (Bacon *et al.* 1966, 1969; Cabib and Bowers 1971). Chitin synthesis is catalyzed by chitin synthase, an enzyme present in the cell membrane in an inactive form (Duran *et al.* 1975) that can be activated *in vitro* by proteinase B (Ulane and Cabib 1976). These authors postulated that proteinase B is responsible *in vivo* for activation of chitin synthase zymogen by fusion of vesicles containing proteinase B with selected sites on the plasma membrane. Hasilik (1974) proposed that, following completion of the primary septum, proteinase B terminates synthesis of chitin by degrading chitin synthase.

The finding that cells bearing nonsense mutations in the proteinase B structural gene, *PRB1*, produce chitin-containing primary septa and divide at normal rates indicates that proteinase B is not required for activaton of chitin synthase zymogen (Zubenko *et al.* 1979). The normal activation process appears to involve proteolysis, but the protease responsible has not been identified (Pilar-Fernandez *et al.* 1982).

## Specific Inactivations

Proteinases have been implicated in a number of specific inactivations of enzymes and oligopeptides. These include the glucose-induced inactivation of several enzymes of gluconeogenesis and the glyoxylate cycle, the maltose and galactose uptake systems, uridine nucleosidase and the vacuolar aminopeptidase; carbon-starvation-induced inactivation of NADP-dependent glutamate dehydrogenase; nitrogen-starvation-induced degradation of NAD-dependent glutamate dehydrogenase and the degradation of $\alpha$-factor by cells of **a** mating type.

When glucose is added to a yeast culture that is using acetate as a source of carbon and energy, the gluconeogenic enzymes hexosediphosphatase, the cytoplasmic isozyme of malate dehydrogenase, phosphoenolpyruvate carboxykinase, and one enzyme of the glyoxylate cycle, isocitrate lyase, are inactivated (Ferguson *et al.* 1967; Gancedo 1971; Gancedo and Schwerzmann 1976; Witt *et al.* 1966; Wolf *et al.* 1979). This process has been termed *catabolite inactivation* (Holzer 1976). It has been proposed that catabolite inactivation involves proteolysis of the target enzymes, for disappearance of cytoplasmic malate dehydrogenase antigen and hexosediphosphatase antigen accompanies the inactivation of these enzyme activities (Hägele *et al.* 1978; Neeff *et al.* 1978; Funayama *et al.* 1980). Proteinase B was suggested as the enzyme responsible for these inactivations (save that of isocitrate lyase) Holzer 1976; Jušik *et al.* 1976; Molano and Gancedo 1974).

By using strains bearing *prb1* nonsense mutations, Zubenko and Jones (1979) have shown that catabolite inactivation of the three gluconeogenic enzymes will occur in the absence of proteinase B activity. Wolf *et al.* (1979) confirmed these results using a mutant lacking proteinase B (the locus bearing the mutation and the type of mutation were not identified) and also found that inactivation of isocitrate lyase proceeded normally in this mutant. Inactivation of the three gluconeogenic enzymes also proceeds normally in the *pra1* mutant, which apparently lacks proteinase A activity (Mechler and Wolf 1981).

Components of the maltose (Robertson and Halvorson 1957; Gorts 1969) and galactose (Matern and Holzer 1977) uptake systems and the vacuolar aminopeptidase (Frey and Röhm 1979b) are susceptible to glucose-induced catabolite inactivation. Inactivation of the aminopeptidase is accompanied by the loss of the antigen (Frey and Röhm 1979b). Studies of these inactivations in proteinase-deficient mutants have not been reported.

The addition of glucose to yeast cultures using ethanol as a carbon source results in a tenfold decrease in the level of uridine nucleosidase (Magni *et al.* 1977). Although there is little evidence that the inactivation is the result of proteolysis *in vivo,* Magni *et al.* (1979) suggested that proteinase A might play a physiological role in this inactivation. Whether this inactivation occurs in mutants lacking proteinase A activity is as yet unreported.

Loss of NADP-dependent glutamate dehydrogenase activity occurs when yeast cells are starved for a source of carbon (Mazon 1978). A parallel loss of NADP-GDH subunit protein accompanies this loss of activity (Hemmings and Mazon 1979; Mazon and Hemmings 1979). Based on the observation that proteinase B inactivates NADP-GDH *in vitro,* Hemmings and Mazon (1979) suggested that proteinase B might be responsible for this inactivation *in vivo.* By employing strains bearing nonsense mutations at *PRB1* and/or *PRC1,* Hemmings *et al.* (1980) demonstrated that neither proteinase B nor carboxypeptidase Y was essential for either the inactivation of the NADP-dependent GDH activity or the associated loss of subunit protein that occurs on carbon starvation. The rates of loss of activity as well as subunit protein observed for these strains as well as for the *pep4-3* mutant (having low levels of proteinases A and B and carboxypeptidase Y) during carbon starvation were not distinguishable from those observed for the wild-type strain. Similar results were reported earlier with mutants lacking proteinase B (Wolf and Ehmann 1979). However, the mutations in the mutants were uncharacterized and one could not be sure that proteinase B activity was absent *in vivo.* Inactivation of NADP-dependent GDH also proceeds normally in *pra1* mutants, which apparently lack activity for proteinase A (Mechler and Wolf 1981).

Nitrogen starvation leads to inactivation and proteolysis of subunit protein of the NAD-dependent glutamate dehydrogenase. Inactivation proceeds at a faster rate than proteolysis and appears to involve phosphorylation (Hemmings 1980). No candidates for the proteolytic enzyme were suggested or tested.

Cells of **a** mating type catalyze cleavage of the pheromone $\alpha$-factor (Maness and Edelman 1978; Ciejek and Thorner 1979; Finkelstein and Strausberg 1979; Ciejek 1980), allowing recovery from $\alpha$-factor-induced cell cycle arrest. The enzyme activity responsible for this cleavage is present in the membranes of **a**, $\alpha$ and **a**/$\alpha$ cells (Ciejek 1980), although the latter two cell types do not catalyze cleavage *in vivo.* The cleavage catalyzed is a specific one and occurs between Leu-6 and Lys-7 of the tridecapeptide (Ciejek 1980). Interestingly, among the cleavages catalyzed by the membrane-located proteinase P of Achstetter *et al.* (1981) is the endoproteolytic cleavage of H-D-Val-Leu-Lys- *p*-nitroanilide. Of interest is whether the susceptible bond is the Leu-Lys bond (as seems likely) and whether proteinase P corresponds to the $\alpha$-factor cleavage activity.

*sst1* and *sst2* (supersensitive to mating factor) mutants are defective in recovery from $\alpha$-factor-induced cell cycle arrest (Chan and Otte 1982a,b). The *sst1* mutant cells degrade $\alpha$-factor at greatly reduced rates. There is some evidence that levels of a membrane-bound protease capable of cleaving $\alpha$-factor may be reduced in these *sst1* mutants (Ciejek 1980). Whether *SST1* encodes the protease and whether the defect in the *sst1* mutants involves an aminopeptidase or an endoprotease is at present unclear. Also unclear is whether the extracellular aminopeptidase (Frey and Röhm 1978) is involved in degradation of $\alpha$-factor. Sprague and Herskowitz (1981) isolated *bar1* mutations, allelic to *sst1* mutations, by an entirely different procedure. See this reference for additional properties of the mutants.

## Degradation of Abnormal Proteins

There is some evidence that abnormal proteins may be labile *in vivo* in yeast. Labile proteins include nonsense fragments, missense proteins, and unassembled subunits of enzyme complexes.

The *HIS4* locus encodes a trifunctional enzyme that catalyzes the second, third, and tenth steps in the histidine biosynthetic pathway. The enzyme is present intracellularly as a dimer of identical 95,000-d subunits (Keesey *et al.* 1979). The *his4C-1176* ochre mutation maps in the distal region of the distal of three cistrons. The 92,000-d nonsense fragment resulting from this mutation is present in very small amounts in cell extracts (Bigelis and Burridge 1978; Bigelis and Fink 1981). Unlike other *his4C* mutations, the *his4C-1176* mutation fails to complement missense mutations in the *HIS4B* region (Fink and Styles 1974). This aberrant complementation pattern is thought to be a result of proteolysis of the nonsense fragment. Studies by Zubenko (1981) indicate that mutations in *PRB1, PRC1,* and *PEP4,* homozygous singly or in combination, do not alter the complementation behavior of the *his4C-1176* allele, making it unlikely that vacuolar proteinases are involved in degradation of this nonsense fragment.

The *ino1-16* mutation lies in the structural gene for inositol-1-phosphate synthase and imparts a temperature-sensitive inositol auxotrophy to cells that bear it (Donahue and Henry 1981). When purified, the enzyme from the mutant is not demonstrably more thermolabile than the normal enzyme. The enzyme is present at low levels in the mutant and is very susceptible to proteolysis during purification (Donahue and Henry 1981). Whether the temperature-sensitive phenotype is a consequence of accelerated proteolysis of the enzyme at the restrictive temperature is unclear, but introduction of a *prb1* nonsense mutation into the *ino1-16* strain does not alleviate the temperature sensitivity (E. Thompson and E. Jones, unpublished observations).

The fatty acid synthase complex is composed of products of two different genes, *fas1* and *fas2* (Schweizer *et al.* 1971). Virtually all noncomplementing mutations in either of the two loci result in failure to produce any cross-reacting material (of either subunit). It has been suggested that both the mutationally altered subunit and the normal subunit may be susceptible to proteolytic degradation because they cannot associate into an enzyme complex (Dietlein and Schweizer 1975; Schweizer *et al.* 1978). However, Schweizer *et al.* (1978) pointed out that the same results would obtain if the synthesis of both were under mutual positive control. Studies involving protease-deficient mutants have not been reported.

## General Proteolytic Processes

### Protein Degradation During Growth and Starvation

Overall protein degradation proceeds at a basal rate of 0.5–1% per hour in growing yeast cells utilizing glucose as a carbon source (Halvorson 1958b; Betz 1976; Lopez and Gancedo 1979). However, this basal level increases to almost 2% per hour when cells are grown on ethanol instead of glucose (Lopez and Gancedo 1979). It has been suggested (Lopez and Gancedo 1979) that the lower rate of proteolysis seen in glucose-grown cells is a result of repression of the vacuolar proteinases by glucose (Hansen *et al.* 1977). Like bacterial and mammalian cells, yeast cells respond to starvation for nitrogen, amino acids, or phosphate with a several-fold increase in the level of protein degradation over the basal level observed for cells growing on glucose as a carbon source (Halvorson 1958a; Betz 1976; Lopez and Gancedo 1979). Unlike these other cell types, however, the inhibition of protein synthesis (in mutants conditionally defective in protein synthesis or by administration of cycloheximide) stimulates proteolysis to the same degree as does starvation for required amino acids (Betz 1976).

Resting cells show higher rates of protein turnover than growing cells (Halvorson 1958a,b) and rates of protein degradation increase two- to threefold as cells enter stationary phase (Bakalkin *et al.* 1976). Possibly the loss of allophanate hydrolase activity in stationary-phase cells is a reflection of this phenomenon (Sumrada and Cooper 1978).

Levels of protein degradation in various proteinase-deficient mutants have been determined for selected conditions, but a complete study has not been carried out. Growing cells of mutants have not been examined for rates of degradation.

Of all the mutants altered in levels of proteinases or their inhibitors, only the *pai1* mutant has been examined for protein degradation in stationary-phase cells. Recall that this mutant shows no derepression of synthesis of the proteinase A inhibitor at stationary phase and, as a consequence, has a high proteinase A:$I^A$ ratio in stationary-phase cells. Rates of overall protein degradation in stationary phase are normal in this mutant (Beck *et al.* 1980). However, tryptophan synthase and $I^B$, known to be rapidly inactivated by proteinase A *in vitro* (Betz *et al.* 1974; Jušik *et al.* 1976), disappear more rapidly than normal when the *pai1* cells enter stationary phase (Beck *et al.* 1980). These results suggest proteinase A as the agent of inactivation of these two proteins. The impact of the *pra1* mutation on these degradations will clearly be of interest.

If cells are transferred from a glucose–ammonia medium to a medium lacking carbon or nitrogen, protein is degraded at a rate of 2.2% per hour. Cells deficient in proteinase B activity show a lower rate of degradation, about 60% of the wild-type rate (Wolf and Ehmann 1979). Cells deficient for carboxypeptidase Y or for carboxypeptidases Y and S show normal rates of degradation when subjected to this regimen (Wolf and Ehmann 1981). If the transfer is made to a medium containing acetate, the cells deficient for proteinase B degrade protein at only 40% of the wild-type rate (Wolf and Ehmann 1979).

If cells growing on acetate as a carbon source are subsequently starved for nitrogen in the presence of acetate, protein is degraded at a rate of 2–3% per hour (Betz and Weiser 1976; Zubenko and Jones 1981). Mutations in the structural gene for carboxypeptidase Y *(PRC1)* decrease degradation by 10–20%; nonsense mutations in the proteinase B *(PRB1)* structural gene reduce degradation by 40–50%. The proteinase B deficiency is epistatic to the carboxypeptidase Y deficiency, implying that proteinase B provides the substrates on which carboxypeptidase Y acts, despite the presence in the cells of proteinase A. The pleiotropic, *pep4-3* mutation, deficient for most, if not all, vacuolar hydrolases, reduces protein degradation under this condition by 60–75% (Zubenko and Jones 1981).

These results clearly implicate proteinase B in the protein degradation triggered by starvation for nitrogen or nitrogen and carbon. Contradictory results were obtained for carboxypeptidase Y, with one report indicating an involvement of the enzyme in nitrogen starvation-induced degradation (Zubenko and Jones 1981) while the other indicates lack of involvement in the carbon and nitrogen starvation-induced degradation (Wolf and Ehmann 1981). Whether this discrepancy is due to the difference in regimen or to strain differences is unclear.

The implication of all of these studies is that vacuolar proteinases participate in the turnover and degradation of protein associated with starvation or entry into stationary phase. Whether this occurs because proteins can enter the vacuole to allow degradation or because the proteinases can enter the cytosol under these conditions is unclear. Evidence that small molecules can depart from the vacuole and induce synthesis of selected enzymes under these conditions has been presented (Whitney *et al.* 1973; Sumrada and Cooper 1978).

## Protein Degradation During Sporulation

Diploid yeast cells of **a**/$\alpha$ mating type are induced to undergo meiosis and sporulation by transfer to an acetate medium free of nitrogen. The process of sporulation in yeast, as in bacteria, is accompanied by massive protein degradation. Sporulating cells have been reported to degrade from 30 to 70% of their vegetative proteins (Hopper *et al.* 1974; Klar and Halvorson 1975; Zubenko and Jones 1981). The differences in extents of degradation are presumed to reflect differences in strains and experimental protocols. That the vacuolar proteinases, A, B, and carboxypeptidase Y, might be responsible for the observed protein degradation and might be essential for sporulation was suggested by Betz and Weiser (1976), for the levels of the three proteinases rise when cells sporulate (Chen and Miller 1968; Klar and Halvorson 1975; Betz and Weiser 1976) and no new proteinases are detected during sporulation (Klar and Halvorson 1975; Betz and Weiser 1976).

The preponderance of the evidence suggests that the protein degradation that accompanies sporulation is not a sporulation-specific proteolysis, but rather a response to nitrogen starvation, for asporogenous diploids of **a**/**a** mating type (and of several proteinase genotypes) (Zubenko and Jones 1981) and haploid cells of **a** mating type (Betz and Weiser 1976) degrade their vegetative proteins in a fashion similar to their **a**/$\alpha$ counterparts. However, Hopper *et al.* (1974) report that **a**/**a** and $\alpha/\alpha$ diploids degrade their proteins at lower rates than their **a**/$\alpha$ counterparts. The basis of the discrepancy is unknown.

Diploids homozygous for *pra1*, and therefore deficient for proteinase A activity, show rates of protein degradation during sporulation only 70% of wild-type rates (Mechler and Wolf 1982). The frequency of ascus formation was only 60% of wild type for these diploids. As the strains tested were not isogenic and as frequencies of sporulation vary widely among strains of different genetic backgrounds, these results are difficult to assess.

Betz (1979) described a recessive mutation that reduced levels of proteinase A to 20% of normal, but concluded it was not a structural gene mutation. This mutation was detected in a *cdc* mutant strain isolated by Hartwell *et al.* (1970) after extensive mutagenesis. Three proteinase A-deficient segregants were backcrossed to the original strain and all three resultant diploids failed to sporulate. From this observation Betz concluded that the proteinase A and sporulation deficiencies are the result of a single mutation. One would obtain the result seen 12.5% of the time even if the two effects were due to distinct muta-tins. Hence these observations contribute little to resolving whether proteinase A activity is required for sporulation.

Diploids homozygous for a nonsense mutation in *PRB1*, and therefore lacking proteinase B activity, show rates of degradation in sporulation medium 50–60% of normal and those lacking carboxypeptidase Y activity show rates of degradation 80–90% of normal. As was seen for asporogenous strains, the degradation deficit caused by deficiency for proteinase B activity is epistatic to that attributable to absence of carboxypeptidase Y activity (Zubenko and Jones 1981). The kinetics and extent of haploidization in diploids bearing *prb1* and/or *prc1* mutations are similar to those of wild-type diploids. Normal asci are produced in the *prc1* homozygote. The asci produced in the *prb1* and *prb1 prc1* homozygotes are abnormal and spores are not readily visible in the light microscope. They are present, however, at normal frequencies, but remain embedded in a matrix within the ascus. Apparently, proteinase B activity is essential for dissolution of the matrix material. The initial report (Zubenko *et al.* 1979) that *prb1* homozygotes do not sporulate stems from the fact that the matrix material renders the spores nearly invisible. The spores and asci produced in *prb1* homozygotes are abnormally small (Zubenko and Jones 1981). These studies utilized isogenic strains.

These results clearly implicate proteinase B and carboxypeptidase Y in the protein degradation that accompanies sporulation and indicate, as well, that proteinase B may play a role in the normal spore dispersal mechanism through dissolution of the matrix material. As the spores resulting from these diploids germinate at normal frequencies, neither proteinase is required for spore germination (Zubenko and Jones 1981).

Wolf and Ehmann (1979) reported that a diploid deficient in proteinase B activity shows lower rates of protein degradation during sporulation and a lower rate and extent of formation of asci. As the strains employed were not isogenic, the significance of this latter observation is unclear.

Wolf and Ehmann (1981) report that diploids lacking carboxypeptidase Y or carboxypeptidases S and Y showed rates of protein degradation during sporulation equivalent to wild-type degradation rates. This observation is clearly at odds with that of Zubenko and Jones (1981). The basis of the discrepancy is unknown. Wolf and Ehmann (1981) indicate that deficiency for carboxypeptidases Y and S has little or no impact on sporulation frequency.

Diploids homozygous for the *pep4-3* mutation, and therefore deficient for proteinases A and B and carboxypeptidase Y, neither undergo meiosis nor sporulate (Zubenko and Jones (1981). As this mutation also reduces activity for other vacuolar hydrolases (Jones *et al.* 1981; Zubenko 1981; Jones *et al.* 1982), one cannot infer the basis of the defect in sporulation.

Trumbly (1980) did not specifically address the question of whether aminopeptidases are required for protein degradation and/or sporulation. However, one can draw some inferences from the strain list presented and the mode of strain construction indicated. Strains homozygous for any three of the four mutations *lap1*, *lap2*, *lap3*, and *lap4* are able to sporulate, although there is no indication of frequency of sporulation and no certainty that these mutations are structural gene mutations. Whether the quadruply mutant diploid will sporulate is unknown. These results imply that the vacuolar aminopeptidase (missing in *lap4* mutants) is not required for sporulation.

Diploids homozygous for the *bin1* mutation, which reduces levels of $I^B$, do not sporulate but those homozygous for *bin2* or *bin3* mutations, which likewise reduce $I^B$ levels, do sporulate (Zubenko 1981). The sporulation deficiency of the *bin1* homozygote cannot be attributed simply to the deficiency for $I^B$, since *bin2* and *bin3* homozygotes sporulate. As the $I^B$ deficiency in these mutants is only partial, possible participation of this inhibitor in sporulation remains untested.

## Peptide Utilization

Yeast possess the ability to take up and utilize a variety of di-, tri-, and oligopeptides (see Becker and Naider 1980 for review). The peptides can supply auxotrophic requirements. Similarly, dipeptides, such as Cbz-Gly-Leu, with blocked $\alpha$-amino groups, can satisy the leucine requirements of Leu- strains (Jones 1977).

Four peptidases, two of which correspond to the aminopeptidases studied by Trumbly (1980), have been detected by their ability to catalyze hydrolysis of various peptides (Rose *et al.* 1979). To which, if any, of the four peptidases the dipeptidase purified by Röhm (1974) corresponds is unknown. Carboxypeptidases Y and S can catalyze cleavage of Cbz-Gly-Leu (Wolf and Weiser 1977).

The impact of proteinase deficiencies on the ability of peptides to satisfy nutritional needs of cells has been examined to a limited extent only. Mutants that lack carboxypeptidases Y and S, *prc1 cps1* double mutants, can no longer utilize Cbz-Gly-Leu as a source of leucine and nitrogen, whereas those lacking carboxypeptidase Y but retaining carboxypeptidase S activity, *prc1* mutants, retain the ability to use the dipeptide (Jones 1977; Wolf and Ehmann 1981). Strains that lack carboxypeptidase S but possess carboxypeptidase Y (*dut1* mutants—whether allelic to *cps1* is unknown) have a reduced ability to utilize the dipeptide as a leucine and nitrogen source (E. Jones, unpublished observations). Possibly the greater importance of carboxypeptidase S in utilization of the dipeptide stems from the fact that the level of the enzyme is greatly elevated when Cbz-Gly-Leu is the only nitrogen source, whereas that of carboxypeptidase Y is unaffected (Wolf and Ehmann 1978b).

Trumbly (1980) reported that a quadruple mutant, *lap1 lap2 lap3 lap4*, having little residual aminopeptidase activity, grows very slowly when leucinamide is the nitrogen source. This mutant retains the ability to utilize Leu-Gly as a leucine source and therefore presumably still possesses dipeptidase activity (Rohm 1974). Mutants lacking the dipeptidase activity or peptidases 3 and 4 of Rose *et al.* (1979) have not been reported.

**Table 6.3.** Mutations that Affect Proteinases, Proteinase Inhibitors, or Proteolytic Processes in *Saccharomyces cerevisiae*

| Mutations | Proteinase, inhibitor, or proteolytic process affected | Comments | References |
|---|---|---|---|
| *pra*$^-$ | Proteinase A | Not in the structural gene | Betz 1979 |
| *pra*$^-$ | Proteinase A | Activity level reflects gene dosage; allelism tests with *pra1* have not been performed | Jones *et al*. 1981 |
| *pra1* | Proteinase A | May be structural gene mutation | Mechler and Wolf 1981 |
| *prb1* | Proteinase B | Structural gene | Zubenko *et al*. 1979 |
| *prb2* | Proteinase B | Not in structural gene | Zubenko *et al*. 1979 |
| *prb3* | Proteinase B | Not in structural gene | Zubenko *et al*. 1979 |
| *prb4* | Proteinase B | Not in structural gene | Zubenko *et al*. 1979 |
| *prb*$^-$ | Proteinase B | Allelism tests with *prb1-prb4* have not been performed | Wolf and Ehmann 1978a, 1979 |
| *prc1* | Carboxypeptidase Y | Structural gene; alleism tests were performed | Wolf and Fink 1975; Wolf and Weiser 1977; Jones 1977 |
| *cps1* | Carboxypeptidase S | May be structural gene | Wolf and Ehmann 1981 |
| *dut1* | Carboxypeptidase S | Allelism tests with *cps1* have not been performed | Zubenko 1981 |
| *pep1-pep3, pep5-pep16, pep18* (see Table 6.2) | Carboxypeptidase Y; some affect levels of proteinases A and B | Not structural gene; pleiotropic; some may affect other vacuole hydrolases | Jones 1977; A. Mitchell, M. Hospodar, and E. Jones, unpublished |

| | | | |
|---|---|---|---|
| *pep4* ≡ *pho9* | Carboxypeptidase Y, proteinases A and B, other vacuole hydrolases | Not structural gene; accumulates enzyme precursors; pleiotropic, activity levels for proteinases A and B reflect gene dosage | Jones 1977; Jones *et al.* 1981; Zubenko 1981; Hemmings *et al.* 1981; Jones *et al.* 1982; Zubenko *et al.* 1981, 1982, 1983; Kaneko *et al.* 1982 |
| *lap1* | Aminopeptidase (LAPI) | | Trumbly 1980 |
| *lap2* | Aminopeptidase (LAPII) | | Trumbly 1980 |
| *lap3* | Aminopeptidase (LAPIII) | | Trumbly 1980 |
| *lap4* | Aminopeptidase (LAPIV) | Vacuolar aminopeptidase | Trumbly 1980 |
| *pai1* | Proteinase A inhibitor | Probably not structural gene | Wolf *et al.* 1979; Beck *et al.* 1980 |
| *bin1* | Proteinase B inhibitor | Probably not structural gene | Zubenko 1981 |
| *bin2* | Proteinase B inhibitor | Probably not structural gene | Zubenko 1981 |
| *bin3* | Proteinase B inhibitor | Probably not structural gene | Zubenko 1981 |
| *sst1* ≡ *bar1* | Degradation of $\alpha$ factor | Cells of **a** mating type slow to recover from $\alpha$-factor arrest | Ciejek 1980; Sprague and Herskowitz 1981; Chan and Otte 1982a,b |
| *ste13* | Processing of $\alpha$ factor precursor; dipeptidylpeptidase IV | *MAT*$\alpha$ (but not *MAT***a**) cells bearing *ste13* are sterile | Sprague *et al.* 1981; D. Julius, L. Blair, A. Brake, and J. Thorner, personal communication |

## Summary and Conclusions

In Table 6.3 is a compilation of mutations known to affect defined proteases, protease inhibitors, or proteolytic processes. It seems probable or certain, as the case may be, that structural genes for proteinase A, proteinase B, carboxypeptidase Y, and carboxypeptidase S have been identified. Evidence in support of structural gene assignments for any other protease or inhibitor is lacking.

Studies employing mutants lacking one or more of the defined proteases of the vacuole indicate clearly that these proteases are not required for any of the proteolytic events that require restricted enzyme specificity. It seems certain that proteinase A and B and carboxypeptidase Y are not required for protein maturation or processing or for any of the specific degradations found in certain cell types or under selected conditions. Catabolite inactivation, degradation of specific enzymes under selected conditions, synthesis and degradation of $\alpha$ factor, activation of chitin synthase zymogen, and processing of precursors of vacuolar hydrolases are examples of proteolytic processes for which vacuolar proteases are not required. It seems equally clear that proteases of the vacuole are involved in the rather nonspecific protein degradation induced by nitrogen starvation, whether or not this degradation is associated with sporulation. The implication of these observations is that the enzymes of the vacuole are brought into play when the only source of building blocks for adaptation or restructuring is entirely within the cell and the cell must adapt or die.

As might be expected, the aminopeptidases and carboxypeptidases are implicated in utilization of peptides. Both types of enzymes are involved in utilization of exogenously supplied peptides as the studies with auxotrophs attest. The epistasis of the *prb1* mutation to the *prc1* mutation in protein degradation implicates carboxypeptidase Y in degradation of peptides produced as a consequence of catalysis by proteinase B.

The lack of involvement of the vacuolar proteinases in the degradation of numerous but selected enzymes upon metabolic shift implies that additional proteolytic systems remain to be discovered. Possibly they are cytoplasmically located.

With the possible exception of the *sst1 (bar1), ste13,* and *pep4* mutations, mutations affecting proteolytic enzymes of highly restricted specificity have not been identified. Identification and characterization of the proteases involved and the genes that control and encode them will undoubtedly lead to fresh insights in the future.

## References

Achstetter, T., Ehmann, C., Wolf, D. (1981). New proteolytic enzymes in yeast. Arch. Biochem. Biophys. 207:445–454.

Aibara, S., Hayashi, R., Hata, T. (1971). Physical and chemical properties of yeast proteinase C. Agric. Biol. Chem. 35:658–666.

Bacon, J., Davidson, E., Jones, D., Taylor, I. (1966). The location of chitin in the yeast cell wall. Biochem. J. 101:36C–38C.

Bacon, J., Farmer, V., Jones, D., Taylor, I. (1969). The glucan components of the cell wall of bakers yeast *(Saccharomyces cerevisiae)* considered in relation to its ultrastructure. Biochem. J. 114:557–565.

Bakalkin, G., Kalnov, S., Zubatov, A., Luzikov, V. (1976). Degradation of total cell protein at different stages of *Saccharomyces cerevisiae* yeast growth. FEBS Lett. 63:218–221.

Barth, R., Wolf, D., Holzer, H. (1978). Studies on the carboxypeptidase Y-inhibitor complex of yeast. Biochim. Biophys. Acta 527:63–69.

Bauer, H., Sigarlakie, E. (1975). Localization of alkaline phosphatase in *Saccharomyces cerevisiae* by means of ultrathin frozen sections. J. Ultrasr. Res. 50:208–215.

Beck, I., Fink, G. Wolf, D. (1980). The intracellular proteinases and their inhibitors in yeast. A mutant with altered regulation of proteinase A inhibitor activity. J. Biol. Chem. 255:4821–4828.

Becker, J., Naider, F. (1980). Transport and utilization of peptides in yeast. In: Microorganisms and Nitrogen Sources, edited by J. Payne. New York: Wiley, pp. 257–275.

Betz, H. (1975). Levels and turnover of the proteinase B inhibitors in yeast. Biochim. Biophys. Acta 404:142–151.

Betz, H. (1976). Inhibition of protein synthesis stimulates intracellular protein degradation in growing yeast cells. Biochem. Biophys. Res. Commun. 72:121–130.

Betz, H. (1979). Loss of sporulation ability in a yeast mutant with low proteinase A levels. FEBS Lett. 100:171–174.

Betz, H., Weiser, U. (1976). Protein degradation and proteinases during yeast sporulation. Eur. J. Biochem. 62:65–76.

Betz, H., Hinze, H., Holzer, H. (1974). Isolation and properties of two inhibitors of proteinase B from yeast. J. Biol. Chem. 249:4515–4521.

Bigelis, R., Burridge, K. (1978). The immunological detection of yeast nonsense termination fragments on sodium dodecylsulfate—polyacrylamide gels. Biochem. Biophys. Res. Commun. 82:322–327.

Bigelis, R., Fink, G. (1981). The *HIS4* multifunctional protein: Immunochemistry of the wild type protein and altered forms. J. Biol. Chem. 256:5144–5152.

Böhni, P., Gasser, S., Leaver, C., Schatz, G. (1980). In: Structure and Expression of the Mitochondrial Genome, edited by A. M. Kroon and C. Sacconi. Amsterdam: North-Holland, pp. 423–433.

Bucking-Throm, E., Düntze, W., Hartwell, L., and Manney, T. (1973). Reversible arrest of haploid yeast cells at the initiation of DNA synthesis by a diffusible sex factor. Exp. Cell Res. 76:99–110.

Bünning, P., Holzer, H. (1977). Natural occurrence and chemical modification of proteinase B inhibitors from yeast. J. Biol. Chem. 252:5316–5323.

Cabib, E., Bowers, B. (1971). Chitin and yeast budding. Localization of chitin in yeast bud scars. J. Biol. Chem. 246:152–159.

Chan, R., Otte, C. (1982a). Isolation and genetic analysis of *Saccharomyces cerevisiae* mutants supersensitive to $G_1$ arrest by **a**-factor and $\alpha$-factor pheromone. Mol. Cell. Biol. 2:11–20.

Chan, R., Otte, C. (1982b). Physiological characterization of *Saccharomyces cerevisiae* mutants supersensitive to $G_1$ arrest by **a**-factor and $\alpha$-factor pheromone. Mol. Cell. Biol. 2:21–29.

Chen, A., Miller, J. (1968). Proteolytic activity of intact yeast cells during sporulation. Can. J. Microbiol. 14:957–963.

Ciejek, E. (1980). Alpha-factor, an oligopeptide pheromone from *Saccharomyces cerevisiae:* Purification, chemical synthesis and cell-mediated proteolysis. Ph.D. Thesis, University of California, Berkeley.

Ciejek, E., Thorner, J. (1979). Recovery of *S. cerevisiae* **a** cells from $G_1$ arrest by $\alpha$-factor pheromone requires endopeptidase action. Cell 18:623–635.

Dietlein, G., Schweizer, E. (1975). Control of fatty acid biosynthesis in *Saccharomyces cerevisiae.* Eur. J. Biochem. 58:177–184.

Doi, E., Hayashi, R., Hata, T. (1967). Purification of yeast proteinases II. Purification and some properties of yeast proteinase C. Agric. Biol. Chem. 31:160–169.

Donahue, T., Henry, S. (1981). Myoinositol-1-phosphate synthase: characteristics of the enzyme and identification of its structural gene in yeast. J. Biol. Chem. 256:7077–7085.

Düntze, W., MacKay, V., Manney, T. (1970). *Saccharomyces cerevisiae:* A diffusible sex factor. Science 168:1472–1473.

Düntze, W., Stötzler, D., Bucking-Throm, E., Kalbitzer, S. (1973). Purification and partial characterization of $\alpha$-factor, a mating-type specific inhibitor of cell reproduction from *S. cerevisiae.* Eur. J. Biochem. 35:357–365.

Duran, A., Bowers, B., Cabib, E. (1975). Chitin synthetase zymogen is attached to the yeast plasma membrane. Proc. Natl. Acad. Sci. USA 72:3952–3955.

Esmon, B., Novick, P., Schekman, R. (1981). Compartmentalized assembly of oligosaccharides on exported glycoproteins in yeast. Cell 25:451–460.

Félix, F., Brouillet, N. (1966). Purification and propriétés de deux peptidases de levure de brasserie. Biochim. Biophys. Acta 122:127–144.

Ferguson, Jr., J., Boll, M., Holzer, H. (1967). Yeast malate dehydrogenase: enzyme inactivation in catabolite repression. Eur. J. Biochem 1:21–25.

Fink, G., Styles, C. (1974). Gene conversion of deletions in the *HIS4* region of yeast. Genetics 77:231–244.

Finkelstein, D., Strausberg, S. (1979). Metabolism of $\alpha$-factor by **a** mating type cells of *Saccharomyces cerevisiae.* J. Biol. Chem. 254:796–803.

Fischer, E., Holzer, H. (1980). Interaction of proteinases and their inhibitors from yeast. Biochim. Biophys. Acta 615:187–198.

Frey, J., Röhm, K. -H. (1978). Subcellular localization and levels of aminopeptidases and dipeptidase in *Saccharomyces cerevisiae.* Biochim. Biophys. Acta 527:31–41.

Frey, J., Röhm, K. -H. (1979a). External and internal forms of yeast aminopeptidase II. Eur. J. Biochem. 97:169–173.

Frey, J., Röhm, K. -H. (1979b). The glucose-induced inactivation of aminopeptidase I in *Saccharomyces cerevisiae.* FEBS Lett. 100:261–264.

Fujishiro, K., Sanada, Y., Tanaka, H., Katunuma, N. (1980). Purification and characterization of yeast proteinase B. J. Biochem. 87:1321–1326.

Funayama, S., Gancedo, J., Gancedo, C. (1980). Turnover of yeast fructosebisphosphatase in different metabolic condition. Eur. J. Biochem. 109:61–66.

Gancedo, C. (1971). Inactivation of fructose 1,6-diphosphatase by glucose in yeast. J. Bacteriol. 107:401–405.

Gancedo, C., Schwerzmann, J. (1976). Inactivation by glucose of phosphoenolpyruvate carboxykinase from *Saccharomyces cerevisiae.* Arch. Microbiol. 109:221–225.

Gorts, C. (1969). Effect of glucose on the activity and the kinetics of the maltose uptake system and of $\alpha$-glucosidase in *Saccharomyces cerevisiae.* Biochim. Biophys. Acta 184:299–305.

Hägele, E., Neeff, J. Mecke, D. (1978). The malate dehydrogenase isoenzymes of *Saccharomyces cerevisiae*. Eur. J. Biochem. 83:67–76.

Halvorson, H. (1958a). Intracellular protein and nucleic acid turnover in resting yeast cells. Biochim. Biophys. Acta 27:255–266.

Halvorson, H. (1958b). Studies on protein and nucleic acid turnover in growing cultures of yeast. Biochim. Biophys. Acta 27:267–276.

Hansen, R., Switzer, R., Hinze, H., Holzer, H. (1977). Effects of glucose and nitrogen source on the levels of proteinases, peptidases, and proteinase inhibitors in yeast. Biochim. Biophys. Acta 196:103–114.

Hartwell, L., Culotti, J., Reid, B. (1970). Genetic control of the cell-division cycle in yeast. I. Detection of mutants. Proc. Natl. Acad. Sci. USA 66:352–359.

Hashimoto, C., Cohen, R., Zhang, W. -J. Ballou, C. (1981). Carbohydrate chains on yeast carboxypeptidase Y are phosphorylated. Proc. Natl. Acad. Sci. USA 78:2244–2248.

Hasilik, A. (1974). Inactivation of chitin synthase in *Saccharomyces cerevisiae*. Arch. Microbiol. 101:295–301.

Hasilik, A. (1980): Biosynthesis of lysosomal enzymes. Trends Biochem. Sci. 4:237–240.

Hasilik, A., Neufeld, E. (1980). Biosynthesis of lysosomal enzymes in fibroblasts. J. Biol. Chem. 255:4946–4950.

Hasilik, A., Tanner, W. (1976a). Biosynthesis of carboxypeptidase Y in yeast. Evidence for a precursor form of the glycoprotein. Biochem. Biophys. Res. Commun. 72:1430–1436.

Hasilik, A., Tanner, W. (1978a). Biosynthesis of the vacuolar yeast glycoprotein carboxypeptidase Y. Conversion of precursor into enzyme. Eur. J. Biochem. 91:567–575.

Hasilik, A., Tanner, W. (1978b). Carbohydrate moiety of carboxypeptidase Y and perturbation of its biosynthesis. Eur. J. Biochem. 91:567–575.

Hasilik, A., Müller, H., Holzer, H. (1974). Compartmentation of the tryptophan-synthetase proteolyzing system in *Saccharomyces cerevisiae*. Eur. J. Biochem. 48:111–117.

Hayashi, R., Aibara, S., Hata, T. (1970). A unique carboxypeptidase activity of yeast proteinase C. Biochim. Biophys. Acta 212:359–361.

Hata, T., Hayashi, R., Doi, E. (1967a). Purification of yeast proteinases I. Fractionation and some properties of these proteinases. Agric. Biol. Chem. 31:150–159.

Hata, T., Hayashi, R., Doi, E. (1967b). Purification of yeast proteinases. III. Isolation and physicochemical properties of yeast proteinase A and C. Agric. Biol. Chem. 31:357–367.

Hayashi, R., Moore, S., and Stein, W. (1973). Carboxypeptidase from yeast. Large scale preparation and the application to COOH-terminal analysis of peptides and proteins. J. Biol. Chem. 248:2296–2302.

Hemmings, B. (1980). Phosphorylation and proteolysis regulate the NAD-dependent glutamate dehydrogenase from *Saccharomyces cerevisiae*. FEBS Lett. 122:297–302.

Hemmings, B., Mazon, M. (1979). Proteolytic degradation of NADP-dependent glutamate dehydrogenase in yeast. In: Limited Proteolysis in Microorganisms, edited by H. Holzer and G. Cohen. DHEW Publ. No. 79-1591, pp. 69–75.

Hemmings, G., Zubenko, G., Jones, E. (1980). Proteolytic inactivation of the NADP-dependent glutamate dehydrogenase in proteinase-deficient mutants of *Saccharomyces cerevisiae*. Arch. Biochem. Biophys. 202:657–660.

Hemmings, B., Zubenko, G., Hasilik, A., Jones, E. (1981). Mutant defective in processing of an enzyme located in the lysosome-like vacuole of *Saccharomyces cerevisiae*. Proc. Natl. Acad. Sci. USA 78:435–439.

Hinze, H., Betz, H., Saheki, T., Holzer, H. (1975). Formation of a complex between yeast proteinases A and B. Hoppe-Seyler's Z. Physiol. Chem. 356:1259–1264.

Holzer, H. (1976). Catabolite inactivation in yeast. Trends Biochem. Sci. 1:178–181.

Hopper, A., Magee, P., Welch, S., Friedman, M., Hall, B. (1974). Macromolecule synthesis and breakdown in relation to sporulation and meiosis in yeast. J. Bacteriol. 119:619–628.

Johnson, M. (1941). Isolation and properties of a pure yeast polypeptidase. J. Biol. Chem. 137:575–586.

Jones, E. (1977). Proteinase mutants of *Saccharomyces cerevisiae*. Genetics 85:23–33.

Jones, E., Zubenko, G., Parker, R., Hemmings, B., Hasilik, A. (1981). Pleiotropic mutations of *S. cerevisiae* which cause deficiency for proteinases and other vacuole enzymes. Alfred Benzon Symposium 16:182–198, edited by D. von Wettstein, J. Friis, M. Kielland-Brandt, A. Stenderup. Copenhagen: Munksgaard.

Jones, E., Zubenko, G., Parker, R. (1982). *PEP4* gene function is required for expression of several vacuolar hydrolases in *Saccharomyces cerevisiae*. Genetics 102:665–677.

Jušik, M., Hinze, H., Holzer, H. (1976). Inactivation of yeast enzymes by proteinase A and B and carboxypeptidase Y from yeast. Hoppe-Seyler's Z. Physiol. Chem. 357:735–740.

Kaneko, Y., Toh-e, A., Oshima, Y. (1982). Identification of the structural gene and a new regulatory gene for the synthesis of repressible alkaline phosphatase in *Saccharomyces cerevisiae*. Mol. Cell. Biol. 2:127–137.

Keesey, Jr., J., Bigelis, R., Fink, G. (1979). The product of the *HIS4* gene cluster in *Saccharomyces cerevisiae:* a trifunctional polypeptide. J. Biol. Chem. 254:7427–7433.

Kenny, A., Booth, A., George, S., Ingram, J., Kershaw, D., Wood, E., and Young, A. (1976). Dipeptidyl peptidase IV, a kidney brush-border serine peptidase. Biochem. J. 157:169–182.

Klar, A., Halvorson, H. (1975). Proteinase activities of *Saccharomyces cerevisiae*. J. Bacteriol. 124:863–869.

Kominami, E., Hoffschulte, H., Holzer, H. (1981a). Purification and properties of proteinase B from yeast. Biochim. Biophys. Acta 661:124–135.

Kominami, E., Hoffschulte, H. Leuschel, L., Maier, K., Holzer, H. (1981b). The substrate specificity of proteinase B from baker's yeast. Biochim. Biophys. Acta 661:136–141.

Kreil, G., Mollay, C., Kaschnitz, R., Haiml, L., Vilas, U. (1980a). Prepromellitin: specific cleavage of the pre and the propeptide *in vitro*. Ann. N.Y. Acad. Sci. 343:338–346.

Kreil, G., Haiml, L., Suchanek, G. (1980b). Stepwise cleavage of the pro part of promellitin by dipeptidylpeptidase IV. Eur. J. Biochem. 111:49–58.

Kuhn, R., Walsh, K., Neurath, H. (1974). Isolation and characterization of an acid carboxypeptidase from yeast. Biochemistry 13:3871–3877.

Kuhn, R., Walsh, K., Neurath, H. (1976). Reaction of yeast carboxypeptidase C with group specific reagents. Biochemistry 15:4881–4885.

Kurjan, J., Herskowitz, I. (1982). Structure of a yeast pheromone gene (MF$\alpha$): a putative α-factor precursor contains four tandem copies of mature α-factor. Cell 30:933–943.

Lenney, J. (1975). Three yeast proteins that specifically inhibit yeast proteases A, B and C. J. Bacteriol. 122:1265–1273.

Lenney, J., Dalbec, J. (1967). Purification and properties of two proteinases from *Saccharomyces cerevisiae*. Arch. Biochem. Biophys. 120:42–48.

Lenney, J., Matile, P., Wiemken, A., Schellenberg, M., Meyer, J. (1974). Activities and cellular localization of yeast proteinases and their inhibitors. Biochem. Biophys. Res. Commun. 60:1378–1383.

Lewin, A., Gregor, I., Mason, T., Nelson, N., Schatz, G. (1980). Cytoplasmically made subunits of yeast mitochondrial $F_1$-ATPase and cytochrome *c* oxidase are synthesized as individual precursors, not as polyproteins. Proc. Natl. Acad. Sci. USA 77:3998–4002.

Löffler, H. -G., Röhm, K. -H. (1979). Comparative studies on the dodecameric and hexameric forms of yeast aminopeptidase I. Z. Naturforsch 34c:381–386.

Looze, Y., Gillet, L., Deconinck, M., Couteaux, B., Polastro, E., Leonis, J. (1979). Protease B from *Saccharomyces cerevisiae*. Int. J. Peptide Protein Res. 13:253–259.

Lopez, S., Gancedo, J. (1979). Effect of metabolic condition on protein turnover in yeast. Biochem. J. 178:769–776.

Maccechini, M. -L., Rudin, Y., Blobel, G., Schatz, G. (1979a). Import of proteins into mitochondria: precursor forms of the extramitochondrially made $F_1$-ATPase subunits in yeast. Proc. Natl. Acad. Sci. USA 76:343–347.

Maccechini, M. -L., Rudin, Y., Schatz, G. (1979b). Transport of proteins across mitochondrial outer membrane. A precursor form of the cytoplasmically made intermembrane enzyme cytochrome *c* peroxidase. J. Biol. Chem. 254:7468–7471.

Magni, G., Santarelli, I., Natalini, P., Ruggieri, S., Vita, A. (1977). Catabolite inactivation of bakers yeast uridine nucleosidase. Isolation and partial purification of a specific proteolytic inactivase. Eur. J. Biochem. 75:77–82.

Magni, G., Pallotta, G., Natalini, P., Ruggieri, S., Santarelli, I., Vita, A. (1978). Inactivation of uridine nucleosidase in yeast. Purification and properties of an inactivating protein. J. Biol. Chem. 253:2501–2503.

Magni, G., Natalini, P., Santarelli, I., Ruggieri, S., Vita, A. (1979). Inactivation of bakers yeast uridine nucleosidase. In: Limited Proteolysis in Microorganisms, edited by H. Holzer, G. Cohen. DHEW Publ. No. 79-1591, pp. 87–96.

Maier, K., Müller, H., Holzer, H. (1979a). Purification and molecular characterization of two inhibitors of yeast proteinase B. J. Biol. Chem. 254:8491–8497.

Maier, K., Müller, H., Tesch, R., Trolp, R., Witt, I., Holzer, H. (1979b). Primary structure of yeast proteinase B inhibitor 2. J. Biol. Chem. 254:12555–12561.

Maier, K., Müller, H., Tesch, R., Witt, I., Holzer, H. (1979c). Amino acid sequence of yeast proteinase B inhibitor 1 comparison with inhibitor 2. Biochem. Biophys. Res. Commun. 91:1390–1398.

Maness, P., Edelman, G. (1978). Inactivation and chemical alteration of mating factor α by cells and spheroplasts of yeast. Proc. Natl. Acad. Sci. USA 75: 1304–1308.

Marriott, M., Tanner, W. (1979). Localization of dolichyl phosphate- and pyrophosphate-dependent glycosyl transfer reactions in *Saccharomyces cerevisiae*. J. Bacteriol. 139:565–572.

Masuda, T., Hayashi, R., Hata, T. (1975). Aminopeptidases in the acidic fraction of the yeast autolysate. Agric. Biol. Chem. 39:499–505.

Matern, H., Holzer, H. (1977). Catabolite inactivation of the galactose uptake system in yeast. J. Biol. Chem. 252:6399–6402.

Matern, H., Betz, H., Holzer, H. (1974a). Compartmentation of inhibitors of proteinases A and B and carboxypeptidase Y in yeast. Biochem. Biophys. Res. Commun. 60:1051–1057.

Matern, H., Hoffman, M., Holzer, H. (1974b). Isolation and characterization of the carboxypeptidase Y inhibitors from yeast. Proc. Natl. Acad. Sci. USA 71:4874–4878.

Matern, H., Barth, R., Holzer, H. (1979). Chemical and physical properties of the carboxypeptidase Y inhibitor from bakers yeast. Biochim. Biophys. Acta 567:503–510.

Matile, P., Wiemken, A. (1967). The vacuole as the lysosome of the yeast cell. Arch. Mikrobiol. 56:148–155.

Matile, P., Wiemken, A., Guyer, W. (1971). A lysosomal aminopeptidase isozyme in differentiating yeast cells and protoplasts. Planta 96:43–53.

Mazon, M. (1978). Effect of glucose starvation on the nicotinamide adenine dinucleotide phosphate-dependent glutamate dehydrogenase of yeast. J. Bacteriol. 133:780–785.

Mazon, M., Hemmings, B. (1979). Regulation of *Saccharomyces cerevisiae* nicotinamide adenine dinucleotide phosphate-dependent glutamate dehydrogenase by proteolysis during carbon starvation. J. Bacteriol. 139:686–689.

McLellan, Jr., W., Lampen, J. (1963). The acid phosphatase of yeast: localization and secretion by protoplasts. Biochim. Biophys. Acta 67:324–326.

Mechler, B., Wolf, D. (1981). Analysis of proteinase A function in yeast. Eur. J. Biochem. 121:47–52.

Mechler, B., Müller, M., Müller, H., Muessdoerffer, F., Wolf, D. (1982). *In vivo* biosynthesis of the vacuolar proteinases A and B in the yeast *Saccharomyces cerevisiae*. J. Biol. Chem. 257:11203–11206.

Metz, G., Röhm, K. -H. (1978). Yeast aminopeptidase I. Chemical composition and catalytic properties. Biochim. Biophys. Acta 429:933–949.

Metz, R., Marx, E., Röhm, K. -H. (1977). The quaternary structure of yeast aminopeptidase I. Z. Naturforsch. 32c:929–937.

Mihara, K., Blobel, G. (1980). The four cytoplasmically made subunits of yeast mitochondrial cytochrome *c* oxidase are synthesized individually and not as a polyprotein. Proc. Natl. Acad. Sci. USA 77:4160–4164.

Molano, J., Gancedo, C. (1974). Specific inactivation of fructose 1,6-bisphosphatase from *Saccharomyces cerevisiae* by a yeast protease. Eur. J. Biochem. 44:213–217.

Mortimer, R., Schild, D. (1982). Genetic map of *Saccharomyces cerevisiae.* In: The Molecular Biology of the Yeast *Saccharomyces.* Metabolism and Gene Expression, edited by J. Strathern, E. Jones, J. Broach. Cold Spring Harbor, N. Y.: Cold Spring Harbor Laboratory, pp. 639–650.

Muessdoerffer, F. (1980). Occurrence of proteinase A isoinhibitors in wild type yeast strains and commercial bakers yeast. Biochem. Biophys. Res. Commun. 97:423–429.

Muessdoerffer, F., Tortora, P., Holzer, H. (1980). Purification and properties of proteinase A from yeast. J. Biol. Chem. 255:12087–12093.

Neeff, J., Hägele, E., Nauhaus, J., Heer, U., Mecke, D. (1978). Evidence for catabolite degradation in the glucose-dependent inactivation of yeast cytoplasmic malate dehydrogenase. Eur. J. Biochem. 87:489–495.

Nelson, N., Schatz, G. (1979). Energy-dependent processing of cytoplasmically made precursors to mitochondrial proteins: Proc. Natl. Acad. Sci. USA 76:4365–4369.

Neupert, W., Schatz, G. (1981). How proteins are transported into mitochondria. Trends Biochem. Sci. 6:1–4.

Novick, P., Ferro, S., Schekman, R. (1981). Order of events in the yeast secretory pathway. Cell 25:461–469.

Nunez de Castro, I., Holzer, H. (1976). Studies on the proteinase A inhibitor $I_3^A$ from yeast. Hoppe-Seyler's Z. Physiol. Chem. 357:727–734.

Perlman, D., Halvorson, H. (1981). Distinct repressible mRNAs for cytoplasmic and secreted yeast invertase are encoded by a single gene. Cell 25:525–536.

Pilar-Fernandez, M., Correa, J., Cabib, E. (1982). Activation of chitin synthetase in permeabilized cells of a *Saccharomyces cerevisiae* mutant lacking proteinase B. J. Bacteriol. 152:1255–1264.

Robertson, J., Halvorson, H. (1957). The components of maltozymase in yeast and their behavior during deadaptation. J. Bacteriol. 73:186–198.

Röhm, K. -H. (1974). Properties of a highly purified dipeptidase from brewers yeast. Hoppe-Seyler's Z. Physiol. Chem. 355:675–686.

Rose, B., Becker, J. Naider, F. (1979). Peptidase activities in *Saccharomyces cerevisiae*. J. Bacteriol. 139:220–224.

Saheki, T., Matsuda, Y., Holzer, H. (1974). Purification and characterization of macromolecular inhibitors of proteinase A from yeast. Eur. J. Biochem. 47:325–332.

Sanada, Y., Fujishiro, K., Tanaka, H., Katanuma, N. (1975). Isolation and characterization of yeast protease B. Biochem. Biophys. Res. Commun. 86:815–821.

Schweizer, E., Kuhn, L., Castorph, H. (1971). A new gene cluster in yeast: the fatty acid synthetase system. Hoppe-Seyler's Z. Physiol. Chem. 352:3277–3284.

Schweizer, E., Werkmeister, K., Jain, M. (1978). Fatty acid biosynthesis in yeast. Mol. Cell. Biochem. 21:95–106.

Skudlarek, M., Swank, R. (1979). Biosynthesis of two lysosomal enzymes in macrophages. J. Biol. Chem. 254:9939–9942.

Sprague, Jr., G., Herskowitz, I. (1981). Control of yeast cell type by the mating type locus. I. Identification and control of expression of the **a**-specific gene, *BAR1*. J. Mol. Biol. 153:305–321.

Sprague, Jr., G., Rine, J., Herskowitz, I. (1981). Control of yeast cell type by the mating type locus. II. Genetic interactions between MAT$\alpha$ and unlinked $\alpha$-specific STE genes. J. Mol. Biol. 153:323–335.

Steiner, D., Quinn, P., Chan, S., Marsh, J., Tager, H. (1980). Processing mechanisms in the biosynthesis of proteins. Ann. N.Y. Acad. Sci. 343:1–16.

Stern, A., Jones, B., Shively, J., Stein, S., Udenfriend, S. (1981). Two adrenal opiod polypeptides: proposed intermediates in the processing of proenkephalin. Proc. Natl. Acad. Sci. USA 78:1962–1966.

Stevens, T. Esmon, B., Schekman, R. (1982). Early stages in the yeast secretory pathway are required for transport of carboxypeptidase Y to the vacuole. Cell 30:439–448.

Stewart, J., Sherman, F., Shipman, N., Jackson, M. (1971). Identification and mutational relocation of the AUG codon initiating transcription of iso-1-cytochrome *c* in yeast. J. Biol. Chem. 246:7429–7445.

Stötzler, D., Düntze, W. (1976). Isolation and characterization of four related peptides exhibiting $\alpha$-factor activity from *Saccharomyces cerevisiae*. Eur. J. Biochem. 65:257–262.

Suarez-Rendueles, M., Schwencke, J., Garcia-Alvarez, N., Gascon, S. (1981). A new X-prolyl-dipeptidyl aminopeptidase from yeast associated with a particulate fraction. FEBS Lett. 131:296–300.

Sumrada, R., Cooper, T. (1978). Control of vacuole permeability and protein degradation by the cell cycle arrest signals in *Saccharomyces cerevisiae*. J. Bacteriol. 136:234–246.

Sutton, D., Lampen, J. (1962). Localization of sucrose and maltose fermenting systems in *Saccharomyces cerevisiae*. Biochim. Biophys. Acta 56:303–312.

Trimble, R., Maley, F. (1977). The use of endo-$\beta$-*N*-acetylglucosaminidase H in characterizing the structure and function of glycoproteins. Biochem. Biophys. Res. Commun. 78:935–944.

Trumbly, R. (1980). A genetic and biochemical study of the aminopeptidases of *Saccharomyces cerevisiae*. Ph.D. Thesis. University of California, Davis.

Ulane, R., Cabib, E. (1974). The activating system of chitin synthetase from *Saccharomyces cerevisiae*. Purification and properties of an inhibitor of the activating factor. J. Biol. Chem. 249:3418–3422.

Ulane, R., Cabib, E. (1976). The activating system of chitin synthetase from *Saccharomyces cerevisiae*. J. Biol. Chem. 251:3367–3374.

Whitney, P., Cooper, T., Magasanik, B. (1973). The induction of urea carboxylase and allophanate hydrolase in *Saccharomyces cerevisiae*. J. Biol. Chem. 248:6203–6209.

Wiemken, A., Schellenberg, M., Urech, K. (1979). Vacuoles: the sole compartment of digestive enzymes in yeast *(Saccharomyces cerevisiae)?* Arch. Microbiol. 123:23–35.

Witt, I., Kronau, R., Holzer, H. (1966). Isoenzyme der malat-dehydrogenase und ihre regulation in *Saccharomyces cerevisiae*. Biochim. Biophys. Acta 128:63–73.

Wolf, D., Ehmann, C. (1978a). Isolation of yeast mutants lacking proteinase B activity. FEBS Lett. 92:121–124.

Wolf, D., Ehmann, C. (1978b). Carboxypeptide S from yeast: regulation of its activity during vegetative growth and sporulation. FEBS Lett. 91:59–62.

Wolf, D., Ehmann, C. (1979). Studies on a proteinase B mutant of yeast. Eur. J. Biochem. 98:375–384.

Wolf, D., Ehmann, C. (1981). Carboxypeptidase S- and carboxypeptidase Y-deficient mutants of *Saccharomyces cerevisiae*. J. Bacteriol. 147:418–426.

Wolf, D., Fink, G. (1975). Proteinase C (carboxypeptidase Y) mutant of yeast. J. Bacteriol. 123:1150–1156.

Wolf, D., Weiser, U. (1977). Studies on a carboxypeptidase Y mutant of yeast and evidence for a second carboxyptidase activity. Eur. J. Biochem. 73:553–556.

Wolf, D., Ehmann, C., Beck, I. (1979). Genetic and biochemical analysis of intracellular proteolysis in yeast. In: Biological Functions of Proteinases, edited by H. Holzer and H. Tschesche. New York: Springer-Verlag, pp. 55–72.

Zubenko, G. (1981). A genetic approach to the study of intracellular proteolysis in *Saccharomyces cerevisiae*. Ph.D. Thesis. Carnegie-Mellon University, Pittsburgh, Pennsylvania.

Zubenko, G., Jones, E. (1979). Catabolite inactivation of gluconeogenic enzymes in mutants of yeast deficient in proteinase B. Proc. Natl. Acad. Sci. USA 76:4581–4585.

Zubenko, G., Jones, E. (1981). Protein degradation, meiosis and sporulation in proteinase-deficient mutants of *Saccharomyces cerevisiae*. Genetics 97:45–64.

Zubenko, G., Mitchell, A., Jones, E. (1979). Septum formation, cell division and sporulation in mutants of yeast deficient in proteinase B. Proc. Natl. Acad. Sci. 76:2395–2399.

Zubenko, G., Mitchell, A., Jones, E. (1980). Mapping of the proteinase B structural gene, *PRB1*, in *Saccharomyces cerevisiae* and identification of nonsense alleles within the locus. Genetics 96:137–146.

Zubenko, G., Park, F., Jones, E. (1981). *PEP4* gene function is required for maturation of vacuolar hydrolases. In: Molecular Biology of Yeast. Cold Spring Harbor, N.Y.: Cold Spring Harbor Laboratory, p. 164.

Zubenko, G., Park, F., Jones, E. (1982). Genetic properties of mutations at the *PEP4* locus in *Saccharomyces cerevisiae*. Genetics 102:679–690.

Zubenko, G., Park, F., Jones, E. (1983). Mutations in the *PEP4* locus of *Saccharomyces cerevisiae* block the final step in the maturation of two vacuolar hydrolases. Proc. Natl. Acad. Sci. USA 80:510–514.

# 7

# Genetic Control of Flocculation[1]

J. R. Johnston* and H. P. Reader**

## Introduction

Flocculation—or conversely, lack of flocculation—is a most important characteristic of yeast strains used in industrial fermentations. Unfortunately, a certain degree of confusion has arisen by use of the term "flocculation" in the scientific literature to describe different phenomena in yeast cell behavior. The term, as used most commonly within the brewing industry in particular, is defined here as "the (reversible) aggregation of dispersed yeast cells into flocs, generally toward the end of fermentation, and subsequent separation of the flocs from the suspending liquid." The definition excludes other forms of yeast aggregation, particularly those of "clumpy growth" and "chain formation," that is, the nonseparation of daughter and mother cells during growth, which are sometimes also referred to as flocculation. The term "nonflocculent" will, therefore, imply the lack of cell aggregation and consequently much slower separation of (dispersed) yeast cells from the liquid medium. Although yeast separation is very often, as expected, by sedimentation, it may also be by flotation because of cell aggregates entrapping bubbles of carbon dioxide, as in the case of "top-fermenting" brewing yeasts.

The technological importance of the flocculence characteristic has been associated predominantly with brewing. Both the time at which flocculation occurs and its intensity (average size of flocs) affect the brewing process and the final product, that is, beer, in several ways. These variables influence the total extent of fermentation (wort attenuation) and consequently alcohol, carbohydrate, and flavor content of the beer. The amount of yeast remaining in suspension

*Yeast Technology Group, Department of Bioscience and Biotechnology, University of Strathclyde, Glasgow G1 1XW, Scotland.
**Cockburn Smithes & Cia, Lda., 4401 Vila Nova de Gaia Codex, Portugal.
[1]We wish to record our gratitude to the Science Research Council and to Allied Breweries (Production) Limited for their financial support, and to P. A. Martin of Allied Breweries, Burton-on-Trent, for his advice and assistance during these investigations.

after primary fermentation affects secondary conditioning and requirements of centrifugation and/or filtering. Differing technologies of brewing demand different flocculation characteristics of the yeasts used if overall efficiency is to be optimized.

It is becoming increasingly evident that the ability to flocculate is a characteristic to be considered in choice of the strain of yeast for other alcoholic fermentations. The ability to flocculate strongly is included among criteria for selection of improved rum yeasts. Yeasts selected for Scotch whisky fermentations are being examined for their flocculation and the significance of this behavior. Certain hybrid strains for production of fuel ethanol from sugarcane juice in Brazil have been developed to include the flocculation characteristic. These relationships between yeast flocculation and industrial fermentations are discussed later in this chapter.

Analysis of the genetic control of flocculation represents one of the few cases of genetic investigation of a yeast characteristic of technological importance. It is, however, also of considerable academic interest, for two principal reasons. First, flocculation is a property associated with structure of the cell wall and most likely with cell wall proteins. The phenomenon, therefore, affords an opportunity to extend our knowledge of the genetic control of structural, rather than enzymatic, proteins. Second, flocculation is a *quantitative* characteristic, varying widely in its degree or intensity and there are still only a few examples of genetic analysis of quantitative characteristics in microorganisms. The major task, with a view to microbial breeding similar to that of plants and animals, of genetic analysis of crucial characteristics such as growth rate, fermentation rate, and product yield, has hardly begun.

## Methodology

A variety of methods for the measurement of flocculation has been described (reviewed in Gilliland 1981). Since yeast strains capable of sporulation often produce some ascospore segregant cultures that exhibit clumpy growth, it is essential to differentiate between this property and "true" flocculation when performing tetrad analysis. A method to allow this has been described by Lewis and Johnston (1974) and Reader (1980) and is illustrated schematically in Figure 7.1. At the end of this procedure, on standing for 10 min, flocculent strains will reaggregate but nonflocculent strains with clumpy growth will remain dispersed. A grading scale of 0 (nonflocculent) to 4 (very flocculent) may conveniently be used. It should be noted that different media and growth conditions have been used in different laboratory groups during genetic analysis of flocculation (Russell *et al.* 1980) and that caution should therefore be exercised in comparing their results, particularly in terms of quantitative determinations.

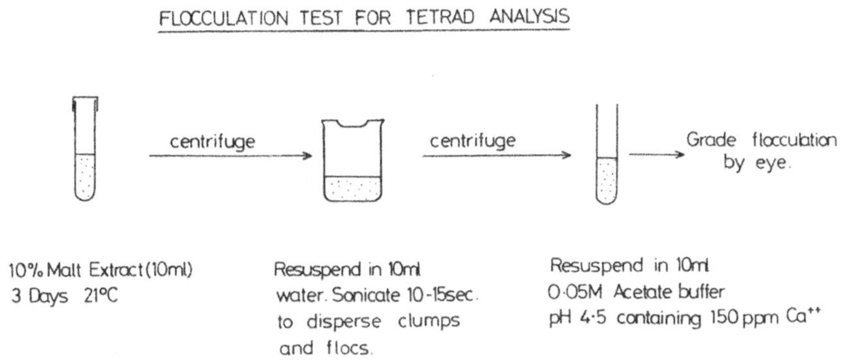


**Figure 7.1.** Flocculation method used in tetrad analysis.

## Origin of Strains

For development of new commercial strains, a primary objective is the identification of genes conferring flocculence in production brewing yeasts. This is generally extremely difficult, however, because of the notoriously poor nature of most brewing strains for genetic studies. The isolation of haploid segregant strains suitable as parent strains in crosses yielding viable tetrads is generally a formidable task. It was indeed fortunate that the first flocculation gene to be clearly identified *(FLO1)* was derived, although indirectly, from a commercial brewing strain. Strain ABX was derived by empirical hybridization of ascospore cultures of brewery production yeast, AB1 (Lewis *et al.* 1976). Rather by chance, the flocculation gene *FLO5* (described later) was also identified in strains derived from a brewery yeast, although of obscure origin. Genes previously designated *FLO2* (Johnston and Lewis 1974) and *FLO4* (Stewart and Russell 1977) were identified in "laboratory" (i.e., genetic stock) strains. These genes, however, are now believed to be allelic with *FLO1* (Russell *et al.* 1980). The recessive gene, *flo3,* was selected as a mutant of a nonflocculent laboratory strain after prolonged continuous culture (Johnston and Lewis 1974). There may also be two further recessive flocculation genes, *flo6* and *flo7*, which have also been identified in laboratory strains (Pearce and Thornton, unpublished results).

Initially, all nonflocculent strains used in genetical investigations were laboratory haploid strains. A nonflocculent mutant strain, however, has been obtained by mutagenesis of a flocculent yeast, although this mutant has been shown to carry a suppressor gene, designated *fsu1* (Holmberg and Kielland-Brandt 1978). Many other nonflocculent strains have now been isolated from crosses of brewery-derived yeasts and laboratory strains.

## Genetic Analysis

It has recently been proposed (Russell *et al.* 1980) that the genes originally described as *FLO1*, *FLO2*, and *FLO4* (see preceding section) are in fact allelic and that this locus be referred to as *FLO1*. The data from which it was concluded that *FLO1* and *FLO2* were allelic genes are shown in Table 7.1. Taken in conjunction with the results of tetrad analysis by Russell and Stewart (Russell *et al.* 1980) that *FLO2/FLO4* crosses gave a PD:NPD:T ratio of 33:0:2, it seems likely that the *FLO1* locus also includes the *FLO4* gene. Additional evidence for this proposal comes from the linkage of *FLO1* to the marker gene *ade1*, for which the evidence is presented in Table 7.2. These data yield a linkage distance between *FLO1* and *ade1* of 37 cM, in close agreement with the mapping of *FLO4* on chromsome I at a distance of 37 cM (Stewart and Russell 1977).

On the basis of the location of marker gene *cys1* on chromosome I, tetrad analysis has been performed on crosses of *FLO1* and *cys1*. Although some difficulty was experienced in the scoring of gene *cys1*, the limited data of PD:NPD:T = 5:4:23 gave no indication of linkage. By pooling all the available tetrad analysis data and using the method of "maximum likelihood" (Snow 1979), Mortimer and Schild (1980) have arrived at a map distance of 38.9 cM for *ade1-FLO1*. Since *cys1* lies outside *ade1-FLO1* and the distance for *cys1-ade1* is 16.8, it is perhaps not surprising that no linkage of *FLO1* to *cys1* was detected from the small number of tetrads analyzed.

In light of the modified conclusion regarding the genes originally described as *FLO1* and *FLO2*, further analysis has been carried out on the gene designated *flo3* (Lewis and Johnston 1976). This gene was originally described in a flocculent mutant isolated from the nonflocculent haploid strain X764-S1 (mutant strain referred to as X764-S1/F). The gene *flo3* was earlier observed to act as a recessive in most crosses but as a semidominant in a minority of crosses (Lewis 1974). This gene, however, requires careful scoring as it normally confers only weak flocculation, which is nevertheless easily dispersed by

**Table 7.1.** Tetrad Analysis for Allelism of Genes Formerly Designated *FLO1* and *FLO2*

| Cross | Hybrid | Number of tetrads PD | NPD | T |
|---|---|---|---|---|
| ABX-1D *(FLO1)* × K2NB-12C *(FLO2)* | XR56 | 23 | 0 | 0 |
| XL110-1A *(FLO1)*[a] × XL86-6A *(FLO2)*[b] | XL116 | 30 | 0 | 0 |
| Pooled total | | 53 | 0 | 0 |

[a]XL110 was cross ABX-2D *(FLO1)* × X764-S1 (nonflocc).
[b]XL86 was cross K2NB-12C *(FLO2)* × S732C (nonflocc).

**Table 7.2.** Tetrad Analysis for Linkage between Genes *FLO1* and *ade1*

| Cross | Hybrid | Number of tetrads | | |
|---|---|---|---|---|
| | | PD | NPD | T |
| ABX-2B *(FLO1)* × S732C *(ade1)* | XR58 | 17 | 1 | 33 |
| K2NB-12C *(FLO1)* × S732C *(ade1)* | XL86 | 9 | 0 | 23 |
| Pooled total | | 26 | 1 | 56 |

sonication and removed by washing with disodium-EDTA. In addition, the recovery of flocculation following sonication is markedly slower than in strains carrying other flocculation genes.

Although strain X764-S1/F *(flo3)* crossed to a nonflocculent haploid (S732C) gave a nonflocculent diploid (XR57) that produced predominantly 2:2 segregations (flocculent:nonflocculent) in tetrads, some segregants showed moderate to strong flocculation. Results of crosses of strains carrying *flo3* to nonflocculent haploid strains are shown in Table 7.3. The mutant strain itself (X764-S1/F) when crossed to three nonflocculent strains produced nonflocculent diploids. confirming *flo3* as a recessive gene. The three more flocculent segregants of the cross XR57, however, produced weakly to moderately flocculent diploids, thereby showing the inheritance of *flo3* as a semidominant gene. The likelihood is that there are modifying factors or modifier genes present in certain crosses that, although not by themselves determining flocculence, influence the *degree* of flocculation conferred in association with the gene *flo3*.

The data presented in Table 7.4 confirmed that *flo3* is a distinct gene from *FLO1*. Tetrad analysis of crosses XR97 show *flo3* and *FLO1* to be nonallelic. The results from crosses XR57 and XR136, indicating that *flo3* is unlinked to *ade1*, are consistent with this conclusion. The low proportion of tetratype asci

**Table 7.3.** Flocculation of Diploids Heterozygous for Gene *flo3*

| Cross | | Flocculation grade[a] | |
|---|---|---|---|
| *flo3* × non-flocc | Hybrid | Parent | Diploid |
| X764-S1/F × S732C | XR57 | 1–2 | 0 |
| X764-S1/F × XL78-2D | XR136 | 1–2 | 0 |
| X764-S1/F × JW4-5C | XR144 | 1–2 | 0 |
| XR57-4D × S1780C | XR62 | 3 | 2 |
| XR57-6D × S1780C | XR63 | 3 | 1–2 |
| XR56-11D × S732C | XR100 | 4 | 2 |

[a]Graded 0 to 4 for nonflocculent to intensely flocculent.

**Table 7.4.** Tetrad Analysis Testing (A) Allelism of *flo3* and *FLO1* and (B) Linkage of *flo3* and *ade1*

| Cross | Hybrid | Number of tetrads PD | NPD | T |
|---|---|---|---|---|
| (A) ABX-2D *(FLO1)* × X764-S1/F *(flo3)* | XR97 | 8 | 14 | 12 |
| XL78-7D *(FLO1)*[a] × X764-S1/F *(flo3)* | XR204 | 1 | 3 | 8 |
| Pooled total | | 9 | 17 | 20 |
| (B) X764-S1/F *(flo3)* × S732C *(ade1)* | XR57 | 3 | 3 | 15 |
| X764-S1/F *(flo3)* × XL78-2D *(ade1)*[a] | XR136 | 3 | 2 | 3 |
| Pooled total | | 6 | 5 | 18 |

[a]XL78 was cross W11 *(FLO1)* × S732C *(ade1)*.

from cross XR97 would appear to indicate that *flo3*, like *FLO1*, is a centromere-linked gene. However, the results from XR204 and particularly the tetratype frequency obtained from crosses with *ade1* (XR57 and XR136) do not confirm this proposition. Overall, the results support the previous conclusion that *flo3* is a non-centromere-linked gene.

During the course of genetic analysis, conducted in cooperation with Stewart and Russell of Labatt Breweries, London, Ontario, it became apparent that there existed a dominant gene for flocculation distinct from *FLO1*. A strain, subsequently referred to as CLX (Reader 1980), derived from the brewery yeast AB1 was initially considered to have the same parentage as hybrid ABX. It was observed, however, that strain CLX flocculated intensely, that is, considerably more strongly than ABX. On ascus dissection, CLX produced only 43% viable ascospores, compared with 99% by ABX. In both cases, all ascospore cultures were flocculent, showing the homozygosity of the parent strains.

From one tetrad of CLX with four viable spores two segregants (CLX-11A and CLX-11B) were crossed to nonflocculent haploids. These crosses were flocculent and gave tetrads showing predominantly 2:2 segregations for flocculent:nonflocculent. When CLX-11A and CLX-11B were crossed together, the resulting hybrid produced 16 tetrads in which all 64 ascospore cultures were flocculent. The high viability of ascospores in all crosses involving CLX-11A and CLX-11B and the 2:2 segregations for several marker genes in some crosses suggest that these two segregant strains of CLX are haploid. It was concluded that they carry the same single dominant gene for flocculation. However, the results presented in Table 7.5 show that this gene is distinct from the gene *FLO1*. Tetrad analysis of cross XR106 shows that the gene carried by CLX-11B is nonallelic to *FLO1*. This gene has been designated *FLO5* (Russell and Stewart 1979; Reader 1980). The results from cross XR112 show that *FLO5* is unlinked to *ade1* and that it is probably not centromere linked.

**Table 7.5.** Tetrad Analysis Testing (A) Allelism of *FLO5* and *FLO1* and (B) Linkage of *FLO5* and *ade1*

| Cross | Hybrid | Number of tetrads PD | NPD | T |
|---|---|---|---|---|
| (A) CLX-11B *(FLO5)* × ABX-1C *(FLO1)* | XR106 | 12 | 4 | 25 |
| (B) CLX-11B *(FLO5)* × S733A *(ade1)* | XR112 | 6 | 8 | 23 |

## Phenotypic Variations

The quantitative phenotypes normally conferred by the genes *FLO1*, *flo3*, and *FLO5* are illustrated in Figure 7.2. Two more recessive genes, considered to be independent and tentatively designated *flo6* and *flo7* (Pearce *et al.*, unpublished results), are also included. However, it has already been pointed out that exceptional segregant strains carrying *flo3* show much stronger flocculation and that *flo3* can be expressed as a semidominant gene (Table 7.3). In this case, modifiers producing increases in the intensity of flocculation appear to be implicated, but suppression of *FLO1* by a specific gene, *fsu1*, has also been reported (Holmberg and Kielland-Brandt 1978).

A laboratory diploid strain, XT25, has been observed to segregate 2:2 for flocculation (R. J. Thornton, personal communication) and two flocculent segregants (S8 and S56) have been used in attempts to identify the gene involved, provisionally referred to as "*floX*." The diploid cross (XR197) of XT25-S8 to XT25-S56 gave 10 tetrads in which all 40 ascospore cultures were flocculent, thereby confirming that S8 and S56 both carried the same single gene for flocculence. However, the many crosses of these two segregants to different nonflocculent haploid strains have produced different results (Thornton, Reader, unpublished results). For example, among 20 such crosses, 11 of

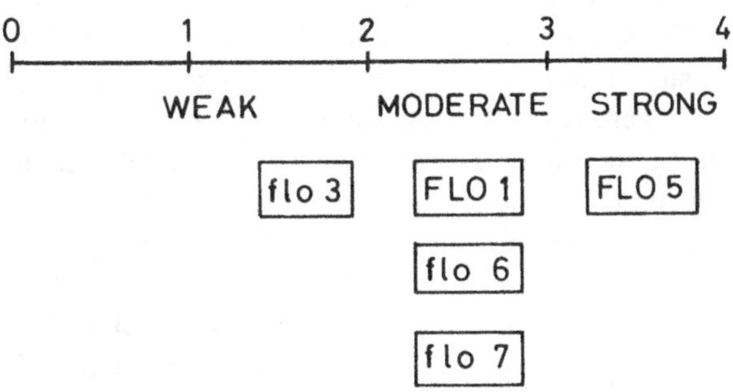


**Figure 7.2.** Quantitative phenotypes of flocculation genes on scales 1–4. Degree of flocculence can also be dependent on the presence of modifier and/or suppressor genes.

the resulting diploids were nonflocculent whereas nine were flocculent (Reader 1980). The gene *floX* is thus similar to *flo3* in showing both recessive and dominant/semidominant patterns of inheritance. Gene *floX* is, however, apparently an independent gene from *flo3* since (1) a cross (XR86) of strains carrying respectively *flo3* and *floX* gave a PD:NPD:T ratio of 2:4:17 and (2) *floX* is apparently linked to *ade1* (PD:NPD:T = 29:4:48).

The latter data suggest that *floX* may be allelic to *FLO1* since the calculated linkage distance of *floX-ade1* is 44 cM. The results of a cross between *floX* and *cys1*, producing a PD:NPD:T ratio of 16:7:31 and a map distance of 68 cM, is consistent with the idea that *floX* is located on the right arm of chromosome I, in close vicinity to *FLO1*. However, this proposal is apprently contradicted by the results of two crosses between strains carrying *floX* and *FLO1*, respectively. The combined PD:NPD:T ratio of 31:8:51 at first sight provides strong evidence against allelism of *floX* and *FLO1* but appears to show that *floX* and *FLO1* are linked, with a distance between them of 55 cM. This apparent linkage distance is, however, not reconcilable with the linkage distance of *floX* to *ade1* and the map location of *FLO1*.

A possible explanation of these apparently inconsistent results is provided by further analysis of the cross, XR94 between *floX* and *FLO1*. All 12 segregants of three tetrads of XR94 that produced 2:2 segregations (i.e., apparent NPD asci) were crossed to nonflocculent haploid strains. With the exception of two crosses (XR171 and XR173), which produced low yields of viable ascospores, the results of tetrad analysis are shown in Table 7.6. On the basis of *floX* and *FLO1* being nonallelic genes, the genotypes of segregants in each tetrad should be *floX FLO1* (flocculent for two segregants) and *FLOX flo1* (nonflocculent for two segregants). The expected result from backcrossing the nonflocculent segregants would then be tetrads of 0:4 (flocculent:nonflocculent) and from backcrossing the flocculent segregants, tetrads of 2:2, 4:0, and 3:1 (flocculent:nonflocculent) as PD, NPD, and T asci, respectively. The results (Table 7.6), however, are completely different from those expected on this basis. All five crosses analyzed from nonflocculent segregants of XR94 (XR154, XR169, XR197, XR159, XR165) produced substantial numbers of flocculent segregants. This result implies that nonflocculent segregants of XR94 do in fact carry a gene (or genes) for flocculation but that the expression of this gene is suppressed. With the minor exception of cross XR154, which is weakly flocculent, the flocculation gene is also suppressed in these diploid crosses. However, its presence is detectable in the proportion of segregants of these crosses where it is not subject to suppression. The results do not indicate the action of a single suppressor gene, since there is a significant lack of 0:4 tetrads, which would be expected as PD asci, and in the case of segregants of crosses XR169 and XR165 suppression is no longer evident. The suppression present in these XR94 segregants, therefore, seems to act in a variable manner when recombined with genes of the second nonflocculent parent in these crosses.

Most of the crosses of flocculent segregants of XR94 (XR200, XR167, XR158, XR160, XR202) show clearly that these segregants lack suppressor factors. The five 1:3 tetrads of XR158 are, however, an exception to this gen-

**Table 7.6.** Tetrad Analysis of Crosses between Segregants of XR94 and Nonflocculent Haploids

| Segregant of XR94 | Flocculation grade | Cross; flocc grade | | No. of tetrads with ratio flocculent:nonflocculent 4:0 | 3:1 | 2:2 | 1:3 | 0:4 | Ratio of segregants flocculent:nonflocculent |
|---|---|---|---|---|---|---|---|---|---|
| 49A | 0 | XR154 | 1 | 0 | 1 | 4 | 2 | 2 | 9:19 |
| 49B | 3 | XR171 | 2 | | | no tetrads | | | |
| 49C | 3 | XR200 | 3 | 0 | 0 | 4 | 0 | 0 | 8:8 |
| 49D | 0 | XR169 | 0 | 0 | 0 | 6 | 0 | 0 | 12:12 |
| 54A | 2 | XR167 | 0 | 0 | 0 | 6 | 0 | 0 | 12:12 |
| 54B | 4 | XR158 | 2 | 1 | 0 | 5 | 5 | 0 | 19:15 |
| 54C | 0 | XR197 | 0 | 0 | 0 | 2 | 10 | 0 | 14:34 |
| 54D | 0 | XR173 | 0 | | | no tetrads | | | |
| 55A | 0 | XR173 | 0 | 0 | 0 | 1 | 4 | 1 | 6:18 |
| 55B | 3 | XR160 | 0 | 3 | 3 | 2 | 0 | 0 | 25:7 |
| 55C | 3 | XR202 | 2 | 0 | 0 | 4 | 0 | 0 | 8:8 |
| 55D | 0 | XR165 | 0 | 0 | 2 | 4 | 0 | 0 | 14:8 |

**Table 7.7.** Tetrad Analysis of Crosses Between Nonflocculent Segregants of XR199

| Segregants of XR199[a] | Cross; flocc. grade | | No. of tetrads with ratio flocculent:nonflocculent | | | | | Ratio of segregants flocculent:nonflocculent |
|---|---|---|---|---|---|---|---|---|
| | | | 4:0 | 3:1 | 2:2 | 1:3 | 0:4 | |
| 1A × 1B | XR219 | 0 | 0 | 0 | 3 | 5 | 0 | 11:21 |
| 1A × 1C | XR221 | 0 | 0 | 0 | 6 | 0 | 0 | 12:12 |
| 1B × 1D | XR222 | 0 | 0 | 0 | 0 | 0 | 8 | 0:32 |
| 1C × 1D | XR223 | 0 | 0 | 0 | 0 | 5 | 1 | 5:19 |
| 11A × 11C | XR227 | 0 | 0 | 0 | 2 | 6 | 2 | 10:30 |
| 11A × 11D | XR228 | 0 | 0 | 4 | 1 | 3 | 0 | 17:15 |
| 11B × 11C | XR229 | 0 | 0 | 0 | 0 | 0 | 7 | 0:28 |
| 11B × 11D | XR230 | 0 | 0 | 0 | 2 | 4 | 0 | 8:16 |

[a]All eight segregants were nonflocculent.

eral result. The evidence of crosses XR200, XR167, and XR202 suggests that the flocculent segregants of XR94 carry only a single gene for flocculation. Alternatively, they could be of genotype *FLO1 floX* if these two genes are closely linked. The results from cross XR160 support this alternative but in this case indicate that the two genes for flocculation are nonlinked.

In an attempt to clarify further the genetic basis of the results obtained with XR94, a cross (XR199) was made between two nonflocculent segregants of XR94 (49A × 55A). The cross XR199 was nonflocculent and 12 tetrads gave the ratios seven 2:2, three 1:3, and two 0:4 (flocculent:nonflocculent). The proportion of flocculent segregants from XR199 was, therefore, approximately one-third ($^{17}\!/_{48}$). This result is similar to those obtained from diploids XR154 and XR159 when the segregant parents XR94-49A and XR94-55A were crossed to other nonflocculent strains. This result from XR199 confirms that segregants 49A and 55A carry a gene for flocculation that is suppressed but not by the same single suppressor gene.

Two tetrads of XR199 were subjected to further analysis. Crosses were constructed between the four spore cultures of each of the two tetrads that gave a 0:4 ratio (flocculent:nonflocculent). These crosses, all of which were nonflocculent, and the results of their tetrad analysis are shown in Table 7.7. These results can be explained in a general manner by the assumption that the parent strains (XR94-49A and XR94-55A) carry different dominant suppressor genes, say SUP1 and SUP2, which suppress the gene *floX*. Crosses in which both parent segregants are either *floX* $SUP1^+$ or $floX^+$ *SUP2* (+ alleles of *SUP1* and *SUP2* are incapable of suppression of flocculation) would give tetrads that are all 0:4. This is the case with XR222 and XR229. Crosses of segregants that are *floX* $SUP1^+$ with $floX^+$ *SUP2,* however, would yield tetrads of 0:4, 2:2, and 3:1 ratios (PD, NPD, and T asci). This expectation is only loosely observed in the case of crosses XR219, XR221, XR223, XR227, and XR230 as there is an overall lack of PD (0:4) compared to NPD (2:2) tetrads. Besides, the production of four 3:1 tetrads by cross XR228 cannot be explained by this model. The genetic basis of suppression is, therefore, more complex than the existence of two suppressor genes, such as *SUP1* and *SUP2*. Interaction between a larger number of genes than it has been possible to identify in these experiments and possibly also the presence of cytoplasmic factors seem to be involved. Nevertheless, the results taken as a whole point to *floX* being either an allele of *FLO1* or, possibly somewhat more likely, an independent gene closely linked to *FLO1*. If the former proposal is correct, however, then *floX* is a distinguishable allele to *FLO1*, since it is suppressible and suppression of *FLO1* itself has not been observed in our strains. The possibility that nonflocculence was due to either cytoplasmic or nuclear *petite* mutations was ruled out by the growth of the nonflocculent segregants on glycerol medium.

## Cell Wall Differences

Two haploid strains carrying gene *FLO1*, XL86-6A and XL112-5A, and their nonflocculent *(flo1)* counterparts, XL86-6B and XR112-5C, were selected for glucan–mannan determinations of their cell walls. Cell walls were prepared by the method described by Work (1971) and glucan and mannan fractionations performed as Mizunaga *et al.* (1971). Mannan and glucan were then determined by a modification (Lewis 1974) of the method of Morris (1948). The results shown in Table 7.8 are mean values of three independent estimations. These results show that it is unlikely that there are significant differences between related flocculent (*FLO1*) and nonflocculent *(flo1)* strains. This conclusion is in agreement with that of Jayatissa and Rose (1976), who used genetically undefined strains and who also noted little difference between phosphate levels of cell walls of flocculent and nonflocculent strains.

The *mnn4* mutation (Ballou *et al.* 1973) results in mannosylphosphate trans-

**Table 7.8.** Glucan and Mannan Contents of Flocculent *(FLO1)* and Nonflocculent Strains

| Strain | Mannan[a] (%) | Glucan[b] (%) | Mannan + glucan (%) | Mannan:glucan |
|---|---|---|---|---|
| XL86-6A *(FLO1)* | 38.9 | 35.9 | 74.8 | 1.08 |
| XL112[c]-5A *(FLO1)* | 41.6 | 37.8 | 79.4 | 1.10 |
| XL86-6B *(flo1)* | 38.6 | 35.4 | 75.1 | 1.09 |
| XL112-5C *(flo1)* | 40.7 | 35.0 | 75.7 | 1.16 |

[a]mg mannose/mg cell wall dry weight.
[b]mg glucose/mg cell wall dry weight.
[c]XL112 was cross ABX-2C × S732C (for XL86, see Table 7.1).

ferase deficiency, and because this mutation is dominant, both haploid and heterozygous diploid cells lack the mannosylphosphate substituent in the side chains of wall mannan (Karson and Ballou 1978). The *mnn4* mutation is considered to be in a regulatory gene and not in the structural gene for mannosylphosphate transferase, thereby explaining its dominance. The latter structural gene is probably defined by the *mnn6* mutation (Cohen and Ballou 1980). A *mnn*4 haploid strain and a related wild-type parent strain, received from C. E. Ballou, were both crossed with flocculent haploids carrying the *FLO5* gene. The intensity of flocculation of these latter strains and of the diploid crosses was measured by the haemocytometric method of Calleja and Johnson (1977). The results, presented in Table 7.9, show no significant difference in flocculation between the two diploids, respectively carrying or lacking the dominant *mnn4* gene, nor between the diploids and their flocculent *(FLO5)* parents. Using the usual visual method of estimating flocculation, all four flocculent strains were assigned grade 4 intensity. It would, therefore, appear that mannosylphosphate groups in the cell wall play no part in the aggregation of cells conferred by the *FLO5* gene.

In recent years, further attempts have been made to identify differences in the cell walls of flocculent and nonflocculent strains. Although surface appendages ("hairs") have been observed on flocculent cells and are apparently absent from nonflocculent cells (Stewart 1981), this morphological difference was not found in a comparison of nearly isogenic flocculent *(FLO1)* and nonflocculent strains (Miki *et al.* 1981). Irreversible deflocculation of flocculent strains has been obtained by treatment with 50 m*M* sodium hydroxide, and such extracts have revealed the presence in flocculent strains of a polypeptide of about 37,000 d, which is absent from nonflocculent strains (Stewart 1981). It is not yet clear, however, whether this polypeptide is a product of a particular flocculation gene. A different polypeptide, composed of 121 amino acids totaling 12,900 d, has been shown to be absent in nonflocculent mutants of a strain carrying the *FLO1* gene (Holmberg 1978). Absence of this polypeptide is apparently correlated with either of the suppressor genes *fsu1* and *fsu2* but is also associated with lack of flocculation due to mitochondrial RD mutation.

On the other hand, no polypeptide unique to flocculent cells could be identified in strains near-isogenic for the presence or absence of the *FLO1* gene

**Table 7.9.** Flocculation of Diploid Strains *(FLO5)* with and without the *mnn4* Gene

| Cross | Hybrid | % Cells in flocs[a] | |
|---|---|---|---|
| | | *FLO5* parent | Diploid |
| CLX-11B *(FLO5)* × LB5-10A *(mnn4)* | XR148 | 99.7 | 99.5 |
| CLX-11A *(FLO5)* × LB5-10B (+) | XR149 | 99.8 | 98.6 |

[a]Mean values of two separate observations on four cultures in each case.

(Miki *et al.* 1981). In this case, expression of the *FLO1* gene was apparently by production of a wall cognor molecule, which interacts with α-branched mannas behaving as surface cognons. Alternatively, the product of the *FLO1* gene could be a polypeptide that acts upon the cognor molecule.

A further possibility is that flocculation depends on the synthesis of enzymes, such as α-mannosidase or amidase, which alter cell wall structure to allow an increase in carboxyl groups for subsequent calcium bridge formation (Beavan *et al.* 1979). Prior to further elucidation of the molecular mechanism of flocculation, the preferred model would seem to be that *FLO* genes code for polypeptides, which act as or are associated with wall cognor molecules and that lytic enzymes, which modify the intensity of flocculation by exposing more carboxyl groups, are specified by various modifier genes.

## Stability of Flocculence

Loss of flocculence, that is, an increase in the proportion of nonflocculent cells, has frequently been observed in laboratory experiments and occasionally also poses problems under production conditions (T. J. B. Webb, personal communication). High mutation rates from dominant to recessive alleles, of the order of $10^{-4}$ per cell division, were first proposed by Thorne (1952) and gross instability of flocculence in a brewing yeast described later (Thorne and Nohr 1963). From similar results, Chester (1963) suggested the additional involvement of cytoplasmic genetic factors. Thorne (1968, 1970) examined the stability of two flocculent beer yeasts during prolonged continuous culture. After periods of 9–12 months, the population of one strain comprised about 15% nonflocculent cells with an additional 6% of isolates reduced in their degree of flocculation. The other strain was relatively more stable, but around 6% of isolates exhibited reduction or loss of flocculence.

Change in the ability to flocculate has sometimes been correlated to mutation to respiration deficiency (probably $rho^-$). Silhankova *et al.* (1970) and others have generally observed RD *(petite)* mutants of flocculent cells to show a loss or reduction in flocculence. Nevertheless, some *petite* mutants of brewing yeasts have exhibited increased flocculating ability, and harvesting of sedimented cells or use of continous fermentation have given these mutants a selective advantage (Gyllang and Martinson 1971; Gilliland 1971). Formaldehyde-induced RD mutants also showed increased flocculation and some of these represented improved production yeasts due to their enhanced sedimentation rates in cylindro-conical fermentation vessels (Cowan *et al.* 1975). Ethidium bromide induced petite mutants of flocculent strains carrying the *FLO1* gene were, however, overwhelmingly either nonflocculent or reduced in their degree of flocculence (Holmberg and Kielland-Brandt 1978). Only one of these latter RD mutants was exceptional in exhibiting increased flocculating ability.

The involvement in some cases of the mitochondrial genome in determining the flocculence phenotype is further evidenced by the results of Wilkie and

Mudd (1981) and Spencer *et al.* (1981). Mitochondrial *petite* ($rho^-$) mutants altered in cell surface characteristics can be detected by their inability to utilize sugars such as galactose and maltose. A fraction of sugar-utilizing revertants of these mutants has been found to be flocculent, apparently as a result of different nuclear mutations with pleiotropic effects (Wilkie and Mudd 1981). Another class of mitochondrial mutants are antibiotic resistant (e.g., to erythromycin, oligomycin, and chloramphenicol). Many of these mutants of a strongly flocculent strain of *S. diastaticus* have been shown to lose, either partially or completely, the flocculence characteristic (Spencer *et al.* 1981). Moreover, upon crossing, a high proportion of segregants that had lost their resistance were observed to have regained the original degree of flocculence.

The stability of three genetically defined strains has been measured (Lewis 1974). Two of these strains (XL112, XL86) were diploid and heterozygous for gene *FLO1* and the other strain was the original haploid *flo3* mutant. These experiments were performed in fermentation tubes containing 10 ml of 10% malt extract medium and static cultures were grown at 25 °C. Serial transfers were made to fresh media twice per day, generally after either 8 or 16 h incubation, with inocula adjusted to initial culture concentrations of $10^6$ cells/ml. Growth rates were determined and gave generation times of 4.5 h and 6.0 h for both diploid strains and the haploid strain. Small numbers of colony isolates were tested at intervals for flocculation and finally 100 colonies of each culture were tested after the complete growth period. The results are presented in Table 7.10. The rates shown should probably be regarded as approximations since they are calculated on the assumption that there is no selective advantage of nonflocculent cells over flocculent cells. Growth rates of flocculent and non-

**Table 7.10.** Rates of Change (Flocculent→Nonflocculent) of *FLO1* Diploids and a *flo3* Haploid

| Strain [genotype] | Culture | Growth period[a] (days) | No. (%) nonflocc. colonies Initial | Final | "Mutation" rate [b] per cell division |
|---|---|---|---|---|---|
| XL112 | 1 | 28 | 0 | 18 | $1.7 \times 10^{-3}$ |
| $\left[\frac{a\ FLO1}{\alpha\ \ +}\right]$ | 2 | 28 | 0 | 29 | $2.7 \times 10^{-3}$ |
| XL86 | 1 | 28 | 0 | 10 | $9.2 \times 10^{-4}$ |
| $\left[\frac{a\ FLO1}{\alpha\ \ +}\right]$ | 2 | 28 | 0 | 2 | $1.8 \times 10^{-4}$ |
| X764-S1/F | 1 | 7 | 6 | 36 | $1.5 \times 10^{-2}$ |
| [α *flo3*] | 2 | 7 | 4 | 43 | $2.0 \times 10^{-2}$ |

[a]Equivalent to 150 generations for XL112, XL86 and 28 generations for X764-S1/F.

[b]Calculated from $2 \log_e 2$ (% final − % initial) ÷ number of generations in growth period.

flocculent cells, however, are unlikely to be equal, at least under the growth conditions used.

Nevertheless, all cultures showed rather rapid increases in the proportion of nonflocculent cells present. Three of the four diploid cultures gave rates of the order of $10^{-3}$ per cell generation, although the fourth culture produced a rate lower by approximately an order of magnitude.

The use of diploid rather than haploid strains was for two reasons. First, they are closer to the nature of brewing yeasts, most of which are triploid or of higher ploidy. Second, the scoring of large numbers of isolates for "true" flocculation is somewhat easier than for haploid strains. The obvious disadvantage, however, is that, without an identical experiment using haploid *FLO1* strains, there is no way of distinguishing between the mechanisms of gene mutation and mitotic segregation. The rate would be exceptionally high for gene mutation but, considering the distance of *FLO1* from its centromere (around 43 cM), is also high compared to available rates of mitotic segregation (Thornton and Johnston 1971).

The equivalent experiment with gene *flo3*, that is, using heterozygous diploid strains, would have measured the change from nonflocculent to flocculent cells because of the (frequency) recessive nature of this gene. Instead of determining the rate of gene mutation in a homozygous *flo3* diploid, it was decided for convenience to use the mutant haploid strain X764-S1/F. This strain was highly unstable when subjected to prolonged growth by serial transfer, producing a rate of flocculent to nonflocculent of $>10^{-2}$. This rate is highly suggestive of a mechanism of cytoplasmic mutation and would seem likely to be due to *petite* mutants that are nonflocculent. Unfortunately, at the time these experiments were performed, nonflocculent colonies were not tested for respiration deficiency. It also seems probable that at least a proportion of nonflocculent cells of the diploid strains XL112 and XL86 was due to their nature as RD mutants. The mechanisms underlying these high rates of change from flocculent to nonflocculent, therefore, await experimental clarification.

An interesting example of this aspect of stability in practice is the case of production strain AB80. Isolation of useful segregant and hybrid strains from this important brewing yeast has been attempted (Reader 1980). The flocculation of 50 colonies of each of the strains AB80 and two of its meiotic segregants, AB80-S1 and AB80-S2, was measured by sedimentation rate or "Chester value" (Chester 1963). These 150 colonies were also tested for respiratory competence. The results presented in Table 7.11 show that 22% of AB80 colonies were respiratory deficient whereas all 100 colonies of the two segregant strains were respiratory competent. The high level of RD mutants of strain AB80 explains the observed instability of this strain with respect to flocculence. The distributions of Chester values of respiratory competent and RD colonies are significantly different. The RD mutants of AB80 are, on average, less flocculent and approximately 10% of them (1 of 11 colonies) are nonflocculent. A significant rate of instability of flocculent to nonflocculent cells in brewing yeasts is by no means uncommon (Gilliland 1980). Both segregant strains of

**Table 7.11.** Sedimentation Rates of Colonies of Strain AB80 and Two of Its Segregants

| | | Number of colonies with | | | | |
|---|---|---|---|---|---|---|
| | | Chester values (min) | | | | |
| Strain | Respiration[a] | 5 | 10 | 15 | 20 | >20 |
| AB80 | +39 | 5 | 29 | 4 | 1 | 0 |
| | −11 | 0 | 6 | 3 | 1 | 1 (>120) |
| AB80-S1 | +50 | 50 | 0 | 0 | 0 | 0 |
| | −0 | | | | | |
| AB80-S2 | +50 | 50 | 0 | 0 | 0 | 0 |
| | −0 | | | | | |

[a] + = respiratory competent; − = respiratory deficient.

AB80 are intensely flocculent and also more stable. These characteristics are important for the use of AB80-S1 and AB80-S2 in the development of hybrid brewing strains.

## Industrial Applications

Several examples of the commercial importance of flocculation have been included in the foregoing sections. It is still true to say that this characteristic is of interest principally within the brewing industry. A trend toward yeast separation by centrifugation, however, means that more nonflocculent strains can be used than hitherto. Nevertheless, many larger breweries use cylindro-conical fermentation vessels and these require selection of a sedimenting (flocculent) strain of yeast (Hoggan 1977). Beer production by continuous fermentation requiring flocculent yeasts has also been described (den Blanken 1974; Seddon 1976). It should be noted, however, that nonflocculent strains can be rendered sufficiently flocculent for use in tower fermenters by their immobilization in calcium alginate gel (White and Portno 1978).

Mutants of brewing yeasts with increased flocculence have been produced. Following mutagenesis with NTG (*n*-methyl-*n*-nitrosonitroguanidine) and 4 weeks continuous culture in a small tower fermenter, three of a total of 50 single colony isolates were more flocculent than the parent strain (Molzahn 1977). In another brewing yeast, RD mutants, predominantly induced by the addition of formaldehyde during mashing, were found to have increased sedimentation rates (Cowan *et al.* 1975). This result shows that mitochondrial mutations can affect the yeast cell wall to cause the change from less flocculent to more flocculent as well as vice versa.

Despite the advent of new methods of recombination in yeast (Johnston and Oberman 1979; Kielland-Brandt 1981), conventional (cell) hybridization remains a useful technique in the development of new industrial strains. Among a large number of hybrids of brewing yeasts bred by Anderson and Martin (1975), only 2% were more flocculent than either parent strain, whereas 13% were less flocculent. An inbred hybrid of a brewing yeast, produced by Reader (1980), had the same degree of flocculence (grade 3) as the parent strain but gave a faster attenuation rate and an increased yeast crop. A selected hybrid bred from the production yeast *(S. carlsbergensis)* of Carlsberg Breweries was more flocculent than the parent strain and produced a very similar beer (Gjermansen and Sigsgaard 1981).

Because it is generally laborious to obtain mating segregants of brewing yeasts and because of the barrier to mating erected by the walls of sporulating diploid and polyploid strains, industrial yeast geneticists have turned to the method of spheroplast fusion. This mechanism also offers the opportunity to hybridize yeasts of different genera that cannot be crossed by conventional breeding. In fusion of brewing yeasts to laboratory haploid strains, recombinants were flocculent regardless of which parent strain was flocculent and which nonflocculent (Stewart 1981). In fusion crosses where both parents were brewing yeasts, one flocculent and the other nonflocculent, hybrids were of two types, either moderately flocculent or very flocculent. Neither type of hybrid in this instance, however, made a suitable beer yeast. These results are consistent with flocculence of the commercial yeasts being conferred by dominant genes *(FLO1, FLO5),* the expression of which is subject to modification.

A disadvantage of fusion hybridization, as with conventional cell mating, is that recombination is between complete parental genomes. The technique of transformation offers the chance to introduce a single, or only a few, specific characteristic(s) into an existing production strain. This is most likely to be successful by use of a transforming vector, but, as far as we are aware, to date flocculation genes have not been incorporated into a suitable plasmid. However, cloning of the genes *FLO1* and *FLO5* has been initiated in one of our laboratories (JRJ). The first results obtained recently with nonvector (native DNA) yeast transformation were highly encouraging (Jensen *et al.* 1979). Transformants for the marker gene *ADE2* were obtained at a frequency of approximately 5 per $10^4$ regenerated spheroplast colonies. Other genes, including *ADE1,* were transferred and incorporated at frequencies around $10^{-5}$, although no transformants were obtained for yet other genes. Further experiments, using DNA from the donor strain ABX-1C (Table 7.5), which carries gene *FLO1,* yielded *ADE1* transformants at a frequency of approximately $10^{-4}$ per regenerated spheroplast colony (Barney *et al.* 1980). Of these, 8% were simultaneously transformed for *FLO1,* showing cotransformation for the linked genes *ADE1* and *FLO1* on chromosome I. Unfortunately, subsequent results have been disappointing and other laboratory groups, including one of ours (JRJ) have been unable to demonstrate nonvector transformation, even for gene *ADE1.*

Flocculation is also an important characteristic to be considered in industrial yeasts other than those for brewing. Strain selection has been carried out for fermentations leading to potable spirits and industrial alcohol. The ability to flocculate well, thereby determining cleaner wash to the still, is listed as one of the criteria for selection by Rose (1976). A flocculent strain of *Saccharomyces uvarum* increased the ethanol yield to 80%, compared to 77% using the former yeast, in a process utilizing high initial concentrations of molasses. Flocculent strains developed by one of us (HPR) have been introduced into hybrid strains used for "gasahol" production in São Paulo state in Brazil (Tavares 1981). This latter hybridization program provides a good example of the use of biometric methods in microbial strain improvement. New hybrid strains have recently been developed for use in Scotch malt whisky fermentations (Sharp and Watson 1979). Some of these, producing around 1% increase in alcohol yield, are flocculent and the importance of this aspect to fermentation performance is being investigated further (Watson 1981). The ability of a yeast, in this case *Torulopsis glabrata,* to flocculate is also critical in optimization of biomass production from low concentration industrial effluents (Meyrath *et al.* 1981).

## References

Anderson, E., Martin, P. A. (1975). Sporulation and mating of brewing yeasts. J. Inst. Brew. 81:242–247.

Ballou, D. L., Kern, K. A., Raschke, W. C. (1973). Genetic control of yeast mannan structure: complementation studies and properties of mannan mutants. J. Biol. Chem. 248:4667–4673.

Barney, M. C., Jansen, G. P., Helbert, J. R. (1980). Use of genetic transformation for the introduction of flocculence into yeast. J. Am. Soc. Brew. Chem. 38(2):71–74.

Beavan, M. J., Belk, D. M., Stewart, G. G., Rose, A. H. (1979). Changes in electrophoretic mobility and lytic enzyme activity associated with development of flocculating ability in *Saccharomyces cerevisiae.* Can. J. Microbiol. 25:888–895.

Calleja, G. B., Johnson, B. F. (1977). A comparison of quantitative methods for measuring yeast flocculation. Can. J. Microbiol. 23:68–74.

Chester, V. E. (1963). A study of changes in flocculence in a single cell culture of a strain of *Saccharomyces cerevisiae.* Proc. Roy. Soc., B 157:223–233.

Cohen, R. E., Ballou, C. E. (1980). Linkage and sequence analysis of mannose-rich glycoprotein core oligosaccharides by proton nuclear magnetic resonance spectroscopy. Biochemistry 19:4345–4358.

Cowan, W. D., Hoggan, J., Smith, J. E. (1975). Introduction of respiratory-deficient mutants in brewery yeast. Tech. Quart. Master Brew. Assoc. Amer. 12:15–22.

den Blanken, J. G. (1974). The tower fermenter for lager production. Brew. Guard. 103:35–39.

Gilliland, R. B. (1971). Stability and variation in brewing yeast. Brew. Guard. October, 29–31.

Gilliland, R. B. (1980). Strain variation in brewing yeasts. In: Stability of Industrial Organisms, edited by B. E. Kirsop. Commonwealth Mycological Institute, Kew, pp. 15–19.

Gilliland, R. B. (1981). Brewing yeast. In: Brewing Science, Vol. 2, edited by J. R. A. Pollock. London: Academic Press. pp. 1–60.

Gjermansen, C., Sigsgaard, P. (1981). Construction of a hybrid brewing strain of *Saccharomyces carlsbergensis* by mating of meiotic segregants. Carlsberg Res. Commun. 46:1–11.

Gyllang, H., Martinson, E. (1971). Properties of some spontaneous mutants from brewers' yeast with increased sedimentation rate. Proc. Eur. Brew. Conv., Estoril, pp. 265–271.

Hoggan, J. (1977). Aspects of fermentation in conical vessels. J. Inst. Brew. 83:133–138.

Holmberg, S. (1978). Isolation and characterisation of a polypeptide absent from nonflocculent mutants of *Saccharomyces cerevisiae.* Carlsberg Res. Commun. 43:401–413.

Holmberg, S., Kielland-Brandt, M. C. (1978). A mutant of *Saccharomyces cerevisiae* temperature-sensitive for flocculation. Influence of oxygen and respiratory-deficiency on flocculence. Carlsberg Res. Commun. 43:37–47.

Jansen, G. P., Barney, M. C., Helbert, J. R., Esposito, M. S. (1979). Nonvector-associated transformation in *Saccharomyces.* J. Appl. Biochem. 1:369–376.

Jayatissa, P. M., Rose, A. H. (1976). Role of wall phosphomannan in flocculation of *Saccharomyces cerevisiae.* J. Gen. Microbiol. 96:165–174.

Johnston, J. R., Lewis, C. W. (1974). Genetic analysis of flocculation in *Saccharomyces cerevisiae* and tetrad analysis of commercial brewing and baking yeasts. In: Second International Symposium on the Genetics of Industrial Microorganisms, edited by K. D. Macdonald. London: Academic Press, pp. 339–355.

Johnston, J. R., Oberman, H. (1979). Yeast genetics in brewing. Prog. Ind. Microbiol. Vol. 15, edited by M. J. Bull. Amsterdam: Elsevier, pp. 151–205.

Karson, E. M., Ballou, C. E. (1978). Biosynthesis of yeast mannan. Properties of a mannylphosphate transferase in *Saccharomyces cerevisiae.* J. Biol. Chem. 253:6484–6492.

Kielland-Brandt, M. C. (1981). The genetics of brewers' yeast. Proc. Eur. Brew. Conv. Copenhagen, pp. 263–276.

Lewis, C. W. (1974). Genetic and related aspects of flocculation in yeast. Ph.D. Thesis, University of Strathclyde, Glasgow.

Lewis, C. W., Johnston, J. R. (1974). Genes controlling flocculation in yeast. Proc. Soc. Gen. Microbiol. 1:73.

Lewis, C. W., Johnston, J. R., Martin, P. A. (1976). The genetics of yeast flocculation. J. Inst. Brew. 82:158–160.

Meyrath, J., Braun, R. (1981). Optimisation of yeast production from effluents using flocculent strains. Adv. Biotechnol. Curr. Dev. Yeast Res., edited by G. Stewart and I. Russell. Toronto: Pergamon Press, pp. 25–31.

Miki, B. C. A., Poon, N. H., James, A. P., Seligy, V. L. (1981). Flocculation in *Saccharomyces cerevisiae:* mechanism of cell–cell interactions. Adv. Biotechnol., Curr. Dev. Yeast Res., edited by G. Stewart and I. Russell. Toronto: Pergamon Press, pp. 193–198.

Mizunagan, T., Kuraishi, H., Aida, K., Uemura, T. (1971). Change of cell wall polysaccharides of yeast in biotin deficiency. J. Gen. Appl. Microbiol. 17:85–96.

Molzahn, S. W. (1977). A new approach to the application of genetics to brewing yeast. J. Amer. Soc. Brew. Chem. 35(2):54–59.

Morris, D. L. (1948). Quantitative determination of carbohydrates with Dreywood's anthrone reagent. Science 107:254–255.

Mortimer, R. K., Schild, D. (1980). Genetic map of *Saccharomyces cerevisiae*. Microbiol. Rev. 44:519–571.

Reader, H. P. (1980). Genetic analysis of flocculation in *Saccharomyces cerevisiae*. Ph.D. Thesis, University of Strathclyde, Glasgow.

Rose, D. (1976). Yeasts for molasses alcohol. Process Biochem. 11(2):10–12, 36.

Russell, I., Stewart, G. G. (1979). Spheroplast fusion of brewer's yeast strains. J. Inst. Brew. 85:95–98.

Russell, I., Stewart, G. G., Reader, H. P., Johnston, J. R., Martin, P. A. (1980). Revised nomenclature of genes that control yeast flocculation. J. Inst. Brew. 86:120–121.

Seddon, A. W. (1976). Continuous fermentation in a tower. The Brewer 62:72–78.

Sharp, R., Watson, D. (1979). Improving yeast strains for whisky production. Brew. Dist. Int. 9(6):50–53.

Silhankova, L., Savel, J., Mostek, J. (1970). Respiratory-deficient mutants of bottom brewers' yeast. II. Technical properties. J. Inst. Brew. 76:280–288.

Snow, R. (1979). Maximum likelihood estimation of linkage and interference from tetrad data. Genetics 92:231–245.

Spencer, J. F. T., Miller, R., Spencer, D. M., and Wilkie, D. (1981). Flocculation and starch utilisation in some antibiotic-resistant mitochondrial mutants of *Saccharomyces diastaticus*. Adv. Biotechnol., Curr. Dev. Yeast Res., edited by G. Stewart and I. Russell. Toronto: Pergamon Press, pp. 33–39.

Stewart, G. G. (1981). The genetic manipulation of industrial yeast strains. Can. J. Microbiol. 27:973–990.

Stewart, G. G., and Russell, I. (1977). The identification, characterisation and mapping of a gene for flocculation in *Saccharomyces* sp. Can. J. Microbiol. 23:441–447.

Tavares, F. C. A. (1981). Genetic engineering for ethanol production. *In:* Genetic Engineering for Biotechnology, edited by O. J. Crocomo, F. C. A. Tavares, D. Sodrzeieski. São Paulo: Promocet, pp. 3–6.

Thorne, R. S.W. (1951). Some aspects of yeast flocculation. Proc. Eur. Brew. Conv., Brighton, England, pp. 21–34.

Thorne, R. S. W. (1968). Some observations on yeast mutation during continuous fermentation. J. Inst. Brew. 76:516–524.

Thorne, R. S. W. (1970). Further observations on yeast mutation during continuous fermentation. J. Inst. Brew. 76:555–563.

Thorne, R. S. W., Nohr, B. (1963). Some observations on the stability of a brewing yeast strain. Brew. Digest 32(2):36–39.

Thornton, R. J., Johnston, J. R. (1971). Rates of spontaneous mitotic recombination in *Saccharomyces cerevisiae*. Genet. Res., Cambridge 18:147–151.

White, F. H., Portno, A. D. (1978). Continuous fermentation by immobilized brewers' yeast. J. Inst. Brew. 84:228–230.

Wilkie, D., Mudd, R. C. (1981). Aspects of mitochondrial control of cell surface characteristics in *Saccharomyces cerevisiae*. Adv. Biotechnol. Curr. Dev. Yeast Res., edited by G. G. Stewart and I. Russell. Toronto: Pergamon Press, pp. 345–349.

Work, E. (1971). Cell walls. In: Methods in Microbiology, Vol. 5A, edited by J. R. Norris, D. W. Ribbons. London: Academic Press, pp. 361–418.

# 8

# Protein and Lipid Composition of the Yeast Plasma Membrane

G. H. Rank* and A. J. Robertson*

## Introduction

Procedures used to isolate the plasma membrane have employed four main strategies. Following the enzymatic removal of the cell wall, osmotically sensitive spheroplasts are lysed to release plasma membrane vesicles subsequently enriched on gradients (Boulton 1965; Longley *et al.* 1968; Christensen *et al.* 1972; Schibeci *et al.* 1973). A recent variation of this approach uses concanavalin A (con A) stabilization of the protoplast (Scarborough 1975) after removal of the cell wall by glusulase (Duran *et al.* 1975; Santos *et al.* 1978) or zymolyase (Bussey *et al.* 1979), resulting in the isolation of ghosts and large membrane sheets. A second method employs cell homogenization followed by differential and density gradient (Matile *et al.* 1967; Matile 1970; Fuhrmann *et al.* 1974; Wehrli *et al.* 1975) or zonal centrifugation (Nurminen *et al.* 1976; Nurminen 1977). Isolation of a cell wall fraction with attached plasma membrane and the subsequent enzymatic removal of the wall is used in a third procedure (Nurminen *et al.* 1970). In addition, a novel procedure was introduced by Fuhrmann *et al.* (1976) that uses adjustment of a vesicle preparation to pH 4 to aggregate and remove nonplasma membrane vesicles (A+); the nonaggregated plasma membrane vesicles are referred to as an A− preparation (Rank *et al.* 1978a).

The principal enzyme used to identify the plasma membrane is a $Mg^{2+}$-stimulated oligomycin-insensitive ATPase with a low pH optimum (Fuhrmann *et al.* 1974; Fuhrmann *et al.* 1976). Membrane vesicles with a high buoyant density (>1.20 g/cm$^3$) are enriched for this enzyme (Schibeci *et al.* 1973; Nurminen *et al.* 1976; Nurminen *et al.* 1977; Delhez *et al.* 1977; Serrano 1978). Preliminary characterizations of plasma membrane ATPase indicated two (Ahlers *et al.* 1978), or three protein components (Willsky 1979). A novel

*Department of Biology, University of Saskatchewan, Saskatoon, Saskatchewan, Canada S7N 0W0.

enzyme marker, chitin synthetase, was shown to be associated with con A stabilized protoplasts in the zymogen form (Duran *et al.* 1975; Duran *et al.* 1979). Cytological identification has used elongated invaginations and hexagonal particle arrangements (Moor *et al.* 1963; Fuhrmann *et al.* 1974; Furhmann *et al.* 1976; Takeo 1976; Kramer *et al.* 1978; Steere *et al.* 1980) associated with plasma membrane fractions; however, these characteristic cytological features are physiologically dependent (Schneider *et al.* 1978). An autoradiographic analysis (Schibechi *et al.* 1973) of labeled protoplasts indicated that following lactoperoxidase-catalyzed iodination, 96% of cellular label was associated with the plasma membrane. Lactoperoxidase-catalyzed iodination was subsequently used to identify membrane fractions (Schibeci *et al.* 1973; Duran *et al.* 1975; Santos *et al.* 1978; Bussey *et al.* 1979) and an A− surface glycoprotein (Rank *et al.* 1978a).

Plasma membrane polypeptides have been resolved into approximately 20 polypeptides by one-dimensional sodium dodecyl sulfate–polyacrylamide gel electrophoresis (SDS-PAGE) (Fuhrmann *et al.* 1974; Wehrli *et al.* 1975; Fuhrmann, 1977; Kramer *et al.* 1978, Santos *et al.* 1978). The utility of two-dimensional SDS-PAGE in further resolving plasma membrane polypeptides is apparent from two recent publications (Bussey *et al.* 1979; Robertson *et al.* 1980). Utilization of crossed immunoelectrophoresis (CIE) has also proven useful in identification of lectin receptors associated with A− preparations (Gerlach *et al.* 1979).

Although numerous lipid analyses have been undertaken in yeast (Rattray *et al.* 1975), there is very little quantitative data on the general lipid composition of the plasma membrane. Longley *et al.* (1968) first reported on the qualitative distribution of the major lipid classes in a poorly identified membrane preparation; they noted the presence of monoglycerides (MG), diglycerides (DG), triglycerides (TG), sterols (S), sterol esters (SE), free fatty acids (FFA), phosphatidyl choline (PC), lysophosphatidyl choline (LPC), phosphatidylethanolamine (PE), phosphatidylinositol (PI), and phosphatidylserine (PS). A quantitative analysis of phospholipid phosphorus (PL-P) indicated plasma membrane enrichment for PE, PI, and PS compared to whole cells. The principal fatty acids of all major lipid classes were palmitic ($C_{16:0}$), palmitoleic ($C_{16:1}$), and oleic ($C_{18:1}$). Suomalainen and Nurminen (1970) confirmed the enrichment for PI and PS the abundance of $C_{16:0}$, $C_{16:1}$, and $C_{18:1}$ in a plasma membrane fraction characterized by a $MG^{2+}$-stimulated ATPase (Nurminen *et al.* 1970). Unfortunately, these preparations may not be highly enriched for the plasma membrane (Dubé *et al.* 1973). Subsequent analyses of these preparations demonstrated the minor presence of glycolipids (Työrinoja *et al.* 1974) and long-chain fatty acids (Nurminen and Suomalainen 1971). Long-chain fatty acids and alkanes were also reported (Blanchardie 1970) in a plasma membrane preparation identified by electron microscopy.

We have confirmed, in quantitative analyses (Rank *et al.* 1978a) of a zymolyase ghost preparation, the presence of the major lipid classes reported by

Longley *et al.* 1968). In addition, the viscosity of zymolyase ghosts, con A ghosts (Duran *et al.* 1975), and gradient purified plasma membrane vesicles (Wehrli *et al.* 1975) was determined (Rank *et al.* 1978a, 1978b) by a fluorescence polarization technique using 1,6-diphenyl1-1,3,5-hexatriene (DPH). The ghost preparations were characterized by a low viscosity (2.3–2.6 poises) and a low S/PL mole ratio ($<0.3$). In sharp contrast, A− had a high viscosity (Rank *et al.* 1978b; 12.5 poises) and a high S/PL mole ratio (2.2). Preliminary lipid analyses of this preparation by Kramer *et al.* (1978) confirmed the presence of a high S/PL mole ratio but failed to quantify major lipid classes and the fatty acid distribution.

We report here on comparative analyses designed to further evaluate the protein and lipid composition of the yeast plasma membrane.

## Cytological Purity of Membrane Preparations

Electron micrographs representative of typical preparations are presented in Figure 8.1. Con A ghosts appear as long sheets with variable entrapment of undefined material. The purity of this preparation varies greatly with the suc-

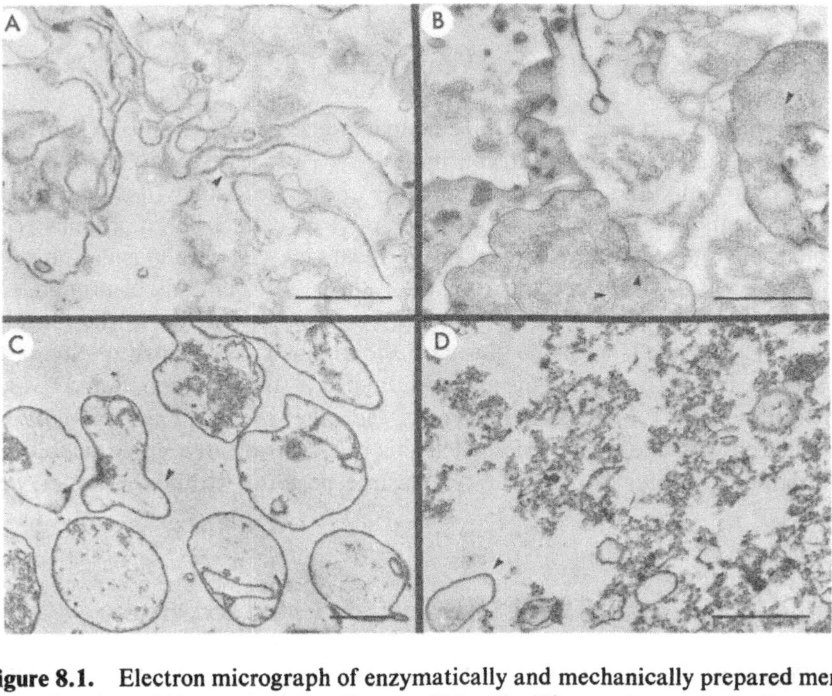


**Figure 8.1.** Electron micrograph of enzymatically and mechanically prepared membrane fractions. (A) con A ghosts (bar = 500 nm). (B) zymolyase ghosts (bar = 1 μm). (C) A− preparation (bar = 500 nm). (D) A+ preparation (bar = 500 nm).

cess in producing protoplasts. Strains of yeast resistant to protoplast formation are often badly contaminated with cell walls. Zymolyase ghosts are composed of membrane sheets and are contaminated with unlysed spheroplasts and cell wall fragments if the strain used is resistant to zymolyase digestion. A− preparations consistently result in vesicle preparations relatively free of entrapped material and cell wall contamination regardless of the strain used, whereas A+ preparations have a high percentage of an undefined clumped material.

## A Vesicle Free Cell Wall Preparation

The common occurrence of cell wall contamination in con A and zymolyase ghosts and the use of adhering cell wall as a method of isolation (Nurminen *et al.* 1970) makes it necessary to evaluate the contribution of cell wall proteins to plasma membrane preparations. Cell walls were homogenized in 400 m*M* KCl at pH4 according to the method for the isolation of A− (Fuhrman *et al.* 1976). Figure 8.2 shows the presence of vesicles in partially purified fractions and the final production of a fraction virtually free of vesicles. The nature of the fibrillar material associated with the inner cell wall is not known (Robertson *et al.* 1980).

## Surface Labeling of Intact Yeast Cells

When intact yeast cells were subjected to lactoperoxidase-catalyzed iodination, there was a 30-fold increase in specific activity ($2.2 \times 10^4$ cpm/mg of protein) over control mixtures ($7.2 \times 10^2$ cpm/mg of protein) lacking lactoperoxidase. This indicated that the surface of the intact yeast cell harbored proteins accessible to lactoperoxidase. Removal of the yeast cell wall with glusulase resulted in spheroplasts with a specific activity of $8.6 \times 10^3$ cpm/mg of protein. Calculations showed that 65% of the total label was released into the glusulase digest. Hypotonic lysis of the glusulase spheroplasts, and centrifugation of the lysate at $100{,}000 \times g$ for 60 min, resulted in a soluble fraction and a crude membrane pellet. Ninety-six percent of the spheroplast $^{125}$I radioactivity was associated with the membrane pellet; this demonstrates the impermeability of the labeling reagent to soluble cytoplasmic proteins within the intact yeast cells.

## One-Dimensional SDS-PAGE Surface Label Experiments

In Figure 8.3A, lanes B, E, and H contain equal amounts (100 $\mu$g) of SDS-solubilized intact cell, total spheroplast, and con A ghost membrane protein. Comparisons of Coomassie blue-stained protein profiles show a striking increase in spheroplast and con A ghost proteins, with apparent molecular

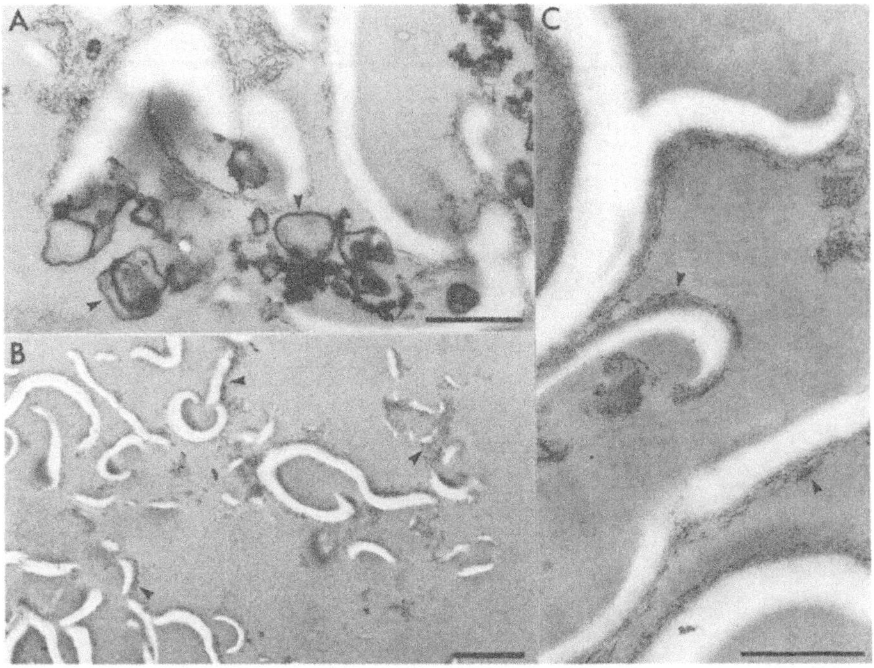


**Figure 8.2.** Electron micrograph of partially purified and purified cell walls. (A) Preparation of partially purified cell walls showing the presence of contaminating vesicular membrane material (arrows). Most of the partially purified wall profiles appeared relatively free of adhering vesicular membrane contamination (bar = 500 nm). (B) Preparation of purified cell walls with associated fibrillar material (arrows). Note the absence of membrane. (bar = 1 $\mu$m). (C) Preparation of purified cell walls with associated fibrillar material (arrows) (bar = 500 nm).

weights greater than 43,000 over intact cells, and a relative decrease in proteins with apparent molecular weights below 43,000, especially in Con A ghost membranes. A major band at 28,000 present in intact cells and spheroplast membranes is virtually absent in con A ghost membranes. Inspection of lanes E and H indicates that spheroplast membranes have several high-molecular-weight protein bands in common with con A ghost membranes, but differ considerably in intensity.

Lanes B, E, and H in Figure 8.3B give the distribution of labeled polypeptides with equal amounts of iodinated intact cell, spheroplast, and con A ghost membrane protein. Spheroplast and con A ghost membranes were isolated from the same iodinated intact cells. A marked increase in the amount of surface-labeled polypeptides during plasma membrane purification is noted in the molecular-weight region greater than 130,000. This correlates with the amount of protein-specific activity found in the analyses of intact cells (lane B, 8.8 cpm/$\mu$g of protein), total spheroplast membranes (lane E, 10.7 cpm/$\mu$g of pro-

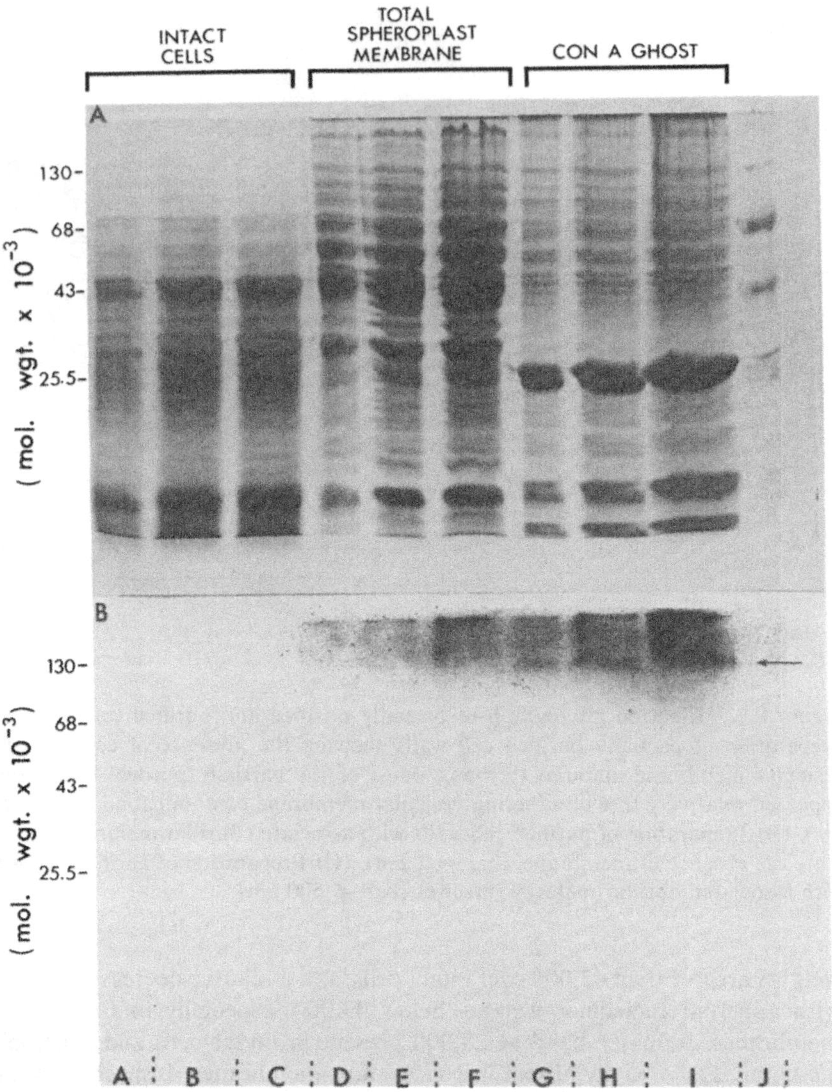


**Figure 8.3.** (A) Coomassie blue-stained one-dimensional SDS–polyacrylamide gels of lactoperoxidase-iodinated intact cells, total spheroplast membranes (100,000 *g* pellet of spheroplast lysate) and con A ghosts isolated from labeled intact cells. (B) This is an autoradiogram of A. The Coomassie blue-stained gel in A was exposed to Kodak X-Omat R x-ray film for 6 weeks at $-80^{\circ}$C with intensifying screens and hypersensitized film. The exposure time was not sufficient for photographic detection of the lower amount of radioactivity in lanes A, B, and C, although a faint darkening of the high-molecular-weight region of the original x-ray film was apparent.

tein), and con A ghosts (lane H, 18 cpm/$\mu$g of protein) during plasma membrane purification. $^{125}$I-labeled polypeptides appeared as diffuse aggregates entering the top of the 10% acrylamide separating gel. A broad, diffuse band appeared to form at an apparent molecular weight of 130,000–140,000 in lanes G, H, and I. Periodic acid–Schiff staining of the preceding preparations gives a positive reaction corresponding exactly to the surface-labeled polypeptides.

## Two-Dimensional SDS-PAGE

### Comparative Analyses

Specific identification of unique membrane polypeptides was not possible in one-dimensional gel analyses because of the difficulty in distinguishing polypeptides migrating with the same apparent molecular weights. Comparing two-dimensional gel patterns of intact cells and derived membrane preparations resulted in the identification of several membrane polypeptides. Membrane proteins were tentatively defined as those components that were consistently enriched in plasma membrane electrophoretograms over intact cell profiles. Conversely, polypeptides greatly reduced or absent after plasma membrane isolation were assumed to be nonmembrane components. Two-dimensional SDS-PAGE profiles containing equal amounts of con A ghosts and A− were compared (Fig. 8.4). Con A ghosts and A− contained approximately 150 polypeptides by Coomassie blue staining. Of the components identified, 80–85% were common to both membrane preparations, suggesting that these two yeast plasma membrane preparations, isolated by differing methods, are enriched for similar membrane proteins. Four polypeptides, labeled 1p, 2p, 5p, and 6p, were greatly enriched in con A ghosts and A− in comparison to intact cells. Polypeptides labeled 3, 4, 7, and 8 were moderately enriched in plasma membrane preparations whereas components 9, 10 and 11 were shown to be soluble polypeptides (Robertson *et al.* 1980). None of the plasma membrane enriched polypeptides were enriched in A+ preparations (Fig. 8.4D). Polypeptides 1p and 2p (often appearing as several spots) are glycoproteins since they gave a positive reaction with periodic acid–Schiff.

### Pronase-Sensitive Surface Polypeptides

Glusulase spheroplasts were incubated with pronase. The specificity of proteolytic digestion was monitored by subjecting lactoperoxidase-iodinated spheroplasts to pronase digestion. This resulted in 80% of the bound $^{125}$I being removed in pronase-treated spheroplasts, whereas control incubations lacking pronase retained 70% of the bound $^{125}$I. Polypeptides 1p, 2p, 3, 4, 7, and 8

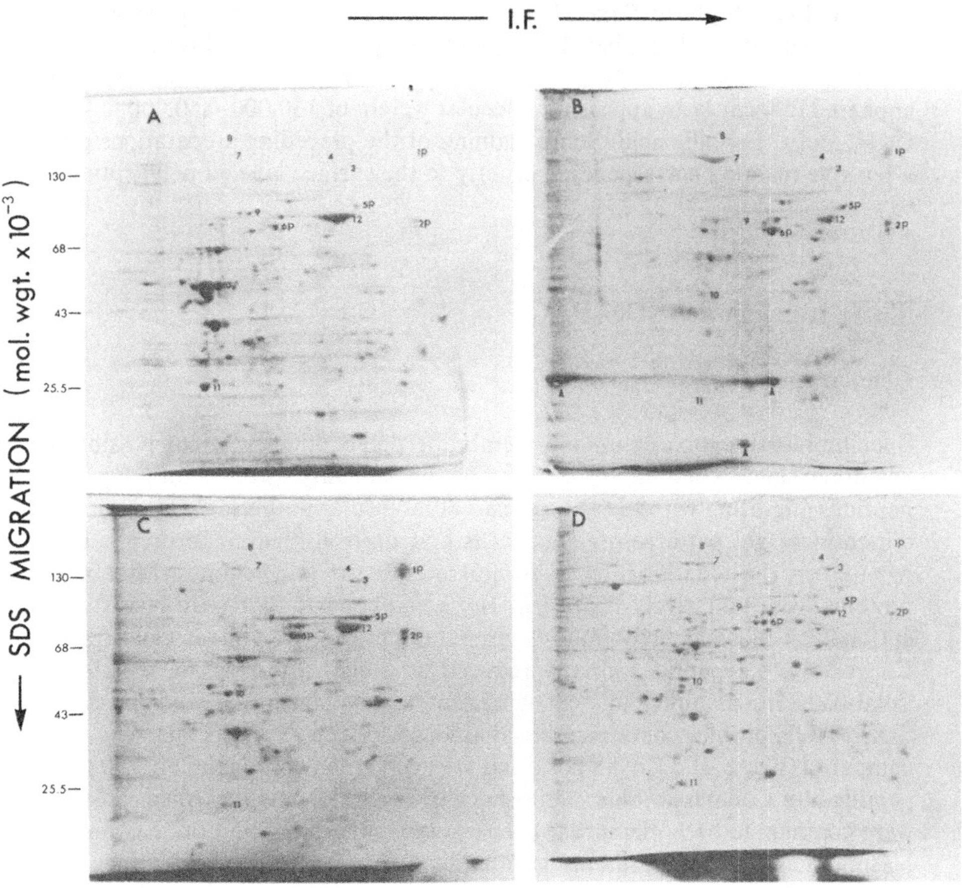


**Figure 8.4.** Two-dimensional separations of intact cell, con A ghost, A− and A+ proteins. (A) Total cell protein; (B) con A ghost; (C) A−; D, A+. The arrow at the top of the figure indicates the direction of the isoelectric focusing (pH 7.0 to 4.8 from left to right) dimension. All preparations were subjected to the same solubilization, electrophoretic, and Coomassie blue-staining conditions. Each isoelectric focusing gel was loaded with 250 μg protein. Spots are referred to as numbered components with the number being placed to the immediate right of the spot. The letter p corresponds to plasma membrane. The arrows in B refer to con A subunits, and the arrow in C designates a series of spots unique to A−.

were virtually removed during pronase digestion of spheroplasts. Other polypeptides such as numbers 5p and 6p were only reduced in intensity. Components 9 to 11, found to be diminished or absent in plasma membrane purified fractions, are examples of polypeptides unaffected by pronase digestion. In general, those polypeptides having apparent molecular weights greater than 70,000 and acidic isoelectric points were most reduced by pronase digestion.

### Surface Label Experiments

When con A ghosts were isolated from lactoperoxidase-iodinated intact cells and subjected to two-dimensional protein analyses, the only $^{125}$I-labeled polypeptide detected was 1p (Fig. 8.5A). Similar results were obtained when intact spheroplasts were surface labeled and subjected to two-dimensional protein analyses.

A− fractions were surface labeled immediately after isolation, in osmotic stabilizer, where a significant percentage of vesicles remain sealed and demonstrate transport properties (Fuhrmann 1977). Autoradiography of two-dimensional profiles of A− proteins detected nine $^{125}$I-labeled components; 1p was the major $^{125}$I-labeled polypeptide (Fig. 8.5B). Minor components are difficult to evaluate for they may represent internal proteins exposed to iodination by the presence of leaky vesicles.

## Cell Wall Polypeptides

Purification of the cell wall as indicated in Figure 8.2 was monitored by two-dimensional SDS-PAGE. Crude and partially purified cell walls were contaminated with approximately 150 polypeptides. However, a single polypeptide, 16w, was shown to be present in significant amounts in purified cell walls (Fig. 8.6). This component appears to be a glycoprotein marker for the cell wall since it gave a positive reaction with periodic acid–Schiff reagent. Since this is the only polypeptide present in significant amounts in cell walls, it appears that the SDS-PAGE profiles of plasma membranes (Fig. 8.4) are not contaminated by cell wall protein and that 16w serves as a cell wall marker polypeptide.

## Comparative Lipid Analyses

### Viscosity

The reported viscosity of plasma membrane preparations (Rank *et al.* 1978a, 1978b) and liposomes is summarized in Table 8.1. In addition, new data on A+ are included. The two ghost preparations and A+ have a similar but lower viscosity than plasma membrane vesicles. Liposomes generated from A+ and A− maintain a large difference (3.2 poises) in viscosiy. Membrane viscosity is increased in the presence of saturated fatty acids and sterols with a 3-$\beta$-hydroxyl group (Demel and De Kruyff 1976). A quantitative analysis of the major lipid classes (Table 8.2) of A+ and A− confirmed the differences in lipid composition suggested by the viscosity determinations. On this basis the high viscosity of A− appears to result from the presence of a high level of free S. Thus the high S/PL mole ratio (2.2) for A− compared to 0.4 for A+ (Table 8.2) is consistent with the 4.2-fold higher viscosity of A−. The

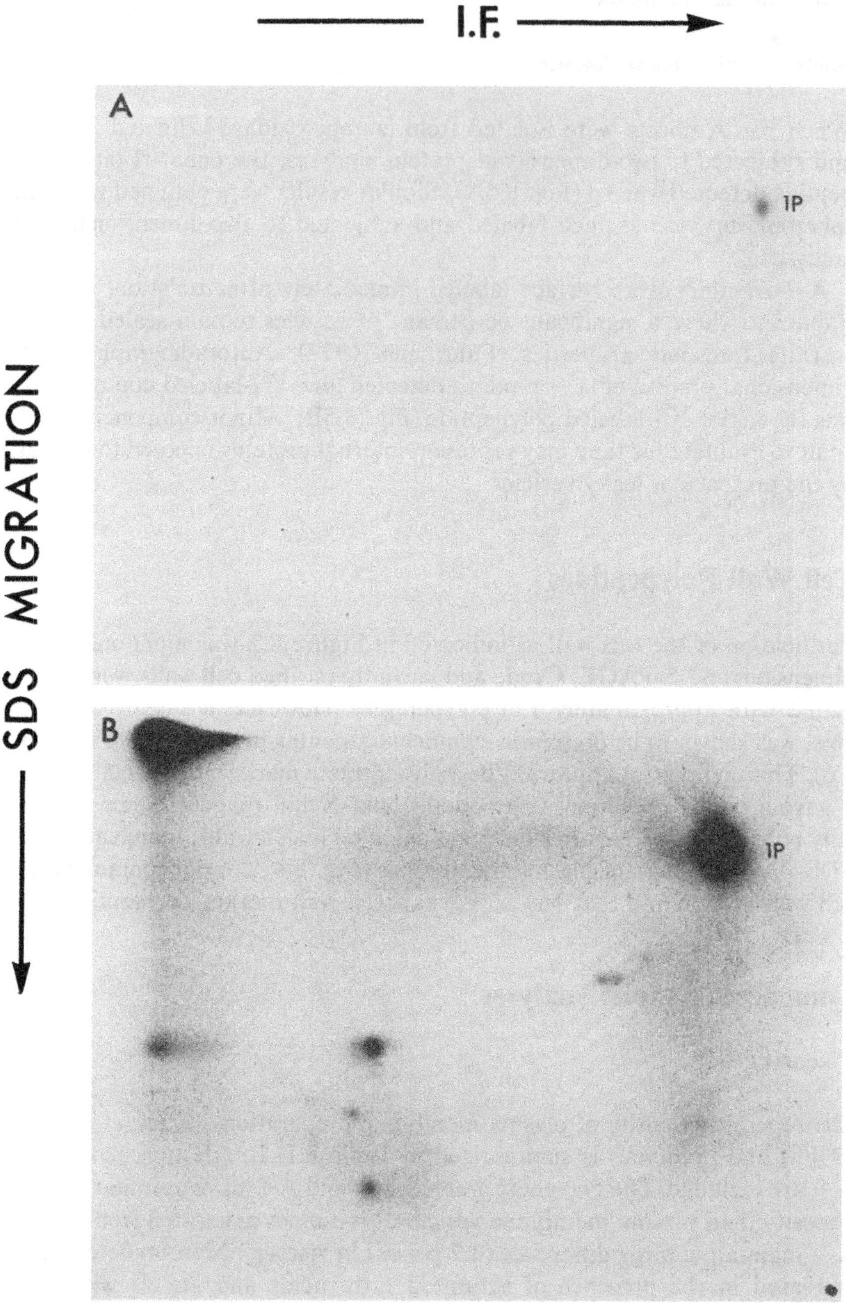


**Figure 8.5.** Autoradiograms of two-dimensional gels from surface-labeled plasma membrane fractions. (A) Autoradiogram of a two-dimensional separation of con A ghost membrane proteins. Con A ghosts were isolated from lactoperoxidase-iodinated intact cells, and 240 μg of protein, with a specific activity of 32 cpm/μg, was placed on isoelectric focusing gels. (B) Autoradiogram of a two-dimensional separation of surface-labeled A− protein. Freshly prepared A− was subjected to lactoperoxidase iodination for 20 min at 25°C prior to electrophoresis. 150 μg of A− protein with a specific activity of 200 cpm/μg was placed onto focusing gels.

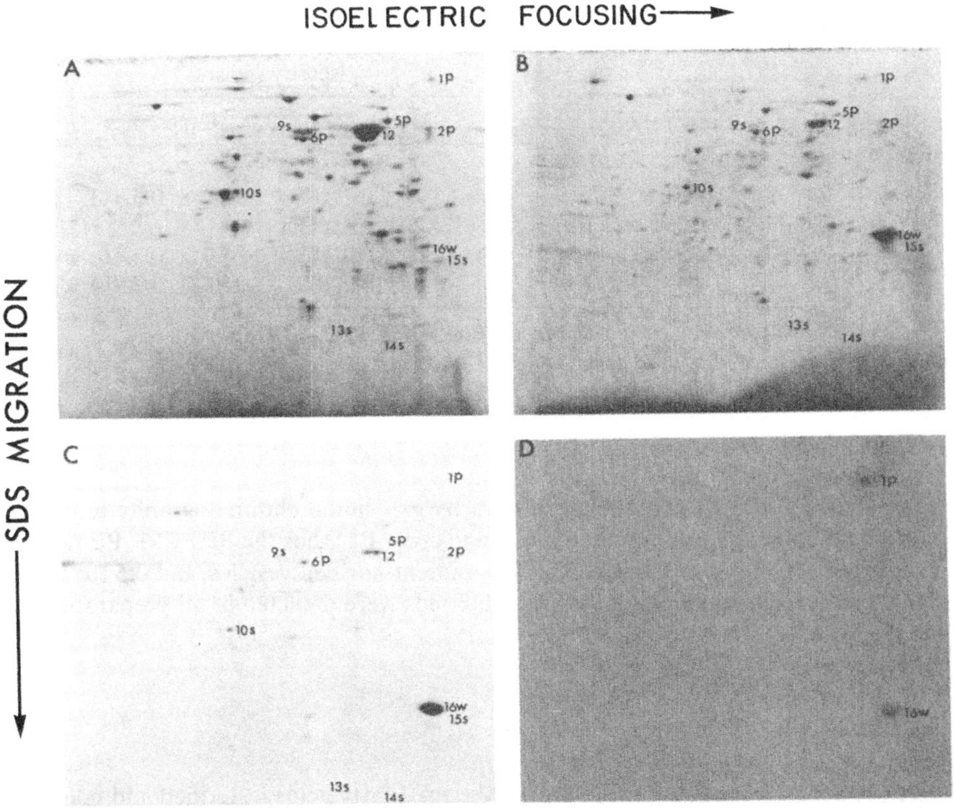


**Figure 8.6.** Comparative two-dimensional SDS–polyacrylamide gels of cell wall fractions. (A) Crude cell walls. (B) Partially purified wall fraction. (C) Purified wall fraction stained for protein with Coomassie blue (D) Purified walls stained for carbohydrate with PAS. If the numbered polypeptides copurify with a particular fraction, the number is followed by a letter denoting the origin of that fraction. The letters p, s, and w correspond to plasma membrane, soluble fraction, and cell wall fraction, respectively (see Robertson *et al.* 1980 for a more detailed discussion).

increased viscosity of A− does not result from saturation of fatty acids since the fatty acid distributions of A+ and A− lipids are similar (Table 8.3).

### Phospholipid Composition

PC, PE, and PI accounted for 89 and 81% of PL-P in con A ghost and A+ preparations (Table 8.4). The overall PL-P distributions of con A ghosts, A+, and zymolyase ghosts were similar. Gradient-purified vesicles were enriched for PI relative to these three preparations. The distribution of A− PL was very different from that in any of the other preparations. 38.8% of PL-P remained at the origin of the two-dimensional thin-layer plate; the composition of this

**Table 8.1.** Viscosity of Membranes and Liposomes[a]

| Preparation | Viscosity (poises)[b] Membranes | Liposomes |
|---|---|---|
| con A ghosts | 2.9 ± 0.31(8)[c] | — |
| Zymolyase ghosts | 3.1 ± 0.21(8) | 2.2 ± 0.37(8) |
| Gradient-purified vesicles | 7.2 + 0.45(4) | — |
| A− | 12.5 ± 0.25(8) | 5.0 ± 0.29(6) |
| A+ | 3.0 ± 0.22(8) | 1.8 ± 0.70(6) |

[a]Data in part summarized from Rank *et al.* 1978a, 1978b.
[b]Mean ± standard error.
[c]Number of preparations.

component is unknown but was shown by gas–liquid chromatography to contain mannose. PI was enriched relative to other PL; thus the PI:PC + PE ratio was 1.8 for A−, 0.3, for A+, 0.6 for gradient-purified vesicles, and 0.5 for con A ghosts. Low amounts of lysophospholipids were detected in all preparations except A−.

## FFA in A−

FFA account for 72% (Table 8.2) of the total fatty acids (esterified and nonesterified) of A− and 16% of the total fatty acids of A+. Lipase activity in A− was not evaluated but could be expected to account for the high level of FFA. Kokke *et al.* (1963) and Van Den Bosch *et al.* (1967) reported on a yeast cell supernatant with phospholipase A and lysophospholipase activities. The combined effect of these enzymes was the production of FFA and glycerophosphorylcholine from PC. Drying of yeast cells for 36 h also resulted in the release of FFA from PL (Van Den Bosch *et al.* 1967). Nurminen and Suom-

**Table 8.2.** Composition of Lipids Extracted from A− and A+[a]

| Source of lipids | Total lipid (μmole/g) Phospolipids | Sterols | Sterol esters | Triglycerides | Diglycerides | Free fatty acids |
|---|---|---|---|---|---|---|
| A− | 140 ± 19 | 314 ± 22 | 6 | 30 ± 6 | 17 ± 5 | 1052 ± 41 |
| A+ | 459 ± 20 | 163 ± 17 | 77 | 55 ± 7 | 49 ± 6 | 210 ± 65 |

[a]Values (except for sterol esters of A+ and A−) are the mean ± standard error of eight preparations. Values for sterol esters of A+ and A− are the average of two preparations. Mole composition was calculated using the following molecular weights: phospholipid (750), sterol (398), sterol ester (650), triglyceride (838), diglyceride (590), free fatty acid (266).

**Table 8.3.** Fatty Acid Distribution[a] of Lipids Extracted from A+ and A−

| | Diglycerides | | Triglycerides | | Fatty acids: Free | | Fatty acids: Esterified | |
|---|---|---|---|---|---|---|---|---|
| | A+ | A− | A+ | A− | A+ | A− | A+ | A− |
| Palmitic | 21 ± 1.8 | 22 ± 1.5 | 16 ± 1.8 | 20 ± 2.8 | 20 ± 2.4 | 19 ± 0.4 | 18 ± 0.8 | 23 ± 1.2 |
| Palmitoleic | 37 ± 1.9 | 32 ± 1.9 | 47 ± 3.9 | 45 ± 5.1 | 41 ± 1.6 | 46 ± 1.8 | 44 ± 3.5 | 37 ± 3.1 |
| Stearic | 6 ± 0.6 | 6 ± 2.0 | 5 ± 1.3 | 5 ± 4.0 | 7 ± 1.0 | 4 ± 0.1 | 4 ± 0.8 | 5 ± 1.2 |
| Oleic | 35 ± 1.7 | 40 ± 1.8 | 32 ± 0.9 | 30 ± 1.9 | 32 ± 2.6 | 32 ± 2.5 | 34 ± 3.2 | 36 ± 2.2 |

[a]Diglycerides, triglycerides, and free fatty acids were isolated on thin-layer plates and fatty acids were quantified as fatty acyl methyl esters after basic (diglycerides and triglycerides) or acidic (free fatty acids) transesterification. Esterified fatty acids were quantified as methyl esters produced by basic transesterification of the complete lipid extract. All values are the mean percent ± standard error from two preparations of four different strains. Greater than 95% of the weight of fatty acids was accounted for by the four fatty acids indicated.

**Table 8.4.** Phospholipid Composition[a]

| Phospholipid | Con A ghosts | Gradient-purified vesicles | A− | A+ |
|---|---|---|---|---|
| Phosphatidylcholine | 35.2(2) | 26.3(2) | 9.5 ± 0.6(4) | 36.8 ± 2.8(4) |
| Phosphatidylethanolamine | 26.1 | 17.4 | 5.5 ± 1.2 | 28.0 ± 0.9 |
| Phosphatidylinositol | 28.1 | 25.7 | 26.8 ± 2.8 | 16.2 ± 2.6 |
| Phosphatidylserine | 1.8 | 4.8 | 8.9 ± 0.8 | 6.9 ± 2.3 |
| Cardiolipin | 1.2 | 2.7 | 3.6 ± 1.2 | 3.3 ± 0.4 |
| Phosphatidic acid | n.d. | 4.8 | 6.9 ± 1.1 | 0.4 ±0.2 |
| Lysophosphatidylcholine | 1.8 | 4.2 | n.d. | 3.1 ± 0.5 |
| Lysophosphatidylethanolamine | 3.2 | 9.3 | n.d. | 4.0 ± 0.9 |
| Phosphatidylglycerol | 2.6 | 4.8 | n.d. | 1.4 ± 0.9 |
| Origin | n.d. | n.d. | 38.8 ± 2.8 | 0.0 |

[a]Data are presented as the average percent phospholipid phosphorus ± standard error. The number of different preparations analyzed is given in brackets. n.d., not detectable in 1500 μg of total lipids.

alainen (1970) identified lysophospholipase and lipase activity in the yeast plasma membrane; the former enzyme was active at pH 4. At pH 4.5 yeast mitochondria showed very high phospholipase activity without the accumulation of lyso derivatives (Vignais *et al.* 1970); PC was degraded rapidly but PI was unchanged. In summary, these observations suggest that PL may be degraded to FFA in A− and that incubation at pH 4, used to isolate A−, may enhance the activity of lipolytic enzymes. Presumably lipases are enriched in A− since A+ is incubated under the same conditions and has a fivefold lower concentration of FFA.

Differences in lipid composition and viscosity between A− and other plasma membrane preparations may result from lipase activity which is activated by incubation at pH 4 and/or excessive contamination of preparations other than A−. Certainly the effect of incubation at pH 4 on the lipid composition of other preparations should be ascertained. However, a massive contamination would be required since surface-labeling techniques have shown that zymolyase ghosts (Bussey *et al.* 1979) and con A ghosts (Duran *et al.* 1975; Santos *et al.* 1978) are clearly enriched for plasma membrane components. Furthermore, the apparent absence of a major uncharacterized A− PL (Table 8.4 origin) in other plasma membrane preparations suggests A− may, in some respects, differ fundamentally from other preparations. One possibility is that A− may represent a plasma membrane domain enriched for surface phosphomannan, which is negatively charged at pH 4 (Rank *et al.* 1978b). Suggestive evidence for domains is seen in the demonstration of plasma membrane fractions of different densities (Schibeci *et al.* 1973; Nurminen *et al.* 1976; Nurminen *et al.* 1977; Blanchardie *et al.* 1977).

## Concluding Remarks

We feel that comparative two-dimensional SDS-PAGE analyses are a useful addition to the standard methods used to assess membrane purity. Polypeptide 1p is a clear marker for the plasma membrane. Polypeptide 1p is well resolved in two-dimensional SDS-PAGE (Fig. 8.4) and is the major surface-labeled component in con A ghosts and A− preparations (Fig. 8.5). In sharp contrast, two-dimensional SDS-PAGE of clean cell walls (Fig. 8.2) does not show major contamination with 1p but is enriched for the unique cell wall polypeptide 16w (Fig. 8.6). The confidence gained from the use of 1p to define A− as a plasma membrane fraction is invaluable in the interpretation placed on the lipid composition of different plasma membrane fractions. The unusual lipid composition of A− (Tables 8.1, 8.2, and 8.4) in comparison to ghosts can be confidently interpreted as a variation related to the type of plasma membrane preparation. The high viscosity and S/PL molar ratio of A− is probably due to the activation of lipase activity at pH 4 utilized in this method of preparation. However, it is also possible that A− is a highly viscous domain of the plasma membrane selected by the method of isolation. Definitive experiments to distinguish between these interpretations should be sought.

The wide range of genetic and physiological manipulations possible in yeast make this species a valuable model to study eukaryotic plasma membrane structure and functional relationships. We are hopeful that future research can assign a function to the major plasma membrane polypeptides resolved by two-dimensional SDS-PAGE and that an understanding of polypeptide–lipid interactions (Esfahani *et al.* 1979) is facilitated by the quantitative lipid analyses presented here.

## References

Ahlers, J., Ahr, E., Seyfarth, A. (1978). Kinetic characterization plasma membrane ATPase from *Saccharomyces cerevisiae. Mol. Cell. Biochem.* 22:39–49.

Blanchardie, P., Carde, J., Cassagne, C. (1977). Occurrence and biosynthesis of very long chain fatty acids and alkanes in plasmalemma enriched fractions from *Saccharomyces cerevisiae. Biol. Cellulaire* 30:127–136.

Boulton, A. A. (1965). Some observations on the chemistry and morphology of the membranes released from yeast protoplasts by osmotic shock. *Exp. Cell Res.* 37:343–359.

Bussey, H., Saville, D., Chevallier, M. P., Rank, G. H. (1979). Yeast plasma membrane ghosts. An analysis of proteins by two-dimensional gel electrophoresis. *Biochim. Biophys. Acta.* 553:185–196.

Christensen, M. S., Cirillio, V. P. (1972). Yeast membrane vesicles: isolation and general characteristics. *J. Bacteriol* 110:1190–1205.

Delhez, J., Dufour, J. P., Thines, D., Goffeau, A. (1977). Comparison of the properties of plasma membrane bound and mitochondria bound ATPase in the yeast *Schizosaccharomyces pombe. Eur. J. Biochem.* 79:319–328.

Demel, R. A., De Kruyff, B. (1976). The function of sterols in membranes. *Biochim. Biophys. Acta* 457:109–132.

Dubé, J., Setterfield, G., Kiss, G., Lusena, C. V. (1973). Fate of the plasma membrane of *Saccharomyces cerevisiae* during cell rupture. *Can. J. Microbiol.* 19:285–290.

Duran, A., Bowers, B., Calib, E. (1975). Chitin synthetase zymogen is attached to the yeast plasma membrane. *Proc. Nat. Acad. Sci. USA* 72:3952–3955.

Duran, A., Cabib, E., Bowers, B. (1979). Chitin synthetase distribution on the yeast plasma membrane. *Science* 203:363–365.

Esfahani, M., Solomon, D. J., Mele, L., Teter, M. N. (1979). Lipid–protein interactions in membranes: effect of the lipid composition on mobility of spin-labeled cysteine residues in yeast plasma membrane. *J. Supramol. Structure* 10:277–286.

Fuhrmann, G. F., Welrli, E., Boem, C. (1974). Preparation and identification of yeast plasma membrane vesicles. *Biochim. Biophys. Acta* 363:295–310.

Fuhrmann, G. F., Boehm, C., Theuvenet, A. P. R. (1976). Sugar transport and potassium permeability in yeast plasma membrane vesicles. *Biochim. Biophys. Acta* 433:585–596.

Fuhrmann, G. F. (1977). The effect of heavy metals on transport functions in plasma membrane of *Saccharomyces cerevisiae. KemiaKemi* 4:616–618.

Gerlach, J. H., Bjerrum, O. J., Rank, G. H., Bøg-Hansen, T. C. (1979). Crossed immunoelectrophoresis of yeast plasma membrane lectin receptors. *Protides of Biological Fluids 27th Colloquim.* Permagon Press, Oxford.

Kokke, R., Hooghwinkel, G. J. M., Booij, H. L., Van Den Bosch, H., Zelles, L., Mulder, E., Van Deenen, L. L. M. (1963). Metabolism of lysolecithin and lecithin in a yeast supernatant. *Biochim. Biophys. Acta* 70:351–354.

Kramer, R., Kopp, F., Niedermeyer, W., and Fuhrmann, G. F. (1978). Comparative studies of the structure and composition of the plasmalemma and the tonoplast in *Saccharomyces cerevisiae. Biochim. Biophys. Acta.* 507:369–380.

Longley, R. P., Rose, A. H., Knights, B. A. (1968). Composition of the protoplast membrane from *Saccharomyces cerevisiae. Biochem. J.* 108:401–412.

Matile, P. (1970). Properties of the purified cytoplasmic membrane of yeast. *FEBS Symp.* 20:39–49.

Matile, P., Moor, H., Muhlthaler, K. (1967). Isolation and properties of the plasmalemma in yeast. *Arch. Microbiol.* 58:201–211.

Moor, H., Muhlethaler, K. (1963). Fine structure in frozen-etched yeast cells. *J. Cell. Biol.* 17:609–628.

Nurminen, T., Oura, E., Soumalainen, H. (1970). The enzymic composition of the isolated cell wall and plasma membrane of baker's yeast. *Biochem. J.* 116:61–69.

Nurminen, T., Suomalainen, H. (1970). The lipolytic activities of the isolated cell envelope fractions of baker's yeast. *Biochem. J.* 118:759–763.

Nurminen, T., Suomalainen, H. (1971). Occurrence of long chain fatty acids and glycolipids in the cell envelope fractions of baker's yeast. *Biochem. J.* 125:963–969.

Nurminen, T., Taskinen, T., Suomalainen, H. (1976). Distribution of membranes, especially of plasma membrane fragments during zonal centrifugations of homogenates from glucose-repressed *Saccharomyces cerevisiae. Biochem. J.* 154:751–763.

Nurminen, T., Taskinen, L., Suomalainen, H. (1977). Distribution of plasma membrane fragments during zonal centrifugation of homogenates from aerobic *Saccharomyces cerevisiae. J. Gen. Microbiol.* 98:301–304.

Rank, G. H., Gerlach, J. H., Robertson, A. J., Van Hoeven, R. P. (1978a). High viscosity vesicles of yeast separated at pH 4 have surface glyprotein. *Nature* 273:682–684.

Rank, G. H., Robertson, A. J., Bussey, H. (1978b). The viscosity and lipid composition of the plasma membrane of multiple drug resistant and sensitive yeast strains. *Can. J. Biochem.* 56:1036–1041.

Rattray, J. B. M., Schibeci, A., Kidby, D. K. (1975). Lipids of yeast. *Bacteriol Rev.* 39:197–231.

Robertson, A. J., Gerlach, J. H., Rank, G. H., Fowke, L. C. (1980). Yeast cell wall, membrane and soluble marker polypeptides identified by comparative two-dimensional electrophoresis. *Can. J. Biochem.* 58:565–572.

Santos, E., Villanveva, J. R., Sentandreu, R. (1978). The plasma membrane of *Saccharomyces cerevisiae:* isolation and some properties. *Biochim. Biophys. Acta* 508:39–54.

Scarborough, G. A. (1975) Isolation and characterization of *Neurosphora crassa* plasma membranes. *J. Biol. Chem.* 250:1106–1111.

Schibeci, A., Rattray, J. B. M., Kidby, D. K. (1973). Isolation and identification of yeast plasma membrane. *Biochim. Biophys. Acta* 311:15–25.

Schneider, H., Fiechter, A., Fuhrmann, G. F. (1978). Plasma membranes from *Candida tropicalis* grown on glucose or hexadecane. I. Isolation, identification, and purification. *Biochim. Biophys. Acta* 512:495–507.

Serrano, R. (1978). Characterization of the plasma membrane ATPase of *Saccharomyces cerevisiae. Mol. Cell. Biochem.* 22:51–63.

Steere, R. L., Erbe, E. F., Moseley, J. M. (1980). Prefracture and cold-fracture images of yeast plasma membranes. *J. Cell Biol.* 86:113–122.

Suomalainen, H., Nurminen, T. (1970). The lipid composition of cell wall and plasma membrane of baker's yeast. *Chem. Phys. Lipids* 4:247–256.

Takeo, K. (1976). Ultrastructural features underlying the hexagonal arrangement of plasma membrane intercalated particles of *Saccharomyces cerevisiae. J. Gen. Microbiol.* 97:331–334.

Työrinoja, K., Nurminen, T., Suomalainen, H. (1974) The cell envelope glycolipids of baker's yeast. *Biochem. J.* 141:133–139.

Van Den Bosch, H., Van Der Elzen, H. M., Van Deenen, L. L. M. (1967). On the phospholipases of yeast. *Lipids* 2:279–280.

Vignais, P. M., Nachbaur, J., Huet, J., Vignais, P. V. (1970). Studies on the phospholipids of yeast mitochondria. *Biochem. J.* 116:42–43.

Wehrli, E., Boem, C., Fuhrmann, G. (1975). Yeast plasma membrane vesicles suitable for transport studies. *J. Bacteriol.* 124:1594–1597.

Willsky, G. R. (1979). Characterization of the plasma membrane $MG^{2+}$-ATPase from the yeast, *Saccharomyces cerevisiae. J. Biol. Chem.* 254:3326–3332.

# 9

# Protoplast Fusion in Yeasts

Richard F. Freeman* and John F. Peberdy**

## Introduction

The yeasts are a group of lower eukaryotic microorganisms showing biological and biochemical diversity. In common usage the term "yeast" is used to describe strains of *Saccharomyces cerevisiae* that have great commercial value in baking, brewing, and distilling. Related yeasts are used in wine making (Snow 1979) and sake brewing and as a source of $\beta$-fructanofuranosidase (invertase; Phaff *et al.* 1978). Recent interest in ethanol as a substitute fuel, or as extender for hydrocarbon fuels for the internal combustion engine, has led to the development of strains specifically for these processes. *S. cerevisiae* is still the organism of choice for fuel alcohol production from sucrose or hydrolyzed starch, although in the longer term its use may be surpassed by *Zymomonas mobilis* (Rogers *et al.* 1980).

In past years some species of *Candida* have been recognized as having some economic potential, especially as food yeasts and as sources of lipid and protein. These processes failed to be economically viable; however, in the current changing times they could be revitalized. In contrast, *Candida albicans* achieves importance, together with several other yeasts, as a common causative agent of mycotic infections.

The value of an understanding of the genetics of these organisms is clearly apparent. In these species that are economically important such knowledge may be applied in breeding programs. In the pathogenic species and in yeasts in general an understanding of the fundamental genetic systems provides the basis for a broader understanding of their biology.

*Patscentre International, Cambridge Division, Melbourn Herts, SG8 6DP England.

**Fungal Genetics and Biochemistry Laboratory, Department of Botany, University of Nottingham, Nottingham, England.

The genetic systems of *S. cerevisiae* and the fission yeast, *Schizosaccharomyces pombe* are the most extensively developed. In the former a compatibility system based on two mating factors is found. Haploid cells of opposite mating types fuse readily to form diploid zygotes that can be induced to undergo meiosis to form four haploid ascospores (a tetrad). On germination the ascospores give rise to haploid cells that exhibit Mendelian reassortment of unlinked genetic markers (Mortimer and Hawthorne 1975). Similar compatibility systems have been described in other yeasts, but some are homothallic, lacking mating types, and in others the underlying control of sexual interactions is unknown.

*S. cerevisiae,* and a few other species, also have a mitochondrial genetic system. This has been extensively studied in the former (Bernardi 1979). Other extranuclear characters have also been described.

Almost all strains of *S. cerevisiae* and several strains of *Sch. pombe* harbor a cryptic dsDNA plasmid, the "2 $\mu$m plasmid." In *S. cerevisiae* about 70 copies are found per cell. It has a molecular weight of $4 \times 10^6$ d and has coding properties for at least three gene products (Hartley and Donelson 1980). The plasmid appears to confer no obvious advantage or disadvantage on the strains that possess it. It incorporates a replicon and when spliced with *E. coli* plasmids serves as an efficient shuttle vector between the two organisms. More recently, the cloning of the centromere for chromosome III (Clarke and Carbon 1980) and also for chromosome XI (Hsiao and Carbon 1981) raises the possibility of constructing plasmids that behave as chromosomes in their stability and segregation.

Most of the industrial strains of *S. cerevisiae* lack the mating reaction of heterothallic laboratory strains. All the yeasts used for brewing, baking, or distilling may be taxonomically classified as *S. cerevisiae.* They include the bottom-fermenting lager yeasts (ex-*S. uvarum* or *S. carlsbergensis*) and the dextrin-hydrolyzing strains that produce amyloglucosidase enzyme (ex-*S. diasticus*). Most of these strains neither mate nor sporulate (Johnston 1965). This is also the case for baker's yeasts (Johnston and Lewis 1976). Various attempts have been made to stimulate industrial strains to behave like laboratory strains (Clayton *et al.* 1972; Anderson *et al.* 1975) and recently a lager yeast was induced to sporulate and the original strain was reassembled from the spores (Gjermansen and Sigsgaard 1981).

The aneuploid or polyploid nature of commercial yeasts is believed to contribute to their observed poor spore viability. It also reduces the range of genetic markers that may be introduced (Kielland-Brandt *et al.* 1979). The *petite* mutation (Ephrussi *et al.* 1949), which results in a nonfunctional cytochrome chain, is, however, ubiquitous to all *S. cerevisiae* strains. The mutation is a consequence of a partial or total lack of mitochondrial DNA (mtDNA) and is therefore irreversible. *Petite* mutants lacking all their mtDNA are readily produced by exposure of normal strains to ethidium bromide (Wilkie, 1975). Phenotypically these mutants are characterized by their inability to grow on nonfermentable carbon sources (e.g., glycerol, lactate, ethanol, and acetate).

The *petite* mutation is a selectable genetic marker that can be readily introduced into commercial yeasts. Crosses may then be carried out using another haploid strain, wild-type for mtDNA ("grande"), but with other counterselectable markers, e.g. auxotrophic markers, as originally described by Gunge *et al.* (1972). The main disadvantage of this technique arises from the phenotypic nature of "grande," which is cytoplasmic. Consequently, prototrophic clones may arise following transfer of mitochondria into the *petite* cytoplasm without nuclear fusion, an event called cytoduction by Zakharov and Yarovoi (1977). However, in a more recent study (Spencer *et al.* 1980), this approach did prove successful. Hybrids were obtained between *S. diastaticus* and a brewer's yeast, and between *S. diastaticus* and a distiller's yeast, using *petite* forms of antibiotic-resistant strains and prototrophic, antibiotic-sensitive strains in the crosses.

The use of protoplasts as tools for genetic recombination in yeasts has proved of value in overcoming some of these problems. Removal of the cell wall, in several yeasts at least, eliminates the cell components associated with mating reactions. The naked cells, spheroplasts or protoplasts, can be manipulated to fuse, normally under the agency of a fusogen, thus bringing together complete genomes and a mixing of the two cytoplasms within a single cell. The consequences of this event have proved to be fairly consistent in a whole range of yeast species, leading to the formation of diploid cells from haploid parent; exceptions have been found in some brewing strains and some pseudomycelial forms.

## Protoplast Isolation, Fusion, and Reversion

The technology of protoplast fusion has been the subject of earlier reviews (Peberdy 1979, 1980; Ferenczy 1981), and only a brief outline will be given here. The isolation of protoplasts and spheroplasts from yeasts has been an established practice for many years following the early work of Eddy and Williamson (1957). Workers investigating the yeast system have been greatly aided by the availability of suitable lytic enzymes from commercial sources. The most widely used has been the preparation of snail digestive juices from the snail, *Helix pomatia,* which is available in a variety of forms (e.g., glusulase). Another product, zymolyase, obtained from *Arthrobacter luteus,* has been used (Kitamura *et al.* 1974). More recently, a new product, Novozym 234, has been shown to be an effective lytic preparation for *S. cerevisiae* (Stephen and Nassim 1981; Hamlyn *et al.* 1981). Strains vary greatly in their susceptibility to lysis and protoplast formation; however, the underlying reasons for this are not understood.

Cells devoid of their wall are osmotically fragile and maintenance of protoplasts and spheroplasts is dependent on external osmotic support. A range of salts, sugars, and sugar alcohols have been used with different yeast strains (Villanueva and Garcia Acha 1971).

Active interest in protoplast fusion as a method for obtaining genetic recombination between strains stemmed from the discovery of the fusogenic activity of the high-molecular-weight polyethylene glycols with plant protoplast systems (Kao and Michayluk 1974). This compounds are now used universally in all microorganisms (Peberdy 1980; Ferenczy 1981). In the majority of published reports, fusion experiments have utilized auxotrophic mutant strains, which after treatment with polyethylene glycol are plated onto a minimal medium forcing the selection of fused protoplasts by nutritional complementation. Dead donor techniques employed with prokaryotes, using heat treatment (Fodor *et al.* 1978), antibiotics (Levi *et al.* 1977), or UV irradiation (Hopwood and Wright 1979, 1981) as lethal agents, should be equally applicable to yeasts. In this approach protoplasts of one strain are rendered nonviable and the selection system used relates to the viable recipient strain. Some success was achieved with commercial yeast strains using such a dead donor system (Freeman, unpublished).

The natural vitamin requirements of a strain may be used to counterselect against one or both parents. Thus all brewing strains of *S. cerevisiae* require biotin, and other strains require additional growth factors as well (e.g., pantothenic acid and *p*-aminobenzoic acid). When this approach has been used, it has proved necessary to reculture sequentially in a vitamin-selective medium after regeneration of the fusion products, since traces of vitamins reduce the effectiveness of the method (Freeman, unpublished).

Successful application of protoplast fusion is dependent on the subsequent culture and reversion of protoplasts to the typical vegetative state. This aspect of the technology has been established practice for many years (Nečas, 1971, 1980). No special medium constituents are necessary other than the inclusion of an osmotic stabilizer. Of all the yeasts, *S. cerevisiae* appears to be the most fastidious in that total wall regeneration occurs only in solidified media. The early events of protoplast reversion in this species were described several years ago (Nečas 1961) and one aspect of the process, repeated nuclear division in the absence of septation, is clearly of significance in considering the consequences of protoplast fusion. Extrapolation of this observation to the fused protoplast condition would suggest that a heterokaryotic state exists for the first two or three cell generations during reversion. The cells recovered as colonies on the selective plates result from an event, or events, that occur in this very early period of reversion.

## The Consequences of Protoplast Fusion

### Intraspecies Protoplast Fusion Involving *Saccharomyces cerevisiae*

Fusion products from crosses in laboratory strains of *S. cerevisiae* of the same and opposite mating types have been genetically analyzed by several groups (Van Solingen and van de Plaat 1977; Yamamoto and Fukui 1977; Maráz *et al.* 1978; Svoboda 1978; Gunge and Tamaru 1978; Christensen 1979). These reports are strikingly in accordance with those anticipated for hybrids constructed by classical mating. Protoplasts from haploid parents were fused to give diploid progeny that contained an equal genetic complement from both parents. Analysis of these progeny revealed that they were indistinguishable from normal sexual hybrids.

It seems clear that for genetically compatible strains, protoplast fusion achieves the same results as mating. The majority of the fusion products recovered are stable and contain a full genetic complement from both parents.

## Protoplast Fusion with Commercial Strains of *S. cerevisiae*

The strains of *S. cerevisiae* used for brewing, distilling, and baking are generally polyploid. The lack of a mating system in these strains has encouraged several groups to employ somatic cell (protoplast) fusion to hybridize these yeasts (Russell *et al.* 1979; Barney *et al.* 1980; Hockney and Freeman 1980; Freeman 1981). The results from these groups generally concur with those experienced with laboratory strains of heterothallic *S. cerevisiae.*

A major difficulty in the manipulation of commercial strains is their lack of good counterselectable markers. The use of dead donor techniques and inherent differences in growth factor requirements of different strains has been used to good effect in this situation (Hockney and Freeman 1980). Effective selective markers to eliminate parental phenotypes from the output of a fusion are central to the success of this technique. If a suitably rigorous selection can be maintained on a hybrid clone for sufficient cell divisions, then stable strains may result through mitotic segregation.

## Protoplast Fusion in Other Genera of Yeasts

Successful crosses have been described for several genera of yeasts. In every case the "conventional" selection system based on nutritional complementation of auxotrophic markers was used. In the majority of crosses the fusion products recovered were diploid, as indicated by DNA estimations and segregation of markers. This situation was found in *Schizosaccharomyces pombe* (Sipiczki 1979), *Rhodosporidium toruloides* (Sipiczki and Ferenczy 1977; Becher and Böttcher 1980; Böttcher *et al.* 1980), *Kluyveromyces lactis* (Morgan *et al.* 1977), *Saccharomycopsis lipolytica* (Stahl 1978), *Hansenula polymorpha* (Savchenko and Kapul'tsevich 1980), *Pichia guillermondii* (Böttcher *et al.* 1980; Klinner *et al.* 1980; Spata and Weber 1980), and *Candida albicans* (Poulter *et al.* 1981; Sarachek *et al.* 1981). In all cases the fusion products exhibited parasexual phenomena and gave rise to segregants either as spontaneous events or as a consequence of induction by haploidizing agents such as benomyl or *p*-fluorophenylalanine. In *C. albicans* (Poulter *et al.* 1981) some of the segregants were presumed to have the same ploidy as the fusion products and it was suggested that these could have arisen as a consequence of mitotic crossing-over, the first such claim for a yeast.

In some of these reports unstable fusion products of variable ploidy were also described, for example, aneuploids in *R. toruloides* (Becher and Böttcher 1980) and triploids and tetraploids in *H. polymorpha* (Savchenko and Kapul'-tsevich 1980). Other phenomena were reported which, with the information available, are difficult to explain, for example, the segregation of only one marker as a spontaneous event.

The only exception known so far to this general uniformity of diploid fusion products was reported in *C. tropicalis*. Two independent groups (Fournier *et al.* 1977; Vallin and Ferenczy 1978) found that the progeny recovered from crosses were heterokaryons. With constant rigorous selection the heterokaryons could be maintained and on occasion gave rise to diploids that were much more stable. Relaxation of selection allowed the parental strains to segregate from the heterokaryons. The occurrence of the heterokaryotic state was attributed to the pseudomycelial form of this yeast. Diploids were also found to arise directly following protoplast fusion; however, independent of their origin, they did yield segregants either spontaneously or following induction. The ploidy of these segregants was not determined and it is probable that they did include aneuploids and diploids, arising as a result of parasexual events.

More recently, protoplast fusion has been used successfully to tackle specific genetic problems in two yeasts. In *K. lactis* matings between compatible *petite*-negative strains yield unstable diploids (Herman and Roman 1966). A recent observation (Morgan 1979) revealed a more extreme situation in crosses between wild and *petite* strains with the recovery of only haploid recombinants from some strain combinations. Using protoplast fusion, it was possible to obtain relatively stable hybrids from strains of the same mating type. Spontaneous or induced haploidization was possible, providing a system for genetic analysis (Morgan *et al.* 1980). In a similar application in *Sch. pombe,* analysis of a sterile mutant, a conditional lethal defective in mitosis, was carried out (Thuriaux *et al.* 1980).

## Interspecies and Intergeneric Hybridization in Yeasts Using Protoplast Fusion

Attempts to generate interspecies hybrids have met with varying degrees of success. Thus the nature of the fusion product derived from *Sch. pombe* and *Sch. octosporus* (Sipiczki 1979) is not clear. The frequency of fusion product recovery was at least two magnitudes higher than the back-mutation frequency for each parent; nevertheless under nonselective conditions the slow-growing hybrids yielded only *Sch. octosporus*. In contrast, a cross between *K. lactis* and *K. fragilis* was more fruitful, yielding several hybrid forms distinguished on the basis of relative growth rates and DNA contents (Whittaker and Leach 1978).

Recently, interesting results have been obtained from several intergeneric crosses. Prototrophic "hybrids" were obtained following fusion of protoplasts from auxotrophic strains of *S. cerevisiae* and *Saccharomycopsis lipolytica* (de van Broock *et al.* 1980). The progeny obtained varied in ploidy, and genetic stability was achieved following prolonged culture on minimal medium.

Using a selection procedure based on the restoration of the *petite* mutation in *S. cerevisiae* and *S. diastaticus,* Spencer and Spencer (1982) obtained hybrids between these strains and *Pichia membranaefaciens, Hansenula capsulata, Saccharomyces rosei,* and *Sch. pombe.* Assimilation tests indicated interactions between the genomes of the different species as well as between the mitochondrial components. In addition, sporulating hybrids were obtained between *H. capsulata* and *S. diastaticus,* which may permit detailed genetic analysis in future.

Also, at the intergeneric level, Provost *et al.* (1978) recovered hybrid fusion products of *Candida tropicalis* and *Saccharomyces fibuligera.* These were mostly unstable, yielding segregants that were like one or the other parent. Characterization of the progeny was based on carbon assimilation patterns and amylo-1,6-glucosidase activities.

## Discussion

It is apparent that protoplast fusion is now an established method for producing hybrids between yeast strains. In many instances, recombination is a clear consequence. In laboratory strains of *S. cerevisiae* the advantage of protoplast fusion over natural conjugation systems is not apparent, the progenies from protoplast fusion crosses being indistinguishable from those obtained by normal matings.

With commercial strains there are clear advantages. Most strains of *S. cerevisiae* that have found commercial application are polyploid; the reason is unclear. Higher ploidy reduces the incidence at which mutations are phenotypically detectable. Most polyploid yeasts do not sporulate, or they produce spores of very low viability. These factors together confer a measure of phenotypic stability on polyploid strains that may have contributed to their selection. Gene dosage phenomena may also contribute to the fitness of such strains compared to haploids or diploids.

Using the approaches described, these difficulties may be overcome, allowing the prospect of constructing strains with increased ploidy as well as hybrids from strains of varying degrees of divergence to reassort desirable characters into a single strain. The genetic manipulation of yeasts for industrial purposes is still in its infancy, so the availability of this technology is timely. The interval between the production of a suitable strain and its adoption into routine production usage is necessarily long. It is essential that all appropriate intermediate-scale trials be carried out to confirm that the "improved" strain does not exhibit some unforeseen disadvantageous characters when used for large-scale fermentations. For this reason the use of genetically manipulated cultures in industrial yeast processes is still uncommon. There is little doubt that the use of industrial strains with modified ploidy, harboring chromosomes from another yeast or cloned DNA from yeasts or other organisms, will soon become commonplace.

For many of the yeasts of no commercial interest the use of protoplast fusion provides for the first time an opportunity to develop a study of their genetics. In species such as *C. albicans* this is of undoubted value in the study of drug resistance and its incidence in natural populations. For some time the ploidy of naturally occurring strains of this yeast has been controversial and an understanding of the genetic system of this organism will resolve this situation.

Clearly, the fusion technique will be applied to more and more species. The demonstration of hybridization between strains and between different species is of interest; however, explanations of the nature of these interactions and subsequent recombination events are frustrated by the lack of basic genetics and the availability of mapped strains. It is hoped that future endeavors will be directed toward these ends, establishing protoplast fusion as an essential component of the methodology of the genetics of some of the lesser known yeasts.

## References


Anderson, E., Martin, P. A. (1975). The sporulation and mating of brewing yeasts. J. Inst. Brew. 81:242–247.

Barney, M. C., Jansen, G. P., Halbert, J. R. (1980). Use of spheroplast fusion and genetic transformation to introduce dextrin utilization into *Saccharomyces uvarum*. Amer. Soc. Brew. Chem. J. 38:1–5.

Becher, D., Böttcher, F. (1980). Hybridization of *Rhodosporidum toruloides* by protoplast fusion. In: Advances in Protoplast Research, edited by L. Ferenczy, G. L. Farkas. Budapest: Akadémiai Kiadó, Oxford: Pergamon Press, pp. 105–111.

Bernardi, G. (1979). Petite mutation in yeasts. Trends Biochem. Sci. 4:197–201.

Böttcher, F., Becher, D., Klinner, U., Samsonova, I. A., Schilowa, B. (1980). Genetic structure of yeast hybrids constructed by protoplast fusion. In: Advances in Protoplast Research, edited by L. Ferenczy, G. L. Farkas. Budapest: Akadémiai Kiadó, Oxford: Pergamon Press, pp. 99–104.

Broock, M. R. de van, Sierra, M. F., Figueroa, L. I. de (1980). Intergeneric fusion of yeast protoplasts. In: Current Developments in Yeast Research, edited by G. G. Steward, I. Russell. Vth Int. Symp. Yeasts, London, Ontario, Canada, July 20–25, 1980.

Clarke, L., Carbon, J. (1980). Isolation of a yeast centromere and construction of functional small circular chromosomes. Nature (London) 287:504–509.

Clayton, E., Howard, G. A., Martin, P. A. (1972). Yeast hybridization. Proc. Ann. Meeting Amer. Soc. Brew. Chem., pp. 78–81.

Christensen, B. E. (1979). Somatic hybridization in *Saccharomyces cerevisiae:* analysis of products of protoplast fusion. Carlsberg Labs Res. Commun. 44:225–233.

Eddy, A. A., Williamson, D. H. (1957). A method of isolating protoplasts from yeast. Nature (London) 179:1252–1253.

Ephrussi, B., Hottinguer, H., Chimenes, C. (1949). Action de l'acriflavine sur les levures. I. La mutation "petite colonie." Ann. d'Institut Pasteur (Paris) 76:351–368.

Ferenczy, L. (1981). Microbial protoplast fusion. Symp. Soc. Gen. Microbiol. 31:1–34.

Fodor, K., Demiri, E. Alföldi, L. (1978). Polyethylene glycol-induced fusion of heat-inactivated and living protoplasts of *Bacillus megaterium*. J. Bacteriol. 135:68–70.

Fournier, P., Provost, A., Bourquignon, C., Heslot, H. (1977). Recombination after protoplast fusion in the yeast *Candida tropicalis*. Arch. Microbiol. 115:143–149.

Freeman, R. F. (1981). Construction of brewing yeasts for production of low carbohydrate beers. Proc. 18th Eur. Brew. Convention Congress. I.R.L. Press, London: pp. 497–504.

Gjermansen, C., Sigsgaard, P. (1981). Construction of a hybrid brewing strain of *Saccharomyces carlsbergensis* by mating of meiotic segregants. Carlsberg Res. Commun. 46:1–11.

Gunge, N., Nakatomi, Y. (1972). Genetic mechanisms of rare matings of the yeast *Saccharomcyes cerevisiae* heterozygous for mating type. Genetics 70:41–58.

Gunge, N., Tamaru, A. (1978). Genetic analysis of products of protoplast fusion in *Saccharomyces cerevisiae*. Jap. J. Genet. 53:41–49.

Hamlyn, P. F., Bradshaw, R. E., Mellon, F. M., Santiago, C. M., Wilson, J. M., Peberdy, J. F. (1981). Efficient protoplast release from fungi using commercial enzymes. Enzyme Microb. Technol. 3:321–325.

Hartley, J. L., Donelson, J. E. (1980). Nucleotide sequence of the yeast plasmid. Nature (London) 286:860–865.

Herman, A., Roman, H. (1966). Allelic specific determinants of homothallism in *Saccharomyces lactis*. Genetics 53:727–740.

Hockney, R. C., Freeman, R. F. (1980). Construction of polysaccharide-degrading brewing yeast by protoplast fusion. In: Advances in Protoplasts Research, edited by L. Ferenczy, G. L. Farkas. Budapest: Akadémiai Kiadó, Oxford: Pergamon Press, pp. 139–144.

Hopwood, D. A., Wright, H. M. (1979). Factors affecting recombinant frequency in protoplast fusion of *Streptomyces coelicolor*. J. Gen. Microbiol. 111:137–143.

Hopwood, D. A., Wright, H. M. (1981). Protoplast fusion in *Streptomyces:* Fusions involving ultraviolet-irradiated protoplasts. J. Gen. Microbiol. 126:21–27.

Hsiao, C. L., Carbon, J. (1981). Direct selection procedure for the isolation of functional centromeric DNA. Proc. Nat. Acad. Sci. USA 78:3760–3764.

Johnston, J. R. (1965). Breeding yeasts for brewing. J. Inst. Brewing 71:130–135.

Johnston, J. R., Lewis, C. W. (1976). Genetic analysis of flocculation in *Saccharomyces cerevisiae* and tetrad analysis of commercial brewing and baking yeasts. In: Second International Symposium on the Genetics of Industrial Microorganisms, edited by K. D. Macdonald. London: Academic Press, pp. 339–355.

Kao, K. N., Michayluk, M. R. (1974). A method for high-frequency intergeneric fusion of plant protoplasts. Planta 115:355–367.

Kielland-Brandt, M. C., Petersen, J. G. L., Michelsen, J. D. (1979). Mutants in the biosynthesis of isoleucine in a non-mating, non-sporulating brewing strain of *Saccharomyces carlsbergensis*. Carlsberg Res. Commun. 44:27–36.

Kitamura, K., Kaneko, T., Yamamoto, Y. (1974). Lysis of viable yeast cells by enzymes of *Arthrobacter luteus*. J. Gen. Appl. Microbiol. 20, 323–344.

Klinner, U., Böttcher, F., Samsonova, I. A. (1980). Hybridization of *Pichia guilliermondii* by protoplast fusion. In: Advances in Protoplast Research, edited by L. Ferenczy, G. L. Farkas. Budapest: Akadémiai Kiadó, Oxford: Pergamon Press, pp. 113–118.

Levi, C., Sanchez Rivas, C., Schaeffer, P. (1977). Further genetic studies on the fusion of bacterial protoplasts. Fed. Eur. Microbiol. Soc. Microbiol. Lett. 2:323–326.

Maráz, A., Kiss, M., Ferenczy, L. (1978). Protoplast fusion in *Saccharomyces cerevisiae* strains of identical and opposite mating types. Fed. Eur. Microbiol. Soc. Microbiol. Lett. 3:319–322.

Morgan, A. J. (1979) Ph.D. Thesis, University of Sussex, Brighton, England.

Morgan, A. J., Heritage, H., Whittaker, P. A. (1977). Protoplast fusion between petite and auxotrophic mutants of the petite-negative yeast, *Kluyveromyces lactis*. Microbios Lett. 4:103–107.

Morgan, A. J., Hall, J. L., Brunner, A., Whittaker, P. A. (1980). Protoplast fusion in the study of mitochondrial genetics in the petite-negative yeast. *Kluyveromyces lactis*. In: Advances in Protoplast Research, edited by L. Ferenczy, G. L. Farkas. Budapest: Akadémiai Kiadó, Oxford: Pergamon Press, pp. 93–98.

Mortimer, R. K., Hawthorne, D. C. (1975). Genetic mapping in yeast. In: Methods in Cell Biology, Vol XI, edited by D. M. Prescott. New York: Academic Press, pp. 221–233.

Nečas, O. (1961). Physical conditions as important factors for the regeneration of naked yeast protoplasts. Nature (London) 192:580–581.

Nečas, O. (1971). Cell wall synthesis in yeast protoplasts. Bacteriol. Rev. 35:149–170.

Nečas, O. (1980). Regeneration of protoplasts. In: Advances in Protoplast Research, edited by L. Ferenczy, G. L. Farkas. Budapest: Akadémiai Kiadó, Oxford: Pergamon Press, pp. 151–161.

Peberdy, J. F. (1979). Fungal protoplasts: isolation, reversion and fusion. Ann. Rev. Microbiol. 33:21–39.

Peberdy, J. F. (1980). Protoplast fusion—a tool for genetic manipulation and breeding in industrial microorganism. Enzyme Microb. Technol. 2:23–29.

Phaff, H. J., Miller, M. W., Mrak, E. M. (1978). The Life of Yeasts. Cambridge, Mass.: Harvard Univ. Press.

Poulter, R., Jeffrey, K., Hubbard, M. J., Shepherd, M. G., Sullivan, P. A. (1981). Parasexual genetic analysis of *Candida albicans* by spheroplast fusion. J. Bacteriol., 146:833–840.

Provost, A., Bourguignon, C., Fournier, P., Ribet, A. M., Heslot, H. (1978). Intergeneric hybridization in yeasts through protoplast fusion. FEMS Microbiol. Lett. 3:309–312.

Rogers, P. L., Lee, K. J., Tribe, O. E. (1980). High productivity ethanol fermentations with *Zymomonas mobilis*. Process Biochem. pp. 7–11.

Russell, I., Stewart, G. G. (1979). Spheroplast fusion of brewers yeast strains. J. Inst. Brew. 85:95–98.

Sarachek, A., Rhoads, D. H., Schwarthoff, R. H. (1981). Hybridization of *Candida albicans* through fusion of protoplasts. Arch. Microbiol. 129:1–8.

Savchenko, G. V., Kapul'tsevich, Yu. G. (1980). Hybridization of yeast from genus *Hansenula* by protoplast fusion. In: Advances in Protoplast Research, edited by L. Ferenczy, G. L. Farkas. Budapest: Akadémiai Kiadó, Oxford: Pergamon Press, pp. 125–130.

Sipiczki, M. (1979). Interspecific protoplast fusion in yeasts. Curr. Microbiol. 3:37–40.

Sipiczki, M., Ferenczy, L. (1977). Fusion of *Rhodosporidium (Rhodotorula)* protoplasts. FEMS Microbiol. Lett. 2:203–205.

Snow, S. R. (1979). Towards genetic improvement of wine yeasts. Amer. J. Horticulture & Viticulture 30:33–37.

Solingen, P. van, Plaat, J. B. van der (1977). Fusion of yeast spheroplasts. J. Bacteriol. 130:946–947.

Spata, L., Weber, H. (1980). A study on protoplast fusion and parasexual hybridization of alkane utilizing yeasts. In: Advances in Protoplast Research, edited by L. Ferenczy, G. L. Farkas, pp. 131–137.

Spencer, J. F. T., Laud, P., Spencer, Dorothy M. (1981). The use of mitochondrial mutants in the isolation of hybrids involving industrial yeast strains. II. Use in isolation of hybrids obtained by protoplast fusion. Mol. Gen. Genet. 178:651–654.

Spencer, J. F. T., Spencer, Dorothy M. (1982). The use of mitochondrial mutants in the isolation of hybrids involving industrial yeast strains. III. Restoration of mitochondiral function in petites of industrial yeast strains by fusion with respiratory-competent protoplasts of other yeast species. Curr. Genet. 4:177–180.

Stahl, U. (1978). Zygote formation and recombination between like mating types in the yeast *Saccharomycopsis lipolytica* by protoplast fusion. Mol. Gen. Genet. 160:111–113.

Stephen, E. R., Nassim, A. (1981). Production of protoplasts in different yeasts by mutantase. Can. J. Microbiol. 27:550–553.

Svoboda, A. (1978). Fusion of yeast protoplasts induced by polyethylene glycol. J. Gen. Microbiol. 109:167–175.

Thuriaux, P., Sipiczki, M., Fantes, P. A. (1980). Genetical analysis of a sterile mutant by protoplast fusion in the fission yeast *Schizosaccharomyces pombe*. J. Gen. Microbiol. 116:525–528.

Vallin, C., Ferenczy, L. (1978). Diploid formation of *Candida tropicalis* via protoplast fusion. Acta Microbiol. Acad. Scientiarum Hungaricae 25:209–212.

Villanueva, J. R., Garcia Acha, I. (1971). Production and use of fungal protoplasts. In: Methods in Microbiology, Vol. 4, edited by C. Booth. London and New York: Academic Press, pp. 665–718.

Whittaker, P. A., Leach, S. M. (1978). Interspecific hybrid production between the yeast *Kluyveromyces lactis* and *Kluyveromyces fragilis* by protoplast fusion. Fed. Eur. Microbiol. Soc. Microbiol. Lett. 4:31–34.

Wilkie, D. (1975). Cytoplasmic inheritance and mitochondrial genetics in yeast. In: Methods in Cell Biology, Vol. XII, edited by D. M. Prescott. New York: Academic Press, pp. 353–372.

Yamamoto, M., Fukui, S. (1977). Fusion of yeast protoplasts. Agricu. Biol. Chem. 41:1829–1830.

Zakharov, I. E., Yarovoy, B. Ph. (1977). Cytoduction as a new tool in studying cytoplasmic heredity in yeast. Mol. Cell. Biochem. 14:15–18.

# 10

# Genetic and Functional Aspects of Yeast Mitochondria[1]

D. Wilkie*

## Introduction: The *Petite* Mutant

The classical example of cytoplasmic inheritance for many years has been the so-called *petite colonie* mutation in *Saccharomyces cerevisiae.* First described by Ephrussi more than 30 years ago, this mutant has been the subject of countless numbers of publications both genetic and biochemical. Nonetheless, the nature of the mutational change—how it arises and how it becomes established in a population, is still not clearly understood. The genetic lesion is in the mitochondrial genome but this is not a mutation in the normal sense: a fragment, usually quite small, of the organellar chromosome (mtDNA) becomes extruded or detached, apparently from any part of the DNA molecule, is reiterated in tandem repeats to become a new circular molecule comparable in size to the normal, and in this state is transmitted to daughter cells during vegetative growth. The mechanism of fragment formation is discussed in detail by Bernardi (1978), who favors the theory that illegitimate pairing and exchange between genomes in their AT-rich regions (see Evans, this volume) is responsible. Whatever the mechanism, the result is a family of bizarre mutants the members of which have several features in common: they are irreversible, they have a nonexistent or otherwise nonfunctional system for protein synthesis (mtDNA codes for all the RNA species of the translation machinery in the organnelle), they are all totally respiratory defective and show a non-Mendelian or cytoplasmic mode of inheritance (Wilkie 1973). Perhaps the

*Department of Botany and Microbiology, University College, London, Gower Street, London, England.

[1]Work in the author's laboratory is currently supported by grants from the Cancer Research Campaign and the Scientific Research Council, which are gratefully acknowledged.

most striking feature of *petite* mutants is the high spontaneous rate with which they arise. In most yeast strains this is about 1% but the mutation can be increased dramatically by intercalating agents, particularly acriflavine and ethidium bromide. This high rate makes their identification straightforward and unmistakable. *Petite* cells plated on standard agar medium containing glucose as energy source are able to grow and give rise to colonies only as long as glucose is available, that is, producing ATP by glycolysis. When glucose is depleted from the medium, colonies stop growing, whereas colonies from normal cells continue to grow by virtue of their respiratory activity. The mutant colonies are thus very much smaller than normal colonies at stationary phase—hence the name *petite colonie.* When transferred to nonfermentable medium containing, for example, glycerol, ethanol, or TCA-cycle intermediates as sole carbon and energy source, *petite* cells are unable to grow, unlike normal cells, which can metabolize these substrates by respiration.

The fact that respiratory-deficient cells can grow and divide more or less normally in glucose medium is a feature of *Saccharomyces* since the organism is a facultative anaerobe. Virtually all other eukaryotic organisms are obligate aerobes in which anaerobiosis or respiratory incapability would greatly curtail or cause cessation of growth. This characteristic has made yeast the organism of choice in studies of mitochondrial mutation and in the detection of drugs that are selectively antimitochondrial in their activity (Wilkie 1972).

The genetics of mitochondria is complicated by the fact that there are usually many copies of the mitochondrial genome in a single cell and estimates of between 50 and 100 have been made for respiring yeast cells (see Evans, this volume). This can vary considerably, depending on growth conditions, and under mitochondrial repression by glucose or anaerobiosis there are probably far fewer copies per cell (Wilkie and Maroudas 1968). Also, individual mitochondria probably contain several copies of the mtDNA molecule, since cells, even those fully adapted to respiration, seem to possess relatively few mitochondria, perhaps six or seven at most (Stevens 1974). The large numbers of mitochondrial profiles often seen in thin sections of cells under the electron microscope are misleading and may not be interpreted as representing individual mitochondria. Apparently the organelle is usually much branched so that profiles may be sections through several branches of a single mitochondrion. In this condition of multiple genomes, it is difficult to see how a mutation, particularly one with a deleterious effect, could ever become established in a population. Not only do *petite* cells arise spontaneously with a very high frequency, but in crosses with normal cells, transmission of the *petite* type can be and frequently is greater than that of the wild type. The frequency of *petite* segregants among the vegetative diploid progeny from zygotes in such crosses (often higher than 90%) depends on the particular *petite* mutant. This characteristic of individual *petites* in regulating their heritability in competition with the normal is traditionally referred to as "suppressivity" from the original suggestion of its discoverers (Ephrussi *et al.* 1955) that *petite* mitochondria somehow can suppress the activity of normal mitochondria and assume dominance in a mixed mitochondrial population. Although this is unlikely, the idea of competition forms the basis of Bernardi's argument: Since the segment of mtDNA that initiated the *petite* condition must have at least one site initiating replication in order that it be perpetuated, it follows that the reconstituted genome (made up of tandem repeats of the segment) of the *petite* mutant must have several, possibly many, replication sites. Presumably, the smaller the original segments, the larger the number of replication sites and the greater the rapidity of replication. This could confer a numerical advantage over wild-type genomes, which have a limited number of replication sites. Assuming that successful transmission depends to a great extent or solely on numerical superiority in genomes, it follows that *petite* mitochondria would compete very well on the whole with wild-type organelles in genome transmission.

From the foregoing arguments and hypotheses, it is obvious that very little is known about the mechanism either of genome or organelle transmission to the daughter cells of zygotes or of their movements in zygotes, where they have been shown to recombine (Thomas and Wilkie 1968a,b). The identification of point mutations in mtDNA, which opened the way for conventional genetic analysis (i.e., the study of segregational data), has provided some insight into the behavioral patterns of mitochondrial genomes.

Before discussing these mitochondrial mutations, a further point may be made with respect to the *petite* mutation. Not all *petite* mutations have mitochondrial genomes, even defective ones. The mitochondria of some *petite* mutations have no detectable mtDNA (so-called $\rho^0$) but apart from this they are phenotypically similar to the *petites* described above ($\rho^-$) in having no respiration activity, failing to develop cytochromes a + $a_3$ and b (because of lack of a mitochondrial protein-synthesizing system), and showing undifferentiated ultrastructure in their mitochondria. However, at the cellular level there are differences between $\rho^-$ and $\rho^0$: Whereas $\rho^-$ *petites* usually have quite good viability and growth rate in YED culture, $\rho^0$ mutants on the whole show poor growth and low viability and are difficult to maintain as stock cultures. These differences in cellular properties are difficult to explain since $\rho^-$ and $\rho^0$ mitochondria are equally respiratory defective and unable to manufacture proteins. It may be tentatively concluded that the elaboration of the inner membrane assembly for the production of ATP by oxidative phosphorylation, important though this undoubtedly is, is not the only function of the mitochondrion. The organelle may have a role to play in the normal development of certain cellular characteristics and $\rho^0$ mitochondria are less able on the whole than $\rho^-$ mitochondria to carry out these functions (evidence that $\rho^-$ mutants can also have concomitant impairment of certain cellular characteristics will be discussed later). However, $\rho^0$ mutants do grow and divide and when they do, the DNA-less organelles are reproduced and transmitted to daughter cells. In zygotes produced in crosses between $\rho^0$ and the wild type, on the other hand, DNA-less mitochondria of the $\rho^0$ parent are not transmitted to daughter cells apparently; the only type of cell in the zygote clones is the $\rho^+$ cell. There is no suppressivity and for this reason the term "neutral *petite*" has been used to distinguish the $\rho^0$ from the suppressive $\rho^-$ mutant. One interpretation of these findings is that only mitochondria with a genome or genomes are capable of being transmitted from zygotes, and a test of this hypothesis would be to obtain zygotes from a $\rho^0 \times \rho^0$ cross; no viable daughter cells would be produced on the assumption that yeast cells (and all other eukaryotic cells for that matter) are incapable of growth and division without mitochondria. (As far as the author is aware, a living eukaryotic cell capable of growth and division but without mitochondria has not been seen.) Another and more likely interpretation is that mitochondria in zygotes either fuse or fragment and reform during which time their respective genomes can recombine genetically. Resulting genomes are then available for transmission within new "hybrid" organelles. In the case of the $\rho^0 \times \rho^+$ cross, the parental $\rho^+$ genomes would be selected automatically, being the only ones available. Some preliminary evidence of mitochondrial disaggregation and subsequent reformation was seen in electron microscopic studies of zygotes during their formation and maturation in a $\rho^+ \times \rho^+$ cross by Smith *et al.* (1972), but this was not clearly seen in similar studies by others using different strains (Waxman, personal communication). Further information on the behavior and transmission of mitochondrial genomes has been provided by genetic studies.

## Recombination in Mitochondria

Evidence is conclusive both from genetic and physical studies that mitochondrial genomes recombine in yeast crosses. This was first demonstrated in crosses between strains carrying mitochondrial drug-resistance markers (Thomas and Wilkie 1968). In identifying the mitochondrial location of drug-resistance genes, two main criteria were used: segregation among diploid (vegetative) progeny of zygotes of resistance (R) and sensitivity (S) in R × S crosses, and failure to transmit resistance in similar crosses in which the resistant parental strain was first converted to the *petite* condition; that is, there was loss of R concomitant with loss of mitochondrial genetic information in *petites*. Segregation in zygote clones of erythromycin resistance ($E^R$) and sensitivity ($E^S$) was studied in $E^R \times E^S$ crosses. Cells sampled from zygote clones and plated gave colonies that were pure with respect to mitochondrial type; that is, constituent cells were either all $E^R$ or all $E^S$ and no heterogeneity was seen in colonies. When cells were sampled from individual zygote clones and drop-inoculated onto medium (necessarily nonfermentable) containing the antibiotic, an indication of the relative frequencies of $E^R$ and $E^S$ segregant cells was obtained (Fig. 10.1). In some clones, $E^R$ cells were very frequent (confluent growth) while in one clone all cells sampled ($\backsim 10^4$) were $E^S$. Because 100 or so genomes are expected in zygotes following fusion of parental cells and because zygotes may produce 10 or so daughter cells, there is clearly a severe restriction in genome transmission for the cells in a zygote clone to be all of one mitochondrial type.

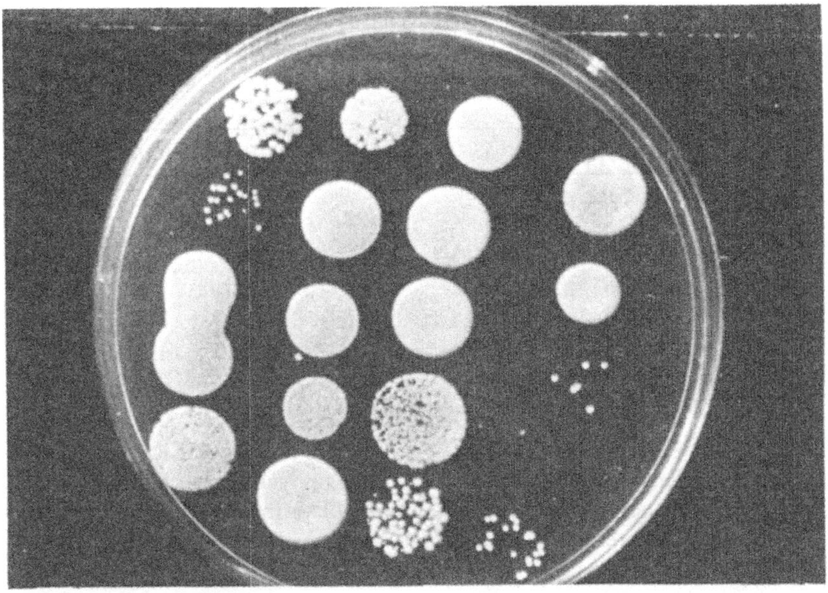

**Figure 10.1.** Drop inocula from individual cell suspensions of 20 zygote clones from the cross $E^R$ (erythromycin resistant) × $E^S$ (erythromycin sensitive). Each drop contained ~$10^4$ cells and medium contained 1 mg/ml erythromycin, inhibitory to sensitive strains. (From Thomas and Wilkie 1968a.)

In demonstrating recombination, a three-point cross $E^RC^RO^S \times E^SC^SO^R$ was set up involving the additional mitochondrial markers $C^R$ (Chloramphenicol resistance) and $O^R$ (oligomycin resistance). From the results in Table 10.1 it can be seen that all possible arrangements of the markers (eight in all) were obtained among random diploids from zygote clones. Furthermore, the frequencies of the different recombinant types were all of the same order, which distinguishes mitochondrial recombination from the meiotic system. In the latter, a centrally located marker would require a double crossover to recombine with flanking markers, and this occurrence would be of a different order of magnitude to single crossover events. These results and numerous similar results obtained subsequently indicated that the situation in zygotes resembles that of a bacteriophage cross in which different genomes engage in rounds of recombinational events making multiple exchanges an effective outcome. The problem of how the genomes from different mitochondria come together to exchange segments remains to be resolved, but a further problem concerns the mechanics of exchange and segregation where circular genomes are concerned. For circular molecules to exchange segments it is clear that at least two crossover events must occur. The consequences of a single crossover or an uneven number of crossovers would be the joining of the two circles into a double-sized one. Subsequent exchanges with other genomes could generate triple-sized or

**Table 10.1.** Genetic Analysis of Mitochondrial Drug-Resistant Mutants: Segregation of Markers Among Random Diploids from the Cross $E^S$ $C^R$ $O^R$ × $E^R$ $C^R$ $O^S$[a]

| Phenotypes | Number |
|---|---|
| S R R | 125 |
| R S S | 82 |
| S R S | 37 |
| R S R | 22 |
| S S R | 9 |
| R R S | 14 |
| S S S | 16 |
| R R R | 12 |

[a] $E^S/E^R$, $C^S/C^R$, $O^S/O^R$ = sensitivity/resistance to erythromycin, chloramphenicol, and oligomycin, respectively. Phenotypes are listed in this order of antibiotics. (Unpublished results of Thomas and Wilkie.)

larger molecules, and, in theory, half of the products of recombination would be of this type. In fact, the contour length of mtDNA molecules from various yeast *(Saccharomyces cerevisiae)* strains seems to be uniformly of the order of 27 μm. Either the multiples cannot compete with the singles in transmission or the mechanism of the recombinational event is not reciprocal in nature and so would not necessarily generate multiples. Recombination could be of the gene conversion type involving heteroduplex DNA and therefore be nonreciprocal. This process has a known tendency to be polarized (Fogel, this volume), which would generate asymmetry in the distribution of respective alleles. Thus if the $E^R$ allele (in the $E^RC^RO^S$ genome) converted to the $E^S$ allele rather than the other way around, a preponderance of genomes of the class $E^SC^RO^S$ would be generated compared to $E^RC^SO^R$. Results in Table 10.1 and in many crosses from other sources confirm this and show that significant deviations from a 1:1 ratio of reciprocal products occur. This point is taken up and discussed in detail by Birky (1978), to which the reader is referred for further information.

A different approach was used by Fonty *et al.* (1978) in their studies of mitochondrial recombination in yeast. They used endonuclease cleavage sites as markers in mitochondrial genomes in crosses, having first established that the two wild-type strains being used were sufficiently different in their cleavage sites with respect to *Hind*II + III, *Hpa*II and *Hae*III restriction enzymes. Thus the mtDNA fragments generated by each parental strain showed recognizably different electrophoretic patterns on agarose–polyacrylamide gels. When individual diploid clones were isolated from the cross, either as random diploids or as buds micromanipulated from zygotes, 75% of them were recombinant, that is, had mtDNA fragment patterns distinct from either parent. Recombination fragment patterns were made up of (1) fragments originating from the mtDNAs of both parental strains due to recombination of cleavage sites, and (2) new fragments not seen in either parent. The most likely explanation for the appearance of new fragments is that unequal crossing-over occurred in the extensive spacer (A + T-rich) regions (see Evans, this volume) of parental genomes.

As well as providing evidence of physical recombination between wild-type mitochondrial genomes, the results also indicated that recombinational events occurred with a high frequency. To obtain further insight into the mechanism of genome transmission, the genome complement of zygote daughter cells can be assessed by isolating daughter cells as they are formed and cloning them. Clones (colonies) can be sampled and plated and presence and frequency of mitochondrial types scored. A diagram of this procedure is presented in Figure 10.2 with results from a three-point cross that underline some of the points raised above. In an extension of these studies of zygote lineages in crosses, Waxman *et al.* (1973) showed that the transmission of mitochondrial type was under nuclear control, and nuclear mutations in the system could give rise to strange but characteristic transmission patterns.

## Antimitochondrial Activity of Drugs

As already pointed out, loss of respiratory function in yeast (as in *petite* mutants) does not significantly impair ability of cells to grow provided glucose is available as a carbon and energy source. On the other hand, mitochondrial incapacity would preclude growth if the carbon source was nonfermentable. A drug that specifically blocked organelle biogenesis (e.g., erythromycin) would likewise arrest growth in nonfermentable but not glycolysable medium. Drugs that are specifically antimitochondrial in their actions are few but toxic compounds that preferentially react with the mitochondrial system are numerous. These compounds have minimum inhibitory concentrations (mic) that are significantly lower in nonfermentable than in glucose media. This is particularly evident in the case of many chemical mutagens and carcinogens (Egilsson *et al.* 1979), which, of course, react with DNA either directly or indirectly. The best examples are acriflavine, ethidium bromide, 5-fluorouracil, daunomycin, cadmium chloride, thioacetamide, 4-nitroquinoline-*N*-oxide, and benzidine. All induce the *petite* mutation but all selectively block transcription of mtDNA, a feature that is detectable in the first instance from absorption spectra of cells in which cytochromes a + $a_3$ and b fail to develop (Fig. 10.3B). The selective nature of the antimitochondrial activity is exemplified in the continuation of cytochrome c synthesis, which requires transcription in the nucleus where the structural gene for cytochrome c is located.

A more direct assessment of inhibition of mtDNA transcription can be demonstrated by monitoring uptake of radioactive uracil into mitochondrial RNA as described by Fukuhara and Kujawa (1970) in yeast. These authors also pointed out that at concentrations that inhibited transcription of mtDNA, the

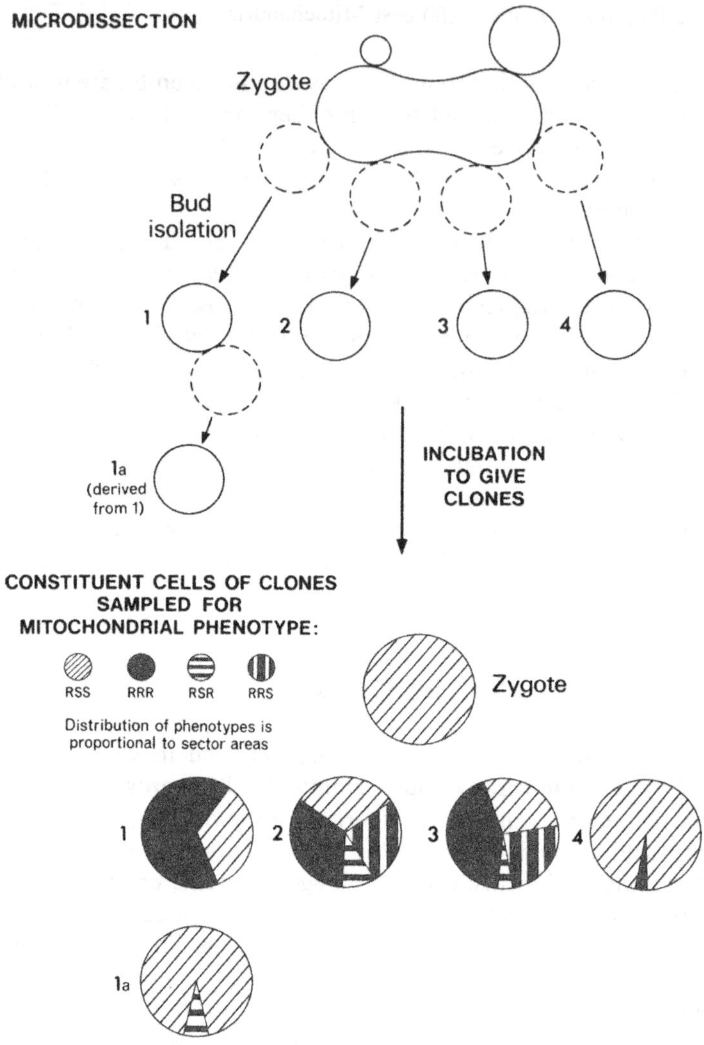


**Figure 10.2.** Zygote cell lineage analysis in the three-point cross RSS (oligomycin resistant, chloramphenicol sensitive, erythromycin sensitive) X SRR (oligomycin sensitive, chloramphenicol resistant, erythromycin resistant). Diploid daughter cells (buds) are removed as they arise from the zygote and cloned. A typical distribution of mitochondrial phenotypes is shown on clone analysis in this particular cross. (From Wilkie 1973.)

**Figure 10.3.** (A) Growth curves of strain D6 in nonfermentable medium (1% yeast extract, 4% glycerol) containing 2,6-diaminoputrine (DAP). (B) Absorption spectra of strain B41 in glucose medium (1% yeast extract, 2% glucose). (a) untreated, control; (b) with 50 $\mu$g/ml DAP; (c) with 100 $\mu$g/ml DAP. Peaks at 603, 562, and 550 nm are $\alpha$ peaks of cytochromes a + $a_3$, b and c respectively, while the $\beta$ peaks of cytochromes b and c occur at 530 and 520 nm, respectively.

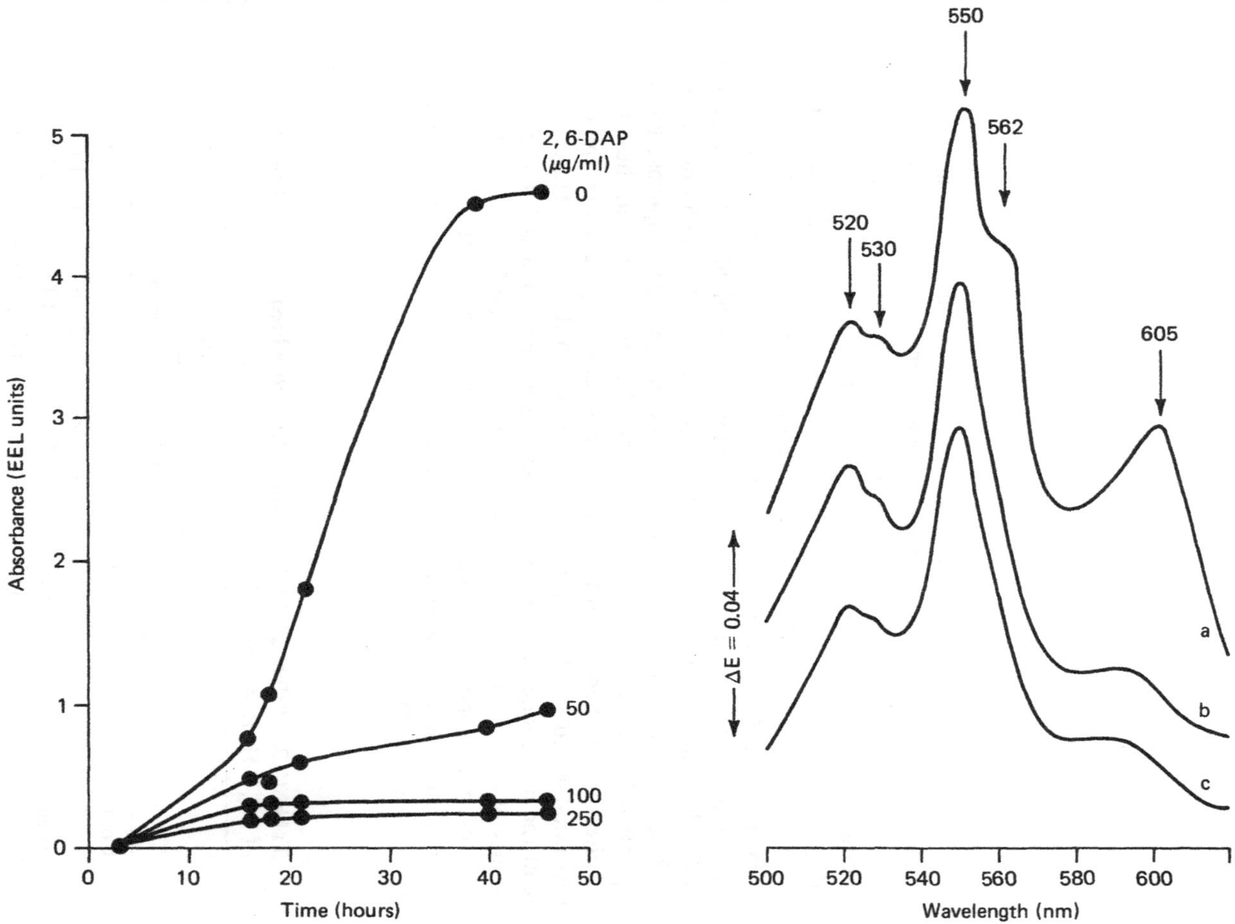
2, 6-DAP
(μg/ml)
0
50
100
250
Absorbance (EEL units)
0
1
2
3
4
5
0
10
20
30
40
50
Time (hours)
550
562
520
530
605
a
b
c
ΔE = 0.04
500
520
540
560
580
600
Wavelength (nm)

(a)

(b)

**Table 10.2.** Effect of Acriflavine and Diaminopurine (DAP) on Incorporation of [$^3$H]UTP into Rat Liver Mitochondria over a 20-min Incubation Period[a]

| | Acriflavin (μg/ml) | | | | | DAP (μg/ml) | | | | |
|---|---|---|---|---|---|---|---|---|---|---|
| | 0 | 10 | 15 | 20 | 25 | 0 | 100 | 150 | 200 | 250 |
| Mitochondria protein, cpm/mg | 524 | 436 | 440 | 415 | 402 | 442 | 382 | 350 | 272 | 192 |
| Percent inhibition | 0 | 12 | 16 | 21 | 23 | 0 | 14 | 21 | 38 | 57 |

[a]From Wallis and Wilkie 1979.

mutagenic action of the dye was low; that is, very few *petite* mutations were induced. In other words, transcription inhibition could be reversible in most cells, which led to the conclusion that the mutagenic action of the dye was not a consequence of this inhibition. This lack of correlation between transcription inhibition and *petite*-inducing ability is well illustrated in our studies with the mutagen and adenine analog 2-6-diaminopurine (Wallis and Wilkie 1979). This drug arrested growth of all yeast strains tested (15) in nonfermentable media (Fig. 10.3A) at about 250 μg/ml but had no significant effect at maximum solubility (5mg/ml) in glucose medium. This highly selective activity against the mitochondria was reflected in the failure of cytochromes a + $a_3$ and b to develop (but not c) in the presence of the drug in glucose medium (Fig. 10.3B). This adenine analog (addition of adenine reversed these inhibitory effects) was then shown to have a direct depressing action on the uptake of tritium-labeled UTP into mitochondrial RNA (Table 10.2). Thus, the failure of the mitochondrially coded cytochromes to develop, with the consequent inability of cells to utilize nonfermentable substrate, can be attributed to inhibition of mitochondrial transcription. To come back to the earlier point, although the drug is efficient in this respect, it is a weak inducer of the *petite* mutation (Table 10.3).

**Table 10.3.** Effect of DAP Treatment on *petite* Frequency in Strains of D6, B41, B/A, and B/B[a]

| | *petite* (%) | | | |
|---|---|---|---|---|
| Treatment | D6 | B41 | B/A | B/B |
| Control | 0.27 (2/748) | 2.03 (26/1280) | 3.54 (24/678) | 0.15 (2/1322) |
| 50 μg/ml DAP | 0.17 (2/1156) | 4.55 (83/1822) | 14.17 (147/1038) | 0.24 (3/1258) |
| 100 μg/ml DAP | 0.22 (2/929) | 3.87 (112/2893) | 4.21 (40/950) | 0.76 (7/922) |

[a]From Wallis and Wilkie 1979.

### Nonmutagenic Drugs

The mitochondrial genetic system seems to be the primary target for a wide variety of chemical mutagens, at least in yeast cells (Egilsson *et al.* 1979), but the organelle is also susceptible to the inhibitory effects of membrane-reacting agents. These affect (usually reversibly) the functional characteristics of the mitochondrion and include such compounds as tricyclic drugs (Linstead and Wilkie 1971), phenylbutazone (Wilkie and Charlwood 1974), the dicarboxylic azeleic and dodecandioic acids (Patel and Wilkie 1982), vitamin A, and aspirin (Bruce and Wilkie 1982a,b). These drugs can be shown to inhibit respiration but in the case of the dicarboxylic acids, vitamin A, and aspirin tested so far, there seems to be a stabilizing effect on the membrane system if the drugs are used in low concentrations. Not only does this result in increased respiratory activity and metabolic rate, but it affords a measure of protection against the antimitochondrial effects of ethidium bromide, Adriamycin, and oligomycin, and this is a particular feature of aspirin (Bruce and Wilkie 1982b). These recent findings broaden the scope of investigations into the functional organization of mitochondrial membranes.

## Mitochondria and Cellular Processes

Lesions in the inner mitochondrial membrane such as those in *petite* mutants lead to deficiencies in respiration but can, at the same time, lead to changes in cell surface characteristics. These include changes in concanavalin A agglutination, in permeation to certain metabolites, in partitioning behavior in a biphasic polymer system, and in flocculating characteristics (Wilkie and Evans 1982). In manifesting these effects, mitochondria do not appear to be contributing cell surface components, but rather the control is achieved by a mitochondrial function that modulates the activity (or the products) of certain nuclear genes. These genes are presumably specifying cell surface proteins that may require modification by mitochondrial action. A possible candidate is $Ca^{2+}$ or other divalent cations that have signaling properties and whose flow is largely controlled by the mitochondrial membrane system. Whatever the mechanism, mitochondrial control of cell surface characteristics has implications for carcinogenesis if the findings can be extrapolated to mammalian cells since the changes listed are similar to those described in neoplastic transformation by carcinogens.

## Conclusions

Mitochondrial genome recombination, replication, and transmission have been briefly discussed together with some of the aspects of biogenesis and function of the organelle. It is clear that major problems remain in this field, but perhaps the most far-reaching of these are the problems relating to the activity of antimitochondrial drugs, particularly since mitochondria appear to have important functions in cellular processes apart from providing ATP. In studies of this kind, the yeast cell makes a useful model of the eukaryotic cell.

## References

Bernardi, G. (1978). The petite mutation in yeast. Trends Biochem. Sci. 4:197–201.

Birky, C. W. (1978). Transmission genetics of mitochondria and chloroplasts. Ann. Rev. Genet. 12:471–512.

Bruce, I. J., Wilkie, D. (1982a). Mitochondrial activity distinguishes aspirin from salicylate and benzoate in yeast cells. Biochem. Pharmacol. (in press).

Bruce, I. J., Wilkie, D. (1982b). Aspirin reverses antimitochondrial activity of ethidium bromide, adriamycin and oligomycin in yeast. Submitted for publication.

Egilsson, V., Evans, I. H., Wilkie, D. (1979). Toxic and mutagenic effects of carcinogens on the mitochondria of *Saccharomyces cerevisiae.* Mol. Gen. Genet. 174:39–46.

Ephrussi, B., Margerie-Hottinguer, H. De, Roman, H. L. (1955). Suppressiveness, a new factor in the genetic determination of the synthesis of respiratory enzymes in yeast. Proc. Natl. Acad. Sci. USA 41:1065–1071.

Fonty, G. Goursot, R., Wilkie, D., Bernardi, G. (1978). The mitochondrial genome of wild-type yeast cells: VII. Recombination in crosses. J. Mol. Biol. 119:213–235.

Fukuhara, H., Kujawa, C. (1970). Selective inhibition of the *in vivo* transcription of mitochondrial DNA by ethidium bromide and acriflavine. Biochem. Biophys. Res. Commun. 41:1002–1008.

Linstead, D., Wilkie, D. (1971). A comparative study of *in vivo* inhibition of mitochondrial function in *Saccharomyces cerevisiae* by tricylic and other centrally-acting drugs. Biochem. Pharmacol. 20:839–846.

Patel, R., Wilkie, D. (1982). Mitochondrial activity of azelaic and dodecanedioic acids in yeast cells. In preparation.

Smith, D. G., Wilkie, D., and Srivastava, K. C. (1972). Ultrastructural changes in mitochondria of zygotes in *Saccharomyces cerevisiae.* Microbios 6:231–238.

Stevens, B. J. (1977). Variation in number and volume of the mitochondria in yeast according to growth conditions. A study based on serial sectioning and computer graphics reconstruction. Biol. Cellulaire 28:37–56.

Thomas, D. Y., Wilkie, D. (1968a). Recombination of mitochondrial drug resistance factors in yeast. Biochem. Biophys. Res. Commun. 30:368–372.

Thomas, D. Y., Wilkie, D. (1968b). Inhibition of respiratory enzyme synthesis by erythromycin: cytoplasmic and nuclear factors controlling resistance. Genet. Res. 11:33–41.

Wallis, C., Wilkie, D. (1979). Mitochondrial activity of 2,6-diaminopurine in *Saccharomyces cerevisiae.* Mol. Gen. Genet. 173:307–313.

Waxman, M. F., Wilkie, D., Eaton, N. (1973). Effect of antibiotics on the transmission of mitochondrial factors in *Saccharomyces cerevisiae.* Mol. Gen. Genet. 127:277–284.

Wilkie, D. (1972). The yeast cell and the antimitochondrial activity of drugs. Med. Biol. Illustr. 22:119–124.

Wilkie, D. (1973). Cytoplasmic genetic system of eucaryotic cells. Brit. Med. Bull. 29:263–268.

Wilkie, D., Charlwood, B. (1974). Effects of mitochondrial inhibition in *Saccharomyces cerevisiae.* In: Biomembranes, edited by L. Packer. New York: Academic Press, pp. 27–34.

Wilkie, D., Evans, I. H. (1982). Mitochondria and the yeast cell surface: implications for carcinogenesis. Trends Biochem. Sci. 7:147–151.

Wilkie, D., Maroudas, N. G. (1968). UV inactivation studies of the cytoplasmic respiratory determinant in yeast. Biochim. Biophys. Acta 166:681–688.

# 11

# Molecular Genetic Aspects of Yeast Mitochondria

Ivor H. Evans*

## General Introduction

Mitochondria are organelles comprising a fundamental and characteristic feature of eukaryotic cytoplasm. The familiar basic structural theme of outer membrane enclosing a folded inner membrane is elaborated into an extensive variety of forms in different eukaryotic organisms and cell types. Inside living cells mitochondrial form is far from static, and harbored within these mobile, plastic, mitochondrial structures is an immense and varied repertoire of biochemical functions.

Three major classes of mitochondrial function can be discerned. First are the well-known energy-linked functions such as conservation of respiratory energy by phosphorylation of ADP to ATP (oxidative phosphorylation), transhydrogenation between the pyridine nucleotides, and ion and solute translocation. A second class of properties is catabolic and anabolic in nature and includes fatty acid oxidation and the synthesis of porphyrin, heme, citrulline, and phospholipids. The third group of mitochondrial functions consists of those concerned with the perpetuation or "biogenesis" of the organelle itself. All these functions, and doubtless others yet undetected, are possessed by the mitochondria of virtually all eukaryotes investigated. Mitochondrial biochemistry or molecular biology is evidently an enormous and highly complex subject, which has nevertheless been grappled with in a number of general reviews (Novikoff 1961; Lehninger 1964; Tandler and Hoppel 1972; Mahler 1973; Lloyd 1974; Munn 1974; Tedeschi 1976), including, very recently, an exposition of the historical development of mitochondrial research by Ernster and Schatz (1981).

*School of Biological Sciences, Queen Mary College, University of London, London E1 4NS England.

The more limited area under consideration here primarily concerns molecular genetic aspects of mitochondrial biogenesis in the yeast *Saccharomyces cerevisiae*. This friendly eukaryotic microbe has become the single most important experimental organism for studies of the mitochondrial genetic system, both because of its great amenability to genetic analysis and because of its unusual capacity to dispense with many mitochondrial functions and so survive mutational or drug-induced mitochondrial lesions that would be lethal to other eukaryotes.

## Mitochondrial DNA

Despite unmistakable evidence of cytoplasmic genes affecting respiration (see Chapter 10 by D. Wilkie) and the discovery of mitochondrial protein synthesis (McLean *et al.* 1958), many biologists were still surprised when in 1962–1963 Nass and Nass presented the first clear evidence for DNA in mitochondria (mtDNA) (Nass and Nass 1962; Nass and Nass 1963a,b). Their ultrastructural study showed that the mitochondria of chick embryo cells contained threadlike structures digestible by DNase but insensitive to RNase, and so inaugurated modern mitochondrial molecular genetics.

The ensuing two decades of intense research activity have culminated in mitochondrial DNA becoming the most thoroughly characterized genetic element of eukaryotic cells. mtDNA is now known to be the center of a semiautonomous genetic system employing an idiosyncratic genetic code and possessing a unique genetic content that is expressed through transcription and translation machinery quite distinct from that of nucleus and cytoplasm, though bearing a resemblance to that of prokaryotes. Although mitochondrial genome size and sequence organization vary strikingly among different eukaryotes, the genetic information is surprisingly constant. Mitochondrial genes encode essentially just a few polypeptides of the respiratory and phosphorylation complexes of the inner membrane plus components of the mitochondrial translation system (mainly ribosomal and transfer RNAs). The vast majority of mitochondrial proteins (~95%), including the enzymes responsible for replicating and transcribing mtDNA, are encoded in the nucleus, are synthesized in the cytoplasm, and must be imported into the mitochondrion.

The logarithmic growth phase of this new field of knowledge has been charted in a profusion of review articles, published symposia, and books, most of which emphasize results obtained with the mitochondrial system of yeast.

*Review articles:* Nass *et al.* 1965; Granick and Gibor 1967; Marchant and Smith 1968; Borst and Kroon 1969; Ashwell and Work 1970; Rabinowitz and Swift 1970; Kuntzel 1971; Linnane and Haslam 1971; Borst 1972; Linnane *et al.* 1972; Borst and Grivell 1973; Mahler 1973a; Tzagoloff *et al.* 1973; Wilkie 1973; Dujon and Michaelis 1974; Gillham 1974; Kasamatsu and Vinograd 1974; Kovac 1974; Schatz and Mason 1974; Butow *et al.* 1975; Clayton and Smith 1975; Linnane and Crowfoot 1975; Mahler *et al.* 1975; Mahler and Raff 1975a,b; Saccone and Quagliariello 1975; Bernardi 1976; Michaelis and Somlo 1976; Schatz 1976; Borst 1977a,b; Borst *et al.* 1977; Nagley *et al.* 1977; Borst and Grivell 1978; Birky 1978; Buetow and Wood 1978; Linnane and Nagley 1978a,b; Bernardi 1979; Borst *et al.* 1979; Butow and Strausberg 1979; Locker and Rabinowitz 1979; Locker *et al* 1979; Tzagoloff *et al.* 1979a,b; Borst 1980a,b; Lloyd and Turner 1980; Perlman *et al.* 1980; Borst 1981a,b; Dujon 1981; Mahler 1981; Stevens 1981; Gray 1982; Gray and Doolittle 1982; Wilkie 1982.

*Symposia:* Slater *et al.* 1968; Miller 1970; Boardman *et al.* 1971; Van den Bergh *et al.* 1972; Kroon and Saccone 1974; Birkey *et al.* 1975; Puiseux-Dao 1975; Bandlow *et al.* 1976; Bücher *et al.* 1976; Saccone and Kroon 1976; Packer and Gomez-Payou 1976; Bandlow *et al.* 1977; Bacila *et al.* 1978a; Cummings *et al.* 1979; Lee *et al.* 1979; Bücher *et al.* 1980; Kroon and Saccone 1980; Von Wettstein *et al.* 1981; Slonimski *et al.* 1982.

*Books:* Roodyn and Wilkie 1968; Sager 1972: Mahler 1973; Gillham 1978.

## The Diversity of Mitochondrial DNAs

Mitochondrial DNAs are highly varied in form, size, and base composition, though broad similarities may be seen in major phylogenetic groups (see Table 11.1; Gillham 1978; Gray 1982; and Gray and Doolittle 1982).

Metazoan mtDNAs tend to be small and circular {5-$\mu$m contour length [= $10 \times 10^6$ daltons (10 megadaltons, or 10 Md)]} or 15–16 kilobase pairs (kbp) (see Borst and Favell 1976).

Higher plant mtDNAs seem to be more heterogeneous and not necessarily all circular, the overall picture being still rather confused (see Leaver and Gray 1982). However, evening primrose (Brennicke 1980) and pea plants (Kolodner and Tewari 1972) apparently contain circular ~30-$\mu$m circumference (~95 kbp or 60 Md) mtDNAs, and the largest mtDNAs yet detected have been found in the muskmelon, a member of the Curcurbitaceae (1600 Md, Ward *et al.* 1981).

Fungal mitochondrial genomes are mostly circular and sizes in yeasts vary between the ~12.8 Md (6 $\mu$m, 18.9 kbp) of *Torulopsis glabrata* (O'Connor *et al.* 1976) and the ~71 Md (34 $\mu$m, 108 kbp) of *Brettanomyces custersii* (Clark-Walker *et al.* 1981), *Saccharomyces cerevisiae* mtDNA consisting of ~50-Md circles (25 $\mu$m, 75 kbp) (Hollenberg *et al.* 1970). One yeast, *Hansenula mrakii,* has recently been shown to possess a linear 18-$\mu$m mtDNA genome (Wesolowski and Fukuhara 1981), whereas the congeneric species *H. wingei* has an 8.2-$\mu$m mtDNA of the more usual, circular form (O'Connor *et al.* 1975). Circular mitochondrial genomes are also found in the filamentous fungi, and contour lengths of ~10 $\mu$m (*Aspergillus nidulans;* Lopez-Perez and Turner 1975) and ~20 $\mu$m (*Neurospora crassa;* Bernard *et al.* 1975) have been recorded.

**Table 11.1.** Size and Conformation of Mitochondrial DNAs

| Taxonomic category | Species | Shape | Length ($\mu$m) | Molecular weight ($\times 10^{-6}$) | Reference |
|---|---|---|---|---|---|
| Protozoa | *Acanthamoeba castellanii* | Circular | 12.7 | 26 | Bohnert and Herrmann (1974) |
| | *Paramecium aurelia* | Linear | 13.8 | 30–35 | Goddard and Cummings (1975) |
| | *Tetrahymena pyriformis* | Linear | ~15 | ~30 | Goldbach *et al.* (1977) |
| | *Trypanosoma mega* | Circular | 9.0 (max)<br>0.74 (min) | 16.1 (max)<br>1.49 (min) | Borst and Hoeijmakers (1979) |
| Slime molds | *Physarum polycephalum* | Circular | 19.1 | 41 | Bohnert (1977) |
| Fungi | *Brettanomyces anomalus* | Circular | 18.2 | ~38 | Clark-Walker and McArthur (1978) |
| Yeasts | *Brettanomyces custersii* | | ~34 | ~71 | Clark-Walker *et al.* (1981b) |
| | *Candida parapsilosis* | Circular | 11.1 | 23.1 | O'Connor *et al.* (1975) |
| | *Hansenula mrakii* | Linear | | 36.4 | Wesolowski & Fukuhara (1981) |
| | *Hansenula wingei* | Circular | 8.2 | 17.3 | O'Connor *et al.* (1975) |
| | *Kloeckera africana* | Circular | 8.3 | ~17.5 | Clark-Walker and McArthur (1978) |
| | *Kluyveromyces lactis* | Circular | 11.4 | 24.0 | O'Connor *et al.* (1975) |
| | *Saccharomyces cerevisiae* | Circular | 21–25 | 46–52 | Gillham (1978) |
| | *Saccharomyopsis lipolytica* | Circular | ~14 | ~29 | Wesolowski *et al.* (1981) |
| | *Schizosaccharomyces pombe* | Circular | 6.0 | 12.5 | O'Connor *et al.* (1975) |
| | *Torulopsis glabrata* | Circular | 6.0 | 12.8 | O'Connor *et al.* (1976) |

| | | | | | |
|---|---|---|---|---|---|
| Filamentous fungi | *Aspergillus nidulans* | Circular | 10–10.6 | 20–21 | Lopez-Perez and Turner (1975) |
| | *Neurospora crassa* | Circular | 18–22 | 41 | Bernardi *et al.* (1975) |
| | *Podospora anserina* | Circular | 31 | 61–63 | Cummings *et al.* (1979b) |
| Metazoa | | | | | |
| Nematoda | *Ascaris lumbricoides* | Circular | 4.8 | ~10.1 | Borst and Flavell (1976) |
| Crustacea | *Artemia salina* | Circular | 5.1 | ~10.7 | Borst and Flavell (1976) |
| Insecta | *Drosophila melanogaster* | Circular | 6.2 | 12–35 | Fauron and Wolstenholme (1976) |
| Echinodermata | *Lytechinus pictus* | Circular | 4.7 | ~9.9 | Borst and Flavell (1976) |
| Pisces | *Carrassius carrassius* | Circular | 5.4 | ~11.3 | Borst and Flavell (1976) |
| Amphibia | *Xenopus laevis* | Circular | 5.8 | ~12.2 | Borst and Flavell (1976) |
| Reptilia | *Terrapene ornata* | Circular | 5.3 | ~11.1 | Borst and Flavell (1976) |
| Aves | *Gallus domesticus* | Circular | 5.4 | ~11.3 | Glaus *et al.* (1980) |
| Mammalia | *Homo sapiens* | Circular | ~5.2 | ~10.9 | Brown *et al.* (1979) |
| Algae | *Chlamydomonas reinhardtii* | Circular | 4.5 | 9.8 | Mahler and Perlman (1979) |
| Embryophyta | | | | | |
| Monocotyledoneae | *Zea mays* | Circular | | ~300 | Ward *et al.* (1981) |
| Dicotyledoneae | *Solanum tuberosum* | Linear | | ~100 | Vedel and Quetier (1974) |
| | *Oenothera berteriana* | Circular | ~31 | ~66 | Brennicke (1980) |
| | *Cucumis melo* | | | ~1600 | Ward *et al.* (1981) |

Perhaps the greatest mtDNA diversity is seen among the protists. The alga *Chlamydomonas reinhardtii* has a modest circular ~5-μm genome (Ryan *et al.* 1978), but other protists have very unusual mtDNAs, such as the linear ~14-μm genomes of *Paramecium aurelia* (Goddard and Cummings 1975), the variably sized linear genomes of *Tetrahymena pyriformis* (Goldbach *et al.* 1977), and the remarkable catenated maxicircles and minicircles of trypanosomes (Borst *et al.* 1980).

Accompanying the diversity in size and conformation of eukaryotic mtDNAs is a considerable variation in base composition. This ranges from approximately 20% (G + C), equivalent to a buoyant density of 1.683, in *Saccharomyces cerevisiae* [even lower values occur in some *petite* mutants (see below), and a buoyant density value of 1.679 has been reported for mtDNA of the avian malarial parasite *Plasmodium lophurae* (Kilejian 1975)] to approximately 60% (G + C) (buoyant density 1.711) in the duck *Anas platyrhynchos* (Borst and Kroon 1969).

## Yeast Mitochondrial DNA

The combined efforts of many laboratories and especially the recent introduction of rapid DNA sequencing techniques have yielded immense quantities of data on mitochondrial genomes. The mtDNAs of three mammals--man (Anderson *et al.* 1981), mouse (Bibb *et al.* 1981), and cow (Anderson *et al.* 1982)—have been completely sequenced and all the known gene–product encoding regions of yeast have been sequenced (see below, and Slonimski *et al.* 1982). Before dealing in some detail with the material relevant to yeast mtDNA, two introductory "signpost" sections are provided as orientation aids.

The first section summarizes key findings on mammalian mtDNA as reference points for comparison with the yeast system. In the second section current knowledge of the yeast mitochondrial genome is briefly outlined. Subsequent sections then consider the gross organization of yeast mtDNA and its base composition, genetic and physical mapping, the sequences of individual yeast genes, and the question of yeast mtDNA introns.

### The Mammalian Mitochondrial Genome

The complete nucleotide sequences of the human (placental) (Anderson *et al.* 1981), mouse (Bibb *et al.* 1981), and bovine (heart) (Anderson *et al.* 1982) mitochondrial genomes are now available (see Anderson *et al.* 1982). The organization of the human and bovine genomes is identical, and they exhibit approximately 70% nucleotide sequence homology.

The genes for the 12S and 16S rRNAs; 22 tRNAs; cytochrome b apoprotein; cytochrome c oxidase subunits I, II, and III; and ATPase subunit 6 have been identified. As in yeast and most other eukaryotes there is no 5S RNA (though higher plant mitochondria contain this species). Eight other possible protein coding genes lacking, as yet, known protein products (unassigned reading frames or URFs) have been identified and are transcribed into stable polyadenylated RNA species (Ojala *et al.* 1980).

The genomes are extremely compact, noncoding bases between the genes, being few or nonexistent. The two DNA strands of mammalian mitochondrial genomes differ sufficiently in base composition to be separable by the resulting density difference into heavy (H) and light (L) strands. The remarkable economy of organization of these genomes is attested by the fact that though most genes are located on (transcribed from) the H strand, several genes (mainly for tRNAs) have been found on the L strand.

The two ribosomal RNA genes lie close together separated only by a single tRNA gene, and indeed a special feature of the organization of mammalian mt genomes appears to be that tRNA and protein-coding sequences are separated or "punctuated" by one or a few tRNA genes. This may have implications for transcript processing.

As in the case of other eukaryotes, including yeast, the genetic code in mammalian mitochondria differs from the "universal" genetic code. UGA codes for tryptophan rather than termination, AUA for methionine and initiation rather than isoleucine, and AGA and AGG for termination rather than arginine. The mitochondrial tRNA anticodons are such that they probably read this genetic code using G:U wobble base pairing (for 2-codon family amino acids) and U:N base pairing (for 4-codon family amino acids).

Another facet of the compression of genetic information in mammalian mtDNAs may be the many reading frames that appear to lack a stop codon, a T or TA following the last sense codon and immediately preceding a tRNA gene or another reading frame. The missing stop codon may be created by polyadenylation of the gene transcript, which would generate a terminal UAA stop codon.

Since mammalian mtDNA appears to be transcribed symmetrically and completely, the genome may possess only a single promoter site (for the initiation of RNA transcription), and this has yet to be identified. It could lie in the "D-loop" region, which has no assigned coding function and apparently contains the origin of replication of the H strand (see Clayton 1982). Although it is the probable site of replication and transcription signals, comparison between human, rat, and bovine sequences shows that this region is the least conserved portion, both in sequence and size, of the mammalian genome.

## The Yeast Mitochondrial Genome: A Synopsis

The circular ($\sim$25-$\mu$m contour length, 50 Md, 75 kbp) mitochondrial genome of *S. cerevisiae* is unusual in being highly (A + T)-rich and in tending to generate a type of deletion mutant (*petite:* see following discussion and Chapter 10) at a very high frequency. Large sections of the genome are so rich in

(A + T) that they must be incapable of specifying proteins and so presumably represent nonsense or "selfish" DNA, though multiple origins of DNA replication do occur in these regions.

As in mammals and most eukaryotes the mtDNA of *S. cerevisiae* encodes just a few essential mitochondrial components: RNA species required for the mitochondrial translation apparatus and polypeptide subunits and apoproteins belonging to the electron transport and phosphorylation complexes of the inner mitochondrial membrane.

Yeast mtDNA bears genes for about 24 mt tRNAs, 2 rRNAs (15S in the small 37–40S mitoribosomal subunit, 21S in the large 50–60S subunit, 5S RNA being absent), and it appears to determine the size of a protein (the *var-1* protein) that is associated with the mitoribosomes. There are also genes for three subunits of cytochrome c oxidase (i.e., subunits COI, II, and III of complex IV), the apoprotein of cytochrome b (part of the b–$c_1$ complex, complex III), and two subunits of the ATP-synthesizing complex (i.e., subunits 6 and 9 of the mitochondrial "ATPase"). A significant difference from mammals is that whereas ATPase subunit 9 is mitochondrially encoded in yeast, the relevant gene does not occur in mammalian mtDNA and so is presumably located in the nucleus. In yeast, unlike mammals, all genes with one exception (a tRNA gene) are located on the same mtDNA strand.

As in the case of the mammalian genome, sequencing of yeast mtDNA has revealed a number of open reading frames for which no corresponding protein product is known (unassigned reading frames, URFs); these are substantial lengths of DNA lacking stop codons in at least one of the possible reading frames and with (usually) an AUG initiator codon toward the 5′ end. Also, as in mammals, the yeast mitochondrial genome provides exceptions to the "universal" genetic code: the UGA termination codon encodes tryptophan, though not all the mammalian mtDNA coding idiosyncrasies are shared (see following discussion and Gray 1982).

No clear logic is evidenced by the arrangement of the genes. The positioning of the genes encoding inner membrane polypeptides (apocytochrome b; COI, II, and III; and ATPase subunits 6 and 9) differs from that in the mammalian genome, but in both cases there is no grouping of genes determining related functions. The genes for the two ribosomal RNA species are more widely separated in yeast than in any other mt genome (see Gray 1982), these genes in mammals being closely apposed. Again, unlike the mammalian mt genome, there is no regular interspersion of rRNA- and polypeptide-coding genes with tRNA genes that, in yeast, are mostly located in the genome quadrant that adjoins the 21S rRNA gene.

Yeast mitochondrial genes differ most sharply from those of mammals in that some of them are split, or "mosaic," containing one or more intervening sequences or introns (see Borst 1980a, 1981). No such inserts are known to occur in the compact mt genomes of mammals. Certain of these yeast mitochondrial introns are absent in some (apparently normal) strains of *S. cerevisiae* and so are nonessential, or "optional" (Borst 1980a). The optional introns do not move around the genome, but occur at fixed sites. In this respect and in the lack of terminal repeat sequences in at least one case (Bos *et al.* 1980; Dujon 1980), yeast mitochondrial introns differ from the translocatable elements or "transposons" of prokaryotes (Kleckner 1981). An optional intron occurs in the 21S rRNA gene and is associated with a polarity-of-recombination phenomenon. The other two mosaic genes known encode polypeptides: the *cob-box* locus specifying apocytochrome b can contain five introns, at least two being optional, and *oxi-3* encoding COI may contain as many as nine introns of which at least four are optional. Intriguingly, the unassigned reading frames so far detected are mostly located in introns.

As the existence of mosaic genes implies, extensive processing and splicing of transcripts occurs in yeast mitochondria, and there is powerful evidence that at least one partially intron-encoded fusion protein—a "maturase" or "trans-acting factor"—has a role in RNA splicing (see Slonimski *et al.* 1982). In contrast with the mammalian mt genome there are probably several promoter sites. Closed-circular RNA molecules are generated during RNA processing by excision of some intron sequences from transcripts.

In common with other mitochondrial systems and with prokaryotes the polypeptide chains of yeast mitochondria are initiated with N-terminal formyl methionine, and mitochondrial protein synthesis is inhibited by, for example, erythromycin and chloramphenicol but is insensitive to cycloheximide, which acts on the cytoplasmic ribosomes of eukaryotes. Mitochondrial protein synthesis is presumably coordinated with cytosolic synthesis of inner membrane proteins, some of which are imported into the organelle as precursor proteins 20–60 amino acids larger than the mature protein and then processed. The ways in which extra- and intraorganellar genetic systems interact to assure a controlled supply of mitochondrial components remain quite unknown, as do the means used to integrate newly synthesized components into the mitochondrial inner membrane. These questions are attracting increasing research attention. It is also being realized that nucleomitochondrial interactions are more subtle and more reciprocal than might be anticipated. Studies of extramitochondrial functions such as carbohydrate fermentation, flocculation, sporulation, and germination have indicated unexpected dependencies on mitochondrial function other than energy provision, and prophesy a wider role in the cell for the mitochondrion and its genome than is usually allowed (see below).

## Size Differences Between the mt Genomes of Different Yeast Strains

Data from DNA–DNA reassociation kinetics and analysis with restriction enzymes usually indicate that the mtDNA of a particular organism is a homogeneous population of molecules with a sequence complexity consistent with genome contour length as measured in the electron microscope; however, close analysis shows that significant differences in mitochondrial genome size can occur within the same species. For example, mtDNAs in different strains of *Tetrahymena pyriformis* vary from 28.2 to 32 kbp in length (Goldbach *et al.* 1977). In the case of *Saccharomyces cerevisiae* size differences such as these have been analyzed in great detail. It is extremely difficult to isolate the ~25-μm circular genomes of *S. cerevisiae* undegraded and routinely measure contour lengths directly, although occasional intact circles have been seen by electron microscopy (Hollenberg *et al.* 1970), so accurate estimates of genome size have come from restriction enzyme analysis. Investigation of a number of yeast (*S. cerevisiae*) strains in this way showed that the mtDNA could vary in size from 68.0 to 78.3 kbp (Prunell *et al.* 1977; Sanders *et al.* 1977a). This size heterogeneity was apparently caused by various combinations of major (1–3 kbp) and minor (~25–50 bp) insertions or deletions, detectable as electrophoretic mobility changes in restriction enzyme fragments and loss or gain of restriction sites, overall gene order being unaffected. Slonimski and Lazowska (1977) surmised that the major insertions could be analogous to the translocatable elements or transposons of bacteria and plasmids, but there is no evidence that these insertions can move to different genomic sites (Borst *et al.* 1977; Heyting *et al.* 1979; Bos *et al.* 1980). Indeed it is now clear that the major insertions are actually mostly located within known mitochondrial genes, and so may be regarded as introns (Borst and Grivell 1978; Grivell *et al.* 1979; Van Ommen *et al.* 1979; and see the preceding "synopsis" section and the following sections on mitochondrial genes and introns).

## Base Composition Aspects of Yeast mtDNA

Yeast (*S. cerevisiae*) mtDNA has one of the lowest (G + C) contents (~18%) of any functional DNA. Early studies using nearest-neighbor analysis showed that the Gs and Cs of yeast mtDNA were highly clustered (Grossman *et al.* 1971). Subsequently, extensive enzymatic analyses by Bernardi and co-workers (Bernardi *et al.* 1972; Piperno *et al.* 1972; Ehrlich *et al.* 1972; Prunell and Bernardi 1974; Bernardi *et al.* 1976; Prunell *et al.* 1977; Prunell and Bernardi 1977; Fonty *et al.* 1978) identified four sequence elements: (1) (A + T)-rich stretches (<5%{G + C}) of average size $4 \times 10^5$ d, making up about 50% of the DNA, and dubbed "spacers," (2) intermediate (G + C)-rich stretches (22–65% {G + C}) of a size similar to that of (1), also making up about 50% of the genome, and termed "genes" (iii) and (iv) short (~50–80 bp) (G + C)-rich segments either containing or lacking sites for the restriction enzymes HaeIII (cleaves at GGCC) and HpaI (cuts at CCGG), and constituting 2–3% of the genome.

It was suggested that the genome was organized as a sequence of tripartite segments: a (G + C)-rich region of regulatory function, possibly corresponding to a promoter, followed by the gene, and finally the spacer, there being a total of about 60 of these elements (Bernardi *et al.* 1976; Prunell and Bernardi

1977). Although it seems likely that the (A + T)-rich segments cannot encode proteins and may well correspond to "selfish" DNA (Doolittle and Sapienza 1980; Orgel and Crick 1980), the existence of transcripts considerably longer than the tripartite elements (Van Ommen *et al.* 1979) implies that "spacers" can be transcribed and that not all (G + C)-rich segments act as promoters. Various possible functions or roles have been ascribed to the (A + T)-rich stretches, and these include

1. The provision of quasi-homologous sequences allowing unequal crossing-over between different genome units and so permitting rapid genome evolution by gross rearrangements (Prunell *et al.* 1977; Fonty *et al.* 1978) (the preservation of gene order in different strains might argue against this); such sequences may also be important in the mechanism of the *petite* mutation (see below).
2. A role in the processing of precursor RNAs (Borst *et al.* 1977).
3. A regulatory role related to the exceptional ability of *S. cerevisiae* to repress mitochondrial biogenesis under anaerobic conditions or in the presence of high glucose concentrations (Borst and Grivell 1978). A "metabolic" hypothesis such as this would presumably not, however, explain the highly (A + T)-rich segments occurring at specific sites in the mt genomes of *Drosophila* species and causing profound differences in size and base composition between them (Shah and Langley 1979; Zakour and Bultmann 1979).

If the (A + T)-rich segments so characteristic of the mtDNA of *S. cerevisiae* are indeed devoid of function or "selfish," an explanation of their existence is still required. A current hypothesis is that yeast mitochondria lack a uracil excision mechanism (see Dujon 1981). Since uracil constantly arises in DNA from the deamination of cytosine, a steady mutagenic drift of GC to AT transitions should occur up to a tolerance limit determined by natural selection.

Sequencing studies of yeast mtDNA have since confirmed some aspects of the Bernardi model and invalidated others (see section on mitochondrial genes, below). Much of the genome indeed contains long stretches of virtually pure (A + T), but while some of these do lie between genes, others are located within split genes as parts of introns. (G + C)-rich sequences have also been found within genes; in the case of the dispensable mini-insert in the 21S rRNA gene (Dujon 1980), a promoter-like or obligate functional role would appear to be definitely precluded. Finally, the total number of genetic elements in the yeast mt genome seems to be substantially less than 60.

## Mapping of Yeast mtDNA

(See Nagley *et al.* 1976; Borst *et al.* 1977; Dujon *et al.* 1977; Linnane and Nagley 1978a,b; Borst *et al.* 1979; Dujon 1981.)

## Genetic Mapping

As discussed in Chapter 10, the first mutations isolated in yeast mtDNA were the *petite* lesions. As these are large, noncomplementing deletions with essentially similar respiratory-deficient phenotypes, they were of little direct use in gene mapping, though in combination with other types of mitochondrial mutation they provide a rapid mapping procedure.

The first mitochondrial point mutations to be isolated were to antibiotic resistance (*ant*$^R$ mutations; see Chapter 10), and it was immediately apparent that determination of frequencies of recombination between these mutations would not alone permit construction of a complete genetic map, since rates of recombination can reach a maximum of 20–25% within only 6000 bp or less (Dujon *et al.* 1974, and see Dujon 1981).

The advent of *mit*$^-$ (defective in one or more mitochondrially coded membrane polypeptides, so respiratory deficient but still capable of mitochondrial protein synthesis) and *syn*$^-$ (defective in a component of the mitochondrial translation system) mutants furnished an extensive new range of widely distributed markers (see, e.g., Tzagoloff *et al.* 1975a,b,c). These three types of localized lesion or point mutation provided, in combination with *petite* mutations, new mapping techniques:

1. Point mutations can be ordered and degree of linkage estimated from frequencies of coretention or separation in random *petite* populations (Molloy *et al.* 1975; Schweyen *et al.* 1976,1978). This approach allowed the circularity of the mitochondrial map to be demonstrated for the first time (Molloy *et al.* 1975).
2. "Libraries" of *petites* derived from a type 1 analysis consist of defined (overlapping) segments of mtDNA covering the entire genome and so can be used to position new mutations on the map by observing "marker rescue" through recombination in crosses (Choo *et al.* 1977).

It was long considered that the rapid marker recombination and mitotic segregation observed in yeast crosses precluded complementation studies. However, Butow and colleagues showed that genes carried on *petite* genomes could be expressed in zygotes arising from crosses with wild-type cells, which provided a functional translation apparatus for the defective *petite* genomes. They termed this technique "zygotic gene rescue" and used it to map mitochondrial genes specifying *var1* (see following discussion) gene products (Butow *et al.* 1977; Strausberg and Butow 1977). Studies of complementation between closely linked but independent *mit*$^-$ mutations have since proved invaluable in analyzing complex "mosaic" genes such as *oxi3* and *cob-box* (Foury and Tzagoloff 1978; Slonimski *et al.* 1978a and see below).

## Physical Mapping

Restriction enzyme analysis of mtDNA has enabled purely physical, circular, maps of the genome to be constructed, giving locations for the many enzyme cleavage sites (Morimoto *et al.* 1976,1977,1978; Sanders *et al.* 1977a,b; Mahler *et al.* 1977; Borst *et al.* 1977; Dujon *et al.* 1977; Linnane and Nagley 1978a,b; Morimoto and Rabinowitz, 1979a,b).

Physical and genetic maps have been correlated by a variety of methods, including

1. Physical mapping of *petite* mtDNAs containing known genetic loci and assignment of the *petite* physical map to that of the $\rho^+$ genome (Morimoto *et al.* 1976; Sanders *et al.* 1976; Borst *et al.* 1979).
2. Hybridization of mtRNA or DNA from genetically defined *petites* to restricted wild-type mtDNA (Sanders *et al.* 1976; Sriprakash *et al.* 1976; Choo *et al.* 1977; Borst *et al.* 1979).
3. Hybridization of known gene products such as mt tRNAs and mt rRNAs to defined segments of mtDNA (Sanders *et al.* 1975,1976; Rabinowitz *et al.* 1976; Van Ommen *et al.* 1977).
4. Electron microscope studies, including partial denaturation mapping and examination of RNA–DNA heteroduplexes (Bos *et al.* 1978a; Faye *et al.* 1979a).
5. Transcription and translation of defined mtDNA segments *in vitro* and identification of the gene product(s) (Grivell and Moorman 1977; Moorman *et al.* 1978). In heterologous systems this approach is vitiated by the problem of premature termination of translation (Moorman *et al.* 1977), which can now be understood as arising from differences between the "universal" genetic code and the mitochondrial version (see below).

As Borst and Grivell (1978) have pointed out, genetic and physical maps obtained by these means agree remarkably well in respect of gene order and overall map distances and suggest that there are no major recombinational hot spots or preferred sites of *petite* induction.

A schematic map of the *S. cerevisiae* mitochondrial genome is given in Figure 11.1.

## Genes on Yeast mtDNA

This section comprises an account of individual yeast mitochondrial genes, focusing mainly on their recently determined nucleotide sequences. Many of the sequences have been provided by the Tzagoloff laboratory using the "short" (~70 kbp) strain D273-10B. This lacks the large inserts—essentially introns—found in strains commonly used in other laboratories. The map positions given below and in Fig. 11.1 are intended only as a guide to relative gene location, but are mainly based on the data of Rabinowitz and co-workers (Morimoto *et al.* 1978; Lewin *et al.* 1978; Morimoto and Rabinowitz 1979a,b), who have constructed detailed physical maps of the mtDNAs of two strains, one being D273-10B (Morimoto and Rabinowitz 1979b). The physical map is divided into 100 units (u), 0/100 being the position of a unique SalI site lying close to the gene for the large ribosomal subunit RNA.

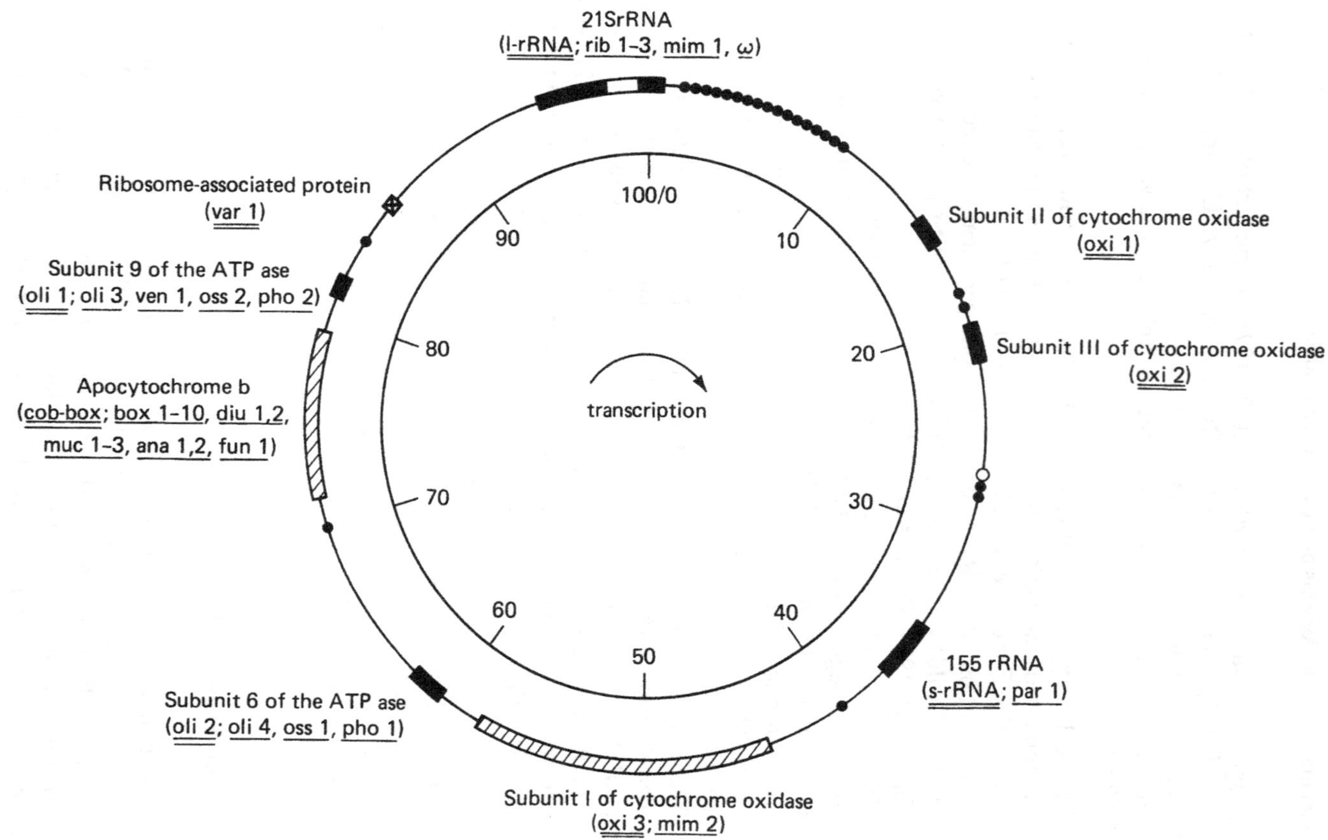
21SrRNA
(l-rRNA; rib 1–3, mim 1, ω)
Subunit II of cytochrome oxidase
(oxi 1)
Subunit III of cytochrome oxidase
(oxi 2)
155 rRNA
(s-rRNA; par 1)
Subunit I of cytochrome oxidase
(oxi 3; mim 2)
Subunit 6 of the ATP ase
(oli 2; oli 4, oss 1, pho 1)
Apocytochrome b
(cob-box; box 1–10, diu 1,2,
muc 1–3, ana 1,2, fun 1)
Subunit 9 of the ATP ase
(oli 1; oli 3, ven 1, oss 2, pho 2)
Ribosome-associated protein
(var 1)
100/0
10
20
30
40
50
60
70
80
90
transcription

## Genes for Ribosomal RNAs

The most striking feature of the organization of the yeast *(S. cerevisiae)* ribosomal RNA genes is the extent of their separation: They are separated (in the direction of transcription, i.e., "clockwise" in Fig. 11.1) by approximately one-third of the total genome (~25 kbp), within which lie many tRNA genes as well as two genes for subunits of cytochrome oxidase. It is therefore very unlikely that the two RNAs are excised from a common precursor transcript, and the question arises of the means used to procure equality of production of the two RNAs.

*Gene Product: (i) The 21S rRNA of the Large (50S) Subunit of the Mitoribosome,*

(ii) possibly the translation product of the unassigned reading frame (URF) of the large (~1100 bp) optional intron.

*Map position:* 94.2–99.4 u.

*Associated genetic loci: rib1* (chloramphenicol resistance, *cap*$^R$), *rib2* (erythromycin resistance, *ery*$^R$, and spiramycin resistance, *spi*$^R$), and *rib3* (*ery*$^R$ and *spi*$^R$); $\omega$ (omega alleles being $\omega^+$, $\omega^n$, $\omega^-$; a locus determining polarity of recombination between flanking markers loci, *rib1*, *2*, and *3*); *mim1* (an allele-specific locus nonspecific suppressor of various *mit*$^-$ mutations).

*Sequence:* Determined by Dujon (1980).

This gene, encoding a ribosomal RNA as primary end product, has several interesting features. It was the first to be genetically located on mtDNA, through the ability of some of its mutant alleles to confer resistance to the

---

**Figure 11.1.** A simplified genetic and physical map of the mitochondrial genome of *Saccharomyces cerevisiae*. This genetic and physical map represents an "average" yeast mitochondrial genome. It is based on many publications, cited in the text, especially the work of Morimoto and colleagues (see, e.g., Morimoto and Rabinowitz 1979a,b), but the gene positions are given only as indications and may be inaccurate for any particular strain. The inner circle represents the physical map, in units from 1 to 100, the 100/0 position being the site of the unique Sal 1 cleavage site. Ten map units equal approximately 7.25 kbp or 2.28 $\mu$m DNA. The outer circle gives the positions of individual mitochondrial genes. Continuous coding sequences for proteins and ribosomal RNAs are represented by black bars. The 21S rRNA gene is shown containing the large intron (white bar) characteristic of $\omega^+$ strains. The *var 1* gene for the ribosome-associated protein is given as a cross-hatched box, since the location of the structural gene for this polypeptide is uncertain (see text). The two mosaic protein-coding genes *cob-box* and *oxi3* contain extensive intronic sequences (see text) and are represented as hatched bars. The tRNA genes are shown as black dots (except for *tRNA*$^{fMet}$, given as an open circle), 17 lying in the *ℓ-rRNA-oxi 1* segment and another eight being widely scattered. In parentheses, below the name of each gene product, is the gene name (double underlined), followed by the (other) associated genetic loci (see also Table 11.2). All mitochondrial genes are transcribed and translated clockwise, except for *tRNA*$_I^{THr}$.

**Table 11.2.** Genes and Gene Products of Yeast mtDNA[a]

| Gene | Associated genetic loci | Mutant phenotypes | Gene product(s) | Mitochondrial component or enzymic complex |
|---|---|---|---|---|
| *l-rRNA* | *rib1, 2, 3* | Resistance to erythromycin, chloramphenicol, and spiramycin | 21 S rRNA[b] | Large ribosomal subunit (21 S rRNA + 38 proteins) |
| | *mim1* | Suppression of *mit*⁻ mutations | | |
| | ω | Polarity of recombination between flanking markers (e.g., *rib* loci) | | |
| *s-rRNA* | *par1*[c] | Paromomycin resistance | 15 S rRNA | Small ribosomal subunit (15 S rRNA + 33 proteins) |
| *var1* | *var1* | Polypeptide size polymorphism | Polypeptide associated with the small mitoribosome subunit | Small ribosomal subunit (15 S rRNA + 33 proteins) |
| tRNA[Aacd] | | Deficiency in mitochondrial protein synthesis (*syn*⁻) | Transfer RNAs (~25 genes specifying ~35 tRNA species) | All known mitochondrial tRNAs (with the possible exception of a single $tRNA^{Lys}$; see R. P. Martin *et al.* 1979). |
| *oxi1* | *oxi1* | Respiratory deficiency (*mit*⁻) | Subunit II of cytochrome oxidase | Complex IV or cytochrome c oxidase (7 nonidentical subunits) |
| *oxi2* | *oxi2* | Respiratory deficiency (*mit*⁻) | Subunit III of cytochrome oxidase | Complex IV or cytochrome c oxidase (7 nonidentical subunits) |
| *oxi3* | *oxi3* | Respiratory deficiency (*mit*⁻) | Subunit I of cytochrome oxidase[e] | Complex IV or cytochrome c oxidase (7 nonidentical subunits) |
| | *mim2* | Suppression of *box 7*⁻ mutations | | |

| | | | | |
|---|---|---|---|---|
| *cob-box* | *box1–10* | Respiratory deficiency (*mit*⁻) | Apocytochrome b[f] | Complex III or the cytochrome b − $c_1$ complex (7 nonidentical subunits) |
| *oli1* | *diu i,2*<br>*muc1,2,3*<br>*ana1,2*<br>*fun1*<br>*oli1,3,5*<br>*ven1*<br>*oss2* | Diuron resistance<br>Mucidin resistance<br>Antimycin resistance<br>Funiculosin resistance<br>Oligomycin resistance<br>Venturicidin resistance<br>Ossamycin resistance | Subunit 9 of the mitochondrial ATPase (the "proteolipid" or "DCCD-binding protein") | The energy-transducing oligomycin-sensitive mitochondrial ATPase complex (10 non-identical subunits) |
| | *pho2* | Phosphorylation deficiency (*mit*⁻) | | |
| *oli2* | *oli2,4*<br>*oss1*<br>*pho1* | Oligomycin resistance<br>Ossamycin resistance<br>Phosphorylation deficiency (*mit*⁻) | Subunit 6 of the mitochondrial ATPase (not yet fully confirmed, see text). | |

[a]Detailed information and references are available in the text. DNA sequencing has revealed at least two intergenic open reading frames of unassigned function (URFs). Several *"ori"* or *"rep"* sequences, each about 300 bp long, and which probably function as origins of DNA replications, have also been detected (see text). Another recently discovered locus lies near *par1* and appears to function in processing tRNA precursors (see text).

[b]In $\omega^+$ strains the 21 S rRNA cistron is split by a large intron containing an open reading frame; it is not known whether this URF is expressed and has function.

[c]*par1* is linked to the 15 S rRNA, but it is not certain whether the locus actually lies within the coding sequence.

[d]Any individual tRNA gene is symbolized $tRNA^{Gly}$, $tRNA^{Val}$, etc.

[e,f]*oxi3* and *cob-box* are "mosaic" or split genes containing variable numbers of introns, depending on the strain. Some of these introns contain URFs which may be expressed with exon sequences as "fusion proteins" with a "maturase" role in splicing or assisting in processing mosaic gene mRNA precursors.

antibiotics chloramphenical and erythromycin. Thus mutations in a ribosomal RNA gene can affect ribosome binding and inhibitory activity of these translation inhibitors, in contrast to the situation in prokaryotes, where this result is usually achieved through mutuational alteration of ribosomal proteins (see Borst and Grivell 1978).

Quite early in the development of yeast mitochondrial genetics, studies of the recombination of antibiotic resistance mutational mapping at the *rib* loci showed that in certain crosses the reciprocal recombinants were very unequal in frequency; that is, there was polarity of recombination (Bolotin *et al.* 1971). This polarity was found to be determined by alleles ($\omega^+$ and $\omega^-$) of the $\omega$ locus, now known to be located in the 21S rRNA. In an $\omega^+ \times \omega^-$ cross, $\omega^+$ and the *rib* alleles to which it is linked are overrepresented in the progeny, at the expense of $\omega^-$ and its linked alleles. A third type of allele, $\omega^n$, showing deficiency in recombination polarity, was isolated later (Dujon *et al.* 1976). Restriction mapping of the *rib1–rib3* region of several strains showed that $\omega^+$ strains possessed a 1.1-kb insert not present in $\omega^-$ strains (Borst *et al.* 1977; Jacq *et al.* 1977; Heyting and Menke, 1979) and heteroduplex analysis located the insertion as an intervening sequence or intron within the structural gene for the 21S rRNA (Bos *et al.* 1978a; Faye *et al.* 1979). Although this large optional intron appears to be necessary for polarity of recombination of neighboring loci—perhaps through a heteroduplex-gene conversion mechanism (Perlman and Birky 1974), it may not be sufficient. The existence of the $\omega^n$ mutants and the recombinational properties of some $\rho^-$ mutants [Dujon and Michel (1976) suggest that the sequences external to the intron itself are also required for polarity to occur.] The intron separates two sites, *rib1A* and *rib1B*, which are close together in $\omega^-$ strains; *rib1A* and *rib1B* interact to determine chloramphenicol resistance or sensitivity, and *rib1B* is the location of $\omega^n$ mutations (see Dujon 1981).

Dujon's sequencing work (Dujon 1980) confirmed that $\omega^+$ strains contained a large 1143-bp intron and showed that $\omega^n$ mutants (derived from $\omega^-$ strains) as well as $\omega^-$ strains lacked the intron. This study also disclosed the presence, in $\omega^+$ but not in $\omega^-$ and $\omega^n$ strains, of a second small insertion, or "mini-insert," of 66 bp, located 156 bp upstream of the major intron and hence lying within an exon of the 21S rRNA gene. It appears likely that the mini-insert is transcribed but not spliced and so persists into the mature rRNA without deleterious effect on its function. Part of the mini-insert sequence (5′-TTCGGGGTTCCGG-3′) is identical to part of a (G + C)-rich sequence found in two different regions of the mitochondrial genome (Cosson and Tzagoloff 1979), and this, together with the existence of 2-bp terminal redundancies flanking both introns, prompted Dujon to suggest that the mini-insert represents a transposable element of the mitchondrial genome. Interestingly, the mini-insert has a strikingly palindromic sequence of 16 bp located almost exactly in the middle of its sequence and able to form a hairpin structure with a 16-bp stem and a 3T loop. The mini-insert as a whole is (G+C)-rich (51% vs. 18% for the entire genome and 24% for the neighboring regions) and so conforms to the definition of the "(G + C) clusters" described by Prunell and

Bernardi (1977), but contrary to prediction, it lies within a gene and, being dispensable or "optional," is not a fundamental unit of the mitochondrial genome.

The major intron contains, in the same DNA strand on the 21S rRNA gene itself, an uninterrupted coding frame (URF) starting with an AUG codon, ending with a UAA, and able to specify a 235 amino acid polypeptide. The hypothetical protein contains a relatively high proportion of basic amino acids (especially lysine) and shows a bias in favor of A or U in the third position of the codon, in good agreement with other mitochondrial genes (Coruzzi and Tzagoloff 1979; Macino and Tzagoloff 1979a,b; Hensgens *et al.* 1979). There is so far no evidence for the existence of such a protein, but the presence of such a long open reading frame in only one out of the six possible frames is very unlikely for a random DNA sequence. The hypothetical protein must obviously have a dispensable function and could conceivably act in the $\omega$ polarity mechanism. Another possibility is that it is a "fossil" gene that is no longer expressed.

It has recently been confirmed that the split gene is transcribed into a precursor RNA from which the intron-specified sequence is excised (Bos *et al.* 1980), splicing occurring at the exon-intron boundaries. Dujon's sequence data (Dujon 1980) make it possible to speculate on those sequence elements of the gene that are presumably necessary for recognition and accurate cleavage and ligation at splice points. Optimization of RNA–RNA base pairing within the intron and adjacent exon sequences of the unprocessed transcript give a variety of secondary structures that may be important in enzyme recognition and splicing, though all the hypothetical structures have relatively low stabilities. Another possibility is that splicing of the 21S rRNA intron occurs after the assembly of the large subunit of the mitochondrial ribosome, in which case interactions with other ribosomal elements such as the ribosomal proteins could contribute to the splicing process.

The complete sequence of the *rrnB* gene of *Escherichia coli* coding for the large rRNA is now available (Brosius *et al.* 1980) and shows regions of important homology to the exon sequences published by Dujon, as does the sequence of the large rRNA gene of the *Chlamydomonas reinhardtii* chloroplast (which also has an intron) (Allet and Rochaix 1979) and the 23S rRNA gene of tobacco chloroplasts (Takaiwa and Sugiura 1982). As Dujon comments, the high sequence homology observed between exons of evolutionary distant species and the fact that the two introns are inserted within the most highly conserved regions, but at two different positions, suggests that these introns were introduced within preexisting exons. It does seem that large ribosomal RNA introns (when known) are always located in the 3′ end of the molecule, which, in the case of *E. coli* at least, appears to be at the ribosome interface (Brosius *et al.* 1980). If splicing does occur after incorporation of the rRNA into the ribosomal subunit, then it is probable that introns at the 3′ end would be more accessible and readily excisable, so accounting for their location at this end of the molecule only.

The high degree of sequence homology found between the 3′ ends of the large rRNAs of phylogenetically distant species suggests that the sequence of this region of the gene is under strong constraints with respect to the function of the gene product. This is also attested to by Dujon's finding that the $\omega^n$ and two *cap*$^R$ *(rib1)* mutations result from single base substitutions. Thus dramatic effects such as suppression of recombinational polarity and conferring of chloramphenicol resistance on the mitoribosome can proceed from comparatively slight mutational change in this highly conserved gene segment.

*Gene Product: The 15S Ribosomal RNA of the Small (37S) Mitoribosome Subunit*

*Map position:* 36–38 3 *u*.

*Associated genetic loci: par1 (par$^R$)* mutants are resistant to the antibiotic paromomycin and map genetically in the region of the small rRNA gene (Schweyen *et al.* 1978). However, $\rho^-$ mutants can retain *par1* but apparently lose all sequences able to hydridize to 15S RNA (Faye *et al.* 1976), so the exact status and location of the *par1* locus remain uncertain, though recent sequencing work in Tzagoloff's laboratory may clarify the situation (see Slonimski *et al.* 1982)).

*Sequence:* Determined by Sor and Fukuhara (1980).

*Coding sequence:* Since the 15S rRNA has not itself been sequenced, the precise boundaries of the gene are not definitely known, but because there are many stretches that are clearly homologous to the established sequence of *E. coli* 16S rRNA the likely beginning and end points of the gene can be located with fair accuracy. The gene is therefore estimated to be 1659 bp long and is apparently continuous (Tabak *et al.* 1979). It is found to contain many (A + T)-rich sequences as well as many short inverted repeats of various lengths capable of generating a number of possible secondary structures through intramolecular base pairing.

In particular, a secondary structure model can be constructed that is strikingly similar to those proposed by Woese *et al.* (1980) and Stiegler *et al.* (1980) for *E. coli* 16S ribosomal RNA. The mitochondrial sequences homologous to *E. coli* 16S ribosomal RNA are often found in the base-paired regions of this model.

The middle of the ribosomal RNA gene contains a (G + C)-rich sequence of about 40 bp that is found in many regions of the mitochondrial genome.

The gene does not contain a typical Shine-Dalgarno sequence (Shine and Dalgarno 1975) that, in prokaryotes, is apparently involved in the recognition of mRNA initiation sites.

Sequencing work in Tzagoloff's laboratory (M. Li, K. Lyon, N. Martin, and A. Tzagoloff: see Slonimski *et al.* 1982) has shown that paromomycin resistance is caused by a small change (loss of a ThaI site) about 200 nucleotides from the putative 3′ end.

*Flanking regions:* A (G + C)-rich sequence similar to that found in the middle of the gene also occurs just before and just after the rRNA gene. The (G + C)-rich sequence beyond the 5′ end is practically identical, over 115 bp to the sequence found near the *oli1* gene for ATPase subunit 9 (Nobrega and Tzagoloff 1980).

## tRNA Genes and the Mitochondrial Genetic Code

Chromatographic analyses of aminoacylated tRNAs have indicated that yeast mitochondria contain 31–35 separable tRNA species, there being at least one species for each of the 20 amino acids (Casey *et al.* 1972; Martin and Rabinowitz 1978; Martin *et al.* 1978; Tzagoloff *et al.* 1978). Some of these [e.g., the four isoaccepting histidine tRNAs (Berlani *et al.* 1980a)] arise from posttranscriptional modification, so that this figure is an overestimate of the actual number of tRNA genes. Hybridization of labeled tRNAs or aminoacylated tRNAs to mitochondrial DNA indicated the existence of 22–23 tRNA genes with a highly nonuniform distribution on the genome (Grivell *et al.* 1979; Wesolowski and Fukuhara 1979), 17–19 of the tRNA genes occurring in the genome segment between the 21S rRNA gene and the *oxi2* locus (encoding COIII).

There are currently considered to be 25 tRNA genes, two each for the amino acids arginine, serine, threonine, and tryptophan and a single gene for each of the remaining amino acids (see Dujon 1981). The DNA sequences of 23 of the genes are known (Bonitz *et al.* 1980b), 22 of the sequences (i.e., excepting $tRNA^{Ile}$, $tRNA^{Tyr}$, and $tRNA_2^{Trp}$ and including some primary tRNA sequences), being found in the following publications: Berlani *et al.* 1980a,b; Bonitz and Tzagoloff 1980; Bos *et al.* 1979; Canaday *et al.* 1980b; Coruzzi *et al.* 1981; Li and Tzagoloff 1979; Martin, N. C., *et al.* 1979a, 1980a,b; Martin, R. P., 1978, 1980; Miller *et al.* 1979, 1980; Newman *et al.* 1980; Nobrega and Tzagoloff, 1980a; Tzagoloff *et al.* 1980).

In agreement with the hybridization studies, the majority of the genes (19) lie in the *rib-oxi2* genome segment (~0–20 u). Four ($tRNA^{Met}$, $tRNA^{Pro}$, $tRNA^{Trp}$ 1,2) are scattered in the 20–50 u genome quadrant, while one each ($tRNA^{Glu}$ and $tRNA^{Ser}$) occur in the remaining two quadrants of the genome. The tRNA genes in the major cluster, in particular, tend to be embedded in highly (A + T)-rich sequences. The genes in the cluster are separated by variable lengths of DNA from 3 bp [$tRNA_2^{Ser}$ (anticodon: 3′-UCG-5′) and $tRNA_2^{Arg}$ {GCA}; Bonitz and Tzagoloff 1980] upward [e.g., 85 bp between $tRNA^{Cys}$ {ACG} and $tRNA^{His}$ {GUG}; Bos *et al.* 1979]. The genes are mostly 70–80 bp long and lack coding sequences for the 3′-CCA tRNA terminus, which is presumably added posttranscriptionally. Unlike some of the yeast nuclear-encoded tRNA genes, the mitochondrial tRNA genes are not split (Johnson *et al.* 1980). The mitochondrial tRNAs have a low (~18–35%) (G + C) content but tend to exhibit the invariant bases and secondary structures generally found in eukaryotic and prokaryotic tRNAs. Thus the tRNAs can be folded into cloverleaf structures with typical T$\psi$C, D, and anticodon loops and acceptor stem. Anomalies occur, however, as in the aspartic acid tRNA, which has unusual D and T$\psi$C loops (Bonitz and Tzagoloff 1980), though these are identical with those of the formylmethionine tRNA of *Neurospora crassa* mitochondria (Heckman *et al.* 1978).

Although sequencing of protein-coding genes has shown that there is a bias for A and U in the third position of amino acid codons (discussed later), most codons appear to be used (see Bonitz *et al.* 1980). This implies that with only 24 or 25 tRNAs/anticodons available, even more bending of base-pairing rules must occur during yeast mitochondrial codon–anticodon interaction than are envisaged in the wobble hypothesis (Crick 1966), which requires a minimum of 32 tRNAs/anticodons.

All tRNAs pertaining to the unmixed codon families (UCH, CCN, GCN etc.) appear to have U in the wobble position of the tRNA anticodon. The implication is that a single tRNA can read all four codons of a given family. One exception to this rule appears to be the arginine tRNA with a 3′-GCA-5′ anticodon, but as none of the 5′-CGN-3′ arginine codons seem to be used in the mitochondrial protein genes (apart from rare occurrence in intron URFs), the tRNA may be functionless (AGA and AGG arginine codons being recognized by the 3′-UCU-5′ anticodon of the second arginine mit tRNA). This two-out-of-three simplified reading method has been discussed by Lagerkvist (1981), who points out that there needs to be reasonably strong base pairing between the bases in the first two positions of the codon–anticodon interaction, which can be, and indeed is, achieved by the presence of at least one G — C base pair.

Most mixed families are read by two tRNAs, so tRNAs recognizing codons ending in U or C have a G in the appropriate anticodon position, thereby permitting normal G — C or wobble G — U base pairing. tRNAs reading codons ending in A or G always have a U in the wobble position of the anticodon. Presumably, there is some mechanism to ensure that these tRNAs do not read all the codons of the mixed family, as do those tRNAs with U in the wobble position, reading all codons of the unmixed families.

One family outside this general pattern for mixed families is the AUN series for isoleucine and methionine. The wobble position of the cognate isoleucine tRNA anticodon is a G (3′-UA*G*-5′), and this is probably modified so that it can pair with all three standard isoleucine codons AUU, AUC, and AUA. This contrasts with the situation in mammalian mitochondria where AUA is apparently recognized as a methionine codon (Barrell *et al.* 1979). The methionine tRNA has a C in the wobble position of the anticodon and so reads only the AUG codon. The yeast mitochondrial formylmethionine (initiator) tRNA also has a C in the wobble position of the anticodon (i.e., the same anticodon UAC as the methionine tRNA). Yeast $tRNA^{fMet}$ has a couple of prokaryotic features not found in the *N. crassa* equivalent—lack of a Watson-Crick base pair at the end of the acceptor stem and a T–ψ–C–A sequence on loop IV—but it cannot be formylated by the *E. coli* transformylase and requires its own mitochondrial enzyme (Canaday *et al.* 1980b).

As has been previously mentioned, yeast mtDNA parts company with the "universal" genetic code, differing from it in two respects. First, the "universal" UGA "opal" terminator is recognized as a tryptophan codon by the product of a recently sequenced $tRNA^{Trp}$ gene (Martin *et al.* 1980b). Second, CUA (and indeed probably the entire CUN family) is recognized as a threonine rather than a leucine codon by a threonine tRNA with a GAU anticodon (Li and Tzagoloff 1979). This is not the case in the mammalian mitochondrial genetic code, where CUN still encodes leucine. In addition to the difference in meaning of the CUA codon just mentioned, the codons AGA and AGG are terminators in the mammalian mitochondrial genome but code for arginine in the yeast mitochondrion (Barrell *et al.* 1980; Dujon 1981).

The distribution of tRNA genes is another major point of contrast between the yeast and mammalian mitochondrial genomes; in yeast there is a striking clustering, whereas in mammals the genes occur regularly between other types of gene, possibly as some form of "punctuation" required during posttranscriptional processing (see earlier section on mammalian mtDNA). The close grouping of 19 of the yeast mt tRNA genes suggests that they may be transcribed as a single unit, especially as in at least one case the genes are separated by as little as 3 bp, thus leaving insufficient room for a promoter sequence for the downstream gene, unless it lies mainly within the structural sequence of the upstream gene. Indeed, it has been found that in some $\rho^-$ mutants retaining the main tRNA gene cluster only, a large transcript is produced from the cluster region but no mature tRNAs are formed (Morimoto *et al.* 1979). If, however, the $\rho^-$ genome includes the *par1* locus as well as the tRNA gene cluster, mature tRNAs are formed, suggesting that a mtDNA-encoded transcript from the *par1*/15S rRNA region is required for processing a tRNA precursor transcript into individual tRNAs.

All yeast mt tRNA genes are certainly transcribed from the same strand and in the same direction *(rib → par1)*, with the one interesting exception of the gene for the threonine tRNA with the leucine (3′-GAU-5′) anticodon: This is transcribed in the opposite direction from the other strand (Li and Tzagoloff 1979). This raises the intriguing question of whether transcription of yeast mtDNA parallels the complete symmetrical transcription of, for example, the HeLa mitochondrial genome to a greater extent than is usually assumed.

To reiterate, the most remarkable feature of the yeast mitochondrial tRNA genes is that there are so few of them. It is not yet possible to say whether this condition and the peculiar characteristics of the mitochondrial genetic code are more primitive or more highly evolved than other systems. However, Tzagoloff and colleagues speculate that it is the very limited number of mitochondrial protein products that has permitted such an extreme simplification of the decoding mechanism, tRNAs in other translation systems often having important secondary roles in, for instance, gene regulation (Bonitz *et al.* 1980b).

## Simple Protein-Coding Mitochondrial Genes

### *Gene Product: Subunit 2 of Cytochrome Oxidase (COII)* mol. wt. 28,480.

*Map position:* 14.2–15.3 u.

*Associated genetic locus: oxi1.*

*Sequence:* Determined by Coruzzi and Tzagoloff (1979) and Fox (1979a) with additional flanking sequences by Coruzzi *et al.* (1981).

*Coding sequence:* The COII coding sequence is 756 nucleotides (251 codons) long, starting with an AUG initiation codon at nucleotide +1 and ending with a UAA terminator at +756. The reading frame is continuous, has a (G + C) content of 27%, and can be identified as the COII gene sequence because it encodes a polypeptide consistent in amino acid composition and molecular weight with what is known of the yeast COII protein (Poyton and Schatz 1975; Poyton *et al.* 1978). The amino acid sequence of the yeast COII protein, as deduced from the DNA sequence, has extensive (49%) homology with the amino acid sequence of subunit 2 of bovine cytochrome oxidase, as determined on the protein itself (Steffens and Buse 1979).

*Codon usage:* Comparison of the sequences of the presumptive yeast gene and the amino acid sequence of the bovine protein confirmed that UGA, an "opal" stop codon in the "universal" code, encodes tryptophan in the yeast mitochondrial genome (Macino *et al.* 1979; Fox 1979b), as it does in the mammalian genome (Barrell *et al.* 1979). The *oxi1* gene also has a high preference for codons terminating in an adenosine or uridine. Only 19 out of the 251 codons have a cytidine or guanosine in the third position, and of these, 5 correspond to AUG, an obligatory codon for methionine. Certain codons, such as CGN and CUN series for arginine and leucine, are not used at all.

*Flanking regions:* The COII gene is preceded by an (A + T)-rich (91%) stretch of 291 nucleotides, within which lies a (G + C)-rich cluster (60% G + C) of 59 nucleotides (−291 to −232). Likewise, the approximately 100 nucleotide sequence following the ochre terminator of the gene is rich in (A + T) (97%).

*Gene Product: Subunit 3 of Cytochrome Oxidase (COIII),* mol. wt. 30,340.

*Map position:* 19.9–21.0 u.

*Associated genetic locus: oxi2.*

*Sequence:* Determined by Thalenfeld and Tzagoloff (1980).

*Coding sequence:* The gene sequence starts with an AUG initiator and continues for 807 nucleotides before ending with two UAA ochre terminators. There are no introns. The reading frame includes nine UGA tryptophan codons and encodes a hydrophobic polypeptide of 269 amino acids. The amino acid composition of the presumptive gene is in fair agreement with analyses of the purified yeast protein (Thalenfeld and Tzagoloff 1980) and of the homologous *N. crassa* (Sebald *et al.* 1973) and bovine (Steffens and Buse 1976) proteins. The molecular weight deduced from the DNA sequence (30,340) is 20% higher than previous estimates determined by SDS-gel electrophoresis of the protein (Poyton and Schatz 1975) but is within the range of errors inherent in molecular-weight determinations of hydrophobic proteins by SDS-gel electrophoresis, which tend toward underestimation (Poyton *et al.* 1978).

*Codon utilization:* As in other mitochondrial genes, UGA apparently codes for tryptophan and CUN for threonine. Only 28 out of the 269 protein codons terminate in a G or C. The CGN series for arginine and GCG for alanine are not used.

*Flanking regions:* The region upstream of the gene is (A + T)-rich (5% G + C) and that downstream averages 16% (G + C). The entire sequenced DNA segment (a $\rho^-$ genome of approximately 4 kbp) contains a number of short (extragenic) sequences with a high (G + C) content (70%). These sequences contain HpaII and HaeIII site clusters and are of a type first recognized by Bernardi and colleagues (Bernardi *et al.* 1976). They are of unknown function and appear to occur randomly in many different regions of yeast mtDNA (Hensgens *et al.* 1979; Prunell and Bernardi 1977; Cosson and Tzagoloff 1979); Tzagoloff *et al.* 1980).

*Gene Product: Subunit 6 (Not Yet Confirmed) of the Oligomycin-Sensitive Mitochondrial ATPase,* mol. wt. 28,257.

*Map position:* 61.5–62.6 u.

*Associated genetic loci: oli2, oli4,* (*oli*$^R$ = oligomycin resistance), *oss1* (oss$^R$ = ossamycin resistance), *pho1* (*pho*$^-$ = phosphorylation deficiency).

*Sequence:* Determined by Macino and Tzagoloff (1980).

*Coding sequence:* The oligomycin resistance-transmitting *petite* genome sequenced by Macino and Tzagoloff (1980) contains a sequence in only one out of the six possible reading frames that is likely to correspond to this gene. The coding sequence of 780 nucleotides is not interrupted, starts with a formylmethionine AUG initiator, and ends with a UAA ochre terminator. The sequence has a (G + C) content of 22% and encodes a protein of 259 amino acid residues with a molecular weight of 28,257. The identity of this protein is still uncertain, because only one of the four (subunits 3, 6, 8, and 9) apparently mitochondrially synthesized yeast ATPase proteins has been purified and characterized (subunit 9, the proteolipid). On the basis of SDS–polyacrylamide gel electrophoresis molecular-weight determinations (Tzagoloff and Meagher, 1971) the *oli2* gene could encode either of the larger mitochondrially synthesized subunits, 3 (mol. wt. ~29,000) or 6 (~22,000), the molecular weight of the smaller protein (subunit 6) probably being underestimated. The amino acid composition bears similarities to that of the *N. crassa* ATPase subunit corresponding to the yeast subunit 6 (Sebald 1977); in addition, *mit*$^-$ mutations in the *oli2* region affect the 22-kd protein (Roberts *et al.* 1979), both of which findings support the identification of *oli2* as the gene for ATPase subunit 6. Certainly, the *oli2* gene encodes a protein of highly hydrophobic composition, with the protein containing three stretches of more than 30 completely nonpolar amino acid residues. The relatively few polar amino acids occur in several clusters. Macino and Tzagoloff (1980) suggest that the *oli2* gene product is an intrinsic membrane protein, with the three hydrophobic domains folded in loops spanning the lipid bilayer, with the polar amino acid clusters projecting from the membrane surface.

*Codon utilization:* There is a preferential use of codons terminating in A or

U in the third position, less than 10% of the 259 amino acid codons terminating in C or G. All five tryptophans in the protein are UGA-encoded.

*Mutations:* Macino and Tzagoloff (1980) sequenced the *oli-2* locus in three different *petite* genomes bearing wild-type, *oli2,* and *oli4* alleles, respectively. Comparison of the sequences suggested that the *oli2* resistance mutation occurred at nucleotide +513 and the *oli4* mutation at +696. Both mutations resulted in a single base change at the third position of a codon, substituting methionine for isoleucine at amino acid residue 171 in the case of *oli2* and phenylalanine for leucine at residue 232 in the case of *oli4*.

*Flanking sequences:* Long A + T-rich sequences occur both upstream and downstream of the *oli-2* gene. A short (G + C)-rich HpaII, HaeIII site cluster occurs downstream at about +1128 to +1152.

*Gene product: Subunit 9 of the Mitochondrial ATPase (the ATPase Proteolipid or the dicyclohexylcarbodiimide {DCCD}-binding Protein)* mol. wt. 7751.

*Map position:* ~81 u.

*Associated genetic loci: oli1, oli3, oli5* (*oli*$^{R}$ = oligomycin resistance), *ven1* (*ven*$^{R}$ = venturicidin resistance), *oss2* (*oss*$^{R}$ = ossamycin resistance), *pho1* (pho$^{-}$ = phosphorylation deficiency).

*Sequence:* Determined by Hensgens *et al.* (1979), Macino and Tzagoloff (1979a,b), and Tzagoloff *et al.* (1980).

*Coding sequence:* The proteolipid amino acid sequence is encoded by a continuous sequence of 228 nucleotides, starting with an AUG initiator codon and terminating with a UAA stop codon. The DNA sequence corresponds almost completely with the amino acid sequence of the protein from a strain that carried oligomycin resistance at this locus (Sebald and Wachter 1978, and see Sebald *et al.* 1979b), with the ambiguity probably being due to an error in the amino acid sequence. The gene sequences of DS401, the strain used by Tzagoloff and colleagues, and RP6, the strain used by Hensgens and co-workers, are identical.

*Codon utilization:* As usual in yeast mitochondrial genes, (A + T)-rich codons are preferred, and the third position of the codon is usually taken by A or T.

*Flanking sequences:* Extensive stretches—100 nucleotides or more—of (A + T)-rich sequences occur at both 5′ and 3′ ends of the gene. The sequences of these (A + T)-rich regions differ somewhat in the two strains/genes that have been sequenced, but as much of this DNA appears to be functionless, this is perhaps not surprising. In the *petite* genome (513-kbp repeat unit) used by the Tzagoloff laboratory for their sequencing work, there are 10 short, apparently partially homologous, sequences averaging 70% of (G + C). They appear to be similar to the (G + C)-rich clusters found elsewhere in the genome (Cosson and Tzagoloff 1979). Inverted repeated sequences occur in some of the clusters, and Tzagoloff and colleagues have proposed (Tzagoloff *et al.* 1980) that they may be important in RNA processing. Most yeast mitochondrial transcripts are much larger than the final messengers (Borst and Grivell 1978;

Levens *et al.* 1979), and considerable processing or maturation of mitochondrial transcripts therefore occurs. Inverted repeats lying outside polypeptide-coding sequences could permit looping or folding of the transcripts into hairpin structures with double-stranded regions that could be recognized and cleaved by double-stranded RNA-specifid RNases.

## Complex Protein-Coding Mitochondrial Genes

*Gene Product(s): (i) Subunit I of Cytochrome Oxidase (COI)* mol. wt. 55,933.

(ii) It is possible that one or more of the several open reading frames in introns of this gene are expressed as protein products and that some of these have RNA splicing activity. There is as yet no direct evidence for the existence of such polypeptides in wild-type cells.

*Map position:* ~41–58 u (this varies, depending on the strain, number of introns, and hence overall gene size).

*Associated genetic loci: oxi3* (*oxi3*$^-$ mutants are respiratory deficient); *mim2* (*mim2* mutants are locus specific, allele nonspecific suppressors of *box7*$^-$ mutants, a class of mutants in the *cob-box* cytochrome b gene).

*Sequence:* Determined by Bonitz *et al.* (1980a).

*Coding sequence:* On SDS–polyacrylamide gel electrophoresis subunit I of yeast cytochrome oxidase migrates with an apparent molecular weight of 40,000 (Rubin and Tzagoloff 1973), but physical and genetic mapping has shown that the *oxi3* locus is much longer than the approximately 1100 nucleotide stretch that would be required to encode such a protein. These considerations, together with the locations or complementation behavior of *oxi3*$^-$ mutants, have indicated that the COI coding sequence is split into several exons, so constituting a "mosaic" gene. The approximately 10-kb *oxi3* region of strain D273-10B contains six open reading frames that together account for the major part of the sequence. Potential protein-coding sequences therefore greatly exceed what should be necessary to encode COI. Since the amino acid sequence of yeast COI has not been determined, the Tzagoloff group used the sequence of the human COI gene (Anderson *et al.* 1981) to locate the likely yeast COI-coding exons within the extensive open reading frames of the *oxi3* locus. On the basis of homology with the mammalian gene the yeast gene comprises seven to eight exons separated by six to seven introns. The uncertainty about the seventh intron arises from the somewhat low homology of the two proteins at the carboxy-terminal end. The seventh and last intron as defined by the Tzagoloff group could indeed encode a small insertion of amino acid residues with no homologue in the human protein. However, codon usage in this sequence favors the intron possibility.

The coding sequence starts with an ATG at nucleotide +1 and ends with the last nucleotide of the TAA ochre terminator at +2279. Thus the *oxi3* locus is nine times longer than the expected length of the gene. This yeast gene, as deduced from the homological approach, is 1533 nucleotides long, encodes 510 amino acid residues averaging 50% homology with the human protein, and has a composition consistent with analytical data for the actual protein (Poyton and Schatz 1975). The molecular weight of 55,933 is some 30% higher than estimated based on SDS–gel electrophoresis of the protein (Mason *et al.* 1973), and although this discrepancy is probably a question of technique (see above), the possibility cannot be excluded that a COI precursor protein is posttranslationally processed to a smaller mature protein.

*Introns:* The exact number of introns in D273-10B—either six or seven ($aI1$-$7_s$)—remains unknown and awaits precise definition of intron-exon boundaries through, for example, mRNA hybridization studies. Several of the *oxi3* exons are part of longer reading frames that extend into or through introns and could encode proteins of 30–80 kd.

The first four exons are of this nature. The longest reading frame starts at nucleotide +169, following the last codon of the first exon and ends at +2502 followed by an ochre terminator. Excluding the COI exon this frame codes for an additional 784 amino acids. The second intron possesses another long reading frame, and introns $aI3_s$ and $aI4_s$ also possess substantial potentially coding sequences. All these sequences seem to be too long to be fortuitous, especially as ochre and amber termination codons occur frequently when they are read in the other two registers. The intron reading frames have a lower (G + C) content than the exons, which means that any protein products would be unusually rich in lysine, asparagine, and tyrosine (i.e., amino acids with either A and/or U in the first 2 codon positions). The intron sequences would also generate proteins more polar than COI. Interestingly, the first and second introns ($aI1_s$ and $aI2_s$) have extensive sequence homology and their reading frames would encode proteins identical in 50% of their amino acid residues. Intron $aI5_s$ is (A + T)-rich and contains stop codons. $aI6_s$ and $aI7_s$ are very short, lack stop codons, and potentially encode 7 and 15 amino acid residues, respectively; poor homology with the human protein in this C-terminal region renders these assignments and exon–intron boundaries uncertain.

The gene organization of *oxi3* varies in a strain-dependent way showing, essentially, a size polymorphism: gene size is determined by the number of inserts or introns present (see Grivell *et al.* 1980). In this context D273-10B is a "short" strain, whereas KL14-4A (see Borst *et al.* 1977), a "long" strain, contains an extra two long introns within the equivalent of the fifth exon.

Curiously, no *oxi3*$^-$ mutants are known to occur in the introns, in strong contrast to the case of the *cob-box* gene. The explanation may be that processing of the *oxi-3* precursor RNA does not require the intervention of a protein splicing factor ("maturase" or "spligase") encoded by one (or more) of its own introns, so differing from the situation apparently obtaining in the *cob-box* gene. Normally, any requirement by *oxi3* for a mitochondrial intron-encoded splicing factor is supplied by the product of a *cob-box* intron ($bI1_s$/$bI4_l$) accounting for the "box" effect, in which *cob-box* intron mutations can block processing of *oxi3* transcript but not vice versa (see the following section on *cob-box*). It must therefore be hardly coincidental that the fourth *oxi3* intron, $aI4_s$, has a striking homology with an intron present in both long and short *cob-box* genes (bI$1_s$/bI$4_l$, see below): the reading frames of these two introns can be aligned over 700 nucleotides and would give proteins that are 70% homologous, the codons differing especially in position 3 of the codon. An obvious if speculative interpretation is that both these essentially similar introns located in different genes specify homologous "maturases"—the *oxi3* aI$4_s$ maturase happens to be inactive or ineffective, so processing of *oxi3* transcripts is dependent on the homologous *cob-box* intron maturase—hence the "box" effect (see following discussion of cob-box introns).

*Codon usage:* In the COI exons as in other yeast mitochondrial genes, UGA encodes tryptophan and CUN, which "universally" encodes leucine, apparently encodes threonine. In addition, codons with A or U in the third position are used preferentially.

The intron reading frames utilize codons that are either absent or infrequent in mitochondrial genes known to give rise to protein products. Thus there are six arginine codons of the CGN series in the first three introns, none of which codons have been found in established yeast mitochondrial genes. Similarly, the established yeast mitochondrion protein-coding genes very rarely use AUA for isoleucine, whereas the intron reading frames use AUA and AUU almost equally. These biases of codon assignment have also been seen in introns of the *cob-box* apocytochrome b gene (see below).

*Intron–exon boundaries and Flanking Sequences:* The sequences at the intron–exon boundaries within both this gene and the mosaic gene (short version) encoding apocytochrome b have been compared (Bonitz *et al.* 1980a). The only significant homologies near presumed splice points have been found between the first two introns of *oxi3* and between the fourth *oxi3* intron and the first intron of the short apocytochrome b gene. All other boundaries appear to have unique sequences.

Netter and colleagues (Netter *et al.* 1982) have recently determined the sequences of two *cis*-dominant mutations in aI4 that block *oxi3* transcript processing. Both mutations affect a very short DNA sequence several hundred bp from intron–exon junctions. An identical sequence is found in the *cob-box* bI$1_s$/bI$4_l$ intron and is critical for excision of the intron sequences. It is therefore possible that this short sequence is a common signal, recognized by the *box 7* mRNA maturase, to splice out introns in both *cob-box* and *oxi3* (see following section on *cob-box*).

The sequence upstream of the first COI exon is highly (A + T)-rich, for at least 1 kbp. Similarly, there is a long (A + T)-rich sequence downstream of the *oxi3* gene (Bonitz *et al.* 1980a).

*Gene Product(s): (i) The Apoprotein of Cytochrome b,* mol. wt. 44,000.

(ii) "Maturases": open reading frames in introns (the number is strain dependent) may be expressed (linked to exon polypeptides) as "fusion" proteins with a role in transcript processing or maturation. There is not yet any direct evidence for the existence of such "maturases" in wild-type cells.

*Map position:* ~71–76 or 81 u (as in *oxi3* the exact position and size

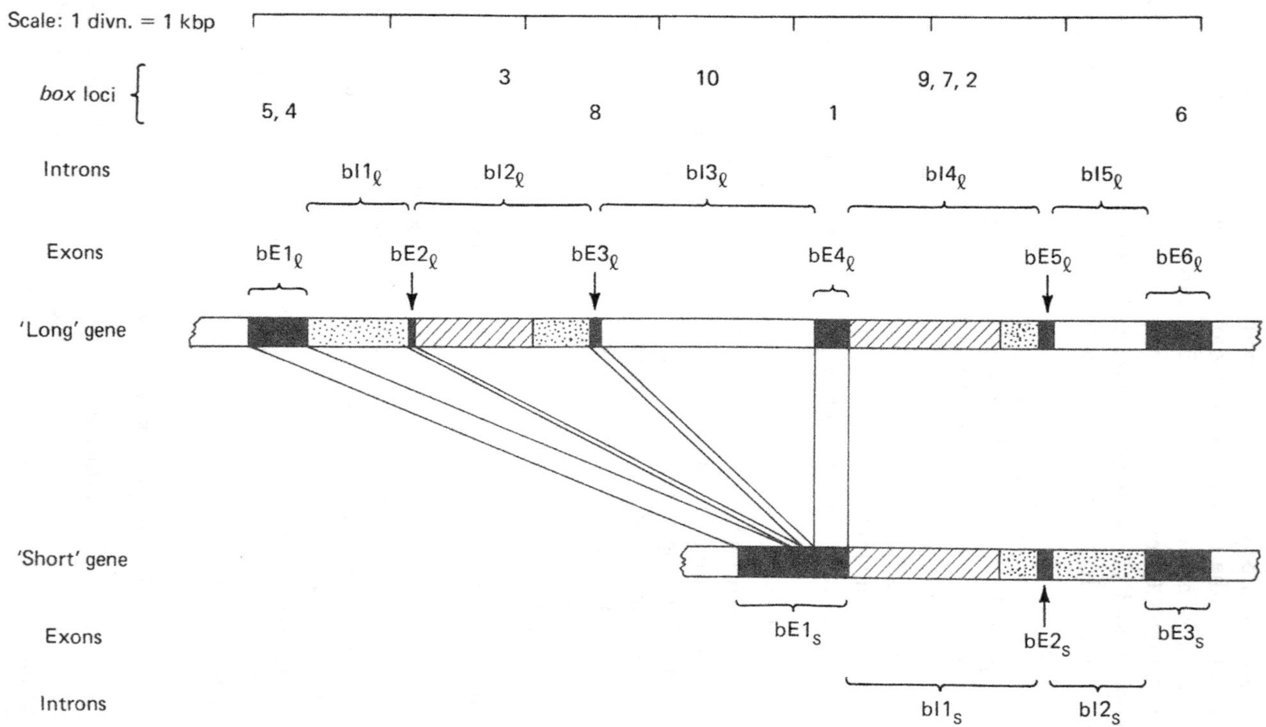

Scale: 1 divn. = 1 kbp
box loci
5, 4
3
8
10
1
9, 7, 2
6
Introns
bI1ℓ
bI2ℓ
bI3ℓ
bI4ℓ
bI5ℓ
Exons
bE1ℓ
bE2ℓ
bE3ℓ
bE4ℓ
bE5ℓ
bE6ℓ
'Long' gene
'Short' gene
Exons
bE1s
bE2s
bE3s
Introns
bI1s
bI2s

depends on the number of introns and hence size of gene, this varying between "short" strains and "long" strains).

*Associated genetic loci:* A series of respiratory-deficient mutants mapping at this locus was isolated from "short" strains such as D273-10B, and termed *cob*$^-$ (cytochrome b deficient; Tzagoloff *et al.* 1975a,b). Subsequently, *mit*$^-$ mutants of this locus were isolated from "long" strains. These showed more complex phenotypes and genetic behavior and were termed *box*$^-$, as some of the mutants were pleiotropic, affecting cytochrome oxidase as well as cytochrome b (Cobon *et al.* 1976; Kotylak and Slonimski 1976,1977; Pajot *et al.* 1976,1977). A total of 10 *box* loci has been identified in the case of the long gene *(box1–10)*. The gene in all versions is now usually referred to as *cob-box*. A number of antibiotic resistance mutations of this gene have also been isolated and map at various loci: *diu1, diu2* (diuron-resistant); *muc1–muc3* (mucidin-resistant); *ana1, ana2* (antimycin-resistant); *fun1* (funiculosin-resistant).

Schematic maps of "short" and "long" *cob-box* genes are given in Figure 11.2.

*Sequence:* The complete sequence for the "short" *cob-box* gene of a "short" strain (D273-10B) has been determined by Nobrega and Tzagoloff (1980b). The *cob-box* intron $bI2_\ell$ (corresponding to the *box 3* locus) and flanking exonic regions of a "long" wild-type strain (777-3A), and of *box 3*$^-$ mutant derivatives have been sequenced by Lazowska and colleagues (Lazowska *et al.* 1980). Also, the sequences of the intron $bI4_\ell$ (the equivalent of the first intron, bI1, in "short" genes/strains) and adjoining regions, in wild type (777-3A) and mutant (*box 1*, *9*, *7* and *2*) "long" strains have recently been established by de la Salle and co-workers (de la Salle *et al.* 1982).

*Coding sequence:* As in the case of *oxi3* a great deal of genetic and physical mapping evidence indicated, before the application of DNA sequencing techniques, that the *cob-box* gene is split (Slonimski *et al.* 1978a; Alexander *et al.* 1979; Haid *et al.* 1979). Sequencing has confirmed that *cob-box* is indeed multiply split, having a mosaic structure of exons and introns (see references given earlier).

Although cytochrome b has been purified from *S. cerevisiae* (Lin *et al.* 1978) the protein has not been sequenced. However, the bovine and human genes

---

**Figure 11.2.** The structure of "long" and "short" *cob-box* genes. The diagram illustrates the structures of a "long" *cob-box* gene from a strain such as KL14-4A and a "short" *cob-box* gene from a strain such as D273-10B (see references in text), though the schemes should not be taken as accurate genetic and physical maps. Exons are represented by black bars; introns are either blank (DNA sequence not fully determined), stippled (all reading frames blocked), or hatched (open reading frame). The positions of the 10 clusters of *box* mutations, isolated and mapped in the "long" gene, are indicated by the figures above the "long" gene diagram; upper-level figures are *box* clusters lying in introns, whereas those at the lower level are *box* clusters located in exons. It can be seen that the two genes differ essentially in the presence of three extra introns—$bI1_\ell$, $bI2_\ell$, and $bI3_\ell$—in the long gene; these introns are all inserted in the equivalent of the $bEI_s$ exon of the short gene.

encoding apocytochrome b have been located and sequenced (Anderson *et al.* 1981; Anderson *et al.* 1982), so allowing the homologous yeast sequences—presumed exons—to be identified within the exon–intron mosaic. This identification depends on there being no introns common to both yeast and bovine genes encoding homologous amino acid sequences and no major insertions or deletions of amino acids in the two proteins. On this basis the *cob-box* gene of strain D273-10B is composed of three exons. The first exon ($bE1_s$) starts with an AUG initiator at +1, extends to +759, and encodes 252/253 amino acid residues. The second exon, $bE2_s$, extends from +2174 to +2227 and specifies an 18-amino acid sequence at most. The third exon, $bE3_s$, specifying the C-terminal portion of the protein, begins at nucleotide +2958, ends with an ochre terminator at +3308, and encodes 115–116 amino acid residues. The entire protein is 385 amino acid residues long and is 50% homologous to bovine cytochrome b, based on the sequence of the bovine gene.

The molecular weight of 44,000 as deduced from the yeast gene sequence is substantially larger than estimates for the actual protein, of 30,000, based on SDS–gel electrophoresis (Lin *et al.* 1978), but cytochrome b behaves anomalously in these gel systems, giving a serious underestimate of the true molecular weight (Beattie *et al.* 1979).

*Introns and splicing:* The mosaic structure of the cytochrome b gene is particularly complex, since the number of introns varies, depending on the strain. The *cob-box* gene sequences by Nobrega and Tzagoloff (1980b) shows a simpler "short gene" organization than that of the "long" gene partially sequenced by Lazowska and colleagues (1980). The short gene organization consists, as indicated earlier, of three exons, $bE1_s$, $bE2_s$, and $bE3_s$, separated by two introns $bI1_s$ and $bI2_s$. In long genes $bE1_s$ is fragmented by two major insertions of approximately 2.0 and 1.4 kbp, with the larger intron containing a "mini-exon" of 14 bp (Lazowska *et al.* 1980). $bE1_s$ thus corresponds to $bE1_\ell$ + $bE2_\ell$ (mini-exon) + $bE3_\ell$ + $bE4_\ell$. Separating these four exons are three optional introns, $bI1_\ell$, $bI2_\ell$, and $bI3_\ell$, which have no counterparts in the short gene (see Figure 11.2).

Two introns are known to contain substantial open reading frames (URFs). These are an optional intron, $bI2_\ell$ (Lazowska *et al.* 1980) and the first intron of the short gene $bI1_s$ (Nobrega and Tzagoloff 1980b), which corresponds to the fourth intron $bI4_\ell$ of long genes (de la Salle *et al.* 1982).

The optional intron URF has been sequenced by Lazowska and co-workers (1980) and encompasses the mini-exon present in long genes. This URF is in the same register as the cytochrome b gene, begins with an AUG, extends for 954 nucleotides, ends with two ochre codons, and is capable of specifying a 318 amino acid polypeptide. The other five possible reading frames of this DNA segment all contain amber and ochre codons, at a frequency consistent with an "average" DNA of the same base composition (18% {G + C}). The possibility that such a long reading frame could occur at random is less than 1 in $10^8$, and therefore its existence does not seem to be fortuitous. *box3* mutations, a cluster of mutations that all impair *cob-box* transcript splicing and that are all complementable in *trans*, are apparently located within this URF, which therefore appears to encode a diffusible *trans*-acting product or factor (TAF) involved in splicing or assisting the splicing of *cob-box* mRNA precursors. In an earlier model this *box3 trans*-acting factor was hypothesized to be an RNA with internal intron sequences complementary to the exon-intron boundaries, which by base pairing could therefore align the sequences to be spliced (Church *et al.* 1979; Halbreich *et al.* 1980). Sequencing of the *box3* locus showed that such complementarity did not exist (Lazowska *et al.* 1980) and, by uncovering the URF, suggested that the *trans*-acting factor (TAF) could be a protein. This view is supported by the demonstration that at least some *box3* mutants are nonsense or missense mutations (Lazowska *et al.* 1980) and that the deficiencies of some *box3* mutants can be alleviated by allele-specific gene nonspecific informational suppressors (Dujardin *et al.* 1980a,b) that presumably act through mitoribosomal misreading. The observation that *box3* mutants can be phenotypically suppressed by paromomycin-induced mitoribosomal misreading also suggests that the TAF is a translation product and, furthermore, implies that when functional, this protein is effective at very low concentrations, since exon mutations are not paromomycin-suppressible (Dujardin *et al.* 1980b). This may explain why a protein corresponding to the TAF has not been detected in wild-type cells.

It appears that splicing out of the first intron ($bI1_\ell$) of long genes does not require a mitochondrially encoded TAF, as splicing occurs in *petites* (*petite* mutants are incapable of mitochondrial protein synthesis, see below). This first splicing event joins the first and second exons ($bE1_\ell$ and $bE2_\ell$) and creates an open reading frame running from $bE1_\ell$ into the URF of the second intron ($bI2_\ell$). It is hypothesized that this precursor to the mature cytochrome b mRNA is itself an mRNA for the *(box3)* maturase or TAF, which is therefore a fusion protein composed of that segment of cytochrome b encoded by exons $bE1_\ell + bE2_\ell$, plus the protein encoded by the URF of intron $bI2_\ell$. The sequence encoded by the cytochrome b portion is the N-terminal section of the apoprotein and is very hydrophobic and nonpolar. By contrast, the segment encoded by the intron URF would be highly basic, rich in lysine as well as in tyrosine and asparagine, though poor in arginine. This fusion protein is presumed to be required for the second splicing event, which excises $bI2_\ell$, and joins $bE1_\ell + bE2_\ell$ to $bE3_\ell$, and so has the curious property of annihilating its own mRNA in order to permit formation of mature apocytochrome b mRNA (Lazowska *et al.* 1980).

While a 38.5-kd polypeptide thought to correspond to this protein can be detected in a *box3* mutant from which *oxi3* has been deleted (so removing the obscuring 41–42 kd COI polypeptide) (Groudinsky, quoted in Lazowska *et al.* 1980), it should be emphasized that this fusion protein has yet to be isolated from wild-type cells and its splicing activities demonstrated.

A speculative scheme for the early splicing events in "long" *cob-box* gene expression is given in Figure 11.3.

It is tempting to surmise that introns other than $bI2_\ell$ might engage in a

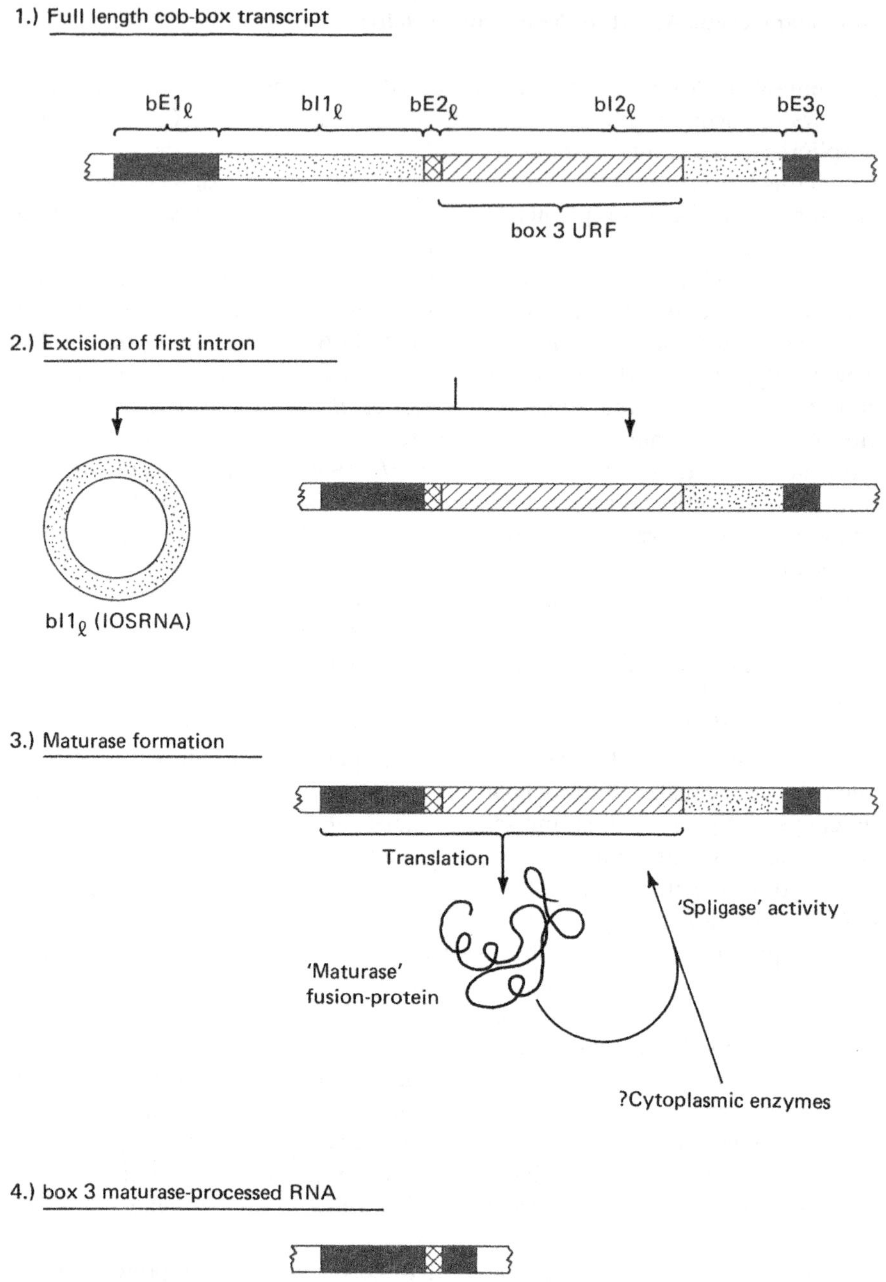


**Figure 11.3.** A model of the initial stages in *cob-box* gene expression (Based on references cited in the text, especially Lazowska *et al.* 1980). (a) Full-length *cob-box* transcript: The entire *cob-box* gene is transcribed into a full-length 7–8-kb precursor. The bar represents (not to scale) the 5′ end of this precursor. The solid and cross-hatched segments are regions of the transcript encoding the apocytochrome b amino acid sequences specified by the first three *cob-box* exons, $bE1_\ell$ (solid), $bE2_\ell$ (cross-hatched), and $bE3_\ell$ (solid). Stippled segments are untranslatable intronic sequences ($bI1_\ell$ and part of $bI2_\ell$), and the hatched section is the open reading frame or URF in

similar form of ingenious and apparently pointless biochemical self-deprecation, but introns such as $bI1_\ell$, the first intron of long *cob-box* genes, clearly do not: This intron sequence contains termination codons in all reading frames and, as already mentioned, its transcribed sequence can be excised in *petites*, so splicing of this intron is quite independent of mitochondrial translation products (see Fig. 11.3).

The large (1700 bp) optional intron $bI3_\ell$ contains the *box10* cluster of *trans*-recessive splicing-defective mutations and so is genetically analogous to the *box3* ($bI2_\ell$) and *box7* ($bI4_\ell$) introns. Partial sequencing by the Slonimski laboratory has revealed that the intron has an open reading frame in the 5′ segment that is uninterrupted for at least 200 bp and that the 3′ end is blocked in all three registers (of the coding strand) for at least 200 bp (Lazowska *et al.* 1981). So in general organization, the possession of a URF at one end and a blocked reading frame at the other, this intron is similar to $bI2_l$ (Lazowska *et al.* 1980) and $bI1_s/bI4_\ell$ (Nobrega and Tzagoloff 1980b; de la Salle *et al.* 1982; and see below). In all likelihood, therefore, $bI3_\ell$ probably also encodes a protein splicing-factor, which could be termed the "*box10* maturase." This yeast intron has some sequence homologies at 5′ and 3′ ends with the single ~1100-bp intron in the apocytochrome b coding gene *(cobA)* of *Aspergillus nidulans* (Waring *et al.* 1981). The mold intron is in exactly the same position in the gene as intron $bI3_\ell$ of the yeast *cob-box* gene. The partial sequence of the *A. nidulans* intron shows that a URF extends for at least 200 bp and that all reading frames are closed at the C-terminus end (Waring *et al.* 1981). The yeast intron is much longer than that of the mold, and it may be significant (in terms of the generality or otherwise of intron functions) that the *Aspergillus cobA* gene lacks introns comparable to $bI4_\ell/bI1_s$ and $bI5_\ell/bI2_s$, the two *cob-box* introns so far found in all yeast strains, long and short. As Lazowska and colleagues point out, experiments effecting exchanges of intron sequences between the two species should of considerable interest.

In the cases of the short-gene intron $bI5_s$ and its long-gene homologue $bI4_\ell$, the sequences are (completely) known (Nobrega and Tzagoloff 1980; de la Salle *et al.* 1982). Both introns contain a long (~1150 bp) reading frame in phase with the preceding exon and ending before the following exon. The two

---

intron $bI2_\ell$, which encompasses the *box3* mutation cluster. (b) Excision of first intron: Cytoplasmic enzyme(s) excise the $bI1_\ell$ sequence, which is released as a single-stranded covalently closed circular 10S RNA species (see Halbreich *et al.* 1980), and ligate $bI1_\ell$ to $bE2_\ell$, so creating the maturase mRNA. (c) Maturase formation: The maturase mRNA is translated to generate a "fusion protein," containing exon- and intron-specified sequences, which, possibly in conjunction with cytoplasmically synthesized enzymes, acts to remove the intron $bI2_\ell$ sequence, so destroying the maturase mRNA. (d) *box3* maturase-processed RNA: The *cob-box* transcript now contains exon sequences $bE1_\ell$, $bE2_\ell$, and $bE3_\ell$, in direct continuity, intron sequences $bI1_\ell$ and $bI2_\ell$ having been removed, and it is ready for further processing into mature apocytochrome b mRNA.

URFs differ by a total of only five base substitutions, which would nevertheless have a significant effect on the amino acid sequences of their putative protein products. Intron $bI1_s$ has an AUG initiation codon one or two amino acids downstream from the first exon, $bE1_s$, and the reading frame that follows could generate a protein of 384 amino acid residues. This hypothetical protein would have an unusual composition, with a very high content of lysine, tyrosine, and asparagine, recalling the putative protein encoded by $bI2_\ell$. Again, as in the case of $bI2_\ell$, $bI4_\ell$ contains a cluster of splicing defective mutations, *box7*. Like *box3* (and *box10*) mutants these can be complemented in *trans*, once again implying the existence of a diffusible intron-encoded product—presumably the URF polypeptide or a protein containing it—necessary for splicing. In contrast with the *box3* product the *box7* TAF (or "spligase" or "maturase") has a wider role, since it is required for the processing of *oxi3* transcripts as well as those of its own gene, *cob-box* (Dhawale *et al.* 1981; Hanson *et al.* 1979). This nonreciprocal interaction between the two genes underlies the "box" effect, in which certain *mit*$^-$ mutations mapping in *cob-box* were found to suppress the expression of COI as well as apocytochrome b (see Dujon 1981 and the section on *oxi3* above). In this context it may be significant that the protein sequences encoded in the URFs of the *cob-box* $bI1_s/bI4_\ell$ intron and the *oxi3* aI4 intron exhibit greater than 50% homology, as do the nucleotide sequences of the adjacent exon–intron boundaries (Bonitz *et al.* 1980a; and see section on *oxi3* above). The hypothetical *box7* maturase, which, as is proposed for the *box3* analogue, may be a fusion protein, also has yet to be isolated and characterized.

*box7* mutants like *box3* mutants accumulate novel polypeptides, and in the case of *box7* these include a 56-kd polypeptide structurally related to cytochrome oxidase subunit I (COI) (Claisse *et al.* 1980). An interesting exception to the normal direction of the "box" effect has been found by Dujardin and colleagues (Dujardin *et al.* 1980a), who have isolated a mutant, *MIM2*, which cures all *box7* mutations but no other mutations and is located in *oxi3*. One interpretation (Lazowska *et al.* 1980) is that the wild-type *oxi3* gene requires an active *box7* maturase in order to process its own transcripts, while a *MIM2* mutated *oxi3* maturase (presumably encoded by an *oxi3* intron and inactive when wild-type) would be able not only to process its own transcripts but also to compensate for the deficiency of the cytochrome b intron $bI4_\ell/bI2_s$-encoded maturase. $aI4_s$ (or its homologues) is the obvious candidate for the *oxi3* intron that encodes the maturase, on account of its homology with the *box7* intron of *cob-box*.

Progress in understanding the mechanism of mitochondrial intron sequence splicing will obviously come in part from detailed analysis of (intronic and boundary) mutations that impair this process. The recent work of de la Salle and colleagues (1982) is relevant here. They sequenced nine mutants in intron $bI4_l$ from four *box* loci (*box 1*, *9*, *7*, and *2*), all blocking transcript processing. One *cis*-dominant *box1* mutant was located extremely close to the exon–intron boundary (just within exon $b4_l$). Other *cis*-dominant mutants of the *box9* and *2* classes were located well within the intron sequence and not at or near intron–exon junctions, as had been anticipated. It was also found that the sequence of wild-type *box9* RNA is complementary to a dodecanucleotide sequence of the small ribosomal RNA, suggesting a possible role of the ribosome in intron splicing (cf. preceding discussion of splicing of the optional intron in the 21S rRNA gene).

The importance of *box9* as a locus for splicing signals is reinforced by the observation that it has a very similar sequence to a region, HOMII, of the $aI4_s$ intron of *oxi3*, an intron in which *cis*-dominant mutations blocking transcript processing occur, this processing being dependent on the *box7* maturase (Netter *et al.* 1982; and see preceding section on *oxi3*).

The (*trans*-recessive) *box7* mutations analyzed all lay in the 3′ half of the intronic URF and three of the mutations caused the introduction of stop codons, all of which reinforces the contention that this URF specifies a protein required in splicing. Overall, the results imply a twofold mitoribosomal involvement in the splicing of mitochondrial introns; first, through translation of intronic URFs into protein splicing factors (maturases), and second, through recognition of signal sequences in precursor RNA, as in the case of the *box9* sequence. It is also clear that sequences important in splicing can be quite distant from intron–exon junctions.

The final *cob-box* intron ($bI2_s/bI5_\ell$) is 733 nucleotides long with (A + T)-rich stretches and no open reading frames of sufficient length to encode a protein. The intron also contains a (G + C)-rich HpaII/HaeIII site cluster, part of which (3′GAGGAAAGCCCCAA-5′) is inversely repeated (5′CTCCTTTCGGGGTT-3′) 500 nucleotides upstream in the same intron (Nobrega and Tzagoloff 1980b). Such sequences are common to many (G + C) clusters in the yeast genome, and it is suggested that they generate secondary structures important as recognition sites for the processing of transcripts (Tzagoloff *et al.* 1980, and see following section on transcription).

*Codon utilization:* As in other yeast mitochondrial genes, most codons terminate in A or U and certain codons are not used. These include the CGN family for arginine, despite the fact that the yeast genome encodes a relevant tRNA gene (S. Bonitz, quoted in Nobrega and Tzagoloff 1980b). Rare codons missing in other yeast mitochondrial genes but found here are UGC for cysteine and GUG for valine.

*Flanking and possible splicing sequences:* In strain D273-10B ("short" genes) the AUG initiator of the first exon is preceded by 139 nucleotides that are almost entirely (A + T). The 2800 nucleotide sequence downstream from the ochre codon terminating the last exon consists of semirepetitive A + T-rich sequences interspersed with numerous G + C clusters containing HpaII and HaeIII sites.

The sequence data of Lazowska and colleagues (1980) show that in the case of the (exon–intron) boundaries of the $bI2_\ell$ intron *(box3)*, the "rule" for nuclear genes (Breathnach *et al.* 1978) is not strictly followed. At the 5′ end the "consensus" GT dinucleotide resides in the exon sequence, the intron beginning with AAT, though the expected AG is found at the 3′ end of the intron.

The first intron of the short *cob-box* gene also finishes with an AG dinucleotide at the 3′ end, the 5′ end starting with AAT and the exon sequence at this juncture terminating in CAA (Nobrega and Tzagoloff 1980b). The second short gene intron lacks both GT and AG in the appropriate positions. Nobrega and Tzagoloff (1980b) have noted that exons $bE2_s$ and $bE3_s$ have a common pentanucleotide sequence, CCTGA, close to their 5′ termini.

The 3′ end of $bI3_l$ ends with the same hexanucleotide sequence as the 3′ end of $bI1_s/bI4_l$, so this sequence constitutes a possible splicing signal:

Intron . . . TAAAAG $\overset{\text{splice}}{\downarrow}$ C . . . Exon

(Lazowska *et al.* 1981).

## An Enigmatic Locus: *var1*

*Map position:* ~86 u.

This locus has been extensively studied by Butow and colleagues (Douglas and Butow 1976; Butow *et al.* 1980,1982) and appears to specify a protein associated with the small (37s) ribosomal subunit (Groot *et al.* 1979).

The *var1* proteins vary in size (40–44 kd); this size variation correlates with the presence or absence of small insertions in a restriction fragment that *petite* deletion mapping identified as carrying the *var1* determinant (Strausberg *et al.* 1978). Thus the protein size polymorphism apparently directly reflected size variations in the *var1* gene itself. In crosses between the different gene forms asymmetric gene conversion occurs through transfer of segments within the *var1* locus between mitochondrial genomes. The polarity of gene conversion within *var1* evinces a similarity to the situation at the $\omega$ locus in the 21S rRNA gene (Strausberg and Butow 1981). Sequencing studies of the *var1* locus have made this interesting locus even more of a puzzle.

It was found (Tzagoloff *et al.* 1980) that the overall (G + C) content of the region was extremely low (~6%), and none of the AUG codons present could generate proteins longer than 37 amino acids—considerably smaller than the ~40 kD *var1* polypeptide. Further complications have arisen through studies of *var1* transcripts. A 16S RNA (~1700 nucleotides) appears to be the *var1* mRNA, showing strain-dependent size variation that correlates with the size of the *var1* protein product. However, only about 250 nucleotides of the *var1* locus hybridize to this RNA, and the RNA itself hybridizes neither with flanking regions nor with other mtDNA regions, nor even with nuclear DNA (see Grivell 1981). It seems that the structural gene for the *var1* polypeptide, or for most of it, remains to be found.

## Unassigned Reading Frames (URFs)

Sequencing studies of regions of mtDNA encoding known genes have uncovered 10 potentially protein-coding open reading frames that do not correspond to known mitochondrial gene products: These are "unassigned" reading frames, or "URFs." Eight of these have already been described (see section on complex genes) and consist of URFs located in mosaic gene introns. One occurs in the optional 21S rRNA intron (Dujon 1980), three have been found in introns of the apocytochrome b *(cob-box)* gene (Tzagoloff *et al.* 1980; Lazowska *et al.* 1980,1981), and four more lie in introns of the gene *(oxi3)* for cytochrome c oxidase subunit I (Bonitz *et al.* 1980a).

At least two URFs have been discovered between genes. One is situated near *oxi1* in the segment of DNA sequenced by Coruzzi and co-workers (1981) that stretches from 10.7 to 17.9 map units. It has an ATG initiator at 15.5 u, is apparently continuous for at least 879 and possibly as much as 1161 nucleotides, and could encode a 386-amino acid protein of molecular weight 48,000. Like intronic URFs, this intergene URF has a low (G + C) content (19%) and is rich in condons for asparagine, lysine and tyrosine. This URF also uses the CGU codon for arginine and the UGG codon for tryptophan. A second (probable) intergenic URF of ~500 bp has also been revealed by sequencing (directed at the *oli2* locus) and lies between *cob-box* and *oli2* (Macino and Tzagoloff 1980).

As yet there is no definitive evidence that any of these URFs are expressed as proteins in wild-type cells lacking blockages of transcript processing, and the only indications of function suggest a role in RNA splicing or, conceivably, mtDNA recombination–gene conversion (see sections on the 21S rRNA gene, complex genes, and introns). It is quite possible that other URFs await discovery; after all, the compact mammalian mitochondrial genome has eight (Anderson *et al.* 1981), and a great deal of yeast mtDNA remains unsequenced. Also certain mitochondrial translation products—such as two of the ATPase subunits (Tzagoloff *et al.* 1979a)—still appear to be proteins in search of genes.

## Introns

Many eukaryotic nuclear genes are now known to be split by insertions of apparently noncoding sequences—introns (Breathnach and Chambon 1981). Introns have also been found in the mitochondrial genes of yeasts and other fungi (see preceding sections on the 21S rRNA gene and complex genes), though they do not seem to occur in mammalian mitochondrial genes. Various general explanations for the existence of intervening sequences have been considered by Borst (1980a and 1981a) in the context of the introns of yeast mtDNA; the discussion below follows his analysis.

1. Introns are needed for the regulation of gene expression in large and complex genomes (Scherrer and Marcaud 1968).

Scherrer and Marcaud suggested—well before the discovery of introns—that regulation of mRNA levels, and hence gene expression, occurs through RNA processing. This suggestion presumably also applies to RNA splicing but is an unattractive explanation of mitochondrial introns for several reasons: The mitochondrial genome is not especially large, and complex, controlled gene expression of some unsplit mitochondrial genes (e.g., *oxi1*, Coruzzi and Tzagoloff 1979) does not appear to require the splicing events demanded by the presence of introns, and indeed, several introns within the split genes are optional, in that they occur in only some yeast strains. Also primary transcripts from unsplit genes may undergo complex multistep maturation processes offering possibilities for regulation without any necessity for splicing (Coruzzi *et al.* 1981).

2. Splicing is required for exit of RNAs from the nucleus (Hamer and Leder, 1979).

This conjecture, arising from studies of expression of SV40–mouse $\beta$ globin gene recombinants, is obviously not applicable to mitochondrial genes.

3. Introns limit homology in duplicated genes: Rapid sequence divergence in introns following a duplication event would reduce homology of the duplicated segments and thereby reduce also the probability of an unequal crossing-over and its possibly damaging consequences (Tiemeyer *et al.* 1978).

Again, this hypothesis is irrelevant, as there are no known duplications (certainly not of split genes) on yeast mtDNA.

4. Splicing is required for the formation of a stable RNA (Gruss *et al.* 1979).

This idea is apparently ruled out by the existence of the same gene (in different strains) with and without introns.

5. Split genes can encode more than one gene product.

Studies with small DNA viruses such as SV40 have shown that more than one protein (e.g., the small-t and large-T antigens) is encoded in the same stretch of DNA. During processing of the primary T/t antigen transcript, two different-sized overlapping segments are excised from within the transcript. This produces a larger mRNA, which contains a translation termination codon missing from the second product, a smaller mRNA. On translation the large mRNA gives a small protein (the t antigen) while the smaller mRNA is translated to yield the much larger T antigen (Fiers *et al.* 1978; Crawford *et al.* 1978). SV40 is able to produce these two proteins from the same gene because there is a substantial stretch of open reading frame downstream from the termination codon used in the synthesis of the small-t antigen. It is the discovery, through DNA sequencing, of extensive lengths of open reading frame in many yeast mtDNA introns—some of them continuous with exon reading frames (see earlier section on URFs)—that has raised the possibility that split genes of yeast mtDNA encode more than one protein product.

Unlike the situation in SV40, however, only a single gene product has so far been found and characterized for each of the split genes (21S rRNA for the 21S rRNA gene, COI for *oxi3* and apocytochrome b for *cob-box*). Nevertheless, a great deal of evidence, mainly obtained with different classes of *mit*$^-$ mutant, suggests that intronic open reading frames are translated and the products ("fusion proteins" or "read-through proteins" containing exon-specified sequences) are required in splicing transcripts (usually just those of the split gene in which the intron is located) (Church *et al.* 1979; Haid, *et al.* 1980; Halbreich *et al.* 1980; Van Ommen *et al.* 1980; Alexander *et al.* 1979, 1980). Thus mutations in introns can cause defective splicing and the accumulation of long, stable transcripts and their presumptive translation products, these being novel polypeptides such as the 42 kD "fusion protein," which contains apocytochrome b (~30 kD) as well as apparently intron-encoded amino acid sequences (Krieke *et al.* 1979; Solioz and Schatz 1979). It is hypothesized that "fusion proteins" such as the 42-kD species are RNA "maturases" splicing out intronic sequences and so destroying the transcripts from which they are translated (their own mRNAs) but leading to the formation of the final, mature, split-gene mRNA for the major gene product, such as apocytochrome b (Slonimski *et al.* 1978a,b; Claisse *et al.* 1978; Jacq *et al.* 1980a,b). Maturase would exist at low steady-state levels in normal, wild-type cells, only rising to readily detectable levels when functionally inactivated by mutations that therefore cause accumulation of maturase mRNA. This attractive hypothesis still lacks direct proof, despite the compelling nature of the circumstantial evidence. Attempts to detect the maturases directly in wild-type cells have so far been unsuccessful (J. Kreike, quoted in Borst 1981a).

Even if the maturase hypothesis is fully validated, the existence of introns in yeast mitochondrial genes remains a puzzle: Teleologically expressed, what is the point of having intervening sequences in genes that simply encode machinery for their own removal from the gene transcripts? The apparent pointlessness of at least some intronic maturases (e.g., the *box3*/bI$2_\ell$ maturase) is emphasized by their dispensability, in the case of the optional introns, and the fact that certain mitochondrial introns can be excised without a requirement for mitochondrial translation products, presumably by the products of nuclear genes. This is the case for the *cob-box* intron bI$1_\ell$ (Halbreich *et al.* 1980) and the large optional intron in the 21S rRNA gene (Faye *et al.* 1974).

6. Introns facilitate reassortment of segments to yield new genes (Gilbert 1978).

As Borst (1980a) has pointed out, the genetic complement of mitochondria seems to have remained remarkably stable through much of evolution—as judged by the similarity of mitochondrial gene products in lower and higher eukaryotes. Nevertheless, new mitochondrial genes may have arisen in the way suggested by Gilbert in an earlier phase of mitochondrial evolution, before the genetic content of modern mitochondrial genomes became consolidated. At least in the case of nuclear genes, Gilbert's hypothesis that exons encode functional segments or domains of proteins has received strong support from studies of immunoglobulin genes and proteins (Sakano *et al.* 1979), where it has been found that each of the six known functional units in the $\alpha$-globulin heavy chain is coded by a separate exon. Whether yeast mitochondrial exons—even those separated by optional introns—have a comparable significance awaits a more detailed characterization of the structures and catalytic activities of the relevant proteins.

Another point made by Gilbert (1978) is that introns increase the length of genes and so should increase overall frequencies of intragenic recombination and gene-conversion events. This could allow more rapid spread and fixation in populations of advantageous alleles, and the exact location of intron–exon boundaries in respect of protein structure would not be important. In this regard it may be significant that split genes do not occur in animal mitochondria, where inheritance of mitochondrial genes is apparently uniparental (matrilineal) (Hutchinson *et al.* 1974), and opportunities for genetically meaningful recombination must be slight. Of course, if a mosaic structure does confer this benefit it may be asked why not all mitochondrial genes are split in organisms where ready recombination of cytoplasmic genes occurs.

7. Introns have no function.

This is probably untrue for yeast mitochondrial introns containing open reading frames and implicated in splicing; at least a definite answer should soon be forthcoming. As for other introns, lack of function would not be a compelling disadvantage once a splicing machinery had been established to excise redundant sequences from transcripts. Such functionless introns would fall into the same selfish DNA category (Orgel and Crick 1980) as many intergenic DNA sequences. Indeed, as Borst again has pointed out (Borst 1980a), if DNA replication machinery has a preference for a certain size of genome [though the very variability of genome sizes even in different yeasts (see following section on evolutionary aspects) might argue against this], then there could be selective pressure for maintenance of noncoding DNA.

In conclusion, there is no single satisfactory explanation for the existence of (yeast) mitochondrial introns. Indeed perhaps a unitary explanation should not be looked for, but at least the rapid progress of yeast mitochondrial genetics promises to dispose of some hypotheses and favor others.

## Mitochondrial Transcription

### In Mammals

The mammalian mitochondrial genome of HeLa cells is transcribed completely and symmetrically (Murphy *et al.* 1975), even though most of the genes are localized on the H strand with only a few scattered genes occurring on the L strand. There is evidence that transcription may be initiated from single promoters on the L and H strands both located in the region of the H-strand replication origin (Cantatore and Attardi 1980). These transcripts are then processed to yield rRNAs, tRNAs, and mRNAs (Amalric *et al.* 1978). Since mRNA protein-coding sequences are flanked immediately on their 5′ sides by tRNA genes (Montoya *et al.* 1981), a "tRNA punctuation" model of RNA processing has been proposed in which it is suggested that the H strand is transcribed into a single polycistronic RNA molecule that is processed later into mature species by precise endonucleolytic cleavages at the borders of tRNA sequences (Ojala *et al.* 1981). Presumably, a similar mechanism could apply to the L strand, although the total information content of this strand—eight tRNAs and one possibly protein-coding open reading frame—is rather low. Control of mitochondrial gene expression in mammals at the transcriptional level is therefore presumably mainly a matter of transcription attenuation, regulation of RNA cleavage events, and the stabilities of the various transcripts and messengers.

### In Yeast

Transcription of yeast mtDNA appears to differ from the mammalian system in a number of important respects, including lack of symmetry, probable use of multiple promoters, and the existence of splicing machinery to process split mitochondrial gene transcripts into functional messengers.

#### Transcripts and Promoters

All thoroughly analyzed yeast mtRNAs have so far proved to be transcribed from the same strand, there being evidence for only one gene, encoding a tRNA, in the "noncoding" strand (Li and Tzagoloff 1979). Hendler and colleagues (Hendler *et al.* 1976) found that some yeast mitochondrial RNA is capable of self-annealing, and Linnane's group (Linnane *et al.* 1980) have observed duplex nucleic acid molecules that they interpret as complete double-stranded transcripts of the entire mtDNA genome. However, transcription of yeast mtDNA *in vitro* by homologous RNA polymerase does not yield significantly self-complementary RNA (Lewis *et al.* 1977), and transcription of quite distinct regions of the genome occurs readily in *petites* (Faye 1976). Also there is now good evidence that at least five different promoters can be used in wild-type mtDNA, since at least five mtRNAs have been shown to carry a di- or triphosphate at the 5′ terminus (Levens *et al.* 1981c).

Although the entire yeast mitochondrial genome may not be transcribed into full-length giant precursor RNA molecules, exceptionally large transcripts certainly do occur, as revealed, for example, by the two-dimensional transcript mapping procedure of Van Ommen and colleagues (Van Ommen *et al.* 1979). The three mosaic genes of the yeast mitochondrial genome (the 21S rRNA gene of $\omega^+$ strains, *oxi3* and *cob-box,* see above) must of course generate primary transcripts substantially larger than needed to encode the final gene product. In the *cob-box* region such transcripts would be at least ~8 kb in the case of "long" genes, and *oxi3* transcripts of 15 kb have been observed by Northern blotting (L. A. M. Hensgens, quoted in Borst 1980a). Also RNA blotting using $^{32}$P-labeled cRNA probes specific for 21S rRNA gene sequences revealed transcripts up to 4.7 kb long (Bos *et al.* 1980). Although split genes containing introns might be expected to give rise to long primary transcripts, lengthy precursor RNAs can also apparently be generated by unsplit genes. Northern blotting experiments have shown that the *oxi1* locus encoding subunit II of cytochrome oxidase is transcribed into a high-molecular-weight RNA, longer than 3 kb, which is processed in discrete steps to a mature 850-nucleotide messenger (Coruzzi *et al.* 1981). The processing involves removal of both 5′ and 3′ sequences, and at least nine discrete species larger than mature mRNA have been detected (Coruzzi *et al.* 1981). The unsplit gene for the small ribosomal RNA also gives rise to a precursor of the 15S rRNA that is larger than the final gene product: Osinga and colleagues (Osinga *et al.* 1981) identified a precursor-RNA bearing approximately 80 extra nucleotides at the 5′ end.

## Splicing

Processing of mosaic gene precursor-RNAs is of course likely to be more complex than that for transcripts from other loci, as not only are cutting (and possibly trimming) events involved, but exonic sequences have to be joined in a "splicing" reaction. Considerable research effort is currently directed toward disentangling the complicated splicing pathways for transcripts from split mitochondrial genes. A possible model for processing RNA containing intervening sequences is provided by those yeast nuclear tRNA genes that are split. Studies by Abelson and colleagues have shown (Peebles *et al.* 1979; Knapp *et al.* 1979; and see Abelson 1979) that *in vitro* maturation of tRNA precursors by soluble cell extracts occurs in two separable stages: in the first, ATP-independent reaction, the precursor is cleaved to yield three linear molecules: two half-tRNA molecules and the intervening sequence. In the second, ATP-dependent reaction, the tRNA fragments are covalently joined to produce mature tRNA. Interestingly, it was found that the first, cleavage reaction generates 5′-hydroxyl and 3′-phosphate termini, in contrast to all previously characterized RNA processing reactions. This also means that the relevant ligase must differ from previously known RNA or DNA ligases, since it joins 3′-phosphate and 5′-hydroxyl termini. Analysis of mutants of SUP4 tRNA suggests that the entire three-dimensional structure of the intron-containing tRNA precursor is likely to be important for recognition by the splicing machinery (see Cooper 1981).

In a discussion of possible RNA-splicing mechanisms, Mahler (1981) distinguished two possible models: (1) "unassisted" or "concerted" and (2) "assisted" or "guided." In type A models exonic or other sequences to be joined after cleavage are held together or in the appropriate spatial orientation either by structural features of the RNA substrate itself, as in base pairing of complementary sequences forming stems and loops (A1), or by the protein complex catalyzing the cleavage and ligation reaction (A2). Maturation of cytosolic tRNA precursors would appear to belong to the A1 category. In type B reactions sequences to be joined are held in association prior to actual covalent linkage by "guide" or "bridge" molecules, presumably either complementary RNAs or (bifunctional) proteins. As discussed earlier, there is a great deal of evidence that open reading frame sequences in at least two *cob-box* introns specify diffusible (*trans*-acting) factors ("TAFs"), probably proteins, that act at particular stages in the maturation of precursor RNA from mosaic genes. Perhaps these factors are type B protein bridge molecules.

Oddly enough, there appears to be no single mechanism for splicing all mosaic gene transcripts: even within the same split *cob-box* gene the *box3* TAF or maturase acts only in the excision of the intron by which it is specified, but not on the *box7* intronic sequence. However, while the *box7* TAF is specific for its own intronic sequence, it is also required for correct splicing of *oxi3* transcripts. Correct processing of mosaic 21S rRNA transcripts is, by contrast, apparently independent of any other mitochondrial DNA sequences and can presumably be effected by an entirely imported cleavage and ligation system.

Evidence has recently emerged that another type of *trans*-acting mitochondrially specified factor is involved in RNA processing and is different from the putative intron maturases in not being a protein. It has been found that a segment of the genome lying between *oxi2* and *par* is required for processing a large transcript from the tRNA gene cluster into mature tRNAs (Morimoto *et al.* 1979). *Petites* containing this "tRNA-processing" locus can zygotically complement *petites* lacking it but retaining the tRNA gene cluster (N. C. Martin, quoted in Grivell, 1981), indicating that the factor cannot be a protein and hence is presumably an RNA. These observations suggest that maturation of mitochondrial tRNAs differs mechanistically from that of the nuclear-coded (split) tRNAs, where there is no evidence for the intervention of an RNA species other than the tRNA precursor/substrate itself. In Mahler's terminology, then, processing of mitochondrial tRNAs may be of the type B, assisted or guided category.

Splicing of the split 21S rRNA transcript does not apparently require a mitochondrially encoded RNA or protein guide, and the sequence of the gene shows that no sequence longer than 6 bp and complementary to the future exon–exon junction occurs in the intron itself, in the flanking exons, or even in the mini-insert (Dujon 1980), apparently excluding a type A1 mechanism. However, Dujon has proposed that a guiding function may still be required for splicing, but that this is provided by the environment of the large ribosomal subunit, excision of the intervening sequence occurring after subunit assembly (Dujon 1980). The sequence data of Lazowska and colleagues (Lazowska *et al.* 1980) have also shown that there is no complementarity between the *box3*/$bI2_{\ell}$ intron of the long *cob-box* gene and the junctions of this intron with its bounding exons, thereby also tending to exclude an A1 mechanism. Indeed, as we have already seen, this intron, possibly like some other introns containing open reading frames, appears to catalyze its own excision by specifying the synthesis of a *trans*-acting protein factor.

## Circles and Hairpins

In the course of processing long (strain KL 14-4A) *cob-box* gene transcripts to produce the 18S (2.2 kb) apocytochrome b mRNA, at least six or seven distinguishable transcript species are generated, excluding the postulated 29S/7.45 kb (or possibly even larger!) primary transcript (Van Ommen *et al.* 1980). It appears that there is no absolute order of splicing, the various transcripts being produced by removal of different combinations of intron sequences; nevertheless all the splicing pathways for long strains probably share the same first step, excision of the first intron, bI1$_{\ell}$. This first excision and splicing event apparently generates the mRNA for the *box3* maturase fusion protein (Lazowska *et al.* 1980; and see the preceding section on *cob-box*). Surprisingly, the bI1$_{\ell}$ intronic sequence is liberated as a single-stranded covalently closed circular 10S (765 nucleotides) RNA species (Halbreich *et al.* 1980). This 10S species is an abundant, stable product, perhaps because circularity confers resistance to enzymic degradation, and, as it is seen in *petites,* it must arise solely through the action of nucleus-encoded enzymes. Conceivably, these RNA circles have some as yet unidentified function, though they are apparently not translated (Van Ommen *et al.* 1980). Circular 11, 19, and 19.2S RNAs are also generated during processing of *oxi3* transcripts (Arnberg *et al.* 1980; Grivell *et al.* 1980).

An indication of how these circular molecules might arise is provided by the tailed circles or "neckties" that have been observed in 25–28S mtRNA fractions (Arnberg *et al.* 1980). These species are not covalently closed circles as they are disrupted by glyoxal treatment, but they withstand 84% formamide/2.8 *M* urea, suggesting the involvement of base pairing between (G + C)-rich sequences. Presumably, these RNA "neckties" are hairpin structures with a single-stranded loop and a double-stranded stem region created by base pairing of complementary (G + C)-rich sequences. Short, 30–70 bp, (G + C)-rich sequences are widespread in the yeast mitochondrial genome (Prunell and Bernardi 1977), and many are homologous with each other (Cosson and Tzagoloff 1979). These observations are the basis of a model of transcript processing proposed by Tzagoloff and colleagues (Tzagoloff *et al.* 1980) in which hairpins are generated by base pairing between homologous (G + C)-rich clusters and the stem sections are recognized and cleaved by a double-stranded RNA-specific RNase [as in the case of *E. coli* 16S rRNA precursor, which forms a hairpin structure whose stem is cleaved by RNase III (Young and Steitz 1978)]. In the Tzagoloff scheme it is envisaged that the cleavage products are linear, but appropriate enzymic machinery with ligating or bond-exchanging activity could presumably liberate circular products (see Borst 1981a).

## Leaders, Caps, and Tails

The 18S mature apocytochrome b mRNA is over twice the size necessary for its amino acid coding function and has an unusually long 5′ leader (~1 kb) with an additional ~400 nucleotides at the 3′ end (Grivell *et al.* 1980). The probable (12S) mRNA for subunit 9 of the ATPase also has a long (~400 nucleotides) leader sequence (Hensgens *et al.* 1979). By contrast, the leader sequence of the presumed mRNA for subunit II of cytochrome oxidase is only 54 nucleotides long and is identical in sequence with the DNA lying upstream of the *oxi1* protein coding region (Coruzzi *et al.* 1981).

Mammalian mitochondrial mRNAs appear to be uncapped (Grohmann *et al.* 1978) and, lacking significant leader sequences (Montoya *et al.* 1981), lack Shine-Dalgarno-type sequences giving complementarity to the 3′ end of the small rRNA, but they do have poly(A) tails (Battey and Clayton 1978). It is not yet clear whether yeast mt mRNAs are also uncapped, and the long leader sequences so prevalent among yeast mtRNAs could conceivably harbor sequences complementary to the small mt rRNA. Alternatively, as pointed out by Barrell and co-workers (Barrel *et al.* 1979), there is also the possibility that ribosomal binding of both lower and higher eukaryotic mRNAs could be effected through base pairing between the small rRNA and sequences *within* the protein-coding region of the mRNA; indeed, there is recent evidence for complementarity between a short sequence internal to the *cob-box* gene and the small mitoribosomal RNA (de la Salle *et al.* 1982). The question of poly(A) tails on yeast mitochondrial mRNAs is still controversial: Hendler and colleagues (Hendler *et al.* 1975) reported binding of mtRNA to poly(U)-sepharose, but other workers (Groot *et al.* 1974; Moorman *et al.* 1978) consider that the binding is merely fortuitous because of long internal (A + U)-rich tracts and that mt mRNAs do not have poly(A) tails.

## RNA Polymerase

There have been a number of reported isolations of DNA-dependent RNA polymerase activity from yeast mitochondria (Tsai *et al.* 1971; Wintersberger 1972; Scragg 1976), but the preparations have differed in properties and tended to exhibit low specific activities. Levens and colleagues have very recently isolated a transcription complex from yeast mitochondria (Levens *et al.* 1981a) and have purified a DNA-dependent RNA polymerase to high specific activity (Levens *et al.* 1981b). This polymerase differs from the equivalent nuclear and bacterial enzymes in a number of respects, such as sensitivity to high ionic strength and insensitivity to $\alpha$ amanitin and rifampicin, but displays similarities to the RNA polymerase from *Xenopus* mitochondria (Wu and Dawid 1972). The enzyme appears to be a ~100–105 kd multimer of a 45-kd monomer, and while mtDNA is its preferred natural template, its most efficient template is poly{d(AT)}. The enzyme appears to transcribe mtDNA *in vitro* nonrandomly, preferentially transcribing rRNA genes, even though its small size compared to other RNA polymerases might suggest a limited ability to recognize and distinguish different start and stop signals. If transcriptional control is fairly rudimentary, then regulation at the level of posttranscriptional processing becomes especially important.

## Mitochondrial Protein Synthesis

### Inhibitors

Mitochondrial protein synthesis has a number of prokaryote-like features. These include the use of formylated methionine in initiation (Bianchetti *et al.* 1977) and sensitivity to various ribosome-binding inhibitors of bacterial protein synthesis such as chloramphenicol and erythromycin, with corresponding insensitivity to inhibitors of cytoplasmic protein synthesis like cycloheximide (see Schatz and Mason 1974; Tzagoloff *et al.* 1979a). These differential responses to inhibitors have been exploited experimentally to allow selective *in vivo* radioactive labeling of mitochondrial translation products. Labeling with inorganic $^{35}SO_4^{2-}$ in the presence of cycloheximide has proved to be a useful and highly sensitive way of examining large numbers of strains (Douglas and Butow 1976; Douglas *et al.* 1979); after *in vivo* labeling the proteins are resolved by polyacrylamide gel electrophoresis and are detected autoradiographically (Groot *et al.* 1978; Douglas *et al.* 1979; Claisse *et al.* 1980). The availability of these techniques has been especially important, since *in vitro* approaches, such as translation of mtRNA in a heterologous system, have been disappointing; these failures can now be understood as a reflection of the unique features of the mitochondrial genetic code (see following discussion).

### Initiator tRNA

The yeast mitochondrial initiator tRNA has recently been sequenced by Canaday and colleagues (Canaday *et al.* 1980b). While the tRNA is formylated by a mitochondrial enzyme extract, it cannot be formylated by the *E. coli* transformylase. Unlike the mt $tRNA^{fMet}$ of *N. crassa* (sequenced by Heckman *et al.* 1978), the yeast initiator tRNA has two typically prokaryotic features: There is no Watson-Crick base pair at the end of the acceptor stem and there is a T–$\psi$–C–A sequence in loop IV. However, both yeast and *N. crassa* mitochondrial initiator tRNAs have a $U_{11}:A_{24}$ base pair in the D stem, unlike prokaryotic initiator tRNAs, which have $A_{11}:U_{24}$. Also both these initiator tRNAs share with bean chloroplast $tRNA^{fMet}$ (Canaday *et al.* 1980a) the highly unusual property of possessing only two G:C pairs next to the anticodon loop. In terms of overall sequence similarities the yeast $tRNA^{fMet}$ differs from both prokaryotic and eukaryotic–cytoplasmic initiator tRNAs, showing highest homology with the initiator tRNA of *N. crassa* mitochondria.

## Ribosomes

As the findings with yeast and *Neurospora* mitochondrial initiator tRNAs suggest, mitochondrial translation systems seem to have unique features and are not closely similar to prokaryotic counterparts. Mitoribosomes differ substantially from those of prokaryotes and of eukaryotic cytoplasm and are quite variable between different phylogenetic groups (see O'Brien and Matthews 1976; Gray 1982).

The yeast *(Saccharomyces cerevisiae)* monomer is 74S and is dissociable into large (50S) and small (37S) subunits containing large and small ribosomal RNAs of 3100 and 1460 nucleotides, respectively (Sriprakash and Clark-Walker 1980). It appears that all the ribosomal proteins (except possibly the *var-1* gene product) are encoded in the nucleus and synthesized outside the mitochondrion. Though they are not sufficiently well characterized for full comparison with those of bacteria and eukaryotic cytoplasm, there do appear to be differences (Faye and Sor 1977; Faye *et al.* 1979b).

Like the mitoribosomes themselves, mitoribosomal RNAs from different eukaryotes are strikingly dissimilar and vary widely in size and base composition (see Mahler 1981; Gray, 1982). Indeed, whereas higher plant mitoribosomes possess a 5S RNA (Cunningham *et al.* 1976), the mitochondria of all other eukaryotes, including yeast, appear to lack this species entirely (Grivell and Reijnders 1974). Bonen (L. Bonen, in Slonimski *et al.* 1982) has found that the 100 nucleotides at the 3′ terminal end of human mitochondrial small subunit rRNA (12S rRNA) exhibits primary and secondary structural homology with 5S rRNA. This sequence can be fitted to 5S rRNA structural models and appears able to fulfil those roles in initiation of protein synthesis that have been proposed for 5S RNA. Thus, in human mitoribosomes at least, 5S rRNA functions may be performed by the 3′ terminus of the small subunit rRNA. Though a similar 5S rRNA homology has not been reported for the 3′ end of yeast mitochondrial 15S rRNA, the sequence of this region of the gene does lack the typical Shine-Dalgarno mRNA-binding sequence seen in prokaryotes (see earlier section on rRNA genes and Sor and Fukuhara 1980). Nevertheless, the sequence contains several stretches that are homologous to *E. coli* 16S rRNA and that fit a secondary structure model very similar to that proposed for the *E. coli* molecule.

An unusual feature of mitoribosomal RNAs is their low content of (post-transcriptionally) modified bases. The 21S rRNA of yeast contains only two methylated ribose residues and one pseudouridine, while the yeast 15S rRNA also contains only one pseudouridine residue and is entirely unmethylated (Klootwijk *et al.* 1975).

## The Genetic Code

The most remarkable feature of mitochondrial translation systems is of course their departure from the "universal" genetic code and their use of what is possibly the absolute minimum number of tRNAs (24–25 in yeast, 22 in man) possible for a functional protein-synthesizing system (see preceding section on tRNA genes and the genetic code). The most general of the mitochondrial exceptions to the usual coding rules appears to be the use of UGA to encode tryptophan instead of termination [in yeast, man, and *Aspergillus nidulans* (Waring *et al.* 1981), but not in maize (Fox and Leaver 1981)], but different mitochondrial systems also exhibit slight coding differences among themselves (e.g., AUA is methionine in mammals but isoleucine in *S. cerevisiae* and the universal code). Reductions in the number of decoding tRNAs required has been achieved by expanding the number of codons read by a tRNA, so that, in many cases, all four codons of a four-codon family can be recognized by the same "adaptor" molecule (e.g., UCN for serine, GGN for glycine, CCN for proline, etc.). This is apparently effected by having U in the first position of the anticodon, so that U:N wobble is possible, though how "two-codon" tRNAs with U in the wobble position are restricted to recognizing codons ending in A or G is still unclear.

The revelation of the unique features of the mitochondrial genetic code has resolved one long-standing problem: the failure of standard *in vitro* translation systems, such as those from *E. coli,* reticulocytes, or wheat germ, to translate mitochondrial mRNAs (Moorman *et al.* 1977,1978). Evidently the widespread tryptophan-encoding UGAs would be interpreted as "stop" signals by the heterologous machinery, leading to the production of the low-molecular-weight products that are in fact seen (Moorman *et al.* 1977,1978). Successful translation of at least the mRNA for cytochrome oxidase subunit II has recently been achieved by using an *in vitro* wheat germ system supplemented by a UGA-suppressor tRNA from *Schizosaccharomyces pombe* (De Ronde *et al.* 1980). However, as pointed out by Borst (Borst 1981a), translation by a similar system of those yeast mitochondrial mRNAs that have exceedingly long (500 nucleotides +) leader sequences has not yet been demonstrated, and development of a truly homologous *in vitro* translation system for mitochondrial mRNAs would be most valuable.

## Protein Import and Membrane Assembly

Although much remains to be learned about (yeast) mitoribosomes, at least the sequences of the ribosomal RNA genes are now known (see above), and future work is likely to follow directions already pioneered for other types of ribosome. However, an exciting and unpredictable area, where the mitochondrion offers special challenges, concerns the fate of the products of the unique mitochondrial translation mechanism. The major enzyme complexes embedded in and contributing to the structure of the inner mitochondrial membrane are composed of polypeptides made both inside and outside the mitochondrion. It may therefore be asked how cytoplasmically synthesized proteins are imported into the mitochondrion (a question also relevant to the many soluble enzymes of the mitochondrial matrix) and how the components of the polymeric enzyme complexes are correctly assembled and appropriately positioned within the membrane. Cognate with this is the question of coordination of protein synthesis in cytoplasmic and mitochondrial compartments so as to provide inner membrane components in the correct stoichiometries and at the correct time, in terms of the cell's overall requirement for mitochondrial biogenesis.

It was at first thought that mitochondrial proteins were discharged through the mitochondrial membranes during the course of polypeptide elongation from ribosomes bound to the mitochondrial membrane, as occurs in the case of the eukaryotic endoplasmic reticulum (Campbell and Blobel 1976). This "vectorial translation" hypothesis (Kellems and Butow 1972) had several implications: (1) that the cytosol should not contain any pools of mitochondrial proteins, (2) that import of polypeptides into mitochondria should be strictly coupled to protein synthesis, and (3) that no discrete large precursors bearing N-terminal "signal sequences" should accumulate in intact cells (Schatz 1979). Recent experiments, by Schatz and colleagues in particular, have shown that these predictions are not fulfilled for a number of cytoplasmically synthesized mitochondrial proteins (Cote *et al.* 1979; Maccecchini *et al.* 1979; Schatz 1979; Lewin *et al.* 1980; Mihara and Blobel 1980). Cytoplasmic precursors of subunits of the cytochrome $bc_1$ complex, cytochrome c oxidase, and F1 ATPase have been found that are all larger than the mature membrane-associated forms by some 2–6 kD. The precursors are transported into the mitochondria in a process that is independent of protein synthesis but requires energy (ATP, Nelson and Schatz 1979) and during which proteolytic cleavage removes the excess (N-terminal leader sequence?) polypeptide material. Cytochrome c peroxidase, an enzyme located in the intermembrane space, is also initially synthesized as a large precursor (Schatz 1979), but, perhaps not surprisingly, no intramitochondrial ATP is necessary for the mature form to appear in its final location (Nelson and Schatz 1979), as the intermembrane space is freely accessible to cytosolic ATP. Although it could be supposed that the extensions of the precursor proteins are recognized by special sites or receptors on the outer mitochondrial membrane and that the proteolytic cleavage effects the translation, such a mechanism cannot be universal. Certain proteins, such as the apoprotein of cytochrome c and the adenine nucleotide translocator, lack extensions or precursor forms, and in these cases the "signal" may be built into the mature protein itself (Neher *et al.* 1980). Also it would appear that processing is not obligatorily coupled to transport, since Boehni and colleagues (Boehni *et al.* 1980) extracted a protease from within mitochondria that converts protein precursors into mature forms.

As usual, discovery—in this context, mitochondrial "preproteins" and processing—simply invites new and interesting questions: Are there different sites or receptors for different types of imported mitochondrial protein, and are these receptors the proteases that cleave the precursors in those cases where such processing occurs? Beyond and including these questions lies the problem still essentially unanswered for any biological system—how do cells make membranes?

## The *Petite* Mutation and mtDNA Replication

The mutation to cytoplasmically inherited respiratory deficiency, known as *petite* ($\rho^-$ or rho$^-$), has been the subject of decades of research and many reviews (see, e.g., Nagai *et al.* 1961; Nagley *et al.* 1977; Locker *et al.* 1979; Bernardi 1979; Whittaker 1979; Dujon 1981 and preceding chapter by Wilkie). Nevertheless, it will be briefly discussed again here because of its special importance in the development of mitochondrial genetics in *Saccharomyces cerevisiae,* it being almost the genetic "trademark" of the species, and because of the insights that studies of this mutation still undoubtedly have to offer into problems of mitochondrial DNA replication and recombination.

### Mechanism

It is now clear that in most spontaneous *petites,* a segment of mtDNA, varying from about 20% to over 99% of the wild-type genome, is excised and lost, and the remaining fragment is used to construct a *petite* mitochondrial genome of approximately normal size and contour length but composed of tandem repetitions of the surviving mtDNA fragment. Further deletions and rearrangements can occur, resulting in a rather more complex sequence organization, and this may be particularly common in *petites* induced by mutagenic treatments. A great range of chemical and physical treatments increase the *petite* frequency above the usual level of one or a few percent, intercalating agents such as ethidium bromide being exceptionally efficient *petite* inducers. Indeed, prolonged treatment with ethidium bromide can cause the elimination of all detectable mtDNA, generating "$\rho^0$" *petites* (Nagley and Linnane 1972b). However, although many of these mutagenic agents or treatments inhibit (mitochondrial) DNA synthesis, inhibition does not seem to be essential to the induction process, since some inhibitors of DNA synthesis are not *petite* inducers (Locker *et al.* 1979). As many *petite* inducers bind to and damage DNA, it seems likely that stimulation of DNA repair is important. If such mtDNA repair involved recombination, then the high frequency of truly or quasi-homologous DNA sequences scattered throughout the mitochondrial genome could result in the illegitimate but site-specific recombination event of the Campbell type (Campbell 1962), which many workers consider the essential step in *petite* formation (Piperno *et al.* 1972; Prunell and Bernardi 1974; Locker *et al.* 1974a,b; Clark-Walker and Miklos 1974). Spontaneous *petite* mutations might arise through a high and probably strain-dependent level of constitutive activity in this (essentially inefficient) repair–recombination pathway. Sequences of quasi-homology providing possible sites for illegitimate recombination include the (G + C)-rich HaeIII, HpaII site clusters occurring in about 60–70 copies per genome unit (Prunell and Bernardi 1977). In accordance with this possibility, examination of a number of spontaneous *petite* genomes has shown that the ends of the repeat unit are formed by clusters of the CCGG and GGCC sequences recognized by HpaII and HaeIII, respectively (Faugeron-Fonty *et al.* 1979). Other *petites,* however, especially those isolated after mutagen treatment, do not have repeat unit ends of this type (Van Kreijl and Bos 1977; Faugeron-Fonty *et al.* 1979). It should be borne in mind that *petite* mutants continue to evolve and that the sequence organization resulting from the original event generating the mutant genome may be obscured by later excisions and rearrangements.

## Origins of Replication

*Petites* that do retain mtDNA are evidently still able to replicate the mutant genome, and they are also able to transmit mitochondrial genetic markers through recombination in crosses with both $\rho^+$ and $\rho^-$ cells. No mitochondrial protein synthesis occurs in *petites* (Schatz and Saltzgaber 1969), presumably because a deletion of even the minimum size observed in *petites* is bound to eliminate or incapacitate one of the genes encoding an essential component of the mitochondrial translation apparatus. This means that replication and recombination of mtDNA are essentially under nuclear control and can certainly proceed in the absence of any mitochondrially synthesized proteins. However, the mitochondrial DNA must presumably possess at least one sequence recognized by mitochondrial DNA polymerase as an origin of replication (an *"ori"* or *"rep"* sequence or locus), and the fact that the repeat units of *petite* genomes can arise from different parts of the wild-type genome suggests that, in contrast to the mammalian situation (Crews *et al.* 1979), there may be several such origins.

The λdv plasmids of *E. coli* (Berg 1974; Chow *et al.* 1974) provide what may be an interesting analogy to the *petite* mutants. These plasmids are deletion mutants of bacteriophage λ. They are circular and consist of an oligomeric series of monomers, dimers, trimers, and higher-order polymers of a basic repeat unit, in various orientations. The repeated unit always arises from the region around the origin of replication. Oligomeric series of circular mtDNAs are also seen in *petite* mutants (Locker *et al.* 1974b; Lazowska and Slonimski 1976). Bernardi and colleagues have examined the DNA sequences of the repeat units of a number of spontaneous *petite* mutants in relation to potential origins of replication (*"ori"* sequences or loci) (Gaillard and Bernardi 1979; de Zamaroczy *et al.* 1979; Gaillard *et al.* 1980; Goursot *et al.* 1980; de Zamaroczy *et al.* 1981). They consider that there are at least seven *ori* sequences distributed unevenly around the wild-type mitochondrial genome, which act as origins of mtDNA replication (de Zamaroczy *et al.* 1981). All *ori* sequences have a common 265-bp stretch with two short (G + C)-rich clusters, A and B, flanking a 23-bp (A + T)-rich palindrome, at one end, separated from C—a long (G + C)-rich cluster at the other end—by an extensive (A + T)-rich sequence. A 300-bp sequence with a somewhat similar organization has been found in ethidium bromide-induced "hypersuppressive" *petites* (see below) by Blanc and Dujon (1980) and also identified as an origin of mtDNA replication. This *"rep1"* sequence is represented only once in the $\rho^+$ genome, but several related sequences also appear to be present, much as in the case of the *ori* sequences of the Bernardi group. Yeast *ori* sequences have homologies of both primary and secondary structure with the HeLa cell origin of mtDNA replication (de Zamaroczy *et al.* 1981), though yeast may be unusual in that its mtDNA can be replicated in the absence of a specific "origin of replication" sequence. Both the Bernardi group and Blanc and Dujon have encountered some *petites* lacking recognizable *ori* or *rep* sequences (i.e., $ori^0$ or $rep^0$; de Zamaroczy *et al.* 1981; Blanc and Dujon 1980). An example of such an $ori^0/rep^0$ *petite* would be the mutant RDIA sequenced by Van Kreijl and Bos (1977), in which the repeat sequence is a mere 68 bp, all (A + T) except for two (G + C) pairs. It is conceivable that *ori*-type sequences do occur in such apparently $ori^0/rep^0$ *petite* genomes but are present in a low copy number and are not stoichiometric with the repeat unit. Alternatively (and more probably), it may simply be that mtDNA polymerase can recognize and replicate virtually any mtDNA sequence, but does so much less efficiently in the absence of *ori/rep* sequences. This view is lent support by the observation that suppressive *petites* always appear to contain *ori/rep* sequences.

## Suppressivity

Suppressivity is a phenomenon that first came to light in the early period of yeast mitochondrial genetics. Suppressive *petites* are those that in crosses with wild-type give rise to a higher frequency of *petite* diploid progeny than is observed in the diploids issuing from the cross between the $\rho^+$ parent of the *petite* and the same wild-type tester strain (Ephrussi *et al.* 1955). The frequency of *petite* diploid offspring is characteristic of the parental suppressive *petite;* neutral *petites,* by contrast, give rise to the same frequency of *petite* diploid offspring as control crosses using the $\rho^+$ parent of the *petite.* Initial attempts to explain the phenomenon at a molecular level were inspired by work on the replication of Q$\beta$ DNA (Mills *et al.* 1967) and postulated that the mitochondrial genomes of suppressive *petites* possessed a replicative advantage over the wild-type genomes (Carnevali *et al.* 1969; Rank 1970). In later hypotheses it was suggested that suppressive $\rho^-$ genomes destroyed wild-type genomes through recombination with them (Michaelis *et al.* 1973; Perlman and Birky 1974; Clark-Walker and Miklos 1974; Slonimski and Lazowska 1977). Recent genetic evidence has shown that, in contradiction to the expectation of the recombination hypothesis, diploid *petites* issuing from a cross between a suppressive *petite* with multiple genetic markers and a wild-type, closely resembled the *petite* parent, lacking markers from the $\rho^+$ parent (Gingold 1981). Also, the results of the Bernardi group and Blanc and Dujon show that at a molecular level the mtDNA of *petite* diploid progeny is actually identical to that of suppressive or "supersuppressive (hypersuppressive)" *petite* parents (Goursot *et al.* 1980; Blanc and Dujon 1980), and the conclusion is drawn that suppressivity reflects a replicative advantage conferred on the *petite* genome by possession of multiple copies of *ori/rep* sequences. The significance of the *ori* sequences is substantiated by the observation that partial or total deletion or rearrangement of *ori* sequences is always accompanied by drastic changes of suppressivity (de Zamaroczy *et al.* 1981). However, this model still does not explain the differences in *petite* frequencies between different zygote clones arising from the same cross of suppressive $\rho^- \times \rho^+$; it may be, as discussed by Gingold (Gingold 1981), that only a very limited number of input mtDNA molecules are able to replicate in the zygote, and the progeny genomes are thus drawn from a small sample of the total input. The effects of media and growth conditions (Ephrussi *et al.* 1966) are other features of suppressiveness still awaiting explanation.

## mtDNA Replication and mtDNA Polymerase

In yeast, mtDNA is usually a rather high proportion of the total cell DNA content, about 10–15% (Hall *et al.* 1976), equivalent to roughly 50 mtDNA molecules per haploid cell (Williamson *et al.* 1977). mtDNA levels are affected by growth conditions and the physiological status of the cell, being reduced in glucose-repressed cells (Goldthwaite *et al.* 1974) and anaerobic cells depleted of ergosterol and unsaturated fatty acids (Nagley and Linnane 1972a) and increased during conjugation (Lee and Johnson 1977). It appears that *in vivo* mtDNA molecules may be associated in aggregates or "chondriolites" (Williamson *et al.* 1977), which may be the heritable units, rather than individual mtDNA molecules, or the fluid, dynamically fusing and fragmenting mitochondrial organelles themselves (see Stevens 1981). Though certainly not manifesting a chromatin-like structure, mtDNA is apparently bound *in vivo* to a lysine- and arginine-rich protein not dissimilar to a histone (Caron *et al.* 1979). Replication of mtDNA continues throughout the cell cycle (Williamson and Moustacchi 1971; Sena *et al.* 1975) and is likewise uncoupled from nuclear DNA replication in zygotes (Sena *et al.* 1976) and during meiosis (Pinõn *et al.* 1974). It is therefore not surprising to find that *cdc* mutants incapable of initiating nuclear DNA synthesis are still able to replicate their mtDNA, though mtDNA synthesis is blocked in *cdc* mutants unable to continue previously initiated nuclear DNA replication during S phase (Newlon and Fangman 1975).

Partial purifications of yeast mtDNA polymerase were achieved more than 10 years ago (Iwashima and Rabinowitz 1969; Wintersberger and Wintersberger 1970). A more recent preparation of higher purity (Wintersberger and Blutsch 1976) was found to lack exonuclease activity and to be particularly effective in gap-filling reactions, preferring a gapped template primer, poly(dA)·$(dT)_{10}$, to nicked DNA. The enzyme consists of a monomer of molecular weight 60,000, which at high salt concentrations forms a dimer of 120,000 and at low salt concentrations forms higher aggregates. Interestingly, in view of the effectiveness of manganese ions in inducing mitochondrial mutations (Putrament *et al.* 1973, 1977), the enzyme was very sensitive to inhibition by $Mn^{2+}$.

The specific activity of the DNA polymerase was found to be dependent on the degree of aeration of the cells and on the carbon source used in the growth medium. It was reduced severalfold by glucose repression and roughly doubled when the yeast cells were grown in the presence of ethidium bromide.

## Comparative Aspects of the *Petite* Mutation

That *petite* mutants are not merely laboratory curiosities, serendipitously available for mitochondrial gene mapping, is shown by the discovery of a naturally occurring respiratory-deficient yeast. *Candida slooffii* is a respiratory-deficient yeast that can be isolated from the anaerobic environment of various mammalian guts. Clark-Walker and colleagues have found that three different isolates of *C. slooffii* contain circular mitochondrial DNAs of different buoyant densities and contour lengths, so resembling the mitochondrial genomes of *petite* mutants of *S. cerevisiae* (Arthur *et al.* 1978). Since the mutant mtDNAs in these *C. slooffii "petites"* can apparently be eliminated by euflavine treatment, without impairing cell viability (Arthur *et al.* 1978), it seems that these strains are harboring and continuing to replicate a functionless "selfish" DNA species.

While the mtDNA of yeast species such as *C. slooffii* and *S. cerevisiae* is apparently entirely dispensable, this may not be the case for other, "*petite*-negative" yeast species, which appear unable to give rise to respiratory-deficient mutants. Bulder (Bulder, 1964a) examined a large number of yeast species and classified them as "*petite*-positive" or "*petite*-negative," depending on whether viable respiratory-deficient mutants could be obtained by acriflavine treatment. Later studies have shown that nuclear or "segregational" respiratory-deficient mutants can in fact be isolated from some *petite*-negative species such as *Schizosaccharomyces pombe, Kluyveromyces lactis,* and *Hansenula wingei* (Heslot *et al.* 1970; Wolf *et al.* 1971; Herman and Griffin 1968; Crandall 1973a,b). In some *petite*-negative species mitochondrial mutagens such as ethidium bromide and acriflavine give rise to what are probably cytoplasmic respiratory-deficient mutants, but these are unstable or inviable (Bulder 1964b; Heslot *et al.* 1970). The question therefore arises of the differences between *petite*-positive and *petite*-negative yeasts. One possibility considered by Clark-Walker and Miklos (Clark-Walker and Miklos 1974) is that the mitochondrial DNAs of *petite*-negative yeasts carry a gene or genes essential for cell growth and division, and certainly, much evidence has accumulated that even in *petite*-positive yeasts such as *Saccharomyces cerevisiae* the mitochondrial genome regulates nuclear gene systems not directly involved in mitochondrial functions (see next section). Alternatively, *petite*-positive yeasts may have special properties that favor the formation of stable deletion mutations in the mitochondrial genome; these might be the nature of the mtDNA replication–recombination–repair enzymes or the sequence organization of the mtDNA itself or, indeed, a combination of both. In analyzing these possibilities Clark-Walker and colleagues have investigated the mitochondrial genomes of a number of different yeasts, both *petite*-positive and negative. Four different species of *petite*-negative yeasts (*Candida parapsilosis, Hansenula wingei, Kluyveromyces lactis,* and *Schizosaccharomyces pombe*) all yielded mtDNAs substantially smaller (12–24 kd) than that of *Saccharomyces cerevisiae* (50 kd) and with higher buoyant densities, corresponding to 37–42% (G + C) (O'Connor *et al.* 1975). This could suggest that major structural features of the mtDNA are important in *petite* positivity. However, the mtDNA of the *petite*-positive yeast *Torulopsis glabrata* has a genome size of only 12 kd—not dissimilar to the 10–12-kd genomes of (*petite*-negative!) mammals—while the buoyant density of *T. glabrata* mtDNA (1.686 $g/cm^3$) implies an (A + T) content near that of *S. cerevisiae,* yet the spontaneous *petite* frequency is a thousandth of that of baker's yeast (O'Connor *et al.* 1976). These observations suggest that large-scale structural features such as mitochondrial genome size and content of (A + T)-rich sequences are not sufficient causes for the high *petite* frequency in *S. cerevisiae.* An examination of spontaneous *petite* frequencies (ranging from 0.003% to 9%) and mtDNA length (~19–108 kbp) in a number of *petite*-positive yeasts showed that the *petite* frequency did generally increase with mtDNA length, but the relationship is far from linear (Clark-Walker *et al.* 1981b). In one group of species, where mtDNA length varied from 56 to 108 kbp, the *petite* frequency was virtually constant, whereas it is well known that in different strains of *S. cerevisiae,* with genome sizes confined to the range 68–76 kbp, the *petite* frequency varies greatly and can be strongly influenced by both mitochondrial (Handwerker *et al.* 1973; Bandlow 1977) and nuclear (Rubin and Blamire 1977, 1979: Backhaus *et al.* 1978; Johnston 1979; Newlon *et al.* 1979; Marmiroli *et al.* 1980) mutations. It would seem likely that factors such as excision–repair–recombination enzymes (all nucleus-specified) may have an overriding role in determining the frequency of occurrence of the *petite* deletion mutation, whereas the viability of the resultant mutants may be a question of whether or not, in any particular species, there is, in the mtDNA, genetic information essential for cell growth or division.

Finally, it may be appropriate to recall that although many aspects of this mutation are still not thoroughly understood, the mutation has been of huge practical utility to mitochondrial geneticists. "Libraries" of *petite* mutants have facilitated genetic mapping of mtDNA (see above) and *petite* mutants have acted as a source of naturally "cloned" DNA for gene sequencing studies, such as those of Tzagoloff and colleagues. Such highly characterized *petite* "gene clones" can now be used as hybridization probes in investigations of mitochondrial genome organization in other yeasts, as well as in more distantly related organisms (see, e.g., Clark-Walker and Sriprakash 1981).

## Genetic Interactions Between Nucleus and Mitochondrion

### General Considerations

Isolating and analyzing mutants are a first, genetic, approach to disentangling what is likely to be an exceedingly complex system: the nucleomitochondrial relationship. At an abstract and simplistic level it is possible to envisage that a mutation in a gene in one "biogenetic" compartment might either increase or decrease the level or functional capacity of a single component in the other compartment or affect in a more or less complex way the levels of a group of two or more components. In the first case the gene could be a structural gene for the component involved or a rather specific regulatory gene. In the second case the gene may be involved in a metabolic pathway or biosynthetic route such that malfunction would cause multiple repercussions—a pleiotropic effect—or the gene may normally exercise a regulatory role over the expression or function of a number of gene products in the other compartment.

#### From Nucleus to Mitochondrion

In the particular case of the mitochondrial and extramitochondrial compartments the interactions in the sense of information flow would appear to be overwhelmingly, but not perhaps entirely, one way: The majority of mitochondrial polypeptides are encoded in the nucleus, but there is no evidence for a single instance of the reverse. The replication and expression of mitochondrial genes is of course also dependent on the nucleus insofar as the mtDNA polymerase, RNA polymerase, and mitoribosomal proteins are specified by nuclear genes. Those nuclear genes affecting the mitochondrial compartment that are likely to be most easily recognized are those that, through mutation, cause a major malfunction of the mitochondrion, usually seen as inability to grow on nonfermentable media (*pet*$^-$ mutations). Such nuclear mutations showing simple Mendelian inheritance ( "segregational" *petites*) could occur in genes encoding a polypeptide component of the respiratory complexes, or in a pathway—such as that of heme biosynthesis—that produces an important prosthetic group required by several enzymes. If the mutation affected mtDNA polymerase, one possible result would be a mutator effect on the mtDNA, whereas mutations impairing mitochondrial gene expression in general might simply produce a

$\rho^-$ *petite* phenocopy. Mutations affecting nuclear regulatory genes could have either simple or complex consequences on the mitochondrial phenotype, as alluded to earlier, and would probably be most easily recognized where there is a fairly specific effect on just one or a few gene products of the mtDNA itself.

Nuclear mutations of these types can be found by obtaining large collections of *pet*$^-$ mutants and subjecting them to detailed analysis (see, e.g., Tzagoloff *et al.* 1975d); the range of possible *pet*$^-$ mutants can be extended even further by isolating temperature-sensitive mutants (see, e.g., Burkl *et al.* 1976). More specific genetic approaches are available where selection can be applied. Although many of the mutations conferring resistance to drugs and antibiotics affecting the mitochondrial system display, entirely cytoplasmic inheritance, nuclear drug resistance mutations can also be isolated. Excluding those cases where resistance is caused by alterations outside the organelle itself, these could provide useful insights into nucleomitochondrial relationships. Nuclear mutations that suppress specific mitochondrial *ant*$^R$ or *mit*$^-$ mutations are also likely to be illuminating. Conversely, the isolation of mitochondrial suppressors of *pet*$^-$ or nuclear *ant*$^R$ mutations could prove fruitful. A third general category of suppressor mutation—mitochondrial suppressors of mitochondrial mutations (*mim* mutations)—is not strictly relevant here, and an example has already been discussed in relation to the intriguing "box effect" (see section on complex mitochondrial genes). The last class of potentially interesting suppressors, nuclear suppressors of nuclear mutations like *pet*$^-$, could obviously include mutations of no interest in the context of nucleomitochondrial relationships, but might also reveal interactions between nucleus-encoded mitochondrial components within the mitochondrion itself, or indeed, between such components and as yet unknown regulatory factors.

## From Mitochondrion to Nucleus

It is to be expected that the mitochondrial genome will interact with those nuclear genes or gene products directly involved in the biogenesis or function of the organelle, but it is not yet clear at what level(s) this might occur. There is no evidence for export of mitochondrially synthesized proteins, and Dujon (1981) considers that the "dogma" of no exchange of RNA molecules between mitochondrion and cytoplasm has never been invalidated experimentally. Small-molecular-weight species, perhaps products of mitochondrial functions, could have a signaling role, though another possibility is that once a particular level of biogenetic activity is "set" (for example, by nuclear control of mtDNA transcription), regulation of component supply from outside the mitochondrion can occur through feedback mechanisms.

As already mentioned, isolation of mitochondrial suppressors of nuclear *pet*$^-$ or *ant*$^R$ mutations represents one approach to the identification and analysis of nucleomitochondrial interactions. Close examination of the impact of particular mitochondrial mutations on levels and functions of nucleus-specified mitochondrial components may also be instructive. It is less easy to prescribe how to discern genetic influences of the mitochondrion on nuclear genes that do not have a direct mitochondrial involvement. Nevertheless, possible examples of this situation have been discovered by observations of unexpected or paradoxical features of the phenotypes of various mitochondrial mutants.

## *pet*$^-$, Mutator, and *cyc*$^-$ Mutants

The first nuclear mutations causing respiratory deficiency were discovered 20 years ago (Sherman 1963; Sherman and Slonimski 1964). At least 51 *pet* loci are now known, and nine of these have been chromosomally mapped (see Broach 1981), but it is clear, in view of the mitochondrion's biochemical complexity, that many more loci must exist: Genetic analysis of a large collection (270) of temperature-sensitive *pet*$^-$ mutants by Burkl and colleagues (Burkl *et al.* 1976) defined 106 complementation groups. Biochemical investigations of a smaller collection of *pet*$^-$ mutants (Tzagoloff *et al.* 1975d) revealed that several of the mutants were specifically deficient in individual mitochondrially synthesized polypeptides, suggestive of a regulatory phenomenon. Similar mutants have been found by Schatz and co-workers. In one of their studies a *pet*$^-$ mutant was found that specifically lacked subunit III of cytochrome oxidase (Ebner *et al.* 1973); as the mutation is suppressible by a nuclear amber suppressor (Ono *et al.* 1975) the gene product causing the deficiency of COIII must itself be a protein. By contrast, two other *pet*$^-$ mutants have been found in which levels of subunits I and II of cytochrome oxidase are severely depressed, but COIII continues to be synthesized (Cabral and Schatz 1978). Many other *pet*$^-$ mutants are known with defects of mitochondrial protein synthesis and a tendency to accumulate $\rho^-$ mutations (Rubin and Blamire 1977a; Schweizer *et al.* 1977; Johnston 1979; Newlon *et al.* 1979; Marmiroli *et al.* 1980); these nuclear mutations may reside in structural genes for components of the mtDNA replication–transcription–translation apparatus. Mutant alleles of specifically "regulatory" genes may occur among mutations like these, but recognizing them is unlikely to be easy. An interesting example of a *pet*$^-$ mutation that does not accumulate $\rho^-$ mutations is *pet9* (also known as $op_1$, Kovac *et al.* 1967; Beck *et al.* 1968). The mutation affects the adenine nucleotide translocator of the inner mitochondrial membrane and, for reasons that are not understood, causes the $\rho^-$ mutation to be lethal. Because of this property the mutation has been extremely useful in isolating limited lesions in the mitochondrial genome of the *mit*$^-$ type (Kotylak and Slonimski 1977).

A number of instances have been encountered of nuclear mutator genes that do not themselves confer respiratory deficiency but that provoke a high frequency of $\rho^-$ formation. In one such case (Evans and Wilkie 1978) the nuclear gene appeared to require a specific mitochondrial genotype and the presence of high glucose in the medium in order to manifest its mutagenic effect in the mitochondrion. Under the appropriate conditions mitochondrial mutations ranging from antibiotic resistance to *petite* deletions were generated at a high frequency, though the mutator effect could be suppressed by caffeine, an inhibitor of error-prone DNA repair. This situation may be compared with that of the *tsm8* mutation: Here a mitochondrial temperature-sensitive mutation mapping in the main tRNA cluster, near $tRNA^{Ile}$ (Monnerot *et al.* 1977), accumulates a high frequency of *petites* at the restrictive temperature (Handwerker *et al.* 1973), but it is not apparently known whether there are nuclear modifier genes.

*CYC* genes are those that specifically affect iso1- and/or iso-2-cytochrome c. At least nine loci have been identified and six of these have been chromosomally mapped (see Broach 1981). *CYC1*, which encodes iso-1-cytochrome c, is now one of the best characterized of all yeast nuclear genes, as it has been both cloned (Montgomery *et al.* 1978) and sequenced (Smith *et al.* 1979).

## Complex Drug-Resistant Mutants

There have been a number of reports of mutants cross-resistant to a variety of drugs, most of which exhibit a complex genetic behavior indicative of nucleocytoplasmic genetic interactions (Rank and Bech-Hansen 1973; Rank *et al.* 1975; Avner and Griffiths 1973; Mitchell *et al.* 1973: Guérineau *et al.* 1974; Howell *et al.* 1974; Lancashire and Griffiths 1975; Carignani *et al.* 1977), both mitotic and meiotic segregations of drug resistance and sensitivity being observed. The genetic models preferred have ranged from essentially single nuclear gene inheritance (Rank and Bech-Hansen 1973; Rank *et al.* 1975) through nucleomitochondrial interactions (Mitchell *et al.* 1973; Howell *et al.* 1974) to the possibility of episomal or plasmid involvement (Guérineau *et al.* 1974; Griffiths *et al.* 1975; Carignani *et al.* 1977). A system very similar to these has recently been investigated by Cohen and Eaton (1979). They isolated a mutant cross-resistant to five drugs—oligomycin, rhodamine 6G, tetracycline, chloramphenicol, and cycloheximide—genetic analysis disclosing that the multiple resistance is determined by a single nuclear gene, *ant1*, and requires three mitochondrial genes for expression of resistance to oligomycin, rhodamine 6G, and tetracycline. *ant1* was located on chromosome VII. This nucleocytoplasmic interaction is very like that analyzed by Mitchell and colleagues (Mitchell *et al.* 1973) in the case of an oligomycin-resistant mutant, and dual nucleocytoplasmic control is also clearly seen in the genetics of a mutant resistant to primaquin, an intercalating dye (Rotman 1975). There is a strong likelihood that an interpretation based on nucleomitochondrial genetic interaction can in fact be applied to others of the investigations cited earlier insofar as two of the relevant nuclear genes (Rank and Bech-Hansen 1973; and Guérineau *et al.* 1975) now appear to be allelic to *ant1* (Cohen 1977).

## Suppressors

Isolation of suppressor mutations provides a more defined, less open-ended genetic approach to the dissection of nucleomitochondrial relations than may mutant isolation and screening. Dujardin and colleagues have recently identified three nuclear suppressor loci that act on specific *mit*$^-$ mutations (Dujardin *et al.* 1980a,b). The suppressors are termed *NAM1* (from "nuclear accommodation of mitochondria"), *NAM2*, and *nam3*. *NAM1* and *nam3* are allele-specific but locus nonspecific, acting on *mit*$^-$ mutations in different genes, and are probably informational suppressors, whereas *NAM2* acts only on *box7*$^-$ mutants. The specificity of action of *NAM2* allows a fairly specific biochemical mechanism to be suggested: It must somehow restore the deficiency of splicing of *cob-box* transcripts that *box7* mutations are to known to impose.

Interestingly, suppressors of antibiotic resistance have been found that act in both directions along the nucleomitochondrial "axis." Colson and co-workers (Colson *et al.* 1974) isolated a single dominant nuclear mutation, conferring resistance to venturicidin and oligomycin as well as increased sensitivity to Synthalin. Introduction of a single mitochondrial oligomycin resistance mutation into diploids containing the nuclear resistance gene suppressed the venturicidin resistance. Conversely, nuclear genes have been found that suppress or modify mitochondrial resistance genes. These include a single dominant nuclear gene (MPR) that modifies paromomycin resistance, and a recessive nuclear gene *(mcr)* able to suppress chloramphenicol resistance; the dominance of *MPR* could suggest that it specifies a ribosomal protein or increases mitochondrial permeability, whereas *mcr*'s recessivity implies it is not encoding a ribosomal protein (Waxman *et al.* 1979). A mitochondrial mutant *tso-r* studied by the Linnane laboratory (Trembath *et al.* 1975) conferred a combination of cold-sensitive growth on nonfermentable media and oligomycin resistance. The cold sensitivity of *tso-r* was found to be suppressible by a nuclear gene, *STS*, but oligomycin resistance was largely unaffected: Trembath and colleagues suggested that *STS* specifies a protein promoting more effective assembly and integration of the mitochondrial ATPase, conceivably one of the F1 proteins or the oligomycin-sensitivity-conferring protein (OSCP).

## Mitochondria and the Regulation of Fermentation Systems

A major regulatory phenomenon in which both nucleocytoplasmic and mitochondrial compartments participate is glucose repression. Appropriate levels of glucose in the medium repress the synthesis of, for example, nonconstitutive dispensable sugar fermentation systems like that for galactose, as well as repressing mitochondrial biogenesis. Derepression of glucose-repressed mitochondrial respiration is probably mediated by cAMP (Fang and Butow 1970; Mahler and Lin 1978) and cAMP will also antagonize glucose repression of extramitochondrial enzymes, such as $\alpha$-glucosidase (Wiseman and Lim 1974).

The observation that chloramphenicol suppresses the effect of cAMP on α-glucosidase activity (Wiseman and Lim 1974) could suggest that the mitochondrial system does not merely passively respond to externally determined cAMP levels but may be involved in the cAMP-mediated glucose derepression mechanism for the cell as a whole. Genetic evidence for a mitochondrial role in glucose repression of sugar fermentation systems has also been provided by Haussmann and Zimmermann (1976) in a study of the effects of the *petite* mutation on the glucose repressibility of invertase synthesis. The evidence for a mitochondrial intervention in derepression–induction of catabolic enzymes extends to several fermentation pathways. Mitchondrial deficiency, stemmiing from either genetic lesion *(petite)* or treatment with antimitochondrial drugs, can block induction of permeases and degradative enzymes for the utilization of galactose, α methylglucoside, and maltose, though expression of constitutive enzymes appears to be unaffected (Kovac and Istenesova 1964; Puglisi and Algeri 1971; Evans and Wilkie 1976a,b; Evans *et al.* 1976; Khan and Greener 1977). Treatment of cells with the permeabilizing agent dimethylsulfoxide bypasses the block on α-glucosidase induction in *petites* unable to ferment α-methylglucoside (Mahler and Wilkie 1978), suggesting that the mitochondrial malfunction causing the sugar-negative phenotype in the *petites* affects cell membranes, either external or internal. The sugar utilization impairment in *petites* can also be overcome by mutation in the nuclear genome (Wilkie and Nudd 1981). This result is not unexpected, as the *petite* sugar utilization phenotype is strain dependent, and crosses between strains differing in the sugar utilization phenotypes of their *petites* have implied a nuclear gene involvement (Evans and Wilkie 1976a; Mahler and Wilkie 1978). These observations also jointly indicate that the block in sugar utilization is not a trivial consequence of respiratory deficiency or lack of mitochondrially generated ATP, *per se.*

In a recent study by Algeri and colleagues, the genetic difference between strains producing *petites* positive or negative for galactose utilization has been analyzed in detail (Algeri *et al.* 1981). They found that a single nuclear gene *IMP1*, and its recessive allele, *imp1*, can determine whether or not *petites* are able to utilize galactose (*IMP1 petites* are galactose-positive). This gene is not allelic with any of the known structural or regulatory genes of the galactose fermentation system. The fact that constitutive mutants of the *GAL80* gene (which probably specifies a repressor) are epistatic to *imp1* implies that the nucleomitochondrial interaction occurs at an early stage of the induction process (Algeri *et al.* 1981). Evidence for a similar situation in the case of maltose utilization has recently been published by Khan (Khan 1982). He found that the maltose utilization phenotype of *petites* from a strain carrying *MAL4* depended on a single nuclear gene, *PMU1*, with only *petites* carrying *PMU1* being maltose-positive.

In sum, the sugar utilization findings indicate that normal mitochondria generate an entity—probably not directly related to ATP production—that on interaction with a nuclear regulatory gene(s) or gene product(s) permits induction of the relevant catabolic enzymes to proceed.

## A Wider Role for Mitochondria?

Unexpected mitochondrial involvement also occurs in the developmental processes of sporulation and spore germination as well as in processes as fundamental as cell proliferation.

It has long been known that petite diploids are unable to sporulate (Ephrussi and Hottinguer 1951), and it has been assumed that this defect is a fairly trivial corollary of defective oxidative phosphorylation. However, Pratje and colleagues (Pratje *et al.* 1979) and Hartig and Breitenbach (1980) found that although some $mit^-$ diploids are unable to sporulate, others, with different $mit^-$ lesions, sporulate well, though they are completely unable to grow on nonfermentable media and have minimal residual respiratory activity. Spore germination exhibits a similar mitochondrial dependency, and respiration has been dissociated from respiratory impairment even more clearly: New types of mitochondrial mutant have been isolated that are almost normal in respiration and growth on nonfermentable substrates but that are unable to pass the first budding stage of germination (Hartig *et al.* 1981; Schroeder and Breitenbach 1981).

A series of experiments by Kovac and colleagues have indicated a mitochondrial role in cell proliferation that extends beyond mere energy provision (Kovac *et al.* 1977). The nuclear $op_l$ *(pet9)* mutation blocks adenine nucleotide translocation across the inner mitochondrial membrane and so renders cells unable to grow on nonfermentable media (Kovac *et al.* 1972). The $op_l$ cells can nevertheless grow well in glucose medium, as can *petite* mutants, but if both mutations, $op_l$ and cytoplasmic *petite,* are present simultaneously in the cell, a few more cell generations occur and then mitosis stops. This implies that there is progressive loss of a factor essential for cell division whose synthesis or activation is blocked when doubly mutant mitochondria are present (Kovacova *et al.* 1968). Not dissimilar observations have been made by Ebner and Schatz (1973), who studied a nuclear $pet^-$ mutant that lacked the F1 mitochondrial ATPase. The mutant had a greatly enhanced—20–30 × wild-type—rate of mutation to $\rho^-$. The cells that contained both nuclear $(pet^-)$ and cytoplasmic $(\rho^-)$ mutations were exceedingly slow growers, suggesting to the authors that yeast cells require mitochondria or some mitochondrial function, even after loss of both oxidative phosphorylation and the ability to synthesize mitochondrial proteins.

Another aspect of what may be the mitochondrion's wider role in the cell is seen in cell surface alterations consequent on genetic lesion in the mitochondrion. It has been found that *petite* mutants are much more susceptible to agglutination by the lectin Concanavalin A than the wild-type parental cells from which they derive (Evans *et al.* 1980). This implies that the *petite* mutation can affect the frequency, distribution, or accessibility of Concanavalin A binding sites—presumably mannan or mannoproteins—in the cell wall or plasmalemma. The same study also showed that *petite* mutants could differ from wild-type cells in electrokinetic characteristics (probably cell surface charge) as manifested by changes in the partitioning properties of the whole cells in immiscible aqueous solutions of polymers [poly(ethylene glycol) and dextran]. More important, these data demonstrated that the nature and degree of this change (affinity for poly(ethylene glycol) phase vs. dextran phase) is strain-dependent, and in one case at least appears to be under single nuclear-gene control, so that these effects do not appear to be a trivial consequence of respiratory deficiency or mitochondrial ATP deprivation.

Flocculence is another cell surface characteristic affected by mitochondrial malfunction: The *petite* mutation can suppress the flocculating ability manifested by the parental respiratory-competent strain (Evans *et al.* 1980). In a comprehensive investigation of the interaction between the *petite* mutation and flocculence, Holmberg and Kielland-Brandt (1978) found that not all *petite* mutants isolated from flocculent strains lose their flocculence, and even increases in flocculence may occur. Moreover, *petites* isolated from the same strain can differ significantly in flocculence. The authors drew conclusions analogous to those of Ebner and Schatz (1973), presuming that some mitochondrial functions other than oxidative phosphorylation and protein synthesis are involved in the expression of flocculence.

The many findings discussed in this section, together indicating that the mitochondrial biogenetic system intervenes in important extramitochondrial affairs, have recently prompted a reconsideration of the mitochondrial role in carcinogenesis. This has been given impetus by the multitudinous and still-growing reports of mitochondrial aberration in cancer cells and by the many observations that, in eukaryotes from yeast to mammals, chemical carcinogens can accumulate in mitochondria and selectively change mtDNA. Extended discussion of these topics and a new model of carcinogenesis through mitochondrial mutation are given in Wilkie and Evans (1982) and Wilkie *et al.* (1982).

## Evolutionary Aspects

The origins of mitochondria and their genomes are a subject of long-standing controversy, but as detailed information—including DNA sequences—proliferates, speculative phylogenetic trees are likely to become less speculative and more firmly rooted.

As is made clear by several recent reviews (Mahler 1981; Gray 1982; Gray and Doolittle 1982; Wallace 1982), two of the most striking features of mitochondrial genomes in general are their enormous heterogeneity of size and sequence organization, apparent even within some closely related organisms, and, in contrast, their surprisingly uniform content of genetic information. This functional conservatism of mitochondria—all characterized major mtDNA gene products being the same, except for subunit 9 of the ATPase (the DCCD-binding proteolipid), could suggest that mitochondria are monophyletic in origin. Hypotheses concerning this origin (see Dodson 1979) fall essentially into two classes: endosymbiont hypotheses (see Margulis 1970), termed exogenous theories by Taylor (1976), and developmental (see Cavalier-Smith 1975) or autogenous (Taylor 1976) hypotheses. The latter class includes the hypothesis of plasmid sequestration (Meyer 1973) developed at length by Mahler and Raff (1975a,b). These hypotheses have been vigorously and extensively debated in the cited reviews and elsewhere, and comment here will be strictly limited. However, it may be said that current evidence and interpreters seem to favor an endosymbiont origin (see Gray 1982; Gray and Doolittle 1982; Wallace 1982). The endosymbiont hypothesis of course proposes that mitochondrial and plastid genomes are descended from the genomes of free-living prokaryotelike (or "eubacterium"-like, see Woese 1981) organisms that entered into a symbiotic association with a proto-eukaryote.

RNA and DNA sequence analysis now strongly supports an endosymbiotic origin of the chloroplast genome from photosynthetic prokaryotes similar to the modern cyanobacteria (Doolittle and Bonen 1981; Gray and Doolittle 1982). The diversity of mitochondrial genomes could obscure similar homologies, but ribosomal RNA genes in particular may prove useful indicators of phylogenetic origins. Thus, on the basis of small ribosomal subunit RNA sequence information, Woese and collaborators have suggested creating a third kingdom, the Archaebacteria, in addition to the true bacteria (Eubacteria) and eukaryotes (Woese and Fox 1977; Fox *et al.* 1980). It is proposed that the archaebacteria, eubacteria, and "urkaryotes" diverged from a universal "progenote," subsequent endosymbiotic colonization of urkaryotes by primitive eubacteria generating eukaryotes containing mitochondria (Woese 1981). Conceivably, more than one colonization occurred: On the basis of cytochrome c data it is suggested that mitochondria originated independently three times, one yielding modern *Tetrahymena* mitochondria; another giving rise to *Crithridia* and *Euglena* mitochondria; and a third yielding fungal, animal, and plant mitochondria (Dayhoff and Schwartz 1981). Support for the basic endosymbiont hypothesis is given by the significant homologies seen between *E. coli* 23S rRNA and the large mitoribosomal subunit RNA of mouse (Van Etten *et al.* 1980), *S. cerevisiae* (Dujon 1980), and tryponosome (I. C. Eperon, quoted in Borst 1980a), and between *E. coli* 16S rRNA and the small mitoribosomal subunit RNA of mouse (Van Etten *et al.* 1980), *S. cerevisiae* (Sor and Fukuhara 1980), *Aspergillus nidulans* (Küntzel *et al.* 1982), wheat (Bonen *et al.* 1977), and trypanosome (I. C. Eperon, in Borst 1980a).

Despite these sequence homologies there are still, in the opinions of some commentators, severe obstacles to the endosymbiont hypothesis. There are, in particular, the various divergences from the "universal" genetic code in different mitochondrial genomes, the expanded codon recognition patterns, enabling use of a minimal number of tRNAs, poor homology between bacterial and mitochondrial tRNAs, and the presence of introns in some mitochondrial genomes. Some of these features, such as the mitochondrial introns, with their requirement for a splicing machinery, could represent archaic survivals from a remote eubacterial past, since lost in modern eubacteria and in the mitochondria of many eukaryotes; as Reanney (1979) has suggested, RNA splicing mechanisms may originally have embodied a primitive means of genetic recombination. Other features, such as the use of a restricted number of tRNAs, divergences in the genetic code, as well as mitochondrial genome diversity in general, may derive from the special situation of mitochondria as populous symbionts living in the protected environment of the "urkaryotic" cell: Natural selection may have been kinder to "experiments" and accidents affecting endosymbiont genomes, especially if the functions of the latter could sometimes be dispensed with, as in the modern yeast *S. cerevisiae* (see Borst 1980a,1981a). Another consideration may be the need of the primitive eukaryotic cell to develop mechanisms preventing translation of mRNAs that have leaked through the mitochondrial membranes into the wrong cell compartments. The transformation of the opal stop codon into a tryptophan codon may have originated as such a translational barrier (see Wallace 1982).

An implication or even requirement of the endosymbiont theory is that in the course of early eukaryotic evolution the majority of the genes required by the originally free-living eubacterial colonist were either lost or transferred to the nucleus. From this viewpoint the current mitochondrial genetic complement is merely the residue of such a transference, a process that may have become blocked by the divergence of mitochondrial and nuclear genetic codes (Borst and Grivell 1978). A clear example of gene transference seems to be the gene for subunit 9 of the ATPase. The product of this gene is mitochondrially specified in yeasts, but in *Neurospora crassa* it is encoded by a nuclear gene on chromosome VII (Sebald 1977; Sebald *et al.* 1977; Sebald *et al.* 1979a) and the gene is absent from mammalian mtDNA (Anderson *et al.* 1981). Since the *Neurospora* protein is apparently synthesized in the cytosol as a preprotein containing an N-terminal leader sequence that is removed during maturation (see Mahler, 1981), the *Neurospora* nuclear gene is likely to contain an extra sequence not present in the yeast mitochondrial gene, and Mahler (1981) has argued that this may constitute evidence for gene transfer in the opposite direction (i.e., from nucleus to mitochondrion) from that usually thought likely. An intriguing new finding is that *N. crassa* mtDNA contains an open reading frame, the putative product of which would be a DCCD-binding protein actually showing more similarity to the DCCD-binding proteins of yeast and bovine heart than to that encoded in the *N. crassa* nucleus (Van den Boogaart *et al.* 1982). Van den Boogaart and colleagues consider that the striking conservation of the primary structure of the (putative) mitochondrially encoded polypeptide argues that the mitochondrial gene is still active (though there is no positive evidence of this), as inactivation of the gene would have resulted in its loss or transformation into a pseudogene (see Little 1982).

While the question of mitochondrial origins is an entertaining problem stemming from events in an extremely remote and inaccessible past, and one that may never be resolved to universal satisfaction, studies on modern yeasts may provide valuable insights into mechanisms of mitochondrial genome diversification. As noted previously (and see Wallace 1982), the yeasts are highly heterogeneous in the size and gross organization of their mtDNAs. Gene location and homology in various yeasts can now be readily studied by hybridization with well-characterized mitochondrial gene probes derived from *S. cerevisiae.* Experiments of this sort have shown that the arrangement of ATPase subunit genes and ribosomal subunit RNA genes in *Kluyveromyces lactis* and *Saccharomycopsis (=Yarrowia) lipolytica* mtDNA is similar to that in *S. cerevisiae,* though the ribosomal rRNA genes are much closer together [~2000 bp in *K. lactis* and ~4000 bp in *Y. lipolytica* (Groot and Van Harten-Loosbraek 1980; Wesolowski *et al.* 1981)]. By contrast, the possibly more distantly related yeasts *Torulopsis glabrata* and *Kloeckera africana* differ in the order of several polypeptide-encoding genes both from *S. cerevisiae* and from each other (Clark-Walker and Sriprakash 1981). These rearrangements and dramatic genome size differences may arise from the existence of the multiple origins of replication observed in intergenic "spacer" regions of *S. cerevisiae* by Bernardi and colleagues and Blanc and Dujon (see preceding section on the *petite* mutation and mtDNA replication). As Bernardi has suggested (see Grivell 1981), multiple origins could permit large mitochondrial genomes to replicate rapidly and could even confer a selective replicative advantage (see preceding discussion of suppressive *petites*); they would also provide regions of homology, dispersed through the genome ("nonhomologous") recombination between which could result in gene rearrangements, thus promoting rapid evolutionary divergence.

## Conclusions and Prospects

The most important single recent development in yeast mitochondrial molecular genetics has probably been the application of rapid DNA-sequencing techniques, especially that of Maxam and Gilbert (1977), to yeast mtDNA. The normal gene sequencing preliminaries, involving bacterial transformation and cloning procedures, have usually been bypassed using the natural *in vivo* gene-cloning system offered by the *petite* mutation. It has thus been possible in the last four or five years to sequence all the known yeast mitochondrial genes, barring one or two genes for tRNAs. These DNA sequences have

1. Given precise amino acid sequences and molecular weights for hitherto poorly characterized polypeptides specified by mtDNA.
2. Confirmed the presence of introns in rRNA- and protein-coding genes and demonstrated that in some yeast strains certain introns are absent, and must therefore be "optional."
3. Revealed open reading frames—mostly in introns—corresponding to no known protein product: unassigned reading frames or "URF"s.
4. Shown that the yeast mitochondrial genome, like that of other eukaryotes, does not adhere to the "universal" genetic code.
5. Tended to support the endosymbiont hypothesis of mitochondrial origins.

The presence of introns in all three split mitochondrial genes—21S rRNA, *cob-box*, and *oxi3*—had been indicated by a considerable weight of evidence before the definitive DNA sequence results. Biochemical techniques such as R looping elegantly revealed the major intron in the 21S rRNA gene of $\omega^+$ strains, and methods such as the construction of restriction fragment maps combined with genetic mapping showed that the *cob-box* and *oxi3* genes are much longer than are required to encode the relevant gene products. On purely genetic grounds—the complementation behavior of different classes of mutant blocking apocytochrome b expression—it was suspected that *cob-box* was split or mosaic—a tribute to the power and sophistication of mitochondrial genetics in *Saccharomyces cerevisiae*.

The existence of multiple introns obviously necessitates a more or less complex splicing machinery to transform primary transcripts into functional ribosomal or messenger RNAs. Once again a problem—in this case, splicing mechanisms—has been illuminated by a study of mitochondrial mutants. The mere fact that in *petites*, which lack mitochondrial protein synthesis, certain introns could be excised, showed that some splicing events, at least, are quite independent of mitochondrial translation products. On the other hand, intense scrutiny of other kinds of mitochondrial mutant, especially those lying within or close to *cob-box* introns, indicated that successful splicing of *cob-box* transcripts requires translation of intronic sequences into diffusible factors capable of complementing intron mutations in *trans*. The fruit of these observations is the remarkable and uniquely mitochondrial "maturase" concept: a "fusion protein" jointly specified by intronic and exonic sequences that acts in the removal of intronic sequences from its own messenger, thereby destroying it, but in the process carrying the maturation of precursor RNA one stage further on. These studies of *cob-box* mutants have also revealed an intriguing interaction between the two mosaic protein-coding genes. It now appears that this interaction, the "box effect," reflects a requirement for splicing of *oxi3* messenger precursor RNA by a *cob-box*-specified maturase.

The mitochondrion is of course the product of the collaboration of two distinct genetic systems, with over 90% of mitochondrial proteins specified in the nucleus and synthesized in the cytoplasm, and it is this fact that makes mitochondrial biogenesis such an attractive and difficult problem. Mechanisms of protein import from the cytoplasm have received particular attention, and it now seems that in many (but not all) cases, imported proteins first appear in the cytoplasm as precursors larger than the mature forms and that transport across the mitochondrial membranes is energy dependent and involves a proteolytic cleavage of the precursor. The question now naturally arises of the nature and distribution of the processing proteases and their relationship to the receptors that are probably necessary for recognizing and binding precursors at the mitochondrial surface. Here, with the problem of membrane assembly, as elsewhere in the mitochondrial field, molecular biology is steadily annexing cell-biological territory.

Although most of the information flow between nucleus and mitochondrion must be one-way, from nucleus to mitochondrion, the possibility of mitochondrial "talk-back" has not been neglected. Interactions essentially concerned only with the mitochondrion and its biogenesis have been probed by the isolation of new nuclear and mitochondrial mutants, especially suppressors. Examination of various mutants has also implicated the mitochondrion in the regulation of nuclear genes concerned with general cell functions such as sugar fermentation, sporulation, and germination.

It is pleasing to record that the organism used in the first mitochondrial genetic crosses has remained the foremost experimental system in the whole field of mitochondrial molecular genetics and that it is still rewarding the faithful with novel phenomena and concepts: maturases, circular splicing products, optional introns, and "teaser-genes" like *var1*. Undoubtedly, *S. cerevisiae* will provide many more insights into mechanisms of mitochondrial biogenesis and the evolution of organellar genomes.

Some likely future developments will now be considered.

Sequencing of the yeast mitochondrial genome will probably soon be completed. This will require tolerance of inordinate numbers of As and Ts on sequencing gels, but more URFs may await discovery as well as more sites of repetitive sequence elements and replication origins. Sequencing of mutations inside and bordering on introns will assist in identification of sequences important to intron functions in the specification of maturase–splicing factors and in polarity–gene conversion events. Similarly, a greater understanding of the sequence requirements of the splicing enzymes should accrue. It may also be possible to investigate some elements of the splicing machinery (e.g., the maturases) by *in vitro* transcription and translation of relevant DNA fragments; in this way the problem of the low steady-state levels in wild-type cells may be overcome, allowing more direct examination of the proteins and their functions. Nuclear genes encoding splicing enzymes—as other nuclear genes affecting mitochondria—are now, of course, potentially accessible through standard genetic engineering techniques, provided mutants are available for complementation; the products of these genes could then also be generated *in vitro* and further characterized.

The range of mitochondrial mutants will expand enormously as "reverse genetics" begins to be applied to the system: recombinant DNA techniques will allow *in vitro* introduction of very precise sequence alterations and deletions into mitochondrial genes and regulatory elements that can then be transformed back into the cell in order to observe the phenotypic effects (see Nagley *et al.* 1980). This will be a way of analyzing origins of replication and promoter sequences as well as of attempting to discover the functions of URFs. Yeast transformation also allows exploration of evolutionary mechanisms and relationships to be extended beyond comparison of DNA sequences: Mitochondrial genes from other yeasts and fungi, or from more distantly related organisms, can be introduced into the mitochondrial genome of *S. cerevisiae* and their *in vivo* performance assessed. Development of transformation systems in other organisms will permit reciprocal experiments.

Suitable *in vitro* systems may also have application in furthering the detailed analysis of mitochondrial protein import and membrane assembly: A great deal will indeed be known about the mitochondrial inner membrane if and when it proves possible to construct the membrane *in vitro* from purified components.

While the last few years have seen the accumulation of a wealth of fundamental knowledge of yeast mitochondria, little of this has been applied to the needs and problems of the food and fermentation industries based on yeast; this may change. Already it has been found that *petite* mutants can be more effective alcohol producers than wild-type ones. This possibility reflects alterations in the glycolytic pathway and in hexokinase activity in particular (Bacila *et al.* 1978b). Of more immediate practical relevance is the use of mitochondrial mutants, especially the easily identified induced *petite* mutation, in hybridizations and protoplast fusions involving industrial yeast strains (Spencer and Spencer 1981).

In a final glance into the farther future we can recall puzzling aspects of mitochondrial behavior, long known to cell biologists, such as apparently ordered mitochondrial movements within the cell and in relation to other organelles (see Novikoff 1961) and wonder how molecular genetics will help here and if yeast will still have a role in finding answers.

**Acknowledgments**

I would first like to record my thanks to Professor David Wilkie and my friends and colleagues in his laboratory for introducing me to yeast and to mitochondrial genes. I am also very grateful to Professor Giorgio Bernardi and his colleagues for the hospitality shown me in his laboratory during the tenure of two EMBO Fellowships and for their patience when initiating me into the arts of preparing mitochondrial DNA. Again, I am very appreciative of the assistance and stimulating discussions afforded me by Professor Bevan and his colleagues in the School of Biological Sciences at Queen Mary College. Finally, thanks are owed to my wife; to Mrs. Joan Stratford, my typist; and to Dr. Frank Spencer, the Senior Editor, for varying degrees of forbearance.

## References

Abelson, J. (1979). RNA processing and the intervening sequence problem. Ann. Rev. Biochem. 48:1035–1069.

Alexander, N. J., Vincent, R. D., Perlman, P. S., Miller, D. H., Hanson, D. K., Mahler, H. R. (1979). Regulatory interactions between mitochondrial genes. 1. Genetic and biochemical characterization of some mutant types affecting apocytochrome b and cytochrome oxidase. J. Biol. Chem. 254:2471–2479.

Alexander, N. J., Perlman, P. S., Hanson, D. K., Mahler, H. R. (1980). Mosaic organization of a mitochondrial gene: evidence from double mutants in the cytochrome b region of *Saccharomyces cerevisiae*. Cell 20:199–206.

Algeri, A. A., Bianchi, L., Viola, A. M., Puglisi, P. P., Marmiroli, N. (1981). *IMP1/impl:* a gene involved in the nucleo-mitochondrial control of galactose fermentation in *Saccharomyces cerevisiae*. Genetics, 97:27–44.

Allet, B., Rochaix, J-D. (1979). Structure analysis at the ends of the intervening DNA sequences in the chloroplast 23S ribosomal genes of *C. reinhardtii*. Cell 18:55–60.

Amalric, F., Merkel, C., Gelfand, R., Attardi, G. (1978). Fractionation of mitochondrial RNA from HeLa cells by high resolution electrophoresis under strongly denaturing conditions. J. Mol. Biol. 118:1–25.

Anderson, S., Bankier, A. T., Barrell, B. G., de Bruijn, M. H. L., Coulson, A. R., Drouin, J., Eperon, I. C., Nierlich, D. P., Roe, B. A., Sanger, F., Schreier, P. H., Smith, A. J. H., Staden, R., Young, I. G. (1981). Sequence and organization of the human mitochondrial genome. Nature 290:457–465.

Anderson, S., Bankier, A. T., Barrell, B. G., de Bruijn, M. H. L., Chen, E., Coulson, A. R., Drouin, J., Eperon, I. C., Nierlich, D. P., Roe, B. A., Sanger, F., Schreier, P. H., Smith, A. J. H., Staden, R., Young, I. G. (1982). Comparison of the human and bovine mitochondrial genomes. In: Mitochondrial Genes. Edited by P. Slonimski, P. Borst, and G. Attardi. Cold Spring Harbor, N.Y.: Cold Spring Harbor Laboratory, pp. 5–43.

Arnberg, A. C., Van Ommen, G-J. B., Grivell, L. A., Van Bruggen, E. F. J., Borst, P. (1980). Some yeast mitochondrial RNAs are circular. Cell 19:313–319.

Arthur, H., Watson, K., McArthur, C. R., Clark-Walker, G. D. (1978). Naturally occurring respiratory deficient *Candida slooffii* strains resembling *petite* mutants. Nature 271:750–752.

Ashwell, M., Work, T. S. (1970). The biogenesis of mitochondria. Ann. Rev. Biochem. 39:251–290.

Avner, P. R., Griffiths, D. E. (1973). Studies of energy-linked reactions. Genetic analysis of oligomycin-resistant mutants of *Saccharomyces cerevisiae*. Eur. J. Biochem. 32:312–321.

Bacila, M., Horecker, B. L., Stoppani, A. O. M. (1978a). Biochemistry and Genetics of Yeast. Pure and Applied Aspects. New York: Academic Press.

Bacila, M., Wanderley-Xavier, A., Horii, J. (1978b). Induction of respiration-deficient mutants of *Saccharomyces* and evaluation of their efficiency for ethanol production. In: Biochemistry and Genetics of yeast. Pure and applied aspects, edited by M. Bacila, B. L. Horecker, and A. O. M. Stoppani. New York: Academic Press, pp. 577–594.

Backhaus, B., Schweyer, R. J., Kaudewitz, F., Dujon, B. (1978). On the formation of $\rho^-$ *petites* in yeast. III. Effects of temperature on transmission and recombination of mitochondrial markers and on $\rho^-$ cell formation in temperature sensitive mutants of *Saccharomyces cerevisiae*. Mol. Gen. Genet. 161:153–173.

Bandlow, W. (1977). Derepression of mitochondrial enzymes in yeast wild type and a mitochondrial conditional mutant, *tsm 8,* retarded in oxidative adaptation. In: Mitochondrial 1977; Genetics and Biogenesis of Mitochondria, edited by W. Bandlow, R. J. Schweyen, K. Wolf, and F. Kaudewitz. Berlin: De Gruyter, pp. 533–541.

Bandlow, W., Schweyen, R. J., Wolf, K., Kaudewitz, F., Eds. (1976). Molecular Biology of Nucleocytoplasmic Relationships. Berlin: De Gruyter.

Bandlow, W., Schweyen, R. J., Wolf, K., Kaudewitz, F., Eds. (1977). Genetics and Biogenesis of Mitochondria. Berlin: De Gruyter.

Barrell, B. G., Bankier, A. T., Drouin, J. (1979). A different genetic code in human mitochondria. Nature 282:189–194.

Barrell, B. G., Anderson, S., Bankier, A. T., De Bruijn, M. H. L., Chen, E., Coulson, A. R., Drouin, J., Eperon, I. C., Nierlich, D. P., Roe, B. A., Sanger, F., Schreier, P. H., Smith, A. J. H., Staden, R., Young, I. G. (1980). Different patterns of codon recognition by mammalian mitochondrial tRNAs. Proc. Natl. Acad. Sci., USA 77:3164–3166.

Battey, J., Clayton, D. A. (1978). The transcription map of mouse mitochondrial DNA. Cell 14:143–156.

Beattie, D. S., Chen, Y.-S., Clejan, L., Lin, L-F. H. (1979). Electrophoretic behaviour of cytochrome *b* in a partially purified preparation and evidence for high molecular weight associated mitochondrial translation products. Biochemistry 18:2400–2406.

Beck, J. C., Mattoon, J. R., Hawthorne, D. C., Sherman, F. (1968). Genetic modification of energy conserving systems in yeast mitochondria. Proc. Natl. Acad. Sci. USA 60:186–193.

Berg, D. E. (1974). Genetic evidence for two types of gene arrangements in new λdv plasmid mutants. J. Mol. Biol. 86:59–68.

Berlani, R. E., Pentella, C., Macino, G., Tzagoloff, A. (1980a). Assembly of the mitochondrial membrane system: Isolation of transfer ribonucleic acid mutants and characterization of transfer ribonucleic acid genes of *Saccharomyces cerevisiae*. J. Bacteriol. 141:1086–1097.

Berlani, R. E., Bonitz, S. G., Coruzzi, G., Nobrega, M., Tzagoloff, A. (1980b). Transfer RNA genes in the cap-oxi 1 region of yeast mitochondrial DNA. Nucl. Acids Res. 8:5017–5030.

Bernard, U., Pühler, A., Mayer, F., Küntzel, H. (1975). Denaturation map of the circular mitochondrial genome of *Neurospora crassa*. Biochim. Biophys. Acta 402:270–278.

Bernardi, G. (1976). Organization and evolution of the mitochondrial genome of yeast. J. Mol. Evol. 9:25–35.

Bernardi, G. (1979). The petite mutation in yeast. Trends Biochem. Sci. 4:197–201.

Bernardi, G., Piperno, G., Fonty, G. (1972). The mitochondrial genome of wild type yeast cells. I. Preparation and heterogeneity of mitochondrial DNAs. J. Mol. Biol. 65:173–189.

Bernardi, G., Prunell, A., Fonty, G., Kopecka, H., Strauss, F. (1976). The mitochondrial genome of yeast: organization, evolution and the *petite* mutation. In: The Genetic Function of Mitochondrial DNA, edited by C. Saccone and A. M. Kroon. Amsterdam: Elsevier/North-Holland, pp. 185–198.

Bianchetti, R., Lucchini, G., Corsti, P., Tortora, P. (1977). Dependence of mitochondrial protein synthesis initiation on formylation of the initiator methionyl-tRNA. J. Biol. Chem. 252:2519–2523.

Bibb, M. J., Von Etten, R. A., Wright, C. T., Walberg, M. W., Clayton, D. A. (1981). Sequence and gene organization of mouse mitochondrial DNA. Cell 26:167–180.

Birky, C. W., Jr. (1978). Transmission genetics of mitochondria and chloroplasts. Ann. Rev. Genet. 12:471–512.

Birky, C. W., Jr., Perlman, P. S., Byers, T. J., Eds. (1975). Genetics and Biogenesis of Mitochondria and Chloroplasts. Columbus, Ohio: Ohio State University Press.

Blanc, H., Dujon, B. (1980). Replicator regions of the yeast mitochondrial DNA responsible for suppressiveness. Proc. Natl. Acad. Sci. USA 77:3942–3946.

Boardman, N. K., Linnane, A. W., Smillie, R. M., Eds. (1971). Autonomy and Biogenesis of Mitochondria and Chloroplasts. Amsterdam: Elsevier/North-Holland.

Boehni, P., Gasser, S., Leaver, C., Schatz, G. (1980). A matrix-localized mitochondrial protease processing cytoplasmically-made precursors to mitochondrial proteins. In: The Organization and Expression of the Mitochondrial Genome, edited by A. M. Kroon and C. Saccone. Amsterdam: Elsevier/North-Holland, pp. 423–433.

Bohnert, H. J. (1977). Size and structure of mitochondrial DNA from *Physarum polycephalum*. Exp. Cell Res. 106:426–430.

Bohnert, H. J., Herrmann, R. G. (1974). The genomic complexity of *Acanthamoeba castellanii* mitochondrial DNA. Eur. J. Biochem. 50:83–90.

Bolotin, M., Coen, D., Deutsch, J., Dujon, B., Netter, P., Petrochilo, E., Slonimski, P. P. (1971). La recombinaison des mitochondries chez *Saccharomyces cerevisiae*. Bull. Inst. Pasteur 69:215–239.

Bonen, L., Cunningham, R. S. , Gray, M. W., Doolittle, W. F. (1977). Wheat embryo mitochondrial 18S ribosomal RNA: evidence for its prokaryotic origin. Nucl. Acids Res. 4:663–671.

Bonitz, S. G., Tzagoloff, A. (1980). Assembly of the mitochondrial membrane system. Sequences of yeast mitochondrial tRNA genes. J. Biol. Chem. 255:9075–9081.

Bonitz, S. G., Coruzzi, G., Thalenfeld, B. E., Tzagoloff, A., Maciono, G. (1980a). Assembly of the mitochondrial membrane system. Structure and nucleotide sequence of the gene coding for subunit I of yeast cytochrome oxidase. J. Biol. Chem. 255:11927–11941.

Bonitz, S. G., Berlani, R., Coruzzi, G., Li, M., Macino, G., Thalenfeld, B. E., Tzagoloff, A. (1980b). Codon recognition rules in yeast mitochondria. Proc. Natl. Acad. Sci. USA 77:3167–3170.

Borst, P. (1972). Mitochondrial nucleic acids. Ann. Rev. Biochem. 41:333–376.

Borst, P. (1977a). Structure and function of mitochondrial DNA. Trends Biochem. Sci. 2:31–34.

Borst, P. (1977b). Structure and function of mitochondrial DNA. In: International Cell Biology 1976–1977, edited by B. R. Brinkley and K. R. Porter. New York: Rockefeller University Press, pp. 237–244.

Borst, P. (1980a). The optional introns in yeast mitochondrial DNA. In: Biological Chemistry of Organelle Formation, edited by T. Bücher, W. Sebald, and H. Weiss. Berlin: Springer-Verlag, pp. 27–41.

Borst, P. (1980b). Control of mitochondrial biogenesis in yeast. In: Alfred Benzon Symposium 16: Molecular Genetics in Yeast. Edited by D. Von Wettstein, J. Friis, M. Keilland-Brandt, A. Stenderup. Copenhagen: Munksgaard, pp. 41–53.

Borst, P. (1981a). The biogenesis of mitochondria in yeast and other primitive eukaryotes. In: International Cell Biology 1980–1981, edited by H. G. Schweiger. Berlin: Springer-Verlag, pp. 239–249.

Borst, P. (1981b). Biosynthesis of mitochondria and chloroplasts. In: Molecular Biology of the Cell, edited by S. Alberts, K. Porter, M. C. Raff, K. Roberts, J. D. Watson, New York: Garland.

Borst, P., Flavell, R. A. (1976). Properties of mitochondrial DNAs. In: Handbook of Biochemistry and Molecular Biology, 3rd ed., edited by G. D. Fasman. Cleveland, Ohio: C. R. C. Press, pp. 363–374.

Borst, P., Grivell, L. A. (1973). Mitochondrial nucleic acids. Biochimie 55:801–804.

Borst, P., Grivell, L. A. (1978). The mitochondrial genome of yeast. Cell 15:705–723.

Borst, P., Hoeijmakers, J. H. J. (1979). Kinetoplast DNA. Plasmid 2:20–40.

Borst, P., Kroon, A. M. (1969). Mitochondrial DNA: physicochemical properties, replication and genetic function. Int. Rev. Cytol. 26:107–190.

Borst, P., Bos, J. L., Grivell, L. A., Groot, G. S. P., Heyting, C., Moorman, A. F. M., Sanders, J. P. M., Talen, J. L., Van Kreijl, C. F., Van Ommen, G. J. B. (1977). The physical map of yeast mitochondrial DNA: Anno 1977. In: Mitochondria 1977: Genetics and Biogenesis of Mitochondria, edited by W. Bandlow, R. J. Schweyen, K. Wolf, and F. Kaudewitz. Berlin: De Gruyter, pp. 213–254.

Borst, P., Sanders, J. P. M., Heyting, C. (1979). Biochemical methods to locate genes on the physical map of yeast mitochondrial DNA. Meth. Enzymol. 56:182–197.

Borst, P., Hoeijmakers, J. H. J., Frasch, A. C. C., Snijders, A., Janssen, J. W. G., Fase-Fowler, F. (1980). The kinetoplast DNA of *Trypanosoma brucei:* structure, evolution, transcription, mutants. In: The Organization and Expression of the Mitochondrial Genome, edited by A. M. Kroon and C. Sacconne. Amsterdam: Elsevier/North-Holland, pp. 7–19.

Bos, J. L., Heyting, C., Borst, P., Arnberg, A. C., Van Bruggen, E. F. J. (1978a). An insert in the single gene for the large ribosomal RNA in yeast mitochondrial DNA. Nature 275:336–338.

Bos, J. L., Van Kreijl, C. F., Ploegaert, F. H., Mol, J. N. M., Borst, P. (1978b). A conserved and unique (AT)-rich segment in yeast mitochondrial DNA. Nucl. Acids Res. 5:4563–4578.

Bos, J. L., Osinga, K. A., Van der Horst, G., Borst, P. (1979). Nucleotide sequence of the mitochondrial structure genes for cysteine tRNA and histidine tRNA of yeast. Nucl. Acids Res. 6:3255–3266.

Bos, J. L., Osinga, K. A., Van der Horst, G., Hecht, N. B., Tabak, H. F., Van Ommen, G.-J. B., Borst, P. (1980). Splice point sequence and transcripts of the intervening sequence in the mitochondrial 21S ribosomal RNA gene of yeast. Cell 20:207–214.

Breathnach, R., Chambon, P. (1981). Organization and expression of eucaryotic split genes coding for proteins. Ann. Rev. Biochem. 50:349–383.

Breathnach, R., Benoist, C., O'Hare, K., Gannon, F., Chambon, P. (1978). Ovalbumin gene: evidence for a leader sequence in mRNA and DNA sequences at the exon–intron boundaries. Proc. Natl. Acad. Sci. USA 75:4853–4857.

Brennicke, A. (1980). Mitochondrial DNA from *Oenothera berteriana.* Plant Physiol. 65:1207–1210.

Broach, J. R. (1981). Appendix III. Genes of *Saccharomyces cerevisiae.* In: The Molecular Biology of the Yeast *Saccharomyces:* Life Cycle and Inheritance, edited by J. N. Strathern, E. W. Jones, and J. R. Broach. Cold Spring Harbor, N.Y.: Cold Spring Harbor Laboratory, pp. 653–727.

Brosius, J., Dull, T. J., Noller, H. F. (1980). Complete nucleotide sequence of a 23S ribosomal RNA gene from *Escherichia coli.* Proc. Natl. Acad. Sci. USA 77:201–204.

Brown, W. M., George, M., Jr., Wilson, A. C. (1979). Rapid evolution of animal mitochondrial DNA. Proc. Natl. Acad. Sci. USA 76:1967–1971.

Bücher, T., Neupert, W., Sebald, W., Werner, S., Eds. (1976). Genetics and Biogenesis of Chloroplasts and Mitochondria. Amsterdam: Elsevier/North-Holland.

Bücher, T., Sebald, W., Weiss, H., Eds. (1980). 31. Colloquium—Mosbach 1980: Biological Chemistry of Organelle Formation. Heidelberg: Springer-Verlag.

Buetow, D. E., Wood, W. M. (1978). The mitochondrial translation system. Subcell Biochem. 5:1–85.

Bulder, C. J. (1964a). Induction of *petite* mutation and inhibition of synthesis of respiration enzymes in various yeasts. Antonie van Leeuwenhoek J. Microbiol. Serol. 30:1–9.

Bulder, C. J. (1964b). Lethality of the *petite* mutation in *petite*-negative yeasts. Antonie van Leeuwenhoek J. Microbiol. Serol. 30:442–454.

Burkl, G., Demmer, W., Holzner, H., Schweizer, E. (1976). Temperature sensitive nuclear "petite" mutants of *Saccaromyces cerevisiae.* In: Genetics, Biogenesis and Bioenergetics of Mitochondria, edited by W. Bandlow, R. J. Schweyen, D. Y. Thomas, K. Wolf, F. Kaudewitz. Berlin: De Gruyter, pp. 39–48.

Butow, R. A., Strausberg, R. L. (1979). Biochemical genetics of mitochondrial biogenesis. Trends Biochem. Sci. 4:110–113.

Butow, R. A., Bennett, W. F., Finkelstein, D. B. Kellems, R. E. (1975). Nuclear cytoplasmic interactions in the biogenesis of mitochondria in yeast. In: Membrane Biogenesis, edited by A. Tzagoloff. New York: Plenum Press, pp. 155–199.

Butow, R. A., Vincent, R. D., Strausberg, R. L., Zanders, E., Perlman, P. S. (1977). Genetic, physical and biochemical analysis of a mitochondrial gene. In: Mitochondria 1977: Genetics and Biogenesis of Mitochondria, edited by W. Bandlow, R. J. Schweyen, K. Wolf, and F. Kaudewitz. Berlin: De Gruyter, pp. 317–335.

Butow, R. A., Lopez, I. C., Chang, H.-P., Farrelly, F. (1980). The specification of *var 1* polypeptide by the *var 1* determinant. In: The Organization and Expression of the Mitochondrial Genome, edited by C. Saccone and A. M. Kroon. Amsterdam: Elsevier/North-Holland, pp. 195–205.

Butow, R. A., Farrelly, F., Zassenhaus, H. P., Hudspeth, M. E. S., Grossman, L. I., Perlman, P. S. (1982). *var 1* determinant region of yeast mtDNA. In: Mitochondrial Genes, edited by P. Slonimski, P. Borst, and G. Attardi. Cold Spring Harbor, N.Y.: Cold Spring Harbor Laboratory, pp. 241–253.

Cabral, F., Schatz, G. (1978). Identification of cytochrome c oxidase subunits in nuclear yeast mutants lacking the functional enzyme. J. Biol. Chem. 253:4396–4401.

Campbell, A. M. (1962). Episomes. Adv. Genet. 11:101–146.

Campbell, P. N., Blobel, G. (1976). The role of organelles in the chemical modification of the primary translation products of secretory proteins. FEBS Lett. 72:215–226.

Canaday, J., Guillemant, P., Weil, J.-H. (1980a). The nucleotide sequences of initiator transfer RNAs from bean cytoplasm and chloroplasts. Nucl. Acids Res. 8:999–1008.

Canaday, J., Dirheimer, G., Martin, R. P. (1980b). Yeast mitochondrial methionine initiator tRNA: characterization and nucleotide sequence. Nucl. Acids Res. 8:1445–1457.

Cantatore, P., Attardi, G. (1980). Mapping of nascent light and heavy strand transcripts on the physical map of HeLa cell mitochondrial DNA. Nucl. Acids Res. 8:2605–2625.

Carignani, G., Lancashire, W. E., Griffiths, D. E. (1977). Extrachromosomal inheritance of rhodamine 6G resistance in *Saccharomyces cerevisiae.* Mol. Gen. Genet. 151:49–56.

Carnevali, F., Morpugo, D., Tecce, G. (1969). Cytoplasmic DNA from petite colonies of *Saccharomyces cerevisiae:* a hypothesis on the nature of the mutation. Science 163:1331–1333.

Caron, F., Jacq, C., Rouvière-Yaniv, J. (1979). Characterization of a histonelike protein extracted from yeast mitochondria. Proc. Natl. Acad. Sci. USA 76:4265–4269.

Casey, J., Cohen, M., Rabinowitz, M., Fukuhara, H., Getz, G. S. (1972). Hybridization of mitochondrial transfer RNAs with mitochondrial and nuclear DNA of grande (wild type) yeast. J. Mol. Biol. 63:431–440.

Cavalier-Smith, T. (1975). The origin of nuclei and of eukaryotic cells. Nature 256:463–468.

Choo, K.-J., Nagley, P., Lukins, H. B., Linnane, A. W. (1977). Biogenesis of mitochondria. 47. Refined physical map of the mitochondrial genome of *Saccharomyces cerevisiae* determined by analysis of an extended library of genetically and molecularly defined mutants. Mol. Gen. Genet. 153:279–288.

Chow, L. T., Davidson, N., Berg, D. E. (1974). Electron microscope study of the structure of λdv DNAs. J. Mol. Biol. 86:69–89.

Church, G. M., Slonimski, P. P., Gilbert, W. (1979). Pleiotropic mutations within two yeast mitochondrial cytochrome genes block mRNA processing. Cell 18:1209–1215.

Claisse, M. L., Spyridakis, A., Wombier-Kluppel, M. L., Pajot, P., Slonimski, P. P. (1978). Mosaic organization and expression of the mitochondrial DNA region controlling cytochrome c reductase and oxidase. II. Analysis of proteins translated from the *box* region. In: Biochemistry & Genetics of Yeast: Pure and Applied Aspects, edited by M. Bacila, B. L. Horecker, A. O. M. Stoppani, New York: Academic Press, pp. 369–390.

Claisse, M. L., Slonimski, P. P., Johnson, J., Mahler, H. R. (1980). Mutations within an intron and its flanking sites: patterns of novel polypeptides generated by mutants in one segment of the *cob-box* region of yeast mitochondrial DNA. Mol. Gen. Genet. 177:375–387.

Clark-Walker, G. D., McArthur, C. R. (1978). Structure and functional relationships of mitochondrial DNAs from various yeasts. In: Biochemistry and Genetics of Yeast: Pure and Applied Aspects, edited by M. Bacila, B. L. Horecker, and A. O. M. Stoppani. New York: Academic Press, pp. 255–272.

Clark-Walker, G. D., Miklos, G. L. G. (1974). Mitochondrial genetics, circular DNA and the mechanism of the *petite* mutation in yeast. Genet. Res. 24:43–57.

Clark-Walker, G. D., Sriprakash, K. S. (1981). Sequence rearrangements between mitochondrial DNAs of *Torulopsis glabrata* and *Kloeckera africana* identified by hybridization with six polypeptide encoding regions from *Saccharomyces cerevisiae* mitochondrial DNA. J. Mol. Biol. 151:367–387.

Clark-Walker, G. D., McArthur, C. R., Sriprakash, K. S. (1981a). Partial duplication of the large ribosomal RNA sequence in an inverted repeat in circular mitochondrial DNA from *Kloeckera africana:* implications for mechanisms of the *petite* mutation. J. Mol. Biol. 147:399–415.

Clark-Walker, G. D., McArthur, C. R., Daley, D. J. (1981b). Does mitochondrial DNA length influence the frequency of spontaneous *petite* mutants in yeasts? Curr. Genet. 4:7–12.

Clayton, D. A. (1982). Replication of animal mitochondrial DNA. Cell 28:693–705.

Clayton, D. A., Smith, C. A. (1975). Complex mitochondrial DNA. Int. Rev. Exp. Pathol. 14:1–67.

Cobon, G. S., Groot-Obbink, D. J., Hall, R. M., Maxwell, R., Murphy, M., Rytka, J., Linnane, A. W. (1976). Mitochondrial genes determining cytochrome b (complex III) and cytochrome oxidase function. In: Genetics and Biogenesis of Chloroplasts and Mitochondria, edited by T. Bucher, W. Neupert, W. Sebald, and S. Werner. Amsterdam: Elsevier/North-Holland, pp. 453–460.

Cohen, J. D. (1977). Genetic analysis of multiple drug cross resistance in *Saccharomyces cerevisiae:* a nuclear-mitochondrial gene interaction. Ph.D. Thesis, City University of New York.

Cohen, J. D., Eaton, N. R. (1979). Genetic analysis of multiple drug cross resistance in *Saccharomyces cerevisiae:* a nuclear-mitochondrial gene interaction. Genetics 91:19–33.

Colson, A. M., Goffeau, A., Briquet, M., Weigel, P., Mattoon, J. R. (1974). Nucleocytoplasmic interaction between oligomycin-resistant mutations in *Saccharomyces cerevisiae.* Mol. Gen. Genet. 135:309–326.

Cooper, T. G. (1981). Tenth international conference on yeast genetics and molecular biology: a review. Curr. Gen. 3:1–4.

Coruzzi, G., Tzagoloff, A. (1979). Assembly of the mitochondrial membrane system DNA sequence of subunit 2 of yeast cytochrome oxidase. J. Biol. Chem. 254:9324–9330.

Coruzzi, G., Bonitz, S. G., Thalenfeld, B. E., Tzagoloff, A. (1981). Assembly of the mitochondrial membrane system. Analysis of the nucleotide sequence and transcripts in the *oxi 1* region of yeast mitochondrial DNA. J. Biol. Chem. 256:12780–12787.

Cosson, J., Tzagoloff, A. (1979). Sequence homologies of (guanosine + cytidine)-rich regions of mitochondrial DNA of *Saccharomyces cerevisiae.* J. Biol. Chem. 254:42–43.

Cote, C., Solioz, M., Schatz, G. (1979). Biogenesis of the cytochrome $bc_1$ complex of yeast mitochondria. J. Biol. Chem. 254:1437–1439.

Crandall, M. (1973a). Comparison of *Hansenula wingei,* a *petite*-negative obligately aerobic yeast, to the *petite*-positive yeast *Saccharomyces cerevisiae.* J. Gen. Microbiol. 75:363–375.

Crandall, M. (1973b). A respiratory-deficient mutant in the obligately aerobic yeast *Hansenula wingei.* J. Gen. Microbiol. 75:377–381.

Crawford, L. V., Cole, C. N., Smith, A. E., Paucha, E., Tegtmeyer, P., Rundell, K., Berg, P. (1978). Organization and expression of early genes of Simian virus 40. Proc. Natl. Acad. Sci. USA 75:117–121.

Crews, S., Ojala, D., Posakony, J., Nishiguchi, J., Attardi, G. (1979). Nucleotide sequence of a region of human mitochondrial DNA containing the precisely identified origin of replication. Nature 277:192–198.

Crick, F. H. C. (1966). Codon–anticodon pairing: the wobble hypothesis. J. Mol. Biol. 19:548–555.

Cummings, D. J., Borst, P., Dawid, I. B., Weissman, S. M., Fox, C. F., Eds., (1979a). Extrachromosomal DNA: ICN-UCLA Symposia on Molecular and Cellular Biology, Vol. 15. New York: Academic Press.

Cummings, D. J., Belcour, L., Grandchamp, C. (1979b). Mitochondrial DNA from *Podospora anserina.* I. Isolation and characterization. Mol. Gen. Genet. 171:229–238.

Cunningham, R. S., Bonen, L., Doolittle, W. F., Gray, M. W. (1976). Unique species of 5S, 18S and 26S ribosomal RNA in wheat mitochondria. FEBS Lett. 69:116–122.

Dayhoff, M. O., Schwartz, R. M. (1981). Evidence on the origin of eukaryotic mitochondria from protein and nucleic acid sequences. Ann. N.Y. Acad. Sci. 361:92–103.

de Deken, R. H. (1961). The dissociation of phenotypic and inheritable effects of euflavin in yeast. Exp. Cell Res. 24:145–148.

de la Salle, H., Jacq, C., Slonimski, P. P. (1982). Critical sequences within mitochondrial introns: pleiotropic mRNA maturase and *cis*-dominant signals of the box intron controlling reductase and oxidase. Cell 28:721–732.

De Ronde, A., Van Loon, A. P. G. M., Grivell, L. A., Kohli, J. (1980). *In vitro* suppression of UGA codons in mitochondrial mRNA. Nature 287:361–363.

de Zamaroczy, M., Baldacci, G., Bernardi, G. (1979). Putative origins of replication in the mitochondrial genome of yeast. FEBS Lett. 108:429–432.

de Zamaroczy, M., Marotta, R., Faugeron-Fonty, G., Goursot, R., Mangin, M., Baldacci, G., Bernardi, G. (1981). The origins of replication of the yeast mitochondrial genome and the phenomenon of suppressivity. Nature 292:75–78.

Dhawale, S., Hanson, D. K., Alexander, N. J., Perlman, P. S., Mahler, H. R. (1981). Regulatory interactions between mitochondrial genes: interactions between two mosaic genes. Proc. Natl. Acad. Sci. USA 78:1778–1782.

Dodson, E. O. (1979). Crossing the procaryote–eucaryote border: endosymbiosis or continuous development? Can. J. Microbiol. 25:651–674.

Doolittle, W. F., Bonen, L. (1981). Molecular sequence data indicating an endosymbiotic origin for plastids. Ann. N. Y. Acad. Sci. 361:248–256.

Doolittle, W. F., Sapienza, C. (1980). Selfish genes, the phenotype paradigm and genome evolution. Nature 284:601–603.

Douglas, M. G., Butow, R. A. (1976). Variant forms of mitochondrial translation products in yeasts: evidence for location of determinants on mitochondrial DNA. Proc. Natl. Acad. Sci. USA 73:1083–1086.

Douglas, M., Finkelstein, D., Butow, R. A. (1979). Analysis of products of mitochondrial protein synthesis in yeast: genetic and biochemical aspects. Meth. Enzymol. 61:58–66.

Dujardin, G., Pajot, P., Groudinsky, O., Slonimski, P. P. (1980a). Long range control circuits within mitochondria and between nucleus and mitochondria. I. Methodology and phenomenology of suppressors. Mol. Gen. Genet. 179:469–482.

Dujardin, G., Groudinsky, O., Kruszewska, A., Pajot, P., Slonimski, P. P. (1980b). Cytochrome b messenger RNA maturase encoded in an intron regulates the expression of the split gene. III. Genetic and phenotypic suppression of intron mutations. In: The Organization and Expression of the Mitochondrial Genome, edited by C. Saccone and A. M. Kroon, Amsterdam: Elsevier/North-Holland, pp. 157–160.

Dujon, B. (1980). Sequence of the intron and flanking exons of the mitochondrial 21S rRNA gene of yeast strains having different alleles at the $\omega$ and rib-1 loci. Cell 20:185–197.

Dujon, B. (1981). Mitochondrial genetics and functions. In: The Molecular Biology of the Yeast *Saccharomyces:* Life Cycle and Inheritance, edited by J. N. Strathern, E. W. Jones, and J. R. Broach. Cold Spring Harbor, N.Y.: Cold Spring Harbor Laboratory, pp. 505–635.

Dujon, B., Michaelis, G. (1974). Extrakaryotic inheritance. Fortschr. Bot. 36:236–246.

Dujon, B., Michel, F. (1976). Genetic and physical characterization of a segment of the mitochondrial DNA involved in the control of genetic recombination. In: The Genetic Function of Mitochondrial DNA, edited by C. Saccone and A. M. Kroon. Amsterdam: Elsevier/North-Holland, pp. 175–184.

Dujon, B., Slonimski, P. P., Weill, L. (1974). Mitochondrial genetics. IX. A model for recombination and segregation of mitochondrial genomes in *Saccharomyces cerevisiae*. Genetics 78:415–437.

Dujon, B., Bolotin-Fukuhara, M., Coen, D., Deutsch, J., Netter, P., Slonimski, P. P., Weill, L. (1976). Mitochondrial genetics. XI. Mutations at the mitochondrial locus $\omega$ affecting the recombination of mitochondrial genes in *Saccharomyces cerevisiae*. Mol. Gen. Genet. 143:131–165.

Dujon, B., Colson, A. M., Slonimski, P. P. (1977). The mitochondrial genetic map of *Saccharomyces cerevisiae:* a compilation of mutations, genes, genetic and physical maps. In: Mitochondria 1977: Genetics and Biogenesis of Mitochondria, edited by W. Bandlow, R. J. Schweyen, K. Wolf, and F. Kaudewitz. Berlin: De Gruyter, pp. 579–669.

Ebner, E., Schatz, G. (1973). Mitochondrial assembly in respiration-deficient mutants of *Saccharomyces cerevisiae*. III. A nuclear mutant lacking mitochondrial adenosine triphosphatase. J. Biol. Chem. 248:5379–5384.

Ebner, E., Mason, T. L., Schatz, G. (1973). Mitochondrial assembly in respiration deficient mutants of *Saccharomyces cerevisiae*. II. Effect of nuclear and extrachromosomal mutations on the formation of cytochrome c oxidase. J. Biol. Chem. 248:5369–5378.

Ehrlich, S. D., Thiery, J.-P., Bernardi, G. (1972). The mitochondrial genome of wild-type yeast cells. III. The pyrimidine tracts of mitochondrial DNA. J. Mol. Biol. 65:207–212.

Ephrussi, B., Hottinguer, H. (1951). Cytoplasmic constituents of heredity: on an unstable cell state in yeast. Cold Spring Harbor Symp. Quant. Biol. 16:75–84.

Ephrussi, B., Margerie-Hottinguer, H., Roman, H. (1955). Suppressiveness: a new factor in the genetic determinism of the synthesis of respiratory enzymes in yeast. Proc. Natl. Acad. Sci. USA 41:1065–1071.

Ephrussi, B., Jakob, H., Grandchamp, S. (1966). Études sur la suppressivité des mutants à déficience réspiratoire de la levure. II. Etapes de la mutation grande en petite provoquée par le facteur suppressif. Genetics 54:1–29.

Ernster, L., Schatz, G. (1981). Mitochondria: a historical review. J. Cell Biol. 91:227s–255s.

Evans, I. H., Wilkie, D. (1976a). Mitochondrial factors in the utilization of sugars in *Saccharomyces cerevisiae*. Genet. Res. 27:89–93.

Evans, I. H., Wilkie, D. (1976b). Mitochondrial role in the induction of sugar utilization in *Saccharomyces cerevisiae*. In: The Genetic Function of Mitochondrial DNA, edited by C. Saccone and A. M. Kroon, Amsterdam: Elsevier/North-Holland, pp. 209–217.

Evans, I. H., Wilkie, D. (1978). Mitochondrial instability in a strain of *Saccharomyces cerevisiae*. Genet. Res. 32:171–182.

Evans, I. H., Egilsson, V., Wilkie, D. (1976). Mitochondrial involvement in the control of cell function in *Saccharomyces cerevisiae*. In: Genetics, Biogenesis and Bioenergetics of Mitochondria, edited by W. Bandlow, R. J. Schweyen, D. Y. Thomas, K. Wolf, and F. Kaudewitz. Berlin: De Gruyter, pp. 125–136.

Evans, I. H., Diala, E. S., Earl, A., Wilkie, D. (1980). Mitochondrial control of cell surface characteristics in *Saccharomyces cerevisiae*. Biochim. Biophys. Acta, 602:201–206.

Fang, M., Butow, R. A. (1970). Nucleotide reversal of mitochondrial repression in *Saccharomyces cerevisiae*. Biochem. Biophys. Res. Commun. 41:1579–1583.

Faugeron-Fonty, G., Culard, F., Baldacci, G., Goursot, R., Prunell, A., Bernardi, G. (1979). The mitochondrial genome of wild-type yeast cells. VIII. The spontaneous cytoplasmic *"petite"* mutation. J. Mol. Biol. 134:493–537.

Fauron, C. M.-R., Wolstenholme, D. R. (1976). Structural heterogeneity of mitochondrial DNA molecules within the genus *Drosophila*. Proc. Natl. Acad. Sci. USA 73:3623–3627.

Faye, G. (1976). Mitochondrial ribosomal subunits from *Saccharomyces cerevisiae* $\rho^-$ mutants. FEBS Lett. 69:167–170.

Faye, G., Sor, F. (1977). Analysis of mitochondrial ribosomal proteins of *Saccharomyces cerevisiae* by two-dimensional polyacrylamide gel electrophoresis. Mol. Gen. Genet. 155:27–34.

Faye, G., Kujawa, C., Fukuhara, H. (1974). Physical and genetic organization of *petite* and grande yeast mitochondrial DNA. IV. *In vivo* transcription products of mitochondrial DNA and localization of 23S ribosomal RNA in *petite* mutants of *Saccharomyces cerevisiae*. J. Mol. Biol. 88:185–203.

Faye, G., Kujawa, C., Fukuhara, H., Rabinowitz, M. (1976). Mapping of the mitochondrial 16S ribosomal RNA gene and its expression in the cytoplasmic *petite* mutants of *Saccharomyces cerevisiae*. Biochem. Biophys. Res. Commun. 68:476–482.

Faye, G., Dennebouy, N., Kujawa, C., Jacq, C. (1979a). Inserted sequence in the mitochondrial 23S ribosomal RNA gene of the yeast *Saccharomyces cerevisiae*. Mol. Gen. Genet. 168:101–109.

Faye, G., Sor, F., Glatigny, A., Lederer, F., Lesquoy, E. (1979b). Comparison of amino-acid compositions of mitochondrial and cytoplasmic ribosomal proteins of *Saccharomyces cerevisiae*. Mol. Gen. Genet. 171:335–341.

Fiers, W., Contreras, R., Haegeman, G., Rogiers, R., Van de Voorde, A., Van Heuverswyn, H., Van Herreweghe, J., Volckaert, G., Ysebaert, M. (1978). Complete nucleotide sequence of SV40 DNA. Nature 273:113–120.

Fonty, G., Goursot, R., Wilkie, D., Bernardi, G. (1978). The mitochondrial genome of wild-type yeast cells. VII. Recombination in crosses. J. Mol. Biol. 119:213–235.

Foury, F., Tzagoloff, A. (1978). Assembly of the mitochondrial membrane system. Genetic complementation of *mit*$^-$ mutations in mitochondrial DNA of *Saccharomyces cerevisiae*. J. Biol. Chem. 253:3792–3797.

Fox, G. E., Stackenbrandt, E., Hespell, R. B., Gibson, J., Maniloff, J., Dyer, T. A., Wolfe, R. S., Balch, W. E., Tanner, R. S., Magrum, L. J., Zablen, L. B., Blakemore, R., Gupta, R., Bonen, L., Lewis, B. J., Stahl, D. A., Luehrsen, K. R., Chen, K. N., Woese, C. R. (1980). The phylogeny of prokaryotes. Science 209:457–463.

Fox, T. D. (1979a). Genetic and physical analysis of the mitochondrial gene for subunit II of yeast cytochrome c oxidase. J. Mol. Biol. 130:63–82.

Fox, T. D. (1979b). Five TGA "stop" codons occur within the translated sequence of the yeast mitochondrial gene for cytochrome c oxidase subunit II. Proc. Natl. Acad. Sci. USA 76:6534–6538.

Fox, T. D., Leaver, C. J. (1981). The *Zea mays* mitochondrial gene coding cytochrome oxidase subunit II has an intervening sequence and does not contain TGA codons. Cell 26:315–323.

Gaillard, C., Bernardi, G. (1979). The nucleotide sequence of the mitochondrial genome of a spontaneous *"petite"* mutant of yeast. Mol. Gen. Genet. 174:335–337.

Gaillard, C., Strauss, F., Bernardi, G. (1980). Excision sequences in the mitochondrial genome of yeast. Nature 283:218–220.

Gilbert, W. (1978). Why genes in pieces? Nature 271:501–502.

Gillham, N. W. (1974). Genetic analysis of the chloroplast and mitochondrial genomes. Ann. Rev. Genet. 8:347–391.

Gillham, N. W. (1978). Organelle Heredity. New York: Raven Press.

Gingold, E. B. (1981). Genetic analysis of the products of a cross involving a suppressive *'petite'* mutant of *S. cerevisiae.* Curr. Genet. 3:213–220.

Glaus, K. R., Zassenhaus, H. P., Fechheimer, N. S., Perlman, P. S. (1980). Avian mtDNA: structure, organization and evolution. In: The Organization and Expression of the Mitochondrial Genome, edited by A. M. Kroon and C. Saccone, Amsterdam: Elsevier/North-Holland, pp. 131–135.

Goddard, J. M., Cummings, D. J. (1975). Structure and replication of mitochondrial DNA from *Paramecium aurelia.* J. Mol. Biol. 97:593–609.

Goldbach, R. W., Arnberg, A. C., Van Bruggen, E. F. J., Defize, J., Borst, P. (1977). The structure of *Tetrahymena pyriformis* mitochondrial DNA. I. Strain differences and occurrence of inverted repititions. Biochim. Biophys. Acta 477:37–50.

Goldthwaite, C. D., Cryer, D. R., Marmur, J. (1974). Effect of carbon source on the replication and transmission of yeast mitochondrial genomes. Mol. Gen. Genet. 133:87–104.

Goursot, R., de Zamaroczy, M., Baldacci, G., Bernardi, G. (1980). Supersuppressive *"petite"* mutants of yeast. Curr. Genet. 1:173–176.

Granick, S., Gibor, A. (1967). The DNA of chloroplasts, mitochondria and centrioles. Progr. Nucl. Acids Res. Mol. Biol. 6:143–186.

Gray, M. W. (1982). Mitochondrial genome diversity and the evolution of mitochondrial DNA. Can. J. Biochem. 60:157–171.

Gray, M. W., Doolittle, W. F. (1982). Has the endosymbiont hypothesis been proven? Microbiol. Rev. 46:1–42.

Griffiths, D. E., Lancashire, W. E., Zanders, E. D. (1975). Evidence of an extrachromosomal element involved in mitochondrial function: a mitochondrial episome? FEBS Letters 53:126–130.

Grivell, L. A. (1981). Mitochondrial genes at Cold Spring Harbor. Curr. Genet. 4:167–171.

Grivell, L. A., Moorman, A. F. M. (1977). A structural analysis of the oxi-3 region on yeast mtDNA. In: Mitochondria 1977: Genetics and Biogenesis of Mitochondria, edited by W. Bandlow, R. J. Schweyen, K. Wolf, and F. Kaudewitz. Berlin: De Gruyter, pp. 371–384.

Grivell, L. A., Reijnders, L. (1974). Yeast mitochondrial ribosomes. Postepy Mikrobiol. 13:67–80.

Grivell, L. A., Arnberg, A. C., Boer, P. H., Borst, P., Bos, J. L., Von Bruggen, E. F. J., Groot, G. S. P., Hecht, N. B., Hensgens, L. A. M., Van Ommen, G. J. B., Tabak, H. F. (1979). Transcripts of yeast mitochondrial DNA and their processing. In:

Extrachromosomal DNA: ICN-UCLA Symposia on Molecular and Cellular Biology, Vol. 15, edited by D. J. Cummings, P. Borst, I. B. Dawid, S. N. Weissman, and C. F. Fox, New York: Academic Press, pp. 305–324.

Grivell, L. A., Arnberg, A. C., Hensgens, L. A. M., Roosendaal, E., Van Ommen, G. J. B., Van Bruggen, E. F. J. (1980). Split genes on yeast mitochondrial DNA: Organization and expression. In: The organization and expression of the mitochondrial genome. Edited by C. Saccone and A. M. Kroon. Amsterdam: Elsevier/North-Holland, pp. 37–49.

Grohmann, K., Amalric, F., Crews, S., Attardi, G. (1978). Failure to detect "cap" structures in mitochondrial DNA-coded poly(A)-containing RNA from HeLa cells. Nucl. Acids. Res. 5:637–651.

Groot, G. S. P., Van Harten-Loosbroek, N. (1980). The organization of the genes for ribosomal RNA on mitochondrial DNA of *Kluyveromyces lactis*. Curr. Genet. 1:133–135.

Groot, G. S. P., Flavell, R. A., Van Ommen, G. J. B., Grivell, L. A. (1974). Yeast mitochondrial RNA does not contain poly(A). Nature 252:167–169.

Groot, G. S. P., Van Harten-Loosbroek, N., Kreike, J. (1978). Electrophoretic behaviour of yeast mitochondrial translation products. Biochem. Biophys. Acta 517:457–463.

Groot, G. S. P., Mason, T. L., Van Harten-Loosbroek, N. (1979). Var 1 is associated with the small ribosomal subunit of mitochondrial ribosomes in yeast. Mol. Gen. Genet. 174:339–342.

Grossman, L. I., Cryer, D. R., Goldring, E. S., Marmur, J. (1971). The *petite* mutations in yeast. III. Nearest-neighbour analysis of mitochondrial DNA from normal and mutant cells. J. Mol. Biol. 62:565–575.

Gruss, P., Lai, C.-J., Dhar, R., Khoury, G. (1979). Splicing as a requirement for biogenesis of functional 16S mRNA of Simian virus 40. Proc. Natl. Acad. Sci. USA 76:4317–4321.

Guérineau, M., Slonimski, P. P., Avner, P. R. (1974). Yeast episome: oligomycin resistance associated with a small covalently-closed non-mitochondrial circular DNA. Biochem. Biophys. Res. Commun. 61:412–419.

Haid, A., Schweyen, R. J., Bechman, H., Kaudewitz, F., Solioz, M., Schatz, G. (1979). The mitochondrial COB region in yeast codes for cytochrome b and is mosaic. Eur. J. Biochem. 94:451–464.

Haid, A., Grosch, G., Schmelzer, C., Schweyen, R. J., Kaudewitz, I. (1980). Expression of the split gene *cob* in yeast mt DNA. Mutational arrest in the pathway of transcript splicing. Curr. Genet. 1:155–161.

Halbreich, A., Pajot, P., Foucher, M., Grandchamp, C., Slonimski, P. P. (1980). A pathway of cytochrome b mRNA processing in yeast mitochondria: specific splicing steps and an intron-derived circular RNA. Cell 19:321–329.

Hall, R. M., Nagley, P., Linnane, A. W. (1976). Biogenesis of mitochondria. XLII. Genetic analysis of the control of cellular mitochondrial DNA levels in *Saccharomyces cerevisiae*. Mol. Gen. Genet. 145:169–175.

Hamer, D. H., Leder, P. (1979). Splicing and the formation of stable RNA. Cell 18:1299–1302.

Handwerker, A., Schweyen, R. J., Wolf, K., Kaudewitz, F. (1973). Evidence for an extrakaryotic mutation affecting the maintenance of the *rho* factor in yeast. J. Bacteriol. 113:1307–1310.

Hanson, D. K., Miller, D. H., Mahler, H. R., Alexander, N. J., Perlman, P. S. (1979).

Regulatory interactions between mitochondrial genes. II. Detailed characterization of novel mutants mapping within one cluster in the COB 2 region. J. Biol. Chem. 254:2480–2490.

Hartig, A., Breitenbach, M. (1980). Sporulation in mitochondrial *OXI 3* mutants of *Saccharomyces cerevisiae.* A correlation with the genetic map. Curr. Genet. 1:97–102.

Hartig, A., Schroeder, R., Mucke, E., Breitenbach, M. (1981). Isolation and characterization of yeast mitochondrial mutants defective in spore germination. Curr. Genet. 4:29–36.

Haussmann, P., Zimmermann, F. K. (1976). The role of mitochondria in carbon catabolite repression in yeast. Mol. Gen. Genet. 148:205–211.

Heckman, J. E., Hecker, L. I., Schwartzbach, S. D., Barnett, W. E., Baumstark, B., RajBhandary, U. L. (1978). Structure and function of initiator methionine tRNA from the mitochondria of *Neurospora crassa.* Cell 13:83–95.

Hendler, F. J., Padmanaban, G., Patzer, J., Ryan, R., Rabinowitz, M. (1975). Yeast mitochondrial RNA contains a short polyadenylic acid segment. Nature 258:357–359.

Hendler, F. J., Halbreich, A., Jakovic, S., Patzer, J., Merten, S., Rabinowitz, M. (1976). Characterization and translation of yeast mitochondrial RNA. In: Genetics and Biogenesis of Chloroplasts and Mitochondria, edited by Th. Bucher, W. Neupert, W. Sebald, and S. Werner, Amsterdam: Elsevier–North-Holland, pp. 679–684.

Hensgens, L. A. M., Grivell, L. A., Borst, P., Bos, J. L. (1979). Nucleotide sequence of the mitochondrial structural gene for subunit 9 of yeast ATPase complex. Proc. Natl. Acad. Sci. USA 76:1663–1667.

Herman, A. L., Griffin, P. S. (1968). Respiratory-deficient mutants in *Saccharomyces lactis.* J. Bacteriol. 96:457–461.

Heslot, H., Louis, C., Goffeau, A. (1970). Segregational respiratory-deficient mutants of a "*petite*-negative" yeast *Schizosaccharomyces pombe* 972h$^-$ J. Bacteriol. 104:482–491.

Heyting, C., Menke, H. H. (1979). Fine structure of the 21S ribosomal RNA region on yeast mitochondrial DNA. III. Physical location of mitochondrial genetic markers and the molecular nature of omega. Mol. Gen. Genet. 168:279–291.

Heyting, C., Meijlink, F. C. P. W., Verbeet, M. Ph., Sanders, J. P. M., Bos, J. L., Borst, P. (1979a). Fine structure of the 21S ribosomal RNA region on yeast mitochondrial DNA. I. Construction of the physical map and localization of the cistron for 21S mitochondrial ribosomal RNA. Mol. Gen. Genet. 168:231–246.

Heyting, C., Talen, J.-L., Weijers, P. J., Borst, P. (1979b). Fine structure of the 21S ribosomal RNA region on yeast mitochondrial DNA. II. The organization of sequences in *petite* mitochondrial DNAs carrying genetic markers from the 21S region. Mol. Gen. Genet. 168:251–277.

Hollenberg, C. P., Borst, P., Van Bruggen, E. F. J. (1970). Mitochondrial DNA. V. A 25-$\mu$ closed circular duplex DNA molecule in wild-type yeast mitochondria. Structure and genetic complexity. Biochim. Biophys. Acta 209:1–15.

Holmberg, S., Kielland-Brandt, M. C. (1978). A mutant of *Saccharomyces cerevisiae* temperature sensitive for flocculation: influence of oxygen and respiratory deficiency on flocculence. Carlsberg. Res. Commun. 43:37–47.

Howell, N., Molloy, P. L., Linnane, A. W., Lukins, H. B. (1974). Biogenesis of mitochondria. 34. The synergistic interaction of nuclear and mitochondrial mutations to produce resistance to high levels of mikamycin in *Saccharomyces cerevisiae*. Mol. Gen. Genet. 128:43–54.

Hutchison, C. A., III, Newbold, J. E., Potter, S. S., Edgell, M. H. (1974). Maternal inheritance of mammalian mitochondrial DNA. Nature 251:536–538.

Iwashima, A., Rabinowitz, M. (1969). Partial purification of mitochondrial and supernatant DNA polymerase from *Saccharomyces cerevisiae*. Biochim. Biophys. Acta. 178:283–293.

Jacq, C., Kujawa, C., Grandchamp, C., Netter, P. (1977). Physical characterization of the difference between yeast mitochondrial DNA alleles omega$^+$ and omega$^-$. In: Mitochondria 1977: Genetics and Biogenesis of Mitochondria, edited by W. Bandlow, R. J. Schweyen, K. Wolf, and F. Kaudewitz, Berlin: De Gruyter, pp. 255–270.

Jacq, C., Lazowska, J., Slonimski, P. P. (1980a). Sur un nouveau mécanisme de la régulation de l'expression génétique. C. R. Acad. Sci. Paris 290:89–92.

Jacq, C., Lazowska, J., Slonimski, P. P. (1980b). Cytochrome b messenger RNA maturase encoded in an intron regulates the expression of the split gene. I. Physical location and base sequence of intron mutations. In: The Organization and Expression of the Mitochondrial Genome, edited by C. Saccone and A. M. Kroon. Amsterdam: Elsevier/North-Holland, pp. 139–152.

Johnson, J. D., Ogdon, R., Johnson, P., Abelson, J., Dembeck, P., Itakura, K. (1980). Transcription and processing of a yeast tRNA gene containing a modified intervening sequence. Proc. Natl. Acad. Sci. USA 77:2564–2568.

Johnston, L. H. (1979). Nuclear mutations in *Saccharomyces cerevisiae* which increase the spontaneous mutation frequency in mitochondrial DNA. Mol. Gen. Genet. 170:327–331.

Kasamatsu, H., Vinograd, J. (1974). Replication of circular DNA in eukaryotic cells. Ann. Rev. Biochem. 43:695–719.

Kellems, R. E., Butow, R. A. (1972). Cytoplasmic type 80S ribosomes associated with yeast mitochondria. I. Evidence for ribosome binding sites on yeast mitochondria. J. Biol. Chem. 247:8043–8050.

Khan, N. A. (1982). Suppression of maltose-negative phenotype by a specific nuclear gene (PMU 1) in the *petite* cells of the yeast *Saccharomyces cerevisiae*. Mol. Gen. Genet. 186:40–43.

Khan, N. A., Greener, A. (1977). Effect of the *petite* mutation on maltose and alpha-methylglucoside fermentation in *Saccharomyces cerevisiae*. Mol. Gen. Genet. 150:107–108.

Kilejian, A. (1975). Circular mitochondrial DNA from the avian malarial parasite *Plasmodium lophurae*. Biochim. Biophys. Acta 390:276–284.

Kleckner, N. (1981). Transposable elements in prokaryotes. Ann. Rev. Genet. 15:341–404.

Klootwijk, J., Klein, I., Grivell, L. A. (1975). Minimal post-transcriptional modification of yeast mitochondrial ribosomal RNA. J. Mol. Biol. 97:337–350.

Knapp, G., Ogden, R. C., Peebles, C. L., Abelson, J. (1979). Splicing of yeast tRNA precursors: structure of the reaction intermediates. Cell 18:37–45.

Kolodner, R., Tewari, K. K. (1972). Physicochemical characterization of mitochondrial DNA from pea leaves. Proc. Natl. Acad. Sci. USA 69:1830–1834.

Kotylak, Z., Slonimski, P. P. (1976). Joint control of cytochrome a and b by a unique mitochondrial DNA region comprising four genetic loci. In: The Genetic Functions of Mitochondrial DNA, edited by C. Saccone and A. M. Kroon. Amsterdam: Elsevier/North-Holland, pp. 143–154.

Kotylak, Z., Slonimski, P. P. (1977). Mitochondrial mutants isolated by a new screening method based upon the use of the nuclear mutation *op. 1*. In: Mitochondria 1977: Genetics and Biogenesis of Mitochondria, edited by W. Bandlow, R. J. Schweyen, K. Wolf, and F. Kaudewitz. Berlin: De Gruyter, pp. 83–89.

Kovac, L. (1974). Biochemical mutants: an approach to mitochondrial energy coupling. Biochim. Biophys. Acta. 346:101–135.

Kovac, L., Istenesova, A. (1964). Inhibition of anabolic processes by dinitrophenol and azide in anaerobically grown yeast. Biochim. Biophys. Acta 82:162–164.

Kovac, L., Lachowicz, T. M., Slonimski, P. P. (1967). Biochemical genetics of oxidative phosphorylation. Science 158:1564–1569.

Kovac, L., Groot, G. S. P., Racker, E. (1972). Translocation of protons and potassium ions across the mitochondrial membrane of respiring and respiration deficient yeast. Biochim. Biophys. Acta 256:55–65.

Kovac, L., Kolarov, J., Subik, J. (1977). Genetic determination of the mitochondrial adenine nucleotide translocation system and its role in the eukaryotic cell. Mol. Cell. Biochem. 14:11–14.

Kovacova, V., Irmlerova, J., Kovac, L. (1968). Oxidative phosphorylation in yeast. IV. Combination of a nuclear mutation affecting oxidative phosphorylation with cytoplasmic mutation to respiratory deficiency. Biochim. Biophys. Acta 162:157–163.

Kreike, J., Beckman, H., Van Hemert, F. J., Schweyen, R. J., Boer, P. H., Kaudewitz, F., Groot, G. S. P. (1979). The identification of apocytochrome b as a mitochondrial gene product and immunological evidence for altered apocytochrome b in yeast strains having mutations in the COB region of mitochondrial DNA. Eur. J. Biochem. 101:607–617.

Kroon, A. M., Saccone, C., Eds. (1974). The Biogenesis of Mitochondria: Transcriptional, Translational and Genetic Aspects. New York: Academic Press.

Kroon, A. M., Saccone, C., Eds. (1980). The Organization and Expression of the Mitochondrial Genome. Amsterdam: Elsevier/North-Holland.

Küntzel, H. (1971). The genetic apparatus of mitochondria from *Neurospora* and yeast. Curr. Top. Microbiol. Immunol. 54:94–118.

Küntzel, H., Köchel, H. G., Lazarus, C. M., Lünsdorf, H., (1982). Mitochondrial genes in *Aspergillus*. In: "Mitochondrial Genes," edited by P. Slonimski, P. Borst, and G. Attardi. Cold Spring Harbor, N.Y.: Cold Spring Harbor Laboratory, pp. 391–403.

Lagerkvist, U. (1981). Unorthodox codon reading and the evolution of the genetic code. Cell 23:305–306.

Lancashire, W. E., Griffiths, D. E. (1975). Studies of energy-linked reactions: genetic analysis of venturidin-resistant mutants. Eur. J. Biochem. 51:403–413.

Lazowska, J., Slonimski, P. P. (1976). Electron microscopy analysis of circular repetitive mitochondrial DNA molecules from genetically characterized *rho*$^-$ mutants of *Saccharomyces cerevisiae*. Mol. Gen. Genet. 146:61–78.

Lazowska, J., Jacq, C., Slonimski, P. P. (1980). Sequence of introns and flanking exons in wild-type and *box 3* mutants of cytochrome b reveals an interlaced splicing protein coded by an intron. Cell 22:333–348.

Lazowska, J., Jacq, C., Slonimski, P. P. (1981). Splice points of the third intron in the yeast mitochondrial cytochrome b gene. Cell 27:12–14.

Leaver, C. J., Gray, M. W. (1982). Mitochondrial genome organization and expression in higher plants. Ann. Rev. Plant Physiol. 33:373–402.

Lee, E.-H., Johnson, B. F. (1977). Volume-related mitochondrial deoxyribonucleic acid synthesis in zygotes and vegetative cells of *Saccharomyces cerevisiae*. J. Bacteriol. 129:1066–1071.

Lee, C. P., Schatz, G., Ernster, L., Eds. (1979). Membrane Biogenesis. New York: Addison-Wesley.

Lehninger, A. L. (1964). The Mitochondrion: Molecular Basis of Structure and Function. Menlo Park, Calif.: Benjamin.

Levens, D., Edward, J., Locker, J., Lustig, A., Merken, S., Morimoto, R., Synenki, R., Rabinowitz, M. (1979). Transcription of yeast mitochondrial DNA. In: Extrachromosomal DNA:ICN-UCLA Symposium on Molecular and Cellular Biology, edited by D. Cummings, P. Borst, I. Dawid, S. Weissman, and C. F. Fox. New York: Academic Press, Vol. 15, pp. 287–304.

Levens, D., Morimoto, R., Rabinowitz, M. (1981a). Mitochondrial transcription complex from *S. cerevisiae*. J. Biol. Chem. 256:1466–1473.

Levens, D., Lustig, A., Rabinowitz, M. (1981b). Purification of mitochondrial RNA polymerase from *S. cerevisiae*. J. Biol. Chem. 256:1474–1481.

Levens, D., Ticho, B., Ackerman, E., Rabinowitz, M. (1981c). Transcriptional initiation and 5′ termini of yeast mitochondrial RNA. J. Biol. Chem. 256:5226–5232.

Lewin, A., Morimoto, R., Merten, S., Martin, N., Berg, P., Christianson, T., Levens, D., Rabinowitz, M. (1977). Physical mapping of mitochondrial genes and transcripts in *Saccharomyces cerevisiae*. In: Mitochondria 1977, edited by W. Bandlow, R. J. Schweyen, K. Wolf, and F. Kaudewitz, Berlin: De Gruyter, pp. 271–289.

Lewin, A., Morimoto, R., Rabinowitz, M., Fukuhara, H. (1978). Restriction enzyme analysis of mitochondrial DNAs of *petite* mutants of yeast: classification of petites and deletion mapping of mitochondrial genes. Mol. Gen. Genet. 163:257–275.

Lewin, A. S., Gregor, I., Mason, T. L., Nelson, N., Schatz, G. (1980). Cytoplasmically made subunits of yeast mitochondrial $F_1$-ATPase and cytochrome c oxidase are synthesized as individual precursors, not as polyproteins. Proc. Natl. Acad. Sci. USA 77:3998–4002.

Li, M., Tzagoloff, A. (1979). Assembly of the mitochondrial membrane system: sequence of yeast mitochondrial valine and an unusual threonine tRNA gene. Cell 18:47–53.

Lin, L.-F. H., Clejan, L., Beattie, D. S. (1978). The synthesis of cytochrome b on mitochondrial ribosomes in baker's yeast. Eur. J. Biochem. 87:171–179.

Linnane, A. W., Haslam, J. M. (1971). The biogenesis of yeast mitochondria. Curr. Top. Cell. Regul. 2:101–172.

Linnane, A. W., Crowfoot, P. B. (1975). Biogenesis of yeast mitochondrial membranes. In: Membrane Biogenesis, edited by A. Tzagoloff. New York: Plenum Press, pp. 99–124.

Linnane, A. W., Nagley, P. (1978a). Structural mapping of mitochondrial DNA. Arch. Biochem. Biophys. 187:277–289.

Linnane, A. W., Nagley, P. (1978b). Mitochondrial genetics in perspective: the derivation of a genetic and physical map of the yeast mitochondrial genome. Plasmid 1:324–345.

Linnane, A. W., Haslam, J. M., Lukins, H. B., Nagley, P. (1972). The biogenesis of mitochondria in microorganisms. Ann. Rev. Microbiol. 26:163–198.

Linnane, A. W., Astin, A. M., Beilharz, M. W., Bingham, C. G., Choo, W. M., Cobon, G. S., Marzuki, S., Nagley, P., Roberts, H. (1980). Expression of the mitochondrial genome of yeast. In: The Organization and Expression of the Mitochondrial Genome, edited by A. M. Kroon and C. Saccone. Amsterdam: Elsevier/North-Holland, pp. 253–263.

Lloyd, D. (1974). The Mitochondria of Micro-organisms. London: Academic Press.

Lloyd, D., Turner, G. (1980). Structure, function, biogenesis and genetics of mitochondria. In: The Eukaryotic Microbial Cell, edited by G. W. Gooday, D. Lloyd, and A. P. J. Trinci. Cambridge: Cambridge University Press, pp. 143–180.

Little, P. F. R. (1982). Globin pseudogenes. Cell 28:683–684.

Locker, J., Rabinowitz, M. (1979). An overview of mitochondrial nucleic acids and biogenesis. Meth. Enzymol. 56:3–16.

Locker, J., Rabinowitz, M., Getz, G. S. (1974a). Tandem inverted repeats in mitochondrial DNA of petite mutants of *Saccharomyces cerevisiae.* Proc. Natl. Acad. Sci. USA 71:1366–1370.

Locker, J., Rabinowitz, M., Getz, G. S. (1974b). Electron microscopic and renaturation kinetic analysis of mitochondrial DNA of cytoplasmic *petite* mutants of *Saccharomyces cerevisiae.* J. Mol. Biol. 88:489–507.

Locker, J., Lewin, A., Rabinowitz, M. (1979). The structure and organization of mitochondrial DNA from *petite* yeast. Plasmid 2:155–181.

Lopez-Perez, M. J., Turner, G. (1975). Mitochondrial DNA from *Aspergillus nidulans.* FEBS Lett. 58:159–163.

McLean, J. R., Cohn, G. L., Brandt, I. K., Simpson, M. V. (1958). Incorporation of labelled amino acids into the protein of muscle and liver mitochondria. J. Biol. Chem. 233:657–663.

Macchecchini, M.-L., Rudin, Y., Blobel, G., Schatz, G. (1979). Import of proteins into mitochondria: precursor forms of the extramitochondrially made $F_1$-ATPase subunits in yeast. Proc. Natl. Acad. Sci. USA 76:343–347.

Macino, G., Tzagoloff, A. (1979a). Assembly of the mitochondrial membrane system. The DNA sequence of a mitochondrial ATPase gene in *Saccharomyces cerevisiae.* J. Biol. Chem. 254:4617–4623.

Macino, G., Tzagoloff, A. (1979b). Assembly of the mitochondrial membrane system. Partial sequence of a mitochondrial ATPase gene in *Saccharomyces cerevisiae.* Proc. Natl. Acad. Sci. USA 76:131–135.

Macino, G., Tzagoloff, A. (1980). Assembly of the mitochondrial membrane system: sequence analysis of a yeast mitochondrial ATPase gene containing the *oli-2* and *oli-4* loci. Cell 20:507–511.

Macino, G., Coruzzi, G., Nobrega, F. G., Li, M., Tzagoloff, A. (1979). Use of the UGA terminator as a tryptophan codon in yeast mitochondria. Proc. Natl. Acad. Sci. USA 76:3784–3785.

Mahler, H. R. (1973a). Biogenetic autonomy of mitochondria. CRC Crit. Rev. Biochem. 1:381–460.

Mahler, H. R. (1973b). Mitochondria: Molecular Biology, Genetics and Development. Reading, Mass.: Addison-Wesley.

Mahler, H. R. (1981). Mitochondrial evolution: organization and regulation of mitochondrial genes. Ann. N.Y. Acad. Sci. 361:53–75.

Mahler, H. R., Lin, C. C. (1978). Exogenous adenosine 3′:5′-monophosphate can release yeast from catabolite repression. Biochem. Biophys. Res. Commun. 83:1039–1047.

Mahler, H. R., Perlman, P. S. (1979). Mitochondrial biogenesis: evolution and regulation. In: Extrachromosomal DNA. ICN-UCLA Symposia on Molecular & Cellular Biology, Vol. 15, edited by D. J. Cummings, P. Borst, I. B. Dawid, S. M. Weissman, and C. F. Fox. New York: Academic Press, pp. 11–33.

Mahler, H. R., Raff, R. A. (1975). The evolutionary origin of the mitochondrion: a non-symbiont model. Int. Rev. Cytol. 43:1–124.

Mahler, H. R., Raff, R. A. (1975b). The symbiont that never was: an inquiry into the evolutionary origin of the mitochondrion. Symp. Soc. Exp. Biol. 29:41–92.

Mahler, H. R., Wilkie, D. (1978). Mitochondrial control of sugar utilization in *Saccharomyces cerevisiae*. Plasmid 1:125–133.

Mahler, H. R., Bastos, R. N., Flury, U., Lin, C. C., Phan, S. H. (1975). Mitochondrial biogenesis in fungi. In: Genetics and biogenesis of mitochondria and chloroplasts, 1st Bioscience Colloquium, 1974, edited by C. W. Birky, P. S. Perlman, and T. J. Byers. Columbus, Ohio: Ohio State University Press, pp. 66–116.

Mahler, H. R., Hanson, D., Miller, D., Bilinski, T., Ellis, D. M., Alexander, N. J., Perlman, P. S. (1977). Structural and regulatory mutations affecting mitochondrial gene products. In: Mitochondria 1977: Genetics and Biogenesis of Mitochondria, edited by W. Bandlow, R. J. Schweyen, K. Wolf, and F. Kaudewitz. Berlin: De Gruyter, pp. 345–370.

Marchant, R., Smith, D. G. (1968). Membranous structures in yeasts. Biol. Rev. 43:459–480.

Margulis, L. (1970). Origin of Eukaryotic Cells. New Haven, Conn.: Yale University Press.

Marmiroli, N., Restivo, F. M., Donnini, C., Bianchi, L., Puglisi, P. P. (1980). Analysis of *rho* mutability in *Saccharomyces cerevisiae*. I. Effects of *mmc* and *pet-ts* alleles. Mol. Gen. Genet. 177:581–588.

Martin, L. P., Schneller, J.-M., Stahl, A. J. C., Dirheimer, A. (1979). Import of deoxyribonucleic acid coded lysine-accepting transfer ribonucleic acid (anticodon C-U-U) into yeast mitochondria. Biochemistry 18:4600–4605.

Martin, N., Rabinowitz, M. (1978). Mitochondrial tRNAs in yeast: Identification of isoaccepting tRNAs. Biochemistry 17:1628–1634.

Martin, N. C., Rabinowitz, M., Fukuhara, H. (1978). Yeast mitochondrial DNA specifies tRNA for 19 amino acids. Deletion mapping of the tRNA genes. Biochemistry 16:4672–4677.

Martin, N. C., Miller, D. L., Donelson, J. E. (1979a). Cloning of yeast mitochondrial DNA in the *E. coli* plasmid pBR322. Identification of tRNA genes. J. Biol. Chem. 254:11729–11734.

Martin, N. C., Miller, D. L., Donelson, J. E., Sigurdson, C., Hartley, J. L., Moynihan, P. S., Dinh, P. H. (1979b). Identification and sequencing of yeast mitochondrial tRNA genes in mitochondrial DNA-pBR322 recombinants. In: ICN-UCLA Symp. Mol. Cell. Biol. Vol. 15, Extrachromosomal DNA, edited by D. J. Cummings, P. Borst, I. B. Dawid, S. M. Weissman, and C. F. Fox. New York: Academic Press, pp. 357–375.

Martin, N. C., Miller, D., Hartley, J., Moynihan, P., Donelson, J. E. (1980a). The tRNA$^{ser}_{AGY}$ and tRNA$^{Arg}_{CGY}$ genes form a gene cluster in yeast mitochondrial DNA. Cell 19:339–343.

Martin, N. C., Pham, H. D., Underbrink-Lyon, K., Miller, D. L., Donelson, J. E.

(1980b). Yeast mitochondrial $tRNA^{Trp}$ can recognize the nonsense codon UGA. Nature 285:579–581.

Martin, R. P., Sibler, A. P., Schneller, J. M., Keith, G., Stahl, A. J. C., Dirheimer, G. (1978). Primary structure of yeast mitochondrial DNA-coded phenylalanine-tRNA. Nucl. Acids Res. 5:4579–4592.

Martin, R. P., Sibler, A. P., Bordonne, R., Canaday, J., Dirheimer, G. (1980). Nucleotide sequence and gene localization of yeast mitochondrial initiator $tRNA^{fMet}$ and UGA-decoding $tRNA^{Trp}$. In: The Organization and Expression of the Mitochondrial Genome, edited by C. Saccone and A. M. Kroon. Amsterdam: Elsevier/North-Holland, pp. 311–314.

Mason, T. L., Poyton, R. O., Wharton, D. C., Schatz, G. (1973). Cytochrome c oxidase from baker's yeast. I. Isolation and properties. J. Biol. Chem. 248:1346–1354.

Maxam, A. M., Gilbert, W. (1977). A new method for sequencing DNA. Proc. Natl. Acad. Sci. USA 74:560–564.

Meyer, R. R. (1973). On the evolutionary origin of mitochondrial DNA. J. Theoret. Biol. 38:647–663.

Michaelis, G., Somlo, M. (1976). Genetic analysis of mitochondrial biogenesis and function in *Saccharomyces cerevisiae.* J. Bioenerg. 8:93–107.

Michaelis, G., Petrochilo, E., Slonimski, P. P. (1973). Mitochondrial genetics. III. Recombined molecules of mitochondrial DNA obtained from crosses between cytoplasmic *petite* mutants of *Saccharomyces cerevisiae:* physical and genetic characterization. Mol. Gen. Genet. 123:51–63.

Mihara, K., Blobel, G. (1980). The four cytoplasmically made subunits of yeast mitochondrial cytochrome c oxidase are synthesized individually and not as a polyprotein. Proc. Natl. Acad. Sci. USA 77:4160–4164.

Miller, D. L., Martin, N. C., Pham, H. D., Donelson, J. E. (1979). Sequence analysis of two yeast mitochondrial DNA fragments containing the genes for $tRNA^{Ser}_{UCR}$ and $tRNA^{Phe}_{UUY}$. J. Biol. Chem. 254:11735–11740.

Miller, D. L., Sigurdson, C., Martin, N. C., Donelson, J. E. (1980). Nucleotide sequence of the mitochondrial genes coding for $tRNA^{Gly}_{GGR}$ and $tRNA^{Val}_{GUR}$. Nucl. Acids Res. 8:1435–1457.

Miller, P. L., Ed. (1970). Control of Organelle Development. Symposia of the Society for Experimental Biology No. 24. Cambridge: Cambridge University Press.

Mills, D. R., Peterson, R. L., Spiegelman, S. (1967). An extracellular Darwinian experiment with a self-duplicating nucleic acid molecule. Proc. Natl. Acad. Sci. USA 58:217–224.

Mitchell, C. H., Bunn, C. L., Lukins, H. G., Linnane, A. (1973). Biogenesis of mitochondria 23. The biochemical and genetic characteristics of two different oligomycin resistant mutants of *Saccharomyces cerevisiae* under the influence of cytoplasmic genetic modification. Bioenergetics 4:161–177.

Molloy, P. L., Linnane, A. W., Lukins, H. B. (1975). Biogenesis of mitochondria: Analysis of deletion of mitochondrial antibiotic resistance markers in *petite* mutants of *Saccharomyces cerevisiae.* J. Bacteriol. 122:7–18.

Monnerot, M., Schweyen, R. J., Fukuhara, H. (1977). Mapping of mutation tsm-8 with respect to transfer RNA genes on the mitochondrial DNA of *Saccharomyces cerevisiae.* Mol. Gen. Genet. 152:307–309.

Montgomery, D. L., Hall, B. D., Gillam, S., Smith, M. (1978). Identification and isolation of the yeast cytochrome c gene. Cell 14:673–680.

Montoya, J., Ojola, D., Attardi, G. (1981). Distinctive features of the 5′-terminal sequence of human mitochondrial mRNAs. Nature 290:465–470.

Moorman, A. F. M., Verkley, F. N., Asselbergs, F. A. M., Grivell, L. A. (1977). Yeast mitochondrial messenger RNA is not correctly translated in heterologous cell-free systems. In: Mitochondria 1977: Genetics and Biogenesis of Mitochondria, edited by W. Bandlow, R. J. Schweyen, K. Wolf, and F. Kaudewitz. Berlin: De Gruyter, pp. 385–399.

Moorman, A. F. M., Grivell, L. A., Lamie, F., Smith, H. L. (1978). Identification of mitochondrial gene products by DNA-directed protein synthesis *in vitro*. Biochim. Biophys. Acta 518:351–365.

Morimoto, R., Rabinowitz, M. (1979a). Physical mapping of the Xba I, Hinc II, Bgℓ II, Xho I, Sst and Pvu II restriction endonuclease cleavage fragments of mitochondrial DNA of *S. cerevisiae*. Mol. Gen. Genet. 170:11–23.

Morimoto, R., Rabinowitz, M. (1979b). Physical mapping of the yeast mitochondrial genome: Derivation of the fine structure and gene map of strain D273-10B and comparison with a strain (MH41-7B) differing in genome size. Mol. Gen. Genet. 170:25–48.

Morimoto, R., Lewin, A., Merten, S., Rabinowitz, M. (1976). Restriction endonuclease mapping and analysis of grande and mutant yeast mitochondrial DNA. In: Genetics and Biogenesis of Chloroplasts and Mitochondria, edited by Th. Bucher, W. Neupert, W. Sebald, and S. Werner. Amsterdam: Elsevier/North-Holland, 519–524.

Morimoto, R., Lewin, A., Rabinowitz, M. (1977). Restriction cleavage map of mitochondrial DNA from the yeast *Saccharomyces cerevisiae*. Nucl. Acids. Res. 4:2331–2351.

Morimoto, R., Mertin, S., Lewin, A., Martin, N. C., Rabinowitz, M. (1978). Physical mapping of genes on yeast mitochondrial DNA: localization of antibiotic resistance loci, and rRNA and tRNA genes. Mol. Gen. Genet. 163:241–255.

Morimoto, R., Locker, J., Synenki, R. M., Rabinowitz, M. (1979). Transcription, processing and mapping of mitochondrial RNA from grande and *petite* yeast. J. Biol. Chem. 254:12461–12470.

Munn, E. A. (1974). The Structure of Mitochondria. London: Academic Press.

Murphy, W. T., Attardi, B., Tu, C., Attardi, G. (1975). Evidence for complete symmetrical transcription *in vivo* of mitochondrial DNA in HeLa cells. J. Mol. Biol. 99:809–814.

Nagai, S., Yanagashima, N., Nagai, H. (1961). Advances in the study of the respiration-deficient (RD) mutation in yeast and other microorganisms. Bacteriol. Rev. 25:404–426.

Nagley, P., Linnane, A. W. (1972a). Cellular regulation of mitochondrial DNA synthesis in *Saccharomyces cerevisiae*. Cell Differ. 1:145–148.

Nagley, P., Linnane, A. W. (1972b). Biogenesis of mitochondria. XXI. Nature of the mitochondrial genome in yeast. Degenerative effects of ethidium bromide on mitochondrial genetic information in a respiration competent strain. J. Mol. Biol. 65:181–193.

Nagley, P., Sriprakash, K. S., Rytka, J., Choo, K. B., Trembath, M. K., Lukins, H. B., Linnane, A. W. (1976). Physical mapping of genetic markers in the yeast mitochondrial genome. In: The Genetic Functions of Mitochondrial DNA, edited by C. Saccone and A. M. Kroon. Amsterdam: Elsevier/North-Holland, pp. 231–242.

Nagley, P., Sriprakash, K. S., Linnane, A. W. (1977). Structure, Synthesis and Genetics of Yeast Mitochondrial DNA. Adv. Microbiol. Physiol. 16:157–277.

Nagley, P., Atchison, B. A., Devenish, R. J., Vaughan, P. R., Linnane, A. W. (1980). Phenotypic and genetic changes in yeast cells transformed with mitochondrial DNA segments joined to 2 micron plasmid DNA. In: The Organization and Expression of the Mitochondrial Genome, edited by A. M. Kroon and C. Saccone. Amsterdam: Elsevier/North-Holland, pp. 75–78.

Nass, M. M. K., Nass, S. (1962). Fibrous structures within the matrix of developing chick embryo mitochondria. Exp. Cell Res. 26:424–427.

Nass, M. M. K., Nass, S. (1963a). Intramitochondrial fibers with DNA characteristics. I. Fixation and electron staining reactions. J. Cell. Biol. 19:593–611.

Nass, M. M. K., Nass, S., Afzelius, B. A. (1965). The general occurrence of mitochondrial DNA. Exp. Cell Res. 37:516–539.

Nass, S., Nass, M. M. K. (1963b). Intramitochondrial fibers with DNA characteristics. II. Enzymatic and other hydrolytic treatments. J. Cell Biol. 19:613–629.

Neher, E. M., Hanney, M. A., Hennig, B., Zimmerman, R., Neupert, W. (1980). Post-translational transport of proteins in the assembly of mitochondrial membranes. In: The Organization and Expression of the Mitochondrial Genome, edited by A. M. Kroon and C. Saccone. Amsterdam: Elsevier/North-Holland, pp. 433–442.

Nelson, N., Schatz, G. (1979). Energy-dependent processing of cytoplasmically made precursors to mitochondrial proteins. Proc. Natl. Acad. Sci. USA 76:4365–4369.

Netter, P., Jacq, C., Carignani, G., Slonimski, P. P. (1982). Critical sequences within mitochondrial introns: cis-dominant mutations of the "cytochrome-b-like" intron of the oxidase gene. Cell 28:733–738.

Newlon, C. S., Fangman, W. L. (1975). Mitochondrial DNA synthesis in cell cycle mutants of *Saccharomyces cerevisiae.* Cell 5:423–428.

Newlon, C. S., Ludescher, R. D., Walter, S. K. (1979). Production of *petites* by cell cycle mutants of *Saccharomyces cerevisiae* defective in DNA synthesis. Mol. Gen. Genet. 169:189–194.

Newman, D., Pham, H. D., Underbrink-Lyon, K., Martin, N. C. (1980). Characterization of tRNA genes in the tRNA region II of yeast mitochondrial DNA. Nucl. Acids Res. 8:5007–5016.

Nobrega, F. G., Tzagoloff, A. (1980). Assembly of the mitochondrial membrane system. Complete restriction map of the cytochrome b region of mitochondrial DNA in *Saccharomyces cerevisiae* D273-10B. J. Biol. Chem. 255:9821–9827.

Nobrega, F. G., Tzagoloff, A. (1980a). Assembly of the mitochondrial membrane system. Structure and location of the mitochondrial glutamic tRNA gene in *Saccharomyces cerevisiae.* FEBS Lett. 113:52–54.

Nobrega, F. G., Tzagoloff, A. (1980b). Assembly of the mitochondrial membrane system. DNA sequence and organization of the cytochrome b gene in *Saccharomyces cerevisiae* D273-10B. J. Biol. Chem. 255:9828–9837.

Novikoff, A. B. (1961). Mitochondria (Chondriosomes). In: The Cell, Biochemistry, Physiology, Morphology, edited by J. Brachet and A. E. Mirsky. New York: Academic Press, pp. 299–421.

O'Brien, T. W., Matthews, D. E. (1976). Mitochondrial ribosomes. In: Handbook of Genetics, edited by R. C. King. New York: Plenum Press, Vol. 5, pp. 535–580.

O'Connor, R. M., McArthur, C. R., Clark-Walker, G. D. (1975). Closed circular

DNA from mitochondrial-enriched fractions of four *petite*-negative yeasts. Eur. J. Biochem. 53:137–144.

O'Connor, R. M., McArthur, C. R., Clark-Walker, G. D. (1976). Respiratory deficient mutants of *Torulopsis glabrata,* a yeast with a circular mitochondrial deoxyribonucleic acid of 6μm. J. Bacteriol. 126:959–968.

Ojala, D., Merkel, C., Gelfand, R., Attardi, G. (1980). The tRNA genes punctuate the reading of genetic information in human mitochondrial DNA. Cell 22:393–403.

Ojala, D., Montoya, J., Attardi, G. (1981). tRNA punctuation model of RNA processing in human mitochondria. Nature 290:470–474.

Ono, B-I., Fink, G., Schatz, G. (1975). Mitochondrial assembly in respiration deficient mutants of *Saccharomyces cerevisiae.* IV. Effects of nuclear amber suppressors on the accumulation of a mitochondrially made subunit of cytochrome c oxidase. J. Biol. Chem. 250:775–782.

Orgel, L. E., Crick, F. H. C. (1980). Selfish DNA: the ultimate parasite. Nature 284:604–607.

Osinga, K. A., Evers, R. F., Van der Laan, C., Tabak, H. F. (1981). A putative precursor for the small ribosomal RNA from mitochondria of *Saccharomyces cerevisiae.* Nucl. Acids Res. 9:1351–1364.

Packer, L., Gomez-Payou, A., Eds. (1976), Mitochondria, bioenergetics, biogenesis and membrane structure. New York: Academic Press.

Pajot, P., Wambier-Kluppel, M. L., Kotylak, Z., Slonimski, P. P. (1976). Regulation of cytochrome oxidase formation by mutations in a mitochondrial gene for cytochrome b. In: Genetics and Biogenesis of Chloroplasts and Mitochondria, edited by T. Bücher, W. Neupert, W. Sebald, and S. Werner. Amsterdam: Elsevier/North-Holland Biomedical Press, pp. 443–451.

Pajot, P., Wambier-Kluppel, M. L., Slonimski, P. P. (1977). Cytochrome c reductase and cytochrome oxidase formations in mutants and revertants in the "box" region of mitochondrial DNA. In: Mitochondria 1977: Genetics and Biogenesis of Mitochondria, edited by W. Bandlow, R. J. Schweyen, K. Wolf, and F. Kaudewitz. Berlin: De Gruyter, pp. 173–184.

Peebles, C. L., Ogden, R. C., Knapp, G., Abelson, J. (1979). Splicing of yeast tRNA precursors: a two-stage reaction. Cell 18:27–35.

Perlman, P. S., Birky, C. W., Jr. (1974). Mitochondrial genetics in baker's yeast: a molecular mechanism for recombinational polarity and suppressiveness. Proc. Natl. Acad. Sci. USA 71:4612–4616.

Perlman, P. S., Alexander, N. J., Hanson, D. K., Mahler, H. R. (1980). Mosaic genes in yeast mitochondria. In: Gene Structure and Expression, edited by D. H. Dean, L. F. Johnson, P. C. Kimbal and P. S. Perlman. Columbus, Ohio: Ohio State University Press, p. 211.

Pinon, R., Salts, Y., Simchen, G. (1974). Nuclear and mitochondrial DNA synthesis during yeast sporulation. Exp. Cell Res. 83:231–238.

Piperno, G., Fonty, G., Bernardi, G. (1972). The mitochondrial genome of wild-type yeast cells. II. Investigations on the compositional heterogeneity of mitochondrial DNA. J. Mol. Biol. 65:191–205.

Poyton, R. O., Schatz, G. (1975). Cytochrome c oxidase from baker's yeast. III. Physical characterization of isolated subunits and chemical evidence for two classes of polypeptides. J. Biol. Chem. 250:752–761.

Poyton, R. O., McKemmie, E., George-Nascimento, C. (1978). The use of orthacryl two-dimensional polyacrylamide gel electrophoresis to identify and compare the subunit polypeptides of bovine heart and yeast cytochrome c oxidase. J. Biol. Chem. 253:6303–6306.

Pratje, E., Schultz, R., Schnierer, S., Michaelis, G. (1979). Sporulation of mitochondrial respiratory deficient *mit*$^-$ mutants of *Saccharomyces cerevisiae.* Mol. Gen. Genet. 176:411–415.

Prunell, A., Bernardi, G. (1974). The mitochondrial genome of wild-type yeast cells. IV. Genes and Spacers. J. Mol. Biol. 86:825–841.

Prunell, A., Bernardi, G. (1977). The mitochondrial genome of wild-type yeast cells. VI. Genome organization. J. Mol. Biol. 110:53–74.

Prunell, A., Kopecka, H., Strauss, F., Bernardi, G. (1977). The mitochondrial genome of wild-type yeast cells. V. Genome evolution. J. Mol. Biol. 110:17–52.

Puglisi, P. P., Algeri, A. A. (1971). Role of the mitochondrion in the regulation of protein synthesis in the eucaryote *Saccharomyces cerevisiae.* Mol. Gen. Genet. 110:110–117.

Puiseux-Dao, S., Ed. (1975). Molecular Biology of Nucleocytoplasmic Relationships. Amsterdam: Elsevier/North-Holland.

Putrament, A., Baranowska, H., Prazmo, W. (1973). Induction by manganese of mitochondrial antibiotic resistance mutations in yeast. Mol. Gen. Genet. 126:357–366.

Putrament, A., Baranowska, H., Ejchart, A., Jachymczk, W. (1977). Manganese mutagenesis in yeast. VI. $Mn^{++}$ uptake mtDNA replication and $E^R$ induction. Comparison with other divalent cations. Mol. Gen. Genet. 151:69–76.

Rabinowitz, R., Swift, H. (1970). Mitochondrial nucleic acids and their relation to the biogenesis of mitochondria. Physiol. Rev. 50:376–427.

Rabinowitz, M., Jakovcic, S., Martin, N., Hendler, F., Halbreich, A., Lewin, A., Morimoto, R. (1976). Transcription and organization of yeast mitochondrial DNA. In: The Genetic Function of Mitochondrial DNA, edited by C. Saccone and A. M. Kroon. Amsterdam: North-Holland, pp. 219–230.

Rank, G. H. (1970). Genetic evidence for "Darwinian" selection at the molecular level. II. Genetic analysis of cytoplasmically-inherited high and low suppressivity in *Saccharomyces cerevisiae.* Can. J. Genet. Cytol. 12:340–346.

Rank, G. H., Bech-Hansen, N. T. (1973). Single nuclear gene inherited cross resistance and collateral sensitivity to 17 inhibitors of mitochondrial function in *Saccharomyces cerevisiae.* Mol. Gen. Genet. 126:93–102.

Rank, G. H., Robertson, A. J., Phillips, K. L. (1975). Modification and inheritance of pleiotropic cross-resistance and collateral activity in *Saccharomyces cerevisiae.* Genetics 80:483–493.

Reanney, D. (1979). RNA splicing and polynucleotide evolution. Nature 277:598–600.

Roberts, H., Choo, W. M., Murphy, M., Marzuki, S., Lukins, H. B., Linnane, A. W. (1979). *mit*$^-$ mutations in the *oli 2* region if mitochondrial DNA affecting the 20,000 dalton subunit of the mitochondrial ATPase in *Saccharomyces cerevisiae.* FEBS Lett. 108:501–504.

Roodyn, D. B., Wilkie, D. (1968). The Biogenesis of Mitochondria. London: Methuen.

Rotman, A. (1975). Genetics of a primaquin-resistant yeast. J. Gen. Microbiol. 80:1–10.

Rubin, B. Y., Blamire, J. (1977). Regulation of yeast mitochondrial DNA synthesis.

I. Analysis of a mutant conditionally deficient in mitochondrial DNA metabolism. Mol. Gen. Genet. 156:41–47.

Rubin, B. Y., Blamire, J. (1979). Regulation of yeast mitochondrial DNA synthesis. II. The kinetics and effect of growth media on the expression of the *tpi* mutation. Mol. Gen. Genet. 169:41–47.

Rubin, M. S., Tzagoloff, A. (1973). Assembly of the mitochondrial membrane system. IX. Purification, characterization and subunit structure of yeast and beef cytochrome oxidase. J. Biol. Chem. 248:4269–4274.

Ryan, R., Grant, D., Chiang, K.-S., Swift, H. (1978). Isolation and characterization of mitochondrial DNA from *Chlamydomonas reinhardtii*. Proc. Natl. Acad. Sci. USA 75:3268–3272.

Saccone, C., Kroon, A. M., Eds. (1976). The Genetic Functions of Mitochondrial DNA. Amsterdam: Elsevier/North-Holland.

Saccone, C., Quagliariello, E. (1975). Biochemical studies of mitochondrial transcription and translation. Int. Rev. Cytol. 43:125–165.

Sager, R. (1972). Cytoplasmic genes and organelles. New York: Academic Press.

Sakano, H., Rogers, J. H., Hüppi, K., Brack, C., Traunecker, A., Maki, R., Wall, R., Tonegawa, S. (1979). Domains and the hinge region of an immunoglobulin heavy chain are encoded in separate DNA segments. Nature 277:627–633.

Sanders, J. P. M., Heyting, C., Borst, P. (1975). The organization of genes in yeast mitochondrial DNA. I. The genes for large and small ribosomal RNA are far apart. Biochem. Biophys. Res. Commun. 65:699–707.

Sanders, J. P. M., Heyting, C., Difranco, A., Borst, P., Slonimski, P. P. (1976). The organization of genes in yeast mitochondrial DNA. In: The Genetic Functions of Mitochondrial DNA, edited by C. Saccone and A. M. Kroon. Amsterdam: Elsevier/North-Holland, pp. 259–272.

Sanders, J. P. M., Heyting, C., Verbeet, M. Ph., Meijlink, F. C. P. W., Borst, P. (1977a). The organisation of genes in yeast mitochondrial DNA. III. Comparison of the physical maps of the mitochondrial DNAs from three wild-type *Saccharomyces* strains. Mol. Gen. Genet. 157:239–261.

Sanders, J. P. M., Verbeet, M. Ph., Meijlink, F. C. P. W., Heyting, C., Borst, P. (1977b). The construction of the physical maps of three different *Saccharomyces* mitochondrial DNAs. Mol. Gen. Genet. 157:271–280.

Schatz, G. (1976). The biogenesis of mitochondria: a review. In: Genetics Biogenesis and Bioenergetics of Mitochondria, edited by W. Bandlow, R. J. Schweyen, D. Y. Thomas, K. Wolf, and F. Kaudewitz. Berlin: De Gruyter, pp. 163–177.

Schatz, G. (1979). How mitochondria import proteins from the cytoplasm. FEBS Lett. 103:203–211.

Schatz, G., Mason, T. L. (1974). The biosynthesis of mitochondrial proteins. Annu. Rev. Biochem. 43:51–87.

Schatz, G., Saltzgaber, J. (1969). Protein synthesis by yeast promitochondria *in vivo*. Biochem. Biophys. Res. Commun. 37:996–1001.

Scherrer, K., Marcaud, L. (1968). Messenger RNA in avian erythroblasts at the transcriptional and translational levels and the problem of regulation in animal cells. J. Cell. Physiol. 72, Suppl. 1:181–212.

Schroeder, R., Breitenbach, M. (1981). The role of the mitochondria in yeast differentiation. In: Sporulation and Germination, edited by H. S. Levinson *et al.*, Proceedings of the 8th International Spore Conference, ASM Publications.

Schweizer, E., Demmer, W., Holner, U., Tahedl, H. W. (1977). Controlled mitochondrial inactivation of temperature-sensitive *Saccharomyces cerevisiae* nuclear *petite* mutants. In: Mitochondria 1977. Genetics and Biogenesis of Mitochondria, edited by W. Bandlow, R. J. Schweyen, K. Wolf, and F. Kaudewitz. Berlin: De Gruyter, pp. 91–105.

Schweyen, R. J., Steyrer, U., Kaudewitz, F., Dujon, B., Slonimski, P. P. (1976). Mapping of mitochondrial genes in *Saccharomyces cerevisiae*. Population and pedigree analysis of retention or loss of four genetic markers in *rho*$^-$ cells. Mol. Gen. Genet. 146:117–132.

Schweyen, R. J., Weiss-Brummer, B., Backhaus, B., Kaudewitz, F. (1978). The genetic map of the mitochondrial genome in yeast: Map positions of drug$^R$ and *mit*$^-$ markers as revealed from population analyses of *rho*$^-$ clones in *Saccharomyces cerevisiae*. Mol. Gen. Genet. 159:151–160.

Scragg, A. H. (1976). The isolation and properties of a DNA-directed RNA polymerase from yeast mitochondria. Biochim. Biophys. Acta 442:331–342.

Sebald, W. (1977). Biogenesis of mitochondrial ATPase. Biochim. Biophys. Acta 463:1–27.

Sebald, W., Wachter, E. (1978). Amino acid sequence of the putative protonophore of the energy-transducing ATPase complex. In: 29th Mosbacher Colloquium on Energy Conservation in Biological Membranes, edited by G. Schäfer and M. Klingenberg. Berlin: Springer-Verlag, pp. 228–236.

Sebald, W., Machleidt, W., Otto, J. (1973). Products of mitochondrial protein synthesis in *Neurospora crassa*. Determination of equimolar amounts of three products in cytochrome oxidase on the basis of amino-acid analysis. Eur. J. Biochem. 38:311–324.

Sebald, W., Sebald-Althaus, M., Wachter, E. (1977). Altered amino acid sequence of the DCCD-binding protein of the nuclear oligomycin-resistant mutant AP-2 from *Neurospora crassa*. In: Mitochondria 1977. Genetics and Biogenesis of Mitochondria, edited by W. Bandlow, R. J. Schweyen, K. Wolf, and F. Kaudewitz. Berlin: De Gruyter, pp. 433–440.

Sebald, W., Graf, Th., Lukins, H. B. (1979a). The dicyclohexylcarbodiimide-binding protein of the mitochondrial ATPase complex from *Neurospora crassa* and *Saccharomyces cerevisiae*. Eur. J. Biochem. 93:587–599.

Sebald, W., Wachter, E., Tzagoloff, A. (1979b). Identification of amino acid substitutions in the dicyclohexylcarbodiimide-binding subunit of the mitochondrial ATPase complex from oligomycin resistant mutants of *Saccharomyces cerevisiae*. Eur. J. Biochem. 100:599–607.

Sena, E. P., Welch, J. W., Halvorson, H. O., Fogel, S. (1975). Nuclear and mitochondrial deoxyribonucleic acid replication during mitosis in *Saccharomyces cerevisiae*. J. Bacteriol. 123:497–504.

Sena, E., Welch, J., Fogel, S. (1976). Nuclear and mitochondrial DNA replication during zygote formation and maturation in yeast. Science 194:433–435.

Shah, D. M., Langley, C. H. (1979). Inter- and intra-specific variation in restriction maps of *Drosophila* mitochondrial DNAs. Nature 281:696–699.

Sherman, F. (1963). Respiration deficient mutants of yeast I. Genetics. Genetics 48:375–385.

Sherman, F., Slonimski, P. P. (1964). Respiration deficient mutants of yeast. II. Biochemistry. Biochim. Biophys. Acta 90:1–15.

Shine, J., Dalgarno, L. (1975). Determinant of cistron specificity in bacterial ribosomes. Nature 254:34–38.

Slater, E. C., Tager, J. M., Papa, S., Quagliariello, E., Eds. (1968). Biochemical Aspects of the Biogenesis of Mitochondria. Bari: Adriatica Editrice.

Slonimski, P. P., Lazowska, J. (1977). Transposable segments of mitochondrial DNA: a unitary hypothesis for the mechanism of mutation, recombination, sequence reiteration and suppressiveness of yeast "*petite* colony" mutants. In: Mitochondria 1977: Genetics and Biogenesis of Mitochondria, edited by W. Bandlow, R. J. Schweyen, K. Wolf, and F. Kaudewitz. Berlin: De Gruyter, pp. 39–52.

Slonimski, P. P., Pajot, P., Jacq, C., Foucher, M., Perrodin, G., Kochko, A., Lamouroux, A. (1978a). Mosaic organization and expression of the mitochondrial DNA region controlling cytochrome c reductase and oxidase. I. Genetic, physical and complementation maps of the *box* region. In: Biochemistry and Genetics of Yeasts: Pure and Applied Aspects, edited by M. Bacila, B. L. Horecker, and A. O. M. Stoppani. New York: Academic Press, pp. 339–368.

Slonimski, P. P., Claisse, M. L., Foucher, M., Jacq, C., Kochko, A., Lamouroux, A., Pajot, P., Perrodin, G., Spyridakis, A., Wambier-Kluppel, M. L. (1978b). Mosaic organization and expression of the mitochondrial DNA region controlling cytochrome c reductase and oxidase. III. A model of structure and function. In: Biochemistry and Genetics of Yeasts: Pure and Applied Aspects, edited by M. Bacila, B. L. Horecker, and A. O. M. Stoppani. New York: Academic Press, pp. 391–401.

Slonimski, P., Borst, P., Attardi, G., Eds. (1982). Mitochondrial Genes. Cold Spring Harbor, N.Y.: Cold Spring Harbor Laboratory.

Smith, M., Leung, D., Gillam, S., Astell, C., Montgomery, D., Hall, B. (1979). Sequence of the gene for iso-1-cytochrome c in *Saccharomyces cerevisiae*. Cell 16:753–761.

Solioz, M., Schatz, G. (1979). Mutations in putative intervening sequences of the mitochondrial cytochrome b gene of yeast produce abnormal cytochrome *b* polypeptides. J. Biol. Chem. 254:9331–9334.

Sor, F., Fukuhara, H. (1980). Séquence nucléotidique du gène de l'ARN ribosomique 15S mitochondrial de la levure. C. R. Acad. Sci. Paris 291:933–936.

Spencer, J. F. T., Spencer, D. M. (1981). The use of mitochondrial mutants in hybridization of industrial yeasts. III. Restoration of mitochondrial function in *petites* of industrial yeast strains by fusion with respiratory-competent protoplasts of other yeast species. Curr. Genet. 4:177–180.

Sriprakash, K. S., Clark-Walker, G. D. (1980). The size of yeast mitochondrial ribosomal RNAs. Biochem. Biophys. Res. Commun. 93:186–193.

Sriprakash, K. S., Molloy, P. L., Nagley, P., Lukins, H. B., Linnane, A. W. (1976). Biogenesis of mitochondria. XLI. Physical mapping of mitochondrial genetic markers in yeast. J. Mol. Biol. 104:485–503.

Steffens, G., Buse, G. (1976). Studies on cytochrome c oxidase. I. Purification and characterization of the enzyme from bovine heart and identification of peptide chains in the complex. Hoppe-Seyler's Z. Physiol. Chem. 357:1125–1137.

Steffens, G. J., Buse, G. (1979). Studies on cytochrome c oxidase. IV. Primary structure and function of subunit II. Hoppe-Seyler's Z. Physiol. Chem. 360:613–619.

Stevens, B. (1981). Mitochondrial Structure. In: The Molecular Biology of the Yeast *Saccharomyces:* Life Cycle and Inheritance, edited by J. N. Strathern, E. W. Jones, and J. R. Broach. Cold Spring Harbor, N.Y.: Cold Spring Harbor Laboratory, pp. 471–504.

Stiegler, P., Carbon, P., Zuker, M., Ebel, J.-P., Ehresmann, C. (1980). Structure sécondaire et topographie du RNA ribosomique 16S d'*Escherischia coli*. C. R. Acad. Sci. Paris 291:937–940.

Strausberg, R. L., Butow, R. A. (1977). Expression of *petite* mitochondrial DNA *in vivo:* Zygotic gene rescue. Proc. Natl. Acad. Sci. USA 74:2715–2719.

Strausberg, R. L., Butow, R. A. (1981). Gene conversion at the *var 1* locus on yeast mitochondrial DNA. Proc. Natl. Acad. Sci. USA 78:494–498.

Strausberg, R. L., Vincent, R. D., Perlman, P. S., Butow, R. A. (1978). Asymmetric gene conversion at inserted segments on yeast mitochondrial DNA. Nature 276:577–583.

Tabak, H. F., Hecht, N. B., Menke, H. H., Hollenberg, C. P. (1979). The gene for the small ribosomal RNA on yeast mitochondrial DNA: physical map, direction of transcription and absence of an intervening sequence. Curr. Genet. 1:33–43.

Tandler, B., Hoppel, C. L. (1972). Mitochondria. New York: Academic Press.

Takaiwa, F., Sugiura, M. (1982). The complete nucleotide sequence of a 23-S rRNA gene from tobacco chloroplasts. Eur. J. Biochem. 124:13–19.

Taylor, F. J. R. (1976). Autogenous theories for the origin of eukaryotes. Taxon 25:377–390.

Tedeschi, H. (1976). Mitochondria: Structure, Biogenesis and Transducing Functions. New York: Springer-Verlag.

Thalenfeld, B. E., Tzagoloff, A. (1980). Assembly of the mitochondrial membrane system. Sequence of the *oxi-2* gene of yeast mitochondrial DNA. J. Biol. Chem. 255:6173–6180.

Tiemeyer, D. C., Tilghman, S. M., Polsky, F. I., Seidman, J. G., Leder, A., Edgell, M. H., Leder, P. (1978). A comparison of two cloned mouse $\beta$-globin genes and their surrounding and intervening sequences. Cell 14:237–245.

Trembath, M. K., Monk, B. C., Kellerman, G. M., Linnane, A. W. (1975). Biogenesis of mitochondria. 40. Phenotypic suppression of a mitochondrial mutation by a nuclear gene in *Saccharomyces cerevisiae.* Mol. Gen. Genet. 140:333–337.

Tsai, M. J., Michaelis, G., Criddle, R. S. (1971). DNA-dependent RNA polymerase from yeast mitochondria. Proc. Natl. Acad. Sci. USA 68:473–477.

Tzagoloff, A. Meagher, P. (1971). Assembly of the mitochondrial membrane system. VI. Mitochondrial synthesis of subunit proteins of the rutamycin-sensitive adenosine triphosphatase. J. Biol. Chem. 247:594–603.

Tzagoloff, A., Rubin, M. S., Sierra, M. F. (1973). Biosynthesis of mitochondrial enzymes. Biochim. Biophys. Acta 301:71–104.

Tzagoloff, A., Akai, A., Needleman, R. B. (1975a). Properties of cytoplasmic mutants of *Saccharomyces cerevisiae* with specific lesions in cytochrome oxidase. Proc. Natl. Acad. Sci. USA 72:2054–2057.

Tzagoloff, A., Akai, A., Needleman, R. B. (1975b). Assembly of the mitochondrial membrane system: Isolation of nuclear and cytoplasmic mutants of *Saccharomyces cerevisiae* with specific defects in mitochondrial functions. J. Bacteriol. 122:826–831.

Tzagoloff, A., Akai, A., Needleman, R. B., Zulch, G. (1975c). Assembly of the mitochondrial membrane system. Cytoplasmic mutants of *S. cerevisiae* with lesions in enzymes of the respiratory chain and the mitochondrial ATPase. J. Biol. Chem. 250:8236–8242.

Tzagoloff, A., Akai, A., Needleman, R. B. (1975d). Assembly of the mitochondrial membrane system: characterization of nuclear mutants of *Saccharomyces cerevisiae* with defects in mitochondrial ATPase and respiratory enzymes. J. Biol. Chem. 250:8228–8235.

Tzagoloff, A., Foury, F., Macino, G. (1978). Genetic loci and complementation groups of the mitochondrial genome. In: Biochemistry and Genetics of Yeast: Pure and Applied Aspects, edited by M. Aeila, B. L. Horecker, and A. O. M. Stoppani. New York: Academic Press, pp. 477–488.

Tzagoloff, A., Macino, G., Sebald, W. (1979a). Mitochondrial genes and translation products. Annu. Rev. Biochem. 48:419–441.

Tzagoloff, A., Macino, G., Nobrego, M. P., Li, M. (1979b). Organization of mitochondrial DNA in yeast. In: Extrachromosomal DNA: ICN-UCLA Symposium on Molecular and Cellular Biology 15, edited by D. Cummings, P. Borst, I. Dawid, S. Weissman, and C. F. Fox. New York: Academic Press, pp. 341–355.

Tzagoloff, A., Nobrega, M., Akai, A., Macino, G. (1980). Assembly of the mitochondrial membrane system. Organization of yeast mitochondrial DNA in the *oli 1* region. Curr. Genet. 2:149–157.

Van den Bergh, S. G., Borst, P., Van Deenen, L. L. M., Riemersma, J. C., Slater, E. C., Tager, J. M., Eds. (1972). Mitochondria: Biogenesis and Bioenergetics; Biomembranes: Molecular Arrangements and Transport Mechanisms. (Federation of European Biochemical Societies, Vol. 28: Eighth Meeting, Amsterdam, 1972). Amsterdam: Elsevier/North-Holland.

Van den Boogaart, P., Samallo, J., Agsteribbe, E. (1982). Similar genes for a mitochondrial ATPase submit in the nuclear and mitochondrial genomes of *Neurospora crassa*. Nature 298:187–189.

Van Etten, R. A., Walberg, M. W., Clayton, D. A. (1980). Precise localization and nucleotide sequence of the two mouse mitochondrial rRNA genes and three immediately adjacent novel tRNA genes. Cell 22:157–170.

Van Kreijl, C. F., Bos, J. L. (1977). The repeating nucleotide sequence in the repetitive mitochondrial DNA from a "low-density" *petite* mutant of yeast. Nucl. Acids Res. 4:2369–2388.

Van Ommen, G.-J. B., Groot, G. S. P., Borst, P. (1977). Fine structure physical mapping of 4S RNA genes on mitochondrial DNA of *Saccharomyces cerevisiae*. Mol. Gen. Genet. 154:255–262.

Van Ommen, G.-J. B., Groot, G. S. P., Grivell, L. A. (1979). Transcription maps of mtDNAs of two strains of *Saccharomyces:* transcription of strain-specific insertions; complex RNA maturation and splicing. Cell 18:511–523.

Van Ommen, G.-J. B., Boer, P. H., Groot, G. S. P., De Haan, M., Roosendaal, E., Grivell, L. A., Haid, A., Schweyen, R. J. (1980). Mutations affecting RNA splicing and the interaction of gene expression of the yeast mitochondrial loci *cob* and *oxi-3*. Cell 20:173–183.

Vedel, F., Quetier, F. (1974). Physico-chemical characterization of mitochondrial DNA from potato tubers. Biochim. Biophys. Acta 340:374–387.

Von Wettstein, D., Friis, J., Keilland-Brandt, M., Stenderup, A., Eds. (1981). Alfred Benzon Symposium 16: Molecular Genetics in Yeast. Copenhagen: Munksgaard.

Wallace, D. C. (1982). Structure and evolution of organelle genomes. Microbiol. Rev. 46:208–240.

Ward, B. L., Anderson, R. S., Bendich, A. J. (1981). The mitochondrial genome is large and variable in a family of plants (Cucurbitaceae). Cell 25:793–803.

Waring, R. B., Davies, R. W., Lee, S., Grisi, E., Berks, M. M., Scazzocchio, C. (1981). The mosaic organization of the apocytochrome b gene of *Aspergillus nidulans* revealed by DNA sequencing. Cell 27:4–11.

Waxman, M. F., Knight, J. A., Perlman, P. S. (1979). Suppression of mitochondrially-determined resistance to chloramphenicol and paromomycin by nuclear genes in *Saccharomyces cerevisiae.* Mol. Gen. Genet. 167:243–250.

Wesolowski, M., Fukuhara, H. (1979). The genetic map of transfer RNA genes of yeast mitochondria: correction and extension. Mol. Gen. Genet. 170:261–275.

Wesolowski, M., Fukuhara, H. (1981). Linear mitochondrial deoxyribonucleic acid from the yeast *Hansenula mrakii.* Mol. Cell. Biol. 1:387–393.

Wesolowski, M., Algeri, A., Fukuhara, H. (1981). Gene organization of the mitochondrial DNA of yeasts: *Kluyveromyces lactis* and *Saccharomycopsis lipolytica.* Curr. Genet. 3:157–162.

Whittaker, P. A. (1979). The *petite* mutation in yeast. Subcell. Biochem. 6:175–232.

Wilkie, D. (1973). Cytoplasmic genetic systems of eukaryotic cells. Brit. Med. Bull. 29:263–268.

Wilkie, D. (1983). Chapter 10 in this volume.

Wilkie, D., Evans, I. H. (1982). Mitochondria and the yeast cell surface: implications for carcinogenesis. Trends Biochem. Sci. 4:197–201.

Wilkie, D., Nudd, R. C. (1981). Aspects of mitochondrial control of cell surface characteristics in *Saccharomyces cerevisiae.* In: Current Developments in Yeast Research: Advances in Biotechnology. Proceedings of the fifth international yeast symposium, London, Canada, 1980, edited by G. G. Stewart and I. Russell. Toronto: Pergamon, pp. 345–349.

Wilkie, D., Evans, I. H., Egilsson, V., Diala, E. S., Collier, D. (1983). Mitochondria, cell surface, and carcinogenesis. Int. Rev. Cytol. (Suppl. 15), edited by J. F. Danielli. New York: Academic Press, pp. 157–189.

Williamson, D. H., Moustacchi, E. (1971). The synthesis of mitochondrial DNA during the cell cycle in the yeast *Saccharomyces cerevisiae.* Biochem. Biophys. Res. Commun. 42:195–201.

Williamson, D. H., Johnston, L. H., Richmond, K. M. V., Game, J. C. (1977). Mitochondrial DNA and the heritable unit of the yeast mitochondrial genome: a review. In: Mitochondria 1977: Genetics and Biogenesis of Mitochondria, edited by W. Bandlow, R. J. Schweyen, K. Wolf, and F. Kaudewitz. Berlin: De Gruyter, pp. 1–24.

Wintersberger, E. (1972). Isolation of a distinct rifampicin-resistant RNA polymerase from mitochondria of yeast, *Neurospora* and liver. Biochem. Biophys. Res. Commun. 48:1287–1294.

Wintersberger, U., Blutsch, H. (1976). DNA-dependent DNA polymerase from yeast mitochondria: dependence of enzyme activity on conditions of cell growth and properties of the highly purified polymerase. Eur. J. Biochem. 68:199–207.

Wintersberger, U., Wintersberger, E. (1970). Studies on DNA polymerase from yeast. 2. Partial purification and characterization of mitochondrial DNA polymerase from wild-type and respiratory-deficient yeast cells. Eur. J. Biochem. 13:20–27.

Wiseman, A., Lim, T.-K. (1974). Removal of glucose repression of α-glucosidase synthesis in yeast protoplasts by adenosine 3′:5′-cyclic monophosphate addition and its prevention by chloramphenicol. Biochem. Soc. Trans. 2:932–935.

Woese, C. R. (1981). Archaebacteria. Sci. Amer. 244:98–122.

Woese, C. R., Fox, G. E. (1977). Phylogenetic structure of the prokaryotic domain: the primary kingdoms. Proc. Natl. Acad. Sci. USA 74:5088–5090.

Woese, C. R., Maniloff, J., Zablen, L. B. (1980). Phylogenetic analysis of the mycoplasmas. Proc. Natl. Acad. Sci. USA 77:494–498.

Wolf, K., Sebald-Althaus, M., Schweyen, R. J., Kaudewitz, F. (1971). Respiration-deficient mutants of *Schizosaccharomyces pombe*. I. Mol. Gen. Genet. 110:101–109.

Wu, G.-J., Dawid, I. B. (1972). Purification and properties of mitochondrial deoxyribonucleic acid dependent ribonucleic acid polymerase from ovaries of *Xenopus laevis*. Biochemistry 11:3589–3595.

Young, R. A., Steitz, J. A. (1978). Complementary sequences 1700 nucleotide apart form a rubonuclease III cleavage site in *Escherichia coli* ribosomal precursor RNA. Proc. Natl. Acad. Sci. USA 75:3593–3597.

Zakour, R. A., Bultmann, H. (1979). Evolution of *Drosophila* mitochondrial DNAs. Analysis of heteroduplex molecules. Biochim. Biophys. Acta 564:342–351.

## Note Added in Proof

Mitochondrial genes are providing no exception to the emerging "rule" that genes are much more mobile than previously thought. A continuous DNA sequence in the yeast nuclear genome has been found to possess extensive homology with non-contiguous yeast mtDNA sequences, one of the latter being *var1*—which has finally been confirmed as a mitochondrial gene [Farrelly, F., Butow, R. A. (1983). Rearranged mitochondrial genes in the yeast nuclear genome. Nature 301: 296–301]. It is proposed by Farrelly and Butow that these nuclear DNA sequences, which cannot be functional and are not transcribed, orginated from the integration of a fragment of *petite* mtDNA into the nuclear genome. The characterization of the *var1* gene has also shown that this protein-coding sequence contains 90% A + T and that the codon ATA encodes methionine, as in mammalian mitochondria (both yeast and mammalian mitochondria therefore employing a different codon assignment from the "standard" genetic code, which is to isoleucine).

Another fungus in which nuclear DNA sequences have been found showing homology to mtDNA is *Podospora anserina* [Fox, T. D. (1983). Mitochondrial genes in the nucleus. Nature 301: 371–372].

Splicing of mtRNA is a very active field. Comparison of sequences of introns from both *S. cerevisiae* and *Aspergillus nidulans* mtDNA has shown that at least seven intron RNAs can form identical secondary structures which bring the ends of each intron together; alignment of the junctions to be cut and joined could, in this way, facilitate the action of the splicing machinery [Davies, R. W., Waring, R. B., Ray, J. A., Brown, T. A., Scazzocchio, C. (1982). Making ends meet: a model for RNA splicing in fungal mitochodria. Nature 300: 719–724]. Nuclear DNA sequences concerned in splicing reactions have also been characterized: a nuclear *pet* gene needed for correct splicing of *oxi3* pre-mRNA has been cloned and sequenced [Faye, G., Simon, M. (1983). Analysis of yeast nuclear gene involved in the maturation of mitochondrial pre-messenger RNA of the cytochrome oxidase submit 1. Cell 32: 77–87]. Its putative product is a protein of 436 amino acids which, interestingly, shows no similarity in terms of amino acid sequence, to the mitochondrial *box7* maturase.

A number of other new developments are summarised in a review of the 1982 EMBO workshop of mitochondrial gene expression [Fox, T. D. (1982). Mitochondrial gene expression at Castro, Italy. Curr. Genet. 6: 1–3].

# 12

# ScV "Killer" Viruses in Yeast

Diane J. Mitchell* and E. Alan Bevan*

## Introduction

Certain strains of the yeast *Saccharomyces cerevisiae* secrete a proteinaceous toxin, the killer toxin, which is lethal to other strains of the same species. Toxin-producing strains are termed *killers* and susceptible strains are termed *sensitives*.

Killer strains possess two species of dsRNA, the L and M genomes, which are separately encapsidated in identical protein coats as the ScV-L and ScV-M viruses.[1] The ScV-L viruses are helper particles that provide the capsid protein and probably also the polymerase activities needed for both viruses. The ScV-M viruses are defective particles having a positive requirement for the presence of both the ScV-L particles and a multiplicity of host gene products for their own maintenance and functional expression. The M genome encodes the killer toxin and most probably the resistance protein that renders killer cells immune to their own toxin.

This review is concerned mainly with the most widely studied system, that of $K_1$ killer strains. The $K_2$ killer system is also discussed but to a lesser extent. Throughout this review K unless otherwise indicated refers to $K_1$ killers. For the sake of clarity, Table 12.1 lists the nomenclature commonly used and recurrent throughout this review. It may be seen from this table that the word *plasmid* is used to denote the M dsRNA genome. Plasmids are generally defined as naked, dispensable chromosomes whereas viruses are organisms

*School of Biological Sciences, Queen Mary College, University of London, London E1 4NS, England.

[1]The term *virus* is used throughout in preference to *viruslike particle* since all the requirements for a virus, save that of infectivity, are met.

**Table 12.1.** Nomenclature

| General | |
|---|---|
| ScV-L | *Saccharomyces cerevisiae* virus encapsidating L genome |
| ScV-M | *Saccharomyces cerevisiae* virus encapsidating M genome |
| L | dsRNA genome—found in most strains killer or not |
| $M_1$[a] | dsRNA genome—encoding $K_1$ toxin and $R_1$ resistance determinant |
| $M_2$ | dsRNA genome—encoding $K_2$ toxin and $R_2$ resistance determinant |
| $K_1$ | {KIL-$k_1$} killer strains producing TOX 1 killer toxin |
| $K_2$ | {KIL-$k_2$} killer strains producing TOX 2 killer toxin |
| 1 | 4.5 kb (full length) mRNA transcript of L genome |
| $1_a$ | 2.3 kb mRNA transcript of L genome |
| m | 1.9 kb (full length) mRNA transcript of $M_1$ genome |
| $m_a$ | 1.2 kb mRNA transcript of $M_1$ genome |
| ScV-P1 | Major capsid polypeptide encoded by L genome |
| L-P1 | |
| L-P2 | Unidentified unique L encoded polypeptides |
| L-P3 | |

| Nuclear Genes | |
|---|---|
| *MAK* | M-Maintenance genes |
| *SPE* | |
| *PET* | |
| *SKI* | M-Regulatory maintenance genes |
| *KRB* | |
| *SEC* | M-Expression genes |
| *KEX* | |
| *REX* | |

Plasmids

| Genotype | Phenotype | dsRNA Genome |
|---|---|---|
| {KIL-$k_1$} | Wild-type $K_1$ killer | $M_1$ |
| {KIL-$k_2$} | Wild-type $K_2$ killer | $M_2$ |
| {KIL-b} | Superkiller; bypasses requirement for some *MAK* genes | Mutant $M_1$ |
| {KIL-d} | Defective maintenance and expression in haploids—diploid dependent | Mutant $M_1$ |
| {KIL-i} {KIL-pH} | Defective in resistance expression—so-called suicide mutants | Mutant $M_1$ |
| {KIL-kd} | No associated phenotype—deletion of nonessential part of {KIL-$k_1$} | Mutant $M_1$ |
| {KIL-n} | Defective in toxin production—neutral mutants | Mutant $M_1$ |
| {KIL-o} | No M plasmid | No M |
| {KIL-s} | Suppressive sensitive—defective interfering mutants—interfere with {KIL-k} replication | Mutant $M_1$ |
| {KIL-sd} | *SKI* dependent plasmid | Mutant $M_1$ |

**Table 12.1** (*Continued*)

| Plasmids | | |
|---|---|---|
| Genotype | Phenotype | dsRNA Genome |
| {KIL-sk} | Superkiller | Mutant $M_1$ |
| {KIL-ts} | Killing is temperature sensitive | Mutant $M_1$ |
| {EXL} | Excludes {KIL-$k_2$}—? ≡ {KIL-$s_2$} | Mutant $M_2$ ? |
| {HOK} | Permits maintenance of {KIL-sd} mutants in SKI host (Helper plasmid) | Mutant L ? |
| {NEX} | Prevents {KIL-$k_2$} exclusion by {EXL} (Nonexclusion plasmid) | Mutant L ? |

[a] If no subscript (e.g., as in $M_1$ or $M_2$) is given here or in the text, then subscript 1, to mean {KIL-$k_1$}, system is implied.

composed of nucleic acid encapsidated in protein coats. It is unusual therefore that the word *plasmid* should ever have been used to describe a viral genome. However, although we disagree with its use in this context, we have had to perpetuate its use, since much of the original literature refers to the M genome as a plasmid.

Figure 12.1 depicts a schematic representation of a $K_1$ killer cell. The *MAK, SKI, PET,* and *SPE* genes are those nuclear genes involved in the maintenance of the ScV-M particles and the *SEC, REX,* and *KEX* genes are those involved in the expression of the M genome.

Killer toxin-producing strains have been found not only in *S. cerevisiae* but also in a wide variety of other genera and species, namely, *Candida, Cryptococcus, Debaromyces, Hansenula, Kluyveromyces, Pichia, Saccharomyces* (non-*cerevisiae*), and *Ustilago* (Bussey and Skipper 1975; Hankin and Puhalla 1971; Kandel and Stern 1979; Maule and Thomas 1973; Naumov and Naumova 1973; Philliskirk and Young 1975; Puhalla 1968).

Rogers and Bevan (1978) undertook a survey of yeast killer strains and found that they could be classified into one of four groups (TOX.1–TOX.4) on the basis of their cross-reactivity. Killer strains of the TOX.1 group are designated $K_1$ killers and these are the *S. cerevisiae* killers, ubiquitous among laboratory stocks, first discovered by Bevan and Makower (1963). $K_2$ killers belong to the TOX.2 group and have been isolated as brewery contaminants (Maule and Thomas 1973; Ouchi and Akiyame 1976; Rogers and Bevan 1978) and from wine yeasts (Naumova and Naumov 1973). They have also been found among the (British) National Collection of Yeast Cultures by Young and Yagui (1978).

Amongst the non-*Saccharomyces* yeasts, perhaps the most interesting killer system is that found in *Kluyveromyces lactis*. The killer toxin production is associated with the presence of two linear DNA plasmids (Gunge *et al.* 1981). Other strains of *Kluyveromyces lactis* that produce a killer toxin of the same

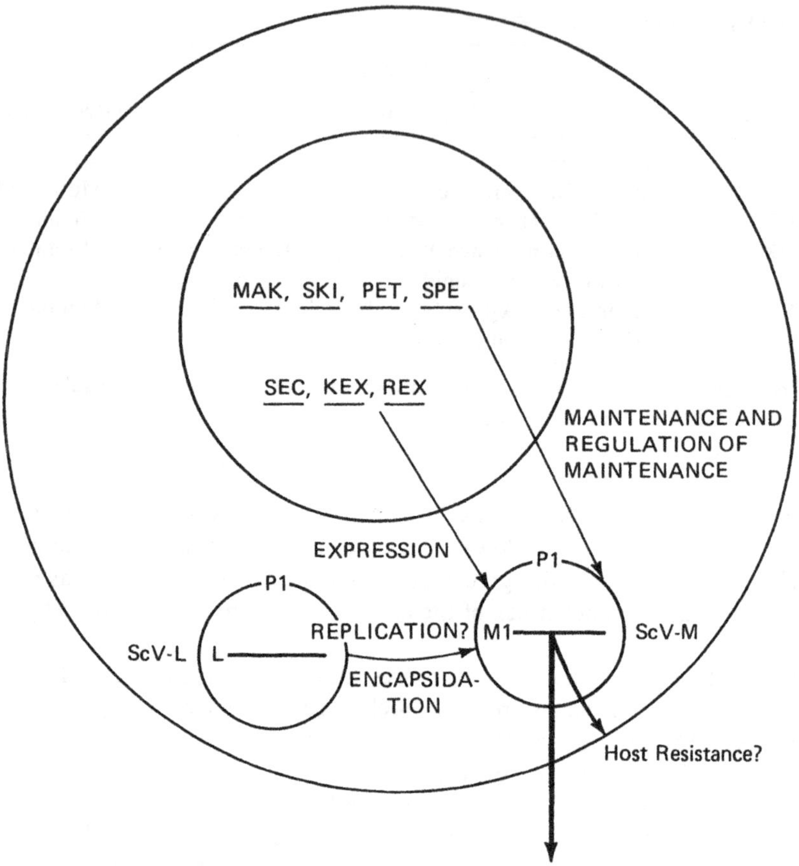


**Figure 12.1** Schematic representation of host–viral and viral–viral interactions in $K_1$ killer cells.

specificity as that described by Gunge and co-workers (1981) have been shown to possess ScV-like viral particles and a dsRNA species of the same molecular weight as the L genome of $K_1$ *S. cerevisiae* strains (authors' unpublished results). This system in particular is fully deserving of further investigation.

## Viral Particles

### Introduction

Adler and Mackenzie (1972) were the first to demonstrate the presence of viral particles in *S. cerevisiae*. The isometric 40-nm particles were isolated from cell-free extracts but were not correlated with either dsRNA or the killer phenotype. The next report (Buck *et al.* 1973) was of spherical viruses, 33–35 nm in diameter, which were shown to be associated with dsRNA but again were not correlated with the killer phenotype.

The first critical experiments, which showed that viruses containing dsRNA were associated with the killer phenotype, were made by Herring and Bevan (1974) and were later confirmed by Adler and co-workers (1976). Isometric, 39-nm particles were correlated with the presence of a dsRNA species that had previously been shown to be associated with the killer character (Berry and Bevan 1972; Bevan *et al.* 1973; Vodkin and Fink 1973).

## Electron Microscopy

An electron micrograph of these icosohedral particles, isolated from a $K_1$ killer strain of *S. cerevisiae,* is shown in Figure 12.2. Particles containing the L dsRNA genome (ScV-L) are indistinguishable in the electron microscope from particles containing the M genome (ScV-M). They do, however, differ in their physical properties, which have been summarized in Table 12.2 and which allow them to be separated on sucrose density gradients (Herring and Bevan 1974).

During attempts to achieve better separation of ScV-L and ScV-M particles, the density gradients were altered from 10 to 35% to 10 to 60% sucrose and the centrifugation times increased. This resulted in the appearance, at the lighter end of the gradients, of novel particles. These particles are ~40 nm in diameter and show some structure, as depicted in Figure 12.3. They appear under two-dimensional transmission electron microscopy to be cuboid, but

**Table 12.2.** Physical Properties of ScV Particles

| Virus | Size (nm) | $S_{20,w}$ | CsCl density (g/c³) | $A_{260}/A_{280}$ | References |
|---|---|---|---|---|---|
| ScV-L | 39[a] | 150–160 | 1.40–1.42 | 1.4–1.6 | Adler *et al.* (1976); Buck *et al.* (1973); Herring and Bevan (1974); Hopper *et al.* (1977); Kane *et al.* (1979); Peitras and Bruenn (1976) |
| ScV-M | 39 | ? | 1.36–1.38 | 1.2 | Herring and Bevan (1974); Kane *et al.* (1979); Pietras and Bruenn (1976) |

[a]Estimates vary from 33 to 41 nm.

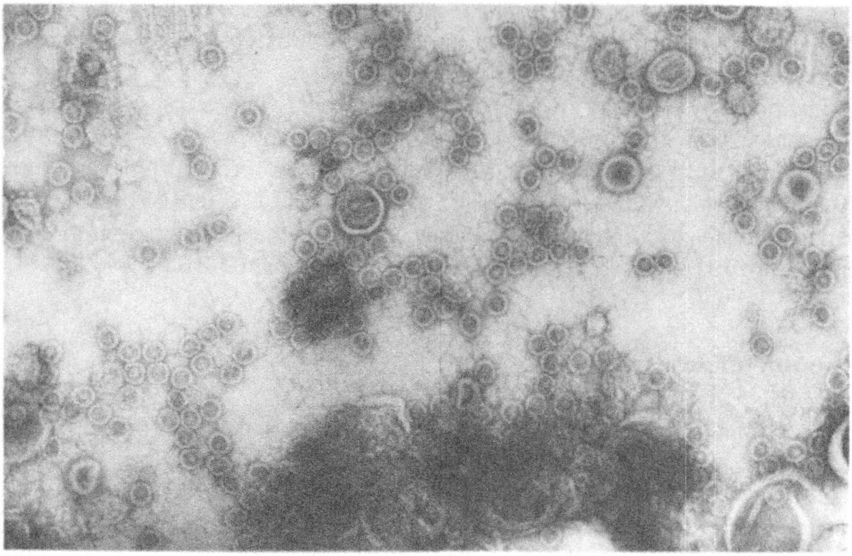

**Figure 12.2** Electron micrograph of ScV-L and ScV-M particles from a $K_1$ killer strain. (× 75000)

obviously, in view of the constraints of viral capsid assembly, must be rods or possibly some other shape. If they are indeed viral particles and not some other contaminant, it remains to be demonstrated whether they are precursors or breakdown products of the ScV particles and whether or not they contain their own nucleic acid species. These particles have so far only been detected in some strains of *S. cerevisiae, Saccharomycopsis lipolytica,* and *Kluyveromyces lactis,* which already possess either ScV-L or ScV-L and ScV-M particles (authors' unpublished observations). Whether or not ScV-free strains possess these novel "particles" is presently unknown.

The minimum number of particles per cell has been estimated at between 100 and 3000 (Bruenn, 1980), with some 90% being ScV-L (Vodkin *et al.* 1974; Wickner and Leibowitz 1976a). An upper limit would be some 30,000 as judged by electron microscopy of thin sections (Weber and Lindner 1975). As to the localization within the cell, the latter observations under the electron microscope showed quite clearly that ScV particles are to be found in the cytoplasm and not within any of the organelles.

## Physical Properties

Table 12.2 summarizes the known physical properties of the ScV particles. Various predictions as to the physical properties of these particles have been

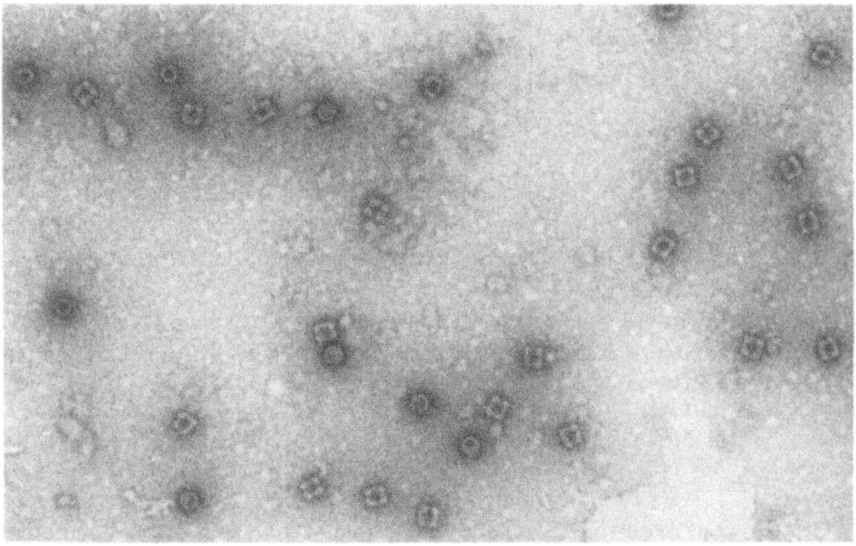

**Figure 12.3** Electron micrograph of novel particles from a $K_1$ killer strain. (× 75000)

made by Bruenn (1980). He estimated the numbers of P1 (capsid) polypeptides in ScV-L particles to be 120, extrapolated this same figure to ScV-M particles, and calculated an expected density for the latter of 1.35–1.36. This is in agreement with observed data (Kane *et al.* 1979; Pietras and Bruenn 1976). He further calculated the particle sizes of ScV-L to be $13.6 \times 10^6$ d (22.8% RNA) and of ScV-M to be $11.7 \times 10^6$ d (10.3% RNA).

Although the ScV particles are unrelated immunologically to the dsRNA viruses of *Aspergillus foetidus*, *Aspergillus niger*, and *Penicillium stoloniferum* (Buck *et al.* 1973), their physical properties (reviewed in Saksena and Lemke 1978) do appear to be similar, namely:

1. Same particle size range: 30–40 nm
2. $S_{20,w}$ in same range: 140–170
3. CsCl densities in same range: 1.3–1.45 g/cm$^3$
4. Molecular weights of particles similar: $\sim 13 \times 10^6$ d
5. All dsRNA species separately encapsidated
6. Molecular weights of component dsRNA genomes in same range: $0.06 \times 10^6$ to $3 \times 10^6$ d
7. Percent of RNA content similar: 10–15%

### Capsids

The capsids for both ScV-L and ScV-M particles may be concluded to be identical based on three findings, namely:

1. They are indistinguishable under the electron microscope (Herring and Bevan 1974).
2. They are both comprised of the same major (at least) polypeptide, estimated to be $75 \times 10^3$ d from SDS–polyacrylamide gel electrophoresis, and accounting for ~90% of the total capsid polypeptide (Herring and Bevan 1975; Oliver *et al.* 1977).[1]
3. Antibodies raised against ScV-L capsid protein (from a strain devoid of ScV-M particles) react against both ScV-L and ScV-M particles (Herring and Bevan 1974).

Minor capsid polypeptides of estimated molecular weights 53,000 d and 37,000 d have also been demonstrated (Oliver *et al.* 1977), but their significance is not known.

More recent experiments on the major capsid polypeptide, designated P1, have shown it to have a molecular weight of ~88,000 d, the minor components being 140,000, 82,000, and 78,000. Furthermore, its amino acid composition has now been determined (Bostian *et al.* 1980b; Hopper *et al.* 1977). Tryptic peptide analysis of the major polypeptide of ScV-M has confirmed that it is identical to that of ScV-L (Bostian *et al.* 1980b).

In conclusion, then, it appears that the major capsid polypeptides for both ScV-L and ScV-M particles are identical. It is, however, possible that differences exist in minor components, and indeed this idea is supported to some extent by the multiplicity of genes required for the maintenance of ScV-M and not ScV-L particles. Some of these genes may code for minor capsid polypeptides to be found in ScV-M but not ScV-L particles.

## ScV Genomes: Molecular Studies

### DsRNA Nature

The evidence for the genomes L and M being dsRNA is unequivocal and has been adequately reviewed elsewhere (Bevan and Mitchell 1979; Bruenn 1980).

### Separate Encapsidation

Separate encapsidation is consistent with both genetic and biochemical data. The genetic data (Mitchell *et al.* 1976a) have been reviewed in Bevan and Mitchell (1979).

---

[1]See footnotes on pages 387 and 397.

The biochemical evidence stems from sucrose density gradient centrifugation of viral preparations. The salient feature of these studies was the appearance of three peaks in the $A_{265}$ absorption profile of a linear gradient. The lightest peak consisted of empty capsids. The middle peak contained predominantly M dsRNA and the heavy peak predominantly L dsRNA (Herring and Bevan 1974). Later work (unpublished results of authors, cited in Bevan and Mitchell 1979) using convex exponential 10–40% sucrose gradients resulted in better separation of these ScV-L and ScV-M peaks such that no cross-contamination occurred and the middle peak contained only M and the heavy peak only L dsRNA. The initial failure to achieve complete separation of the peaks is due to their very similar particle sizes (see above).

## Size of L and M

Size determinations of the L and M molecules vary from $2.5 \times 10^6$ to $3.4 \times 10^6$ for L and from $1.1 \times 10^6$ to $1.7 \times 10^6$ for M (Bevan *et al.* 1973; Bruenn and Kane 1978; Herring and Bevan 1977; Sweeney *et al.* 1976; Vodkin *et al.* 1974; Wickner and Leibowitz 1976b). These differences arise in part from the different RNA standards used, molecular-weight estimates only being as accurate as the extent to which the molecular weights of the markers have been accurately determined, and in part from the different values used for the internucleotide distance of dsRNA molecules. Molecular-weight determinations can be accurately made only on sequenced molecules or by reference to sequenced molecules in the electron microscope. This has yet to be accomplished for L and M. A more detailed discussion of the reasons for these size discrepancies is given by Bruenn (1980).

## Structure

Both dsRNA molecules are simple colinear duplexes having equal strand lengths and with no nicks, single-stranded regions of any significant length (as originally postulated by Vodkin *et al.* 1974), or side loops. These conclusions are based on denaturation studies (Bevan *et al.* 1973) and electron microscope observations (Bruenn and Brennan 1980; Fried and Fink 1978; Holm *et al.* 1978; and Sweeney *et al.* 1976).

It was originally suggested (Shalitin and Fischer 1975) that L and M dsRNAs have poly(A) tracts. This work is open to criticism on the grounds that nowhere did these authors show that total cellular mRNA in yeast does not possess poly(A) tracts, which was crucial to their investigations. Furthermore, Bruenn (unpublished, cited in Bruenn 1980) has now demonstrated the absence of any poly(A) tract in excess of 10 nucleotides in either L or M on the basis of $T_1$ fingerprint analysis and pancreatic RNAse fingerprint analysis.

It has been suggested that M but not L dsRNA possesses a 200-bp AU-rich region revealed and located in the middle of the M genome by electron microscopy under partially denaturing conditions (Fried and Fink 1978). This result is somewhat difficult to reconcile with the fingerprint analysis unless this 200-bp AU-rich region is interspaced by GC base pairs at least every 10 nucleotides.

The 3′ ends of both strands of L and M have been sequenced (Bruenn and Brennan 1980) and are shown in Figure 12.4. The (+) strands have been designated from the data accumulating from the *in vivo* and *in vitro* transcription and translation studies, which are discussed fully later, under "M-Gene Expression." The 3′ ends of the (+) strands of both L and M show some sequence homology, shared bases being underlined in Figure 12.4. The 5′ ends of both L and M are pppGp (Bruenn and Keitz 1976), which is in contrast to the capped 5′ ends both of reovirus (Chow and Shatkin 1975; Miura *et al.* 1974) and of higher eukaryotic mRNAs (Adams and Cory 1975). The 3′ ends of the (−) strands of L and M, and hence the 5′ ends of the (+) strands, show sequence heterogeneity. Both L and M have single unpaired A residues at each 3′ termini (Bruenn and Brennan 1980).

Despite the partial sequence homology at these 3′ (+) strand ends, L and M do not have any overall sequence homology. This has been determined by heteroduplex analysis (Fried and Fink 1978) and RNAse T1 fingerprinting (Bruenn and Kane 1978).

## Replication

The mode of replication of the dsRNA, L and M, genomes has not been fully elucidated. Various workers have made comparisons with other systems, in particular the Reoviridae, and postulated models for replication (Bostian *et al.* 1982a; Bruenn 1980; Buck 1980).

In outline, L dsRNA replication appears to occur by the formation of ss,

| DsRNA | Sequence | Strand |
|---|---|---|
| L | HO-ACUUUUUAAAAAUUUAAG | (+) |
| M | HO-ACUUUUUAUUUCUUUACUG | (+) |
| L | GUGUGCGAGUGGAAAAAGCA-OH | (−) |
| M | GAAACACCCAUCA-OH | (−) |

**Figure 12.4** 3′ end sequences of L and M genomes.

probably (+) strand, intermediates that serve as templates for the doubling-up ss → ds process. These two main processes are discussed separately below followed by a consideration of dsRNA replication in the host cell cycle.

### *Ds → ss*

ScV-L and ScV-M particles have been shown to possess particle-associated ssRNA polymerase (transcriptase) activities that produce full-length, noncomplementary (+) strand copies of the L and M dsRNA genomes (Bruenn *et al.* 1980; Herring and Bevan 1977; Welsh and Leibowitz 1980; Welsh *et al.* 1980). It has been suggested that initiation of replication requires binding of the capsid protein to the dsRNA (Brennan *et al.* 1981b). In the case of the L genome at least this binding may occur at the 3′ terminal sequence of the (+) strand, which has a similar stem and loop structure to alfalfa mosaic virus, where capsid binding is an apparent requisite for replication initiation (Koper-Zwarthoff and Bol 1980).

The (+) strands of L and M dsRNAs are both uncapped, the 5′ termini being pppGp (Bruenn and Keitz 1976), and so is the ssRNA full-length *in vitro* transcript (Bobek and Brennan, cited in Bruenn 1980). This is quite unlike the situation in the Reoviridae, where both the ssRNA transcripts and the (+) strand of the dsRNA are capped, having 7MeG ppp Gm 5′ termini (Furuichi *et al.* 1975a,b).

There is some *in vitro* evidence that these ss intermediates, once formed, are extruded from their parent particles (Welsh and Leibowitz 1978).

### *Ss → ds*

The RNA polymerase activity necessary for doubling the ss to dsRNA has also been demonstrated for ScV-L. It has been shown to be associated with particles, isolated from log-phase cells, that sediment more slowly than L-dsRNA-containing particles. The end product of the reaction is full-length L dsRNA (Bevan and Herring 1976).

### *Cell Cycle*

Presumably, the ScV replication is controlled somehow in line with host cell division such that intracellular numbers are limited. A mechanism for ScV copy number must exist to ensure both host and viral survival and is indicated by the strain specific, even under different physiological conditions, copy number observations (Oliver *et al.* 1977). These authors demonstrated a relationship between mitochondrial metabolism and the replication of the L genome.

Whether or not this is related to the effects of the *PET.18* or *MAK.10* gene loci in controlling the maintenance of both mitochondria and M genomes (Bevan *et al.* 1973; Wickner 1977; Wickner and Leibowitz 1976b) is not known. Further consideration of these gene loci is given later under "Viral Maintenance." Clare and Oliver (1979) have shown that L synthesis is coordinated with host cell protein synthesis and that during nitrogen starvation the replication of ScV-L is subordinate to the host cell requirements. However, the ScV-L particles are never reduced to zero levels.

In conclusion, then, this control mechanism has not yet been elucidated and the only reported studies to date show that ScV components seem to replicate throughout the cell cycle (Bostian *et al.*, cited in Bostian *et al.* 1982a; Newman *et al.* 1981) and that dsRNA in particular replicates at all stages except S (Zakian *et al.* 1981).

## Provirus Hypothesis

The idea that genomic copies of the dsRNA genome may exist is not without some precedence. The literature offers many well-documented examples of the existence of proviruses. For example, it is well established that temperate DNA bacteriophages such as phage λ may exist in the prophage state. Other examples include certain of the animal DNA tumor viruses such as polyoma virus or SV40 (Botcham *et al.* 1976) and the RNA retroviruses (Hill *et al.* 1974).

The possibility that a proviral form of M dsRNA exists is raised by the observation (Young and Yagiu 1978) that some strains produce killer toxin when they apparently lack dsRNA. This occurrence could be explained were these strains to possess a proviral form of M dsRNA. Such a proviral form, being a DNA copy incorporated into the nuclear genome, would necessarily imply that sequence homology, between the M dsRNA genome and chromosomal DNA, would exist.

Shalitin and Fischer (1975) and Vodkin (1977) have reported the existence of sequence homology between ScV dsRNA and yeast nuclear DNA. Hastie and co-workers (1978) have reported their failure to detect such homology. In view of these contradictory reports and of the implications any such homology would suggest, these reports will be considered further.

In essence, the report of Shalitin and Fischer (1975) was based on their finding that homology exists between yeast nuclear DNA and poly(A)-containing mRNA isolated from yeast, after showing that ScV dsRNA contains poly(A) sequences. The salient feature of these experiments is that nowhere did the authors demonstrate that yeast cellular RNAs do not contain poly(A) sequences, this being crucial to their work. More detailed comments as to the validity of these experiments have already been made (Bevan and Mitchell 1979).

Vodkin's (1977) report, that homology between L dsRNA and DNA exists, was based on the results of experiments in which it was shown that the rate of conversion of denatured labeled dsRNA to the double-stranded form was increased by the addition of denatured yeast DNA. The implication here is that the dsRNA was forming DNA–RNA hybrids. Evidence for the existence of such hybrids came from the appearance of a novel labeled band in isopycnic caesium sulfate gradients.

The reassociation kinetics of L dsRNA in the presence of denatured yeast DNA, prepared from one of the same strains used by Vodkin (1977), was also measured by Hastie *et al.* (1978). These authors reported no effect of the yeast DNA on the reassociation kinetics of the dsRNA. Welsh and Leibowitz (1982), using similar methods, have also failed to detect any sequence homology between L dsRNA and genomic DNA.

The weight of the available evidence seems to point to the conclusion that probably no such sequence homology exists. In view of this conclusion, Vodkin's (1977) results need to be explained. Several possibilities exist that would explain this author's results, namely:

1. The addition of host DNA may have resulted in the appearance of networks that would have resulted in the steric hindrance of the ribonuclease (as suggested in Buck 1980).
2. The genomic DNA may have been contaminated with polysaccharide that would have resulted in the same effects as point (1).
3. The genomic DNA may have been contaminated with metal ions—known to be capable of affecting rates of hybridization.
4. The dsRNA preparation may have been contaminated with rRNA; this, however, is disclaimed to some extent by Vodkin (1977).

These possibilities were suggested by and are more fully discussed by McFadden (1982).

Wyse, in collaboration with us, has now almost unequivocally demonstrated that no proviral forms of the ScV genomes exist (unpublished results). Two experiments were undertaken, the first using DNA as the probe and the second using RNA as the probe. In the first preliminary experiments, which were based on the Northern blotting procedure (Alwine *et al.* 1977), purified denatured L and L + M dsRNA were transferred to DBM paper. These were challenged with highly labeled nick-translated purified genomic DNA from three strains, namely:

1. a killer strain L M
2. a sensitive strain L M-o
3. a sensitive strain L-o M-o

Some hybridization was achieved showing similar levels to those of Vodkin's (1976) experiments. Different levels of hybridization were achieved: L M > L M-o > L-o M-o, but with L-o M-o hybridization still being significant. These results indicated that some homology may exist, so a more crucial experiment was performed. Highly labeled ss mRNA *in vitro* transcripts of L dsRNA were used as a probe and challenged to restricted genomic DNA using the Southern blotting method (Southern 1975). No bands lit up, which suggested two things: first, that no proviral forms of the L genome appear to exist, and second, that the hybridization observed in the preliminary experiments, and also those of Vodkin (1977), were probably due to aspecific binding.

## Transmission

Despite numerous attempts by many workers, no *de novo* infection of yeast cells by ScV particles has yet been unequivocally established. The only such reports, those of Lhoas (1972) and Nesterova (1974), are open to serious criticisms, which have been discussed in detail elsewhere (Bevan and Mitchell 1979).

The only mechanism, therefore, to ensure the continued existence of these particles is their transmission through the cell lines during mitotic divisions and during mating and subsequent meiotic divisions (Bevan *et al.* 1973; Mitchell *et al.* 1976a; Somers and Bevan 1969). Detailed consideration of the mode of inheritance of these particles is given later under "Viral Maintenance."

## Curing

Strains of *S. cerevisiae* possessing ScV-L and ScV-M particles can be cured of the ScV-M particles under a variety of different conditions, namely:

1. Heat (Wickner 1974a; Young and Yagiu 1978)
2. Cycloheximide (Fink and Styles 1972; Young and Yagiu 1978)
3. 5-Fluorouracil (Mitchell *et al.* 1976b)

In no instance of elevated temperature or cycloheximide curing was the loss of the ScV-L particles seen. In only one strain (K7) treated with 5-FU were the ScV-L particles lost in addition to the ScV-M particles, and this strain had previously been shown to behave in crosses as if it possessed a mutant L genome (Mitchell *et al.* 1976a). In the case of the 5-FU cured strains it was shown quite clearly that the loss of killer activity was due to curing of the M genome *per se* and not to a nuclear maintenance gene mutation: Crossing "cured" sensitives to killer strains resulted in progeny all showing a 4M (killer) to $OM^-$ (sensitive) segregation.

In these curing experiments, the parent was L M and the progeny were either L M-o or L-o M-o. This result is taken as further evidence to support the hypothesis that M is dependent on L: No L-o M progeny were detected. A similar analysis on the same (K7) strain has recently been performed by Sommer and Wickner (1982). Whereas the previous authors (Mitchell *et al.* 1976b) isolated spontaneous mutants and 5-FU induced mutants, Sommer and

Wickner obtained mutants using elevated temperatures. Both sets of results are qualitatively identical, the mutants being either L M-o or L-o M-o, and in one case (Mitchell *et al.* 1976b) these mutants were back-crossed to discount nuclear gene mutations. From their data, Sommer and Wickner (1982) reached the same conclusion as these authors have made, that this is further evidence of M being dependent on L.

## Viral Genomes

The available evidence indicates that the ScV system in *S. cerevisiae* is one of a helper (ScV-L) and a defective (ScV-M) virus. In this system the defective virus utilizes the coat protein (at least) of the helper virus for encapsidation and also possesses its own genetic information not present in the helper genome. The defective virion is thus analogous to the RNA-5 of cucumber mosaic virus (Mossop and Franki 1978).

Considerably more information is available as to the organization and expression of the M as opposed to the L genome. For this reason the L and M genomes will be considered separately with respect to what is known about their defined and/or postulated gene functions, genomic organization, gene expression, and control of expression.

### ScV-L Genome

#### Gene Products

For a dsRNA virus to be able to promote its own self-replication in a host, there is a positive minimum requirement for that virus to encode a capsid polypeptide and any necessary RNA-dependent RNA polymerase(s). It is now known that ScV-L encodes the capsid polypeptide for both ScV-L and ScV-M particles and is generally assumed to encode at least one RNA polymerase.

##### *Capsid Polypeptide: L-P1*

*In vitro* translation of denatured L dsRNA in the wheat germ cell-free system yields L-P1, an 88,000-kd polypeptide that has been shown to be identical to the major ScV-L capsid polypeptide ScV-P1 on the basis of three criteria, namely:

1. Co-migration in SDS-gel electrophoresis
2. Immunological cross-reactivity
3. Identical tryptic peptide patterns

(Hopper *et al.* 1977).

*A priori* it would seem reasonable to conclude that the M genome utilizes the L-encoded capsid polypeptide for its own encapsidation, since (a) ScV-P1 by itself exceeds the coding capacity of M and (b) M has actually been demonstrated to encode the exotoxin and a glycosylated protein thought to be the resistance determinant (see following discussion).

That the capsid polypeptides of ScV-L and ScV-M are indeed identical has now been demonstrated. The polypeptides have the same molecular weight, are serologically related (Herring and Bevan 1975), and show no differences in peptide mapping (Bostian *et al.* 1980b).

### *RNA-Dependent RNA Polymerase*

Two lines of reasoning lead to the conclusion that the L genome probably encodes an RNA-dependent RNA polymerase activity. First, there is sufficient coding capacity ($\sim$100 kd) on the L genome after subtracting the capsid polypeptide requirement (88 kd) from the total coding capacity calculated at 189 kd assuming one-strand full-length coding with no overlapping sequences, for example (Buck 1980). Second, ScV particles have associated ds → ss and ss → ds RNA polymerase activities (for references and fuller discussion see under "Replication"). However, it should be pointed out that this is by no means unequivocal proof of L-encoded polymerase.

It is entirely possible, and indeed there is some evidence to suggest, that more than one polymerase may be involved in replication. This evidence stems from the sequence analysis of the termini of the L and M genomes (Bruenn and Brennan 1980). Although the 3′ termini of the (+) strands of L and M dsRNA (initiation sites for the ds → ss transcription) show significant sequence homology, and 3′ termini of the (−) strands, although both GC rich, lack such homology (see Fig. 12.4). Such differences in sequence imply different polymerase recognition sequences, and since the ds → ss and ss → ds events are apparently temporally separated, it may be concluded that more than one polymerase may be involved.

Both the number of polymerases involved and the question as to whether or not they are encoded by the L genome remain important unresolved issues.

### *L-P2 and L-P3*

*In vitro* translation of denatured L dsRNA and of hybrid selected *in vivo* mRNAs in the reticulocyte system yields four products, namely, L-P1, L-P2, L-P3, and L-P4 (see Table 12.3). L-P1, as previously shown, is the capsid polypeptide. L-P4 reacts, like L-P1, to anticapsid IgG and is concluded to be a truncated L-P1 product. L-P2 and L-P3 do not react with anticapsid IgG but do react with anti-VLP IgG and hence may be unique L-encoded VLP com-

**Table 12.3.** *In vitro* Translation of L or L encoded mRNAs[a]

| Substrate | System | Products | Comments |
|---|---|---|---|
| Denatured L dsRNA<br>*In vivo* mRNAs (native total cellular RNAs)<br>*In vivo* mRNAs (hybrid selected transcripts—include 1 and $1_a$) | *In vitro* translation (wheat germ) | L-P1<br>L-P2a<br>L-P4 | ≡ ScV-P1<br>L-P1 premature termination products (L-P2a, L-P4) |
| Denatured L dsRNA<br>*In vivo* mRNAs (hybrid selected transcripts—include 1 and $1_a$) | *In vitro* translation (reticulocyte) | L-P1<br>L-P2<br>L-P3<br>L-P4 | Unique L-encoded VLP components (L-P2, L-P3) |
| *In vivo* 1 + $1_a$ mRNAs | *In vitro* translation (wheat germ) | L-P1<br>L-P$2_a$<br>L-P4 | |
| *In vivo* 1 mRNA<br>*In vivo* $1_a$ mRNA | *In vitro* translation (wheat germ) | L-P1 | |

[a]Data from Hopper *et al.* 1977; Bostian *et al.* 1982b.

ponents (RNA polymerase?). Reports of preliminary peptide mapping of L-P2 show that it is different from L-P1 (Bostian *et al.* 1982a).[1]

## Genome Organization

Very little is known about the organization of the L as opposed to the M genome. Although considerable homology, between L and M, exists at the 3′ (+) strand termini, 12 out of 18 bp being identical, no such homology exists at the other, distal to transcription initiation, 3′ ends, as depicted in Figure 12.4 (Bruenn and Brennan 1980). No region comparable to the 200-bp AU-rich "bubble" found in M (see later) has been detected in L (Fried and Fink 1978; Welsh and Leibowitz 1982).

Bostian and co-workers (1982a) have suggested that the half-length L transcript $1_a$ may be a 5′ terminal fragment of 1 [i.e., both transcripts initiating at the 3′ (+) strand end of L dsRNA] (see below). If this is the case then, since both 1 and $1_a$ have L-P1 (ScV-P1 ≡ capsid polypeptide) activity, this would locate the capsid gene at the 3′ (+) strand end of L.

Other gene functions, including control interaction sites, remain to be defined and located on the L genome.

[1]Please see footnote on page 397. Preliminary results on the *in vitro* translation of isolated $L_1$ and isolated $L_2$ genomes has revealed that $L_1$ encodes L-P2 and L-P3, whilst $L_2$ encodes L-P1 and L-P4 (El-Sherbeini, Bostian, Tipper, Bevan and Mitchell, unpublished results).

## Gene Expression

L dsRNA is 4.5 kb long. Two *in vivo* transcripts, a full-genome-length 4.5-kb transcript, 1 and a 2.3-kb transcript, $1_a$ have been identified (Bostian *et al.* 1982a). Both 1 and $1_a$ were shown to be ssRNA transcripts of the (+) strand copy (by definition, since they have mRNA activity) of the L dsRNA. These same authors conclude on the basis of poly(U)-sepharose binding that 1 but not $1_a$ is polyadenylated. The actual data, however, show that only a trace of 1 binds to poly(U)-sepharose and that most does not bind. This is in agreement with Haylock and Bevan's (1981) findings on a full-length mRNA of L dsRNA. These authors show that their full-length transcript does not bind to oligo-(dT)-cellulose and therefore conclude that it does not possess poly(A) tails or oligo(A) stretches. They do state, however, that their transcript is a (−) strand copy, but one of us (EAB) and Haylock (personal communication) have stated that (−) is a typographical error and that it should read (+) strand.

Both 1 and $1_a$ exhibit L-P1 activity in wheat germ *in vitro* translations (see Table 12.3), with no data being given on L-P2 activity (Bostian *et al.* 1982a). The $1_a$ transcript at 2.3 kb, which is roughly half-genome-length, is only just large enough to encode the activity it shows, that of L-P1 (≡ ScV-P1, 88 kd). Since Bostian and co-workers (1982b) conclude that 1 is probably the *in vivo* replicative intermediate and $1_a$ the message, then either $1_a$ is a single species that codes for L-P1 and another as yet unidentified mRNA encodes L-P2 (and/or L-P3?) or else $1_a$ is a mixture of two half-length transcripts only one of which encodes L-P1. It should be possible to settle this problem by the *in vitro* translation of the separated $1_a$ in the reticulocyte system, to determine whether L-P2 and/or L-P3 activities are encoded.

Various *in vitro* translation experiments have been performed using either the wheat germ or the reticulocyte systems. Denatured L dsRNA, native total cellular RNAs (*in vivo* mRNAs), and hybrid selected transcripts (*in vivo* mRNAs including 1 and $1_a$) have all been translated in both systems. In addition, the *in vivo* 1 and $1_a$ transcripts have been translated in the wheat germ system. The results from these experiments (Bostian *et al.* 1982a) have been summarized in Table 12.3.

The evidence that L-P1 is the 88 kd ScV-P1 capsid polypeptide has already been presented. L-P2a (85 kd) and L-P4 (77 kd) have been shown to have peptide maps identical to L-P1 and are assumed to be premature termination products. L-P4, like L-P1, reacts to anticapsid IgG and has therefore also been assumed to be a truncated L-P1 product. The nature of L-P2 and L-P3 has already been discussed in the previous section: It would seem that these are unique L-encoded VLP components.

The *in vivo* functions of L-P2 and L-P3 remain to be identified, as does the transcript(s) that encodes them. The hypothesis that 1 is an intermediary replication transcript and $1_a$ the *in vivo* message also needs to be demonstrated.

### Gene Mutations Affecting Expression

To date, no plasmid mutations or nuclear gene mutations affecting the expression of the L genome have been identified. There are two obvious reasons as to why no such mutants have been isolated, namely:

1. Such mutations could well be lethal to the ScV-L particles carrying them.
2. Such mutants would probably not confer any readily detectable phenotypic change.

## ScV-M Genome

### Gene Products

Early presumptive evidence that M encodes both resistance to and production of the toxin came from their co-segregation with the M genome in crosses (Mitchell *et al.* 1976a). This conclusion has now been almost fully substantiated by the *in vitro* translation experiments of Bostian and co-workers (described later). The M genome has been shown to code for the exotoxin itself and a glycosylated protein assumed to be the resistance determinant (Bostian *et al.* 1980a; 1982a,b).

#### *Killer Toxin*

The $K_1$ killer toxin is a small protein with a molecular weight of 11,470 d (Palfree and Bussey 1979). The active form is monomeric, being a single peptide chain of some 109 amino acid residues. The toxin is heat-labile, being stable only $\sim$22°C and active only within the narrow pH range 4.2–4.6 (Woods 1966).

The toxin is secreted into the growth medium of producing strains during exponential phase (Palfree and Bussey 1979). Sensitive (KIL-o) cells show a rapid decrease in viability when exposed to the toxin, maximum killing being attained after only two or three hours (Bussey 1972; Makower 1964; Skipper and Bussey 1977). A wide range of events takes place during the killing event. After an initial lag peroid of some 40 min there is a coordinate cessation of macromolecule synthesis (Bussey and Sherman 1973). This coincides with damage to the plasma membrane, concluded from the demonstration that potassium ion and ATP leakage occurs (Bussey and Sherman 1973; Skipper and Bussey 1977). Although damaged cells shrink in volume, they do not lyse, nor do they contain large pores, since macromolecule leakage is not observed (Bussey 1974). It has been suggested by Bussey (1981) that the plasma membrane is the primary site of action of the toxin, but there is no direct evidence to support this.

Toxin is concluded to bind initially to receptor sites on the cell wall. The evidence to support this comes from the study of two unlinked gene mutations *kre.1* and *kre.2*. Mutant cells are resistant to the toxin and exhibit a lower binding of toxin to the cell wall: A same-site reversion at the *kre.1* locus results in the expression of normal toxin binding ability and sensitivity (Al-Aidroos and Bussey 1978; Bussey *et al.* 1973). Other studies, including quantitative measurements of binding to the receptors of sensitive strains, indicate that the number of receptor sites is around $1.1 \times 10^7$ per cell and that a single receptor species is present (Bussey *et al.* 1979). Studies on wild-type and *kre* mutant strains have led to the conclusion by Bussey (1981) that the receptor molecule is a $\beta$-(1→6)-D-glucan.

There is some evidence to suggest that second-site receptors exist, probably on the plasma membrane. This evidence has been reviewed and a model for toxin action has been presented by Bussey (1981). However, the nature of the interaction of toxin with the plasma membrane remains the major unsolved problem in the action of the killer toxin.

Other killer toxins—$K_2$ from {KIL-$k_2$} cells, $K_3$ from {KIL-$k_3$} cells, and $K_4$ from {KIL-$k_4$} cells—exist. In general, other killer toxins, insofar as they have been studied, appear to act like $K_1$ in causing membrane damage, but the possibility still exists that they have different modes of action (Rogers 1976; Skipper and Bussey 1977).

### *Resistance Protein*

It is probable that the M-encoded 30-kd glycosylated protein is the resistance determinant (see following discussion). However, this has not been unequivocally established. The actual mechanism of immunity is also unknown, but it has been shown that immunity does not operate at the cell-wall receptor site: {KIL-$k_1$} and {KIL-o} cells display equivalent amounts of wall receptor (Bussey *et al.* 1979). Bussey (1981) has suggested that immunity probably occurs at the membrane level.

## Genome Organization

A model for the functional sequence of $M_1$ dsRNA has been put forward by Bostian *et al.* (1982a). This model has been modified slightly and is shown as Figure 12.5. This figure should be referred to as an illustration of the points made in the following text.

Sequence analysis of M dsRNA has revealed the presence of single unpaired 3′ terminal A residues at both ends (Bruenn and Keitz 1976; Bruenn and Brennan 1980). This analysis also reveals that there is considerable homology within the AU-rich termini between the L and M dsRNAs. It has been proposed that this region is the site of *in vitro* transcription initiation (Brennan *et*

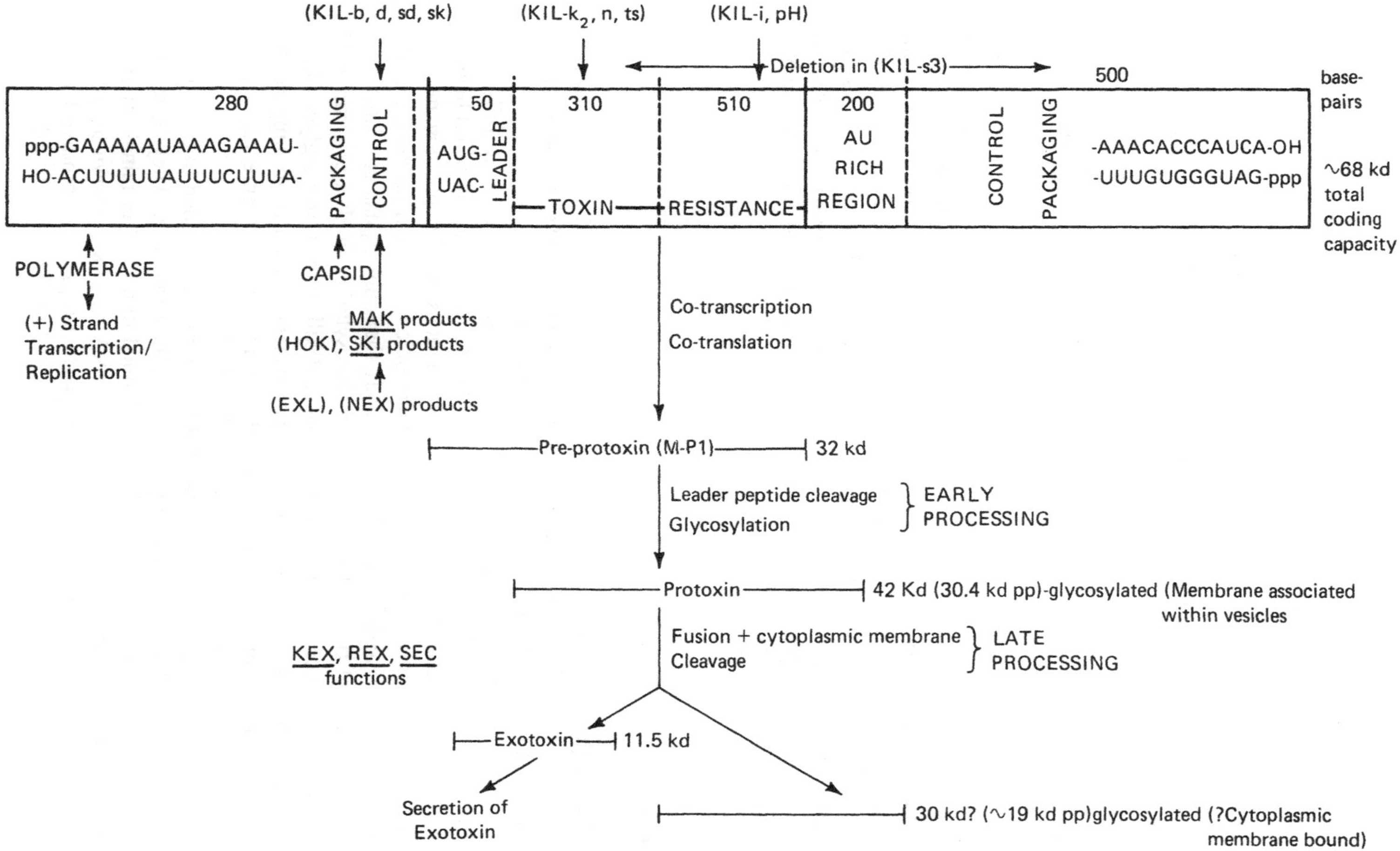


**Figure 12.5** Functional model of $M_1$ dsRNA. (After Bostian *et al.* 1982b.)

*al.* 1981b). The plus strand has been identified as the 3′ → 5′ strand starting at the AU-rich terminus (Brennan *et al.* 1981a).

Electron microscopy under partially denaturing conditions has revealed the presence of a 200-bp AU-rich region located at the middle of the M genome (Fried and Fink 1978). This 200-bp AU-rich region presumably originates at the C terminal end of the putative resistance gene for the following reason: $M_1$ dsRNA is selectively cleaved by micrococcal S1 nuclease at 65 °C, resulting in the removal of this 200-bp AU-rich region, leaving a 1.0-kb dsRNA fragment. This fragment retains the AU-rich terminus and full $M_1$-P1 mRNA, which is the dicistronic message for toxin and resistant functions, activity (Welsh and Leibowitz 1982).

The sequence leader, toxin gene, resistance gene has been determined from transcriptional/translational studies (Bostian *et al.* 1982a), and although no formal proof is available, this sequence is presumably preceded by the ribosome binding site. The two remaining, as yet unaccounted for, regions must contain capsid interaction sites for packaging and presumably also control regions with which the various maintenance gene *(MAK, SKI)* products interact. As suggested by Bostian *et al.* (1982a), these control regions probably also contain the sites for the plasmid mutations {KIL-b, d, sd, sk} that affect M genome maintenance (see later).

## Gene Expression

The biochemistry of M genome expression has been studied in depth, with the result that a well-supported scheme for the steps involved in expression has now been put forward by Bostian and co-workers (1982b).

In outline, the 1.9-kb $M_1$ genome is transcribed to give a 1.9-kb full-length transcript, m, which possibly functions as the ss intermediate in M replication, and a 1.2-kb transcript, $m_a$, which possibly functions as the expression message. Both m and $m_a$ give rise to a primary dicistronic translation product, the $M_1$-P1 32-kd pre-protoxin, but it is thought likely that only $m_a$ has this function *in vivo*. Removal of the leader sequence and glycosylation yields a protoxin of 42 kd, which is processed into a nonglycosylated killer toxin and a glycosylated protein assumed to be the resistance component. Evidence to support this scheme is given later, and the scheme itself is included in Figure 12.5 (reproduced in altered form from Bostian *et al.* 1982a).

Both m and $m_a$ are single-stranded RNAs that are assumed to be polyadenylated on the basis of their binding to poly(U)-sepharose. They have been identified as *in vivo* products and $m_a$ is assumed to be a 5′ terminal fragment of m. Both m and $m_a$ and denatured $M_1$ dsRNA have been shown to give rise to the 32-kd putative primary translation product $m_1$-P1 (pre-protoxin) in the *in vitro* cell free translation system (see Fig. 12.5). The $M_1$-P1 mRNA has been demonstrated to be membrane associated *in vivo*, which is consistent with a co-translational conversion of $M_1$-P1 itself to a glycosylated protoxin. This postulated protoxin molecule has been isolated. It is 42 kd, unstable, glycosylated, and membrane associated within membrane vesicles. When membrane vesicle preparations are added to the *in vitro* translation of the pre-protoxin $M_1$-P1 or the $M_1$-P1 mRNA, conversion of $M_1$-P1 to a 42-kd product, resembling the protoxin, occurs. This is consistent with the loss of around 1600 peptides, assumed to be the terminal leader sequence, and subsequent glycosylation (Bostian *et al.* 1982a,b). Both the *in vivo* membrane association of the $M_1$-P1 mRNA and the requirement for membrane vesicles in the co-translational conversion of $M_1$-P1 to the glycosylated protoxin conform with the scenario proposed for secreted proteins (Blobel and Dobberstein 1975).

Loss of 1.6 kd from the 32-kd pre-protoxin and glycosylation gives rise to the 42-kd glycosylated protoxin of which some 30.4 kd is polypeptide. The exotoxin itself has an estimated molecular weight of ~11.5 kd and is nonglycosylated (Palfree and Bussey 1979), which leaves a glycosylated protein of 30 kd having a polypeptide component of around 19 kd. It is known that the $M_1$ genome codes not only for the toxin but also for the membrane-located resistance determinants (see previous discussions), and it is therefore logical to conclude that the remaining 30-kd glycosylated protein is that resistance component.

The final steps, involving cleavage of protoxin into the exotoxin and a glycosylated protein (resistance component?) and the secretion of exotoxin, remain to be elucidated. The *in vivo* nature of $M_1$-P1 pre-protoxin also remains to be established.

The polyadenylation of M dsRNA transcripts is still a contentious issue. As previously stated, both m and $m_a$ bind to poly(U)-sepharose with an affinity that suggests a poly(A) length of about 90, which is greater than most yeast mRNAs. Bruenn (cited in Bruenn 1980) has shown no oligo(A) greater than about 10 in M dsRNA despite the internal 200-bp AU-rich bubble (Fried and Fink 1978). Nevertheless, the poly(A) nature of the mRNA may still be attributable to this 200-bp AU-rich region. If this is the case, then similar extrapolations to the L transcripts cannot be made. If indeed 1 is polyadenylated (see above), then the polyadenylation cannot be accounted for by the presence of such an AU-rich region in the L genome because from the results of Fried and Fink (1978) apparently no such region exists.

## Gene Mutations Affecting Expression

A range of plasmid and host gene mutations has been identified, all of which affect the expression of the M genome. The lesions and their effects on each mutant are discussed, each mutation being considered individually below.

### *Nuclear Gene Mutations*

Three classes of nuclear gene mutations have so far been identified, namely, (1) *sec,* (2) *kex,* and (3) *rex.*

1. *Sec* mutations (*sec.*1 to *sec.*23) confer a temperature-sensitive inability to secrete normally secreted proteins, e.g., acid phosphatase and invertase (Novick *et al.* 1980). Several of these *sec* mutants accumulate 80–100-nm membrane vesicles, and Bussey (cited in Bostian *et al.* 1982b) has shown that one of these, a *sec.1* mutant, also accumulates protoxin at the nonpermissive temperature. Bostian and co-workers (1982b) have concluded that *sec.1* is presumably defective in fusion of the membrane vesicles with the cytoplasmic membrane. Since such fusion is essential for the processing of protoxin into exotoxin and resistance protein, the *sec.1* mutation apparently blocks this processing.

2. *Kex* mutations also affect protein secretion, but, unlike *sec* mutations, are not lethal. Two *kex* mutations have so far been identified; namely, *kex.1* and *kex.2,* both of which affect toxin production but have no effect on the resistance phenotype (Wickner 1974b; Wickner and Leibowitz 1976a). *Kex.1* mutants do not produce the killer exotoxin, but are apparently wild-type for production of other exoproteins. *Kex.2* mutants, however, are known to be defective in many exocellular proteins and glycoproteins, all of which show altered electrophoretic mobility. It has been suggested (Bostian *et al.* 1982b) that *kex.2* mutants are defective in some late processing step that is essential for function of some (e.g., killer exotoxin and $\alpha$-pheromone), but not for secretion (Rogers *et al.* 1979). The results of *in vivo* transcription and labeling experiments show that transcription and subsequent translation are apparently normal up to the protoxin stage, in both *kex.1* and *kex.2* mutants (Bostian *et al.* 1982b). Hence *kex* mutations like *sec.1* apparently affect the late processing of protoxin to yield mature toxin.

3. *Rex* mutants are defective in resistance expression (Wickner 1979a). No information is as yet available as to which step in resistance expression is blocked, whether at transcriptional or translational level, but in view of the findings with *sec* and *kex* mutants, the *rex* mutants are probably also defective in late processing.

### *Plasmid Gene Mutations*

Mutations of the M genome itself range from the putative point mutations in the neutral {KIL-n} mutants to large deletions in the suppressive sensitive {KIL-s} mutants.

1. {KIL-b, d, sd, sk} mutations all affect the maintenance of the M genome and presumably are all located within the putative control region of the M genome (see Fig. 12.5). They will be discussed in detail later, under "Viral Maintenance."

2. {KIL-$K_2$} killer mutants have been isolated from wine (Naumova and Naumov 1973; Naumov *et al.* 1973) and brewery (Maule and Thomas 1973; Ouchi and Akiyama 1976; Rogers and Bevan 1978) fermentations and have been found among the (British) National Collection of Yeast Cultures (Young and Yagiu 1978). {KIL-$k_2$} mutants are variants of the {$KIL_1$} strains in that their M genomes encode toxins of different specificities (i.e., $K_1$ killer toxin kills $K_2$ toxin-producing strains and vice versa). The above-named workers have shown such differences as altered pH optimum for toxin stability; Rogers and Bevan (1978) have determined the pattern of cross-reaction between strains of different sources and have shown that differences between toxin activities/specificities allow the division of those toxins into four groups, $K_1$, $K_2$, $K_3$, and $K_4$, so at least four plasmid genotypes, {KIL-$k_1$} to {KIL-$k_4$}, exist. Such mutations presumably lie within the toxin gene itself (see Fig. 12.5).

3. {KIL-i} and {KIL-pH} are the so-called suicide mutants. {KIL-i} mutants were originally isolated by Sweeney and co-workers (1976) and {KIL-pH} mutants were isolated by Bussey and co-workers (1982). These mutants produce toxin but not resistance determinants, and in the case of the {KIL-i} strains, cells show complex dsRNA patterns (Sweeney *et al.* 1976). The lesions involved in these "suicide" mutants may be presumed to reside within the resistance gene (see Fig. 12.5), but there is, to date, no formal evidence to support this hypothesis.

4. The {KIL-n} mutation was originally described by Bevan and Makower (1963) and later Bevan and Somers (1969). {KIL-n} neutral mutants do not produce a functional toxin but are still resistant to toxin. The $M_n$ genome has the same molecular weight as the M genome (Bevan *et al.* 1973), so the mutation is assumed to be a very small alteration or point mutation rather than a deletion and to reside within the toxin gene (see Fig. 12.5). *In vitro* translation of denatured $M_n$ dsRNA produces apparently normal pre-protoxin (Bostian *et al.* 1980a), yet no protein-resembling toxin, on the basis of peptide cross-reactivity with antitoxin IgG, is secreted by the N2 neutral strain (Bussey 1981; Bussey, cited in Bostian *et al.* 1982b). Further *in vitro* translation experiments (Bussey *et al.* 1982) show that $M_n$ dsRNA is transcribed, translated, and processed, apparently normally, up to the stage of protoxin. It has therefore been concluded (Bostian *et al.* 1982b) that the {KIL-n} mutation affects either the stability of the toxin itself or the processing of the protoxin.

5. {KIL-s} mutant strains possess defective interfering, ScV-S, particles and were isolated from killer {KIL-k} and neutral {KIL-n} strains. These mutants were originally designated suppressive sensitives on the basis of their behavior in crosses (Somers 1973). In a cross between a killer {$KIk_1$} strain and a suppressive L-sensitive {KIL-$s_1$} strain most of the progeny of meiosis and most of the progeny of mitotic segregation in the diploid are {KIL-$s_1$} mutants. Suppressive sensitive mutants have been shown to possess M genomes ($M_s$) that have a lower than normal molecular weight and that apparently replace the M genomes in heterozygous $M/M_s$ diploids or diploid progeny (Sweeney *et al.* 1976; Tzen *et al.* 1974; Vodkin 1977; Vodkin *et al.* 1974).

Several different {KIL-s} genomes have been analyzed. They appear to fall into two size classes, 0.7–0.8 kbp (S3, S14, and S733) and 1.3–1.5 kbp (S1, S2, S4, and S17) (Bruenn and Kane 1978; Bruenn and Brennan 1980; Fried and Fink 1978; Kane *et al.* 1979; Sweeney *et al.* 1976). The $M_s$ genomes all appear to be derived from the M genome on the basis of $T_1$ fingerprint analysis, electron microscope heteroduplex analysis, and 3′ end sequence analysis (Bruenn and Brennan 1980; Bruenn and Kane 1978; Fried and Fink 1978 and Kane *et al.* 1979). These same authors showed that $M_s$ genomes appear to be internal deletion mutants of M possessing identical sequences at both ends: The 3′ termini are identical for at least 23 nucleotides at one end and 34 at the other. The S3 M genome deletion mutant has been estimated to retain the first 500-bp and the last 280-bp fragments of M. Translation of S3 M dsRNA produces S-M, an 8-kd polypeptide that retains the complete leader sequence and part of the $M_1$-P1 toxin sequence (Bostian *et al.* 1982b). On the basis of this information the translation initiation site for M has been placed at around 280 bp from the 3′ terminus of the plus strand and the S3 deletion located (see Fig. 12.5).

Although {KIL-$s_1$} mutants cause the elimination of the {KIL-$k_1$} genomes, they in no way interfere with the L genome or with each other, i.e., different {KIL-$s_1$} mutant genomes do not eliminate each other and stable strains possessing more than one {KIL-$s_1$} mutant genome have been constructed (Sweeney *et al.* 1976).

Analogies can be made between these suppressive sensitive mutants and the defective interfering (DI) particles of animal viruses. DI viral RNAs have been shown to be mutant on the basis of either internal deletions (Guild and Stollar 1977; Johnson and Lazzarini 1977; Keene *et al.* 1977; Kennedy 1976) or deletions at either, 5′ or 3′, end of the parent genomes (Huang and Baltimore 1977; Keene *et al.* 1978; Leppert *et al.* 1977). No suppressive sensitive mutations of the M genome, which are the result of terminal deletions, have been isolated. This may be because of a positive requirement for these termini for replication and if they were to be deleted it would prove lethal for the genome. Nothing is known about the mechanism by which $M_s$ genomes cause the elimination of the M genomes. However, the fact that the termini, thought to be required for replication in the ScV system, are always retained suggests that $M_s$ genomes interfere at the replication level.

Suppressive sensitive {KIL-$s_2$} mutants of the {KIL-$k_2$} killer system have not been formally identified. However, the {EXL} mutants (see later) may well prove to be such {KIL-$s_2$} mutants.

## Viral Maintenance

A multiplicity of gene loci control, either directly or indirectly, the maintenance of the M plasmid. However, no nuclear gene mutation has ever been recorded that results in the loss of the L plasmid. There are two, possibly three, postulated mutants of the L and several identified mutants of the M plasmids known to be defective or altered in some aspect of M plasmid maintenance. A consideration of the chromosomal maintenance genes and of these plasmid mutants follows, together with some discussion on plasmid–plasmid interactions.

## Nuclear Gene Control of L Maintenance[1]

To date, with the exception of *mak.10* in the K7 strain (see below), no nuclear gene mutation has been identified that results in the loss of the L plasmid. Of the numerous gene loci demonstrated to affect the maintenance of the $M_1$ plasmid (see later), only one locus, *MAK.3*, has any effect on L. Strains possessing the recessive allele, *mak.3*, have an approximately tenfold reduction in L copy number (Wickner and Leibowitz 1979). However, it has recently been suggested (Wickner and Toh-e 1982) that this result may have been due to the presence of {HOK} in the cross.

## Putative L Plasmid Mutants

There are no defined L plasmid mutants as such, possibly for the reasons already suggested (see earlier), although there is evidence to suggest that both the {HOK} and {NEX} elements may be mutant forms of L. However, recent results from this laboratory have shown that most killer strains possess two different forms of L, one of which is dependent on the *MAK.10* maintenance gene (Mitchell and Sherbeni, manuscript in preparation). Indirect evidence that L plasmid mutants exist comes from the 3′ end sequence analysis of Bruenn and Brennan (1980): These authors found marked 3′ (−) strand sequence heterogeneity.

### {HOK}

{HOK} plasmids allow the replication of defective M mutants, {KIL-sd}, in a $SK1^+$ host (see below) and are thus defined *H*elper of *K*iller (Wickner and

---

[1]Nucleic acid extracted from killer strains of *Saccharomyces cerevisiae* and run on agarose gels under sub-optimal conditions for M dsRNA yielded two species of L dsRNA ($L_1$ and $L_2$) which migrated at slightly different rates. $L_2$ dsRNA was shown to be dependent upon MAK3, MAK10 and PES1 but not MAK1 nuclear genes for its maintenance. $L_1$ was shown to be independent of all of these genes.

{HOK} a non-Mendelian genetic element which enables a replication-defective M genome, {KIL-$sd_1$}, to be maintained in a $SKI^+$ host (Wickner and Toh-e 1982) was found to be present in an M-o derivative of the strain K7 which possesses $L_2$ but not $L_1$. Curing of $L_2$ at elevated temperatures resulted in a co-curing of {HOK} and it is therefore concluded that either $L_2$ is equivalent to or else it encodes {HOK} (El-Sherbeini *et al.* 1982).

Table 12.4. Plasmid Gene Requirements for Nuclear Maintenance Genes[a]

| | MAK genes | | | | | | | Other maintenance genes | | |
|---|---|---|---|---|---|---|---|---|---|---|
| Plasmid | 1 | 3 | 4<br>6<br>27 | 5<br>14,15<br>19<br>26 | 7–9<br>11–13<br>17,18<br>20–25<br>28 | 10 | 16 | PET.18 | SPE.2 | MKT 1 |
| L | X | Partial[b] | X | X | X | X | X | X | X | X |
| $M_1$ ({KIL-$k_1$}) | ✓ | ✓ | ✓ | ✓ | ✓ | ✓ | ✓ | ✓ | ✓ | X (even if {NEX} present) |
| $M_2$ ({KIL-$k_2$}) | ? | ✓ | ? | ? | ? | ✓ | ✓ | ✓ | ? | ✓ (if NEX present) |
| {EXL} | ✓ | ✓ | ? | ? | ? | ✓ | ? | ✓ | ? | ? |
| {NEX} | ? | ✓ | X | ? | ? | ? | ? | ? | ? | X |
| {HOK} | X | ✓ | X | X | ? | ✓ | X | ✓ | ? | X |
| $L_{K7}$ | ? | ? | ? | ? | ? | ✓ | ? | ? | ? | ? |

[a]Key: ✓ indicates requirement for dominant allele; X indicates no such requirement; ? indicates not tested.
[b]See text under "Nuclear Control of L Maintenance."

Toh-e 1982). Since {HOK} plasmids segregate in a non-Mendelian manner and are transferred during cytoduction, they may be concluded to be extrachromosomal elements. The data from various experiments point to the conclusion that {HOK} is either a mutant L genome or a helper plasmid for L, that is,

1. In the absence of {HOK}, L copy numbers decrease by about fourfold.
2. {HOK} depends on *MAK.3, MAK.10,* and *PET.18* for its maintenance, which are some of those genes needed for M maintenance, and those genes bypassed or suppressed by the *skil.1* mutations (see Table 12.4).
3. {HOK} supplies a function to a replication-defective M mutant, {KIL-sd}.
4. Strains tested and shown to be L-o are also {HOK-o}, although the reverse is not necessarily true.

These data accumulate from the work of Wickner and Toh-e (1982) and Lund (1982). Wickner and Toh-e (1982) suggested that either {HOK} is an altered form of L or else it is a helper plasmid for both {KIL-sd} and L genome replication. Recent experiments on the co-curing of {HOK} {NEX} lend further support to the hypothesis that {HOK} is a form of L.

## {*NEX*}

Wickner (1980) has provided evidence that {NEX}, which prevents {EXL} from excluding {KIL-$k_2$}, is a non-Mendelian genetic element. He also showed that {KIL-$k_2$} in the presence of {NEX} becomes dependent on the product of the MKT.1 gene for maintenance, whereas {KIL-$k_2$} is independent of this gene under any conditions. The replication of {NEX} was shown to be dependent on *MAK.3,* but independent of *MAK.4, MAK.6, MAK.27,* and MKT.1. The dependency of {NEX} on *MAK.3,* and its helper (nonexcluder) functions, suggests that it is a dsRNA plasmid. If this is the case then {NEX} may be a form of L, but it is not a mutant M genome: {KIL-$k_2$}-o, {NEX} strains lack any detectable M dsRNA. The interaction of {NEX} with other plasmids is discussed below under "Plasmid Exclusion/Interference Phenomena."

The data derived from different sources suggest that {HOK} {NEX} and L are interrelated and may indeed all be forms of the same species:

1. {HOK} requires the *MAK.3, MAK.10,* and *PET.18* genes for maintenance (Wickner and Toh-e 1982). {NEX} requires at least *MAK.3* for its maintenance (Wickner 1980).
2. The majority of strains that are {HOK} are also {NEX}. Only two strains have been found that are {HOK} {NEX-o}, and none have been found that are {HOK-o} {NEX} (Sommer and Wickner 1982).
3. An L {HOK} {NEX} strain was treated with elevated temperatures, ethanol or 2-propanol. This resulted in all ($>100$) cured progeny being both {HOK-o} {NEX-o}. When tested for L, these progeny showed a fourfold reduction in L copy number. When {HOK} {NEX-o} strains were cured of

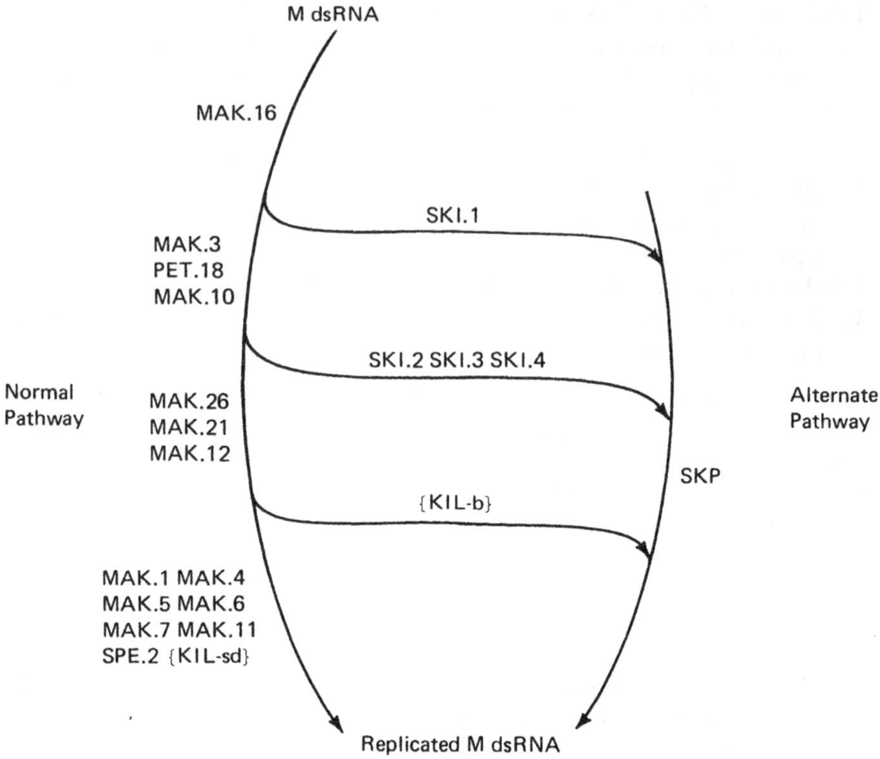


**Figure 12.6** Model of M dsRNA replication. This model incorporates all of the available information on the interactions of the *MAK* and *SKI* gene products on M and on the M plasmid mutants {KIL-b} and {KIL-sd}. The model is reproduced from Wickner (1981) who predicted the existence of *skp* genes whose products are needed for the alternate pathway but not for the normal pathway.

{HOK}, they exhibited the same reduction in L copy number (Sommer and Wickner 1982).

4. Following EMS mutagenesis, strains that had lost L (and M) were both {NEX-o} and {HOK-o} (Lund 1982)

The conclusion from these data is that {NEX} may be a mutant L genome and that it seems to require {HOK} activity or that {NEX} and {HOK} are both functions of the same (L?) plasmid. If {HOK} and {NEX} are mutant forms of L, then it is to be noted that they require *MAK.3, MAK.10,* and *PET.18* and (at least) *MAK.3* gene activities, respectively. L has been shown to be independent of these functions except in one instance, that of the L genome in the K7 strain. This L genome shows abnormal segregation patterns and appears to be under the control of the *MAK.10* gene (Mitchell *et al.* 1976a). This implies that the L of K7 is mutant, but it has been shown (Wickner and

Toh-e 1982) that K7 does not possess {HOK}. Therefore, L of K7 and {HOK} may both be different, mutant forms of the normal L genome. It is not known whether the L of K7 requires the dominant allele of *MAK.3* or *PET.18* for maintenance; it would be interesting to test this and to test whether {NEX} requires *MAK.10* and *PET.18* in addition to *MAK.3*.

That {HOK} requires *MAK.3, MAK.10,* and *PET.18* is remarkable *per se:* These genes are precisely those that are bypassed or suppressed by the *ski.1-1* mutation (Toh-e and Wickner 1980). Perhaps L is ordinarily replicated along the alternate pathway (see Fig. 12.6) and {HOK} is a mutant L that can only gain access to that pathway after the point at which *ski.1-1* permits access. If this is the case then it is somewhat surprising that {HOK} does not, as shown by Wickner and Toh-e (1982), require *MAK.16*. It is of interest then to determine whether or not *ski.1-1* mutants can maintain {HOK} in the presence of a *mak.3, mak.10,* or *pet.18* mutation.

## Nuclear Gene Control of $M_1$ Maintenance

The dominant alleles of 29 nuclear genes, designated *MAK, PET,* and *SKI,* are all required for the maintenance of the $M_1$ genome. Other genes, designated *SKI* and *KRB,* are involved in the regulation of maintenance. These genes are all discussed separately below, and the $M_1$ and other plasmid requirements for these genes are depicted in Table 12.4.

### *MAK* Genes

A mutation at any one of *MAK.1* or *MAK.3-28* sites results in a loss of the $M_1$ but not the L genome (Guerry-Kopecko and Wickner 1980; Mitchell *et al.* 1976a; Somers and Bevan 1969; Wickner 1974b, 1978, 1979a,b; Wickner and Leibowitz 1976b, 1979).

The *MAK* gene products have not been identified, nor is their function known. However, it may be assumed that most if not all of these loci actually determine some host-specific function other than the maintenance of the M genomes in view of the findings in analogous (plasmid/bacterial) systems. For example, it has been shown that *mak.3, mak.15, mak.17, mak.20, mak.22,* and *mak.27* mutants all result in a temperature-independent reduction in host growth rate (Wickner and Leibowitz 1979). Furthermore, *mak.1* and *mak.16,* like *PET.18* (see following discussion), result in a temperature sensitivity for growth, although the loss of $M_1$ is temperature independent (Leibowitz and Wickner 1978; Wickner and Leibowitz 1976b, 1979).

Some of the *MAK* gene functions may be bypassed by the *ski* mutations: This is discussed later, under *SKI*.

## *PET.18 (PES.1)* Gene

The *pet.18* and the *pes.1* mutations are almost certainly the same. *pes.1* mutants were shown to control the maintenance of both functional mitochondria (rho) and the M genome (killer phenotype) by Bevan and co-workers (1969). *pes.1* was shown to map close to the **a**/$\alpha$ mating type locus and the centromere on chromosome III (Theivendirarajah 1969). The *pet.18* mutation has the same pleiotrophic effects as *pes.1* and has been located to the same position on the genetic map (Fink and Styles 1972; Leibowitz and Wickner 1978; Wickner and Leibowitz 1976b). The function of this gene, like that of the *MAK* genes, has yet to be defined. What is known is that the two known functions of the *PET.18* gene are separable in that the *KRB.1* gene restores the ability of $M_1$ but not the maintenance in *pet.18* mutants (Wickner and Leibowitz 1977).

## *SPE.2* Gene

Cohn and co-workers (1978a) have demonstrated that *SPE.2* codes for an enzyme, adenosylmethionine decarboxylase, involved in the biosynthetic pathway for the polyamines spermidine and spermine. *spe.2* mutants have been shown to be completely deficient in spermine and spermidine, to grow at only one-sixth of the normal growth rate, to be unable to sporulate, and to lose their M genomes. All of these *spe.2*-induced effects are presentable by the addition of either spermidine or spermine (Cohn *et al.* 1978a,b).

## *SKI* Genes

*SKI* genes and the *KRB.1* gene are considered to be host regulatory genes that control the replication of the M genome. The nuclear *ski*$^-$ mutants were originally isolated as "superkillers" and were mutant in that they exhibited enhanced toxin activity (Toh-e and Wickner 1980). The latter activity could have resulted in one of three ways, namely:

1. As a result of an increase in M copy number
2. As a result of an increase in expression rate of the M genome
3. As a result of an alteration to the M gene product, the toxin, in that the toxin itself is more stable

The third suggestion can be considered to be the least likely in that it would involve the *ski* gene playing some role in the processing of toxin or toxin precursors. Furthermore, the toxin produced by *ski* and *SKI* has been shown (a) to exhibit the same heat lability when heat inactivation rates were measured

(Toh-e and Wickner; cited in Toh-e and Wickner 1980) and (b) to have the same specificity (Ouchi and Wickner; cited in Wickner 1981).

There is no evidence either for or against the second suggestion, but there is evidence to support the first hypothesis, i.e., that *ski* mutations result in the superkiller phenotype as a result of increasing the M genome replication rate and therefore copy number. This evidence is threefold, namely:

1. *ski* mutants have increased M genome copy numbers.
2. *ski* mutations bypass or suppress certain *mak* mutations.
3. {KIL-sd} mutant M genomes can be maintained in the presence of *ski* mutations.

Each line of evidence will be discussed separately:

1. *ski* mutants have increased M genome copy numbers. The data to support this conclusion are rather variable but nonetheless suggest that *ski.2*, *ski.3*, and *ski.4* mutations all result in an increase in the amount of M dsRNA. *skil.1* mutations, which in addition to resulting in a superkiller phenotype also decrease cellular growth rate, do not seem to increase M levels (Toh-e *et al.* 1978).

2. *ski* mutations bypass or suppress certain *mak* mutations. Different *ski* mutations allow the bypassing or suppression of certain *MAK* gene encoded maintenance functions, i.e., certain *ski mak* double mutants are able to maintain M genomes and hence remain killers. *ski.1-1* bypasses all the *mak* mutations tested with the exception of *mak16.1*. *ski.2*, *ski.3*, and *ski.4* mutations are alike in that they do not permit the suppression of *mak.16*, *mak.3*, *mak.10*, or *pet.18* but allow the bypass or suppression of all other *mak* genes tested (Toh-e and Wickner 1980). These authors, on the basis of their data, have put forward two models to explain the bypass of the *mak* mutations. These data, together with the available information on the {KIL-b} and {KIL-sd} plasmid mutants, have been integrated to produce a model for their interactions (Toh-e and Wickner 1980; Wickner 1981). This model has been reproduced as Figure 12.6. It is entirely conceivable that an alternate pathway for replication of M could result in an increased M copy number. However, this is not consistent with the observation already made, that *skil.1* mutations result in no such increase.

3. {KIL-sd} mutant M genomes can be maintained in the presence of *ski* mutations. {KIL-sd} mutants are defective in some aspect of their replication that is complemented by *ski* mutations (Toh-e and Wickner 1979). They are discussed in some detail later.

As noted under item 2, immediately preceding, these results are consistent with M replication occurring via either of two alternate pathways (Fig. 12.6). This model predicts that M replicates along the normal (*MAK*-controlled) pathway but that {KIL-sd} mutants and M in *mak* mutants replicate via the alternate (*SKI*-controlled) pathway.

### *KRB.1* Gene

A dominant chromosomal mutation KRB.1 results in a bypass of the requirements for both the *MAK.7* and the *PET.18* gene products for $M_1$ maintenance (Wickner and Leibowitz 1977). These authors further showed that KRB.1 only complemented one of the functional defects of the *pet.18* pleiotrophic mutation, that of maintaining the M genome: *KRB.1 pet.18* double mutants were still temperature sensitive for growth and still (rho)°. *KRB.1* mutants, unlike the *ski* mutants, are not superkillers, nor are they translational suppressors. Wickner (1981) has suggested that they may be constitutive for an enzyme that is needed by M for replication and that is normally induced by *PET.18* or *MAK.7*.

## Maintenance-Affected $M_1$ Plasmid Mutants

In this section only those $M_1$ plasmid mutants that show an alteration either in copy number or in their nuclear gene requirements for maintenance are considered.

### {KIL-b}

{KIL-b} plasmids (*k*iller *b*ypass) have the same resistance specificities as the {KIL-$k_1$} plasmids. They have been shown to confer a superkiller phenotype on their hosts and to be independent of the *MAK.4, MAK.7, MAK.11,* and *MAK.17* genes yet dependent on all other M-maintenance genes tested, namely, *MAK.3, 10, 12, 16, 21, 26,* and *PET.18*. {KIL-b} plasmids therefore bypass or suppress some but not all of the *MAK* genes bypassed by all of the *ski* mutations. For this reason the authors of this work, Toh-e and Wickner (1980), have stated that {KIL-b} mutants may be considered to be plasmid-encoded *ski* mutants.

### {KIL-d}

{KIL-d} mutants are dependent on host diploidy for stable maintenance and expression: The {KIL-d} plasmids show no killer phenotype and are rapidly lost in haploid strains (Wickner 1976b). This author has suggested that these {KIL-d} diploid-dependent mutants may be analogous to the chromosomal *cyc.7-2* mutants that overproduce iso-2 cytochrome c in any haploid but not in $a/\alpha$ diploids. The *cyc.7-2* locus has obviously come under the control of the mating-type locus, since homozygous diploids, *a/a* or $\alpha/\alpha$, behave like haploids (Rothstein and Sherman 1980). It remains to be seen whether {KIL-d} mutants behave in homozygous diploids as they do in haploids.

### {KIL-sd}

{KIL-sd} mutants are defined by their inability to be maintained in wild-type *(SKI)* strains: They are dependent on a *ski* mutation for their maintenance (Toh-e and Wickner 1979). These mutants are obviously defective in some maintenance function that is bypassed by the *ski* mutation, presumably in a manner similar to the bypass of *MAK* gene requirements by *ski* gene products (see earlier, under "SKI Genes"). {KIL-sd} mutant genomes have the same molecular weight as the wild-type M genomes (Toh-e and Wickner 1979) and the lesion presumably occurs within the postulated M genome control region (see Fig. 12.5). If, as discussed previously, the *ski* gene mutations result in the replication of M via an alternate pathway to that defined by the maintenance genes *MAK, SPE.2,* and *PET.18,* and if {KIL-sd} plasmids are mutant in their interaction with one or more of these maintenance gene products, then they would have an absolute requirement for a *ski* mutation for their replication.

### {KIL-s}

Suppressive sensitive {KIL-s} mutants are internal deletion mutants and their structure has already been considered in some detail, under "M Plasmid Gene Mutations" in the "Viral Genomes" section. {KIL-s} mutants interfere with the replication of the wild-type M {KIL-k} genomes but not with each other nor with the L genomes (Bruenn and Kane 1978; Fried and Fink 1978; Somers 1973; Sweeney *et al.* 1976; Vodkin *et al.* 1974). Some {KIL-s} plasmids are present in about twice the copy number at which M is maintained (Vodkin *et al.* 1974). The mechanism by which they interfere with {KIL-k} replication is not yet known.

### {KIL-sk}

{KIL-sk} mutants confer a superkiller phenotype on their host. In one strain at least, the superkiller phenotype is thought to be due to a two- to threefold increase in M genome copy number and an increased toxin stability (Palfree and Bussey 1979). {KIL-sk} mutations could, however, result from an increased efficiency of transcription. Bostian *et al.* (1982a) have suggested that of the two mRNA molecules found (1.9 kb and 1.2 kb), the larger, full-length ones may be involved in replication and the smaller ones are partial transcripts resulting from premature termination or cleavage. If this is indeed the case, then these authors may well be correct in suggesting that {KIL-sk} mutations act at this processing site.

### Nuclear Gene Control of $M_2$ Maintenance

Strains of the $K_2$ killer system have L and $M_2$ dsRNA. The wild-type $M_2$ genome has a designated genotype {KIL-$k_2$} and encodes a TOX.2 killer toxin.

Of the nuclear genes that control the maintenance of $M_1$, *MAK.3, MAK.10, MAK.16,* and *PET.18* are also required for $M_2$ maintenance (Naumov and Naumova 1973; Wickner 1980, 1981). None of the other maintenance genes has been tested (see Table 12.4).

The dominant allele of MKT.1 has been shown to be necessary for the maintenance of $M_2$ under certain conditions (see following discussion) but is not needed for $M_1$ maintenance (Wickner 1980).

### Maintenance-Affected $M_2$ Plasmid Mutants

No $M_2$ plasmid mutants have been formally identified. This is probably a reflection of the lack of work on the {KIL-$K_2$} system, as opposed to the {KIL-$k_1$} system, rather than a statement about their prepondence. {EXL}, considered here, is the only known candidate for being an M plasmid mutant.

#### {EXL}: A Putative {KIL-$s_2$} Mutant

{EXL} has been demonstrated to be a non-Mendelian element: It segregates in a non-Mendelian manner and can be transferred during cytoduction (Wickner 1980). {EXL} excludes {KIL-$k_2$} from cells into which both are introduced, in presumably the same way that {KIL-$s_1$} excludes {KIL-$k_1$}. On the basis of these exclusion properties, and the fact that it requires *MAK.1, MAK.3, MAK.10,* and *PET.18* for maintenance, Wickner (1980) has concluded that it is probably a form of dsRNA and that it may in fact be a {KIL-$s_2$} plasmid. However, no dsRNA species has yet been shown to be associated with the {EXL} plasmid.

### Plasmid Exclusion/Interference Phenomena

Three aspects of plasmid–plasmid interactions will be discussed here, namely:

1. Incompatibility
2. Exclusion
3. Helper functions

## Incompatibility

Zygotes that are {KIL-$k_1$} {KIL-$n_1$} quite rapidly segregate mitotically into pure killer {KIL-$k_1$} and pure neutral {KIL-$n_1$} clones, the numbers of each type being strain dependent (Bevan and Somers 1969). Thus {KIL-$k_1$} and {KIL-$n_1$} plasmids are obviously incompatible in the same host. A similar phenomenon occurs in bacterial–plasmid systems, where Ishii *et al.* (1978) and Novick and Hoppensteadt (1978) have explained it on the basis of fluctuations in the random selection of genomes for replication followed by segregation.

## Exclusion

The phenomenon in which zygotes possess two different plasmids, the same one of which is always lost from the progeny, is defined as exclusion. Several examples of plasmid exclusion in this system exist:

1. {KIL-$k_1$} excludes {KIL-$k_2$} (Naumov and Naumova 1973; R. B. Wickner (cited in Wickner 1981) and authors' unpublished observations).
2. {KIL-$s_1$} excludes both {KIL-$k_1$} and {KIL-$k_2$} (Somers 1973; Wickner 1980).
3. {EXL} excludes {KIL-$k_2$} but not, it is to be noted, {KIL-$k_1$} (Wickner 1980).

The basis of these exclusion phenomena is unknown, but it seems possible that it may be because some plasmids are better candidates than others for efficient replication.

## Helper Functions

Where one plasmid assists another in its maintenance it is defined as providing a helper function. Two such examples are known:

1. {HOK} prevents {KIL-sd} loss in a *SKI* host (Wickner and Toh-e 1982).
2. {NEX} prevents {KIL-$k_2$} exclusion by {EXL}, but only if *MKT.1* is present. {NEX} does not, however, prevent {KIL-$k_2$} exclusion by {KIL-$k_1$} or {KIL-$s_1$} (Wickner 1980).

In the first example {HOK} is assumed to supply some function that complements or bypasses the lesion in the {KIL-sd} mutant genomes. The fact that in the preceding item 3 of the "Exclusion" section it was shown that {EXL} excludes {KIL-$k_2$} but not {KIL-$k_1$} and that in item 2 of the present section it is seen that {NEX} prevents {KIL-$k_2$} exclusion by {EXL} but does not prevent its exclusion by {KIL-$k_1$} or {KIL-$s_1$} could be interpreted as evidence for differences in replication or control of $M_1$ and $M_2$ plasmids. Contradictory evidence is presented by the facts that (a) {KIL-$k_2$} requires at least two of the maintenance genes for {KIL-$k_1$} (see Table 12.4) and (b) {KIL-$k_2$} is excluded by both {KIL-$k_1$} and {KIL-$s_1$}, thereby implying some similarities in replication control.

## Other Cytoplasmic Elements

The dsRNA genomes L, M, and their mutant forms, including the putative L mutants {HOK} and {NEX} and the possible $M_2$ mutant {EXL}, are not the only extrachromosomal elements in yeast. *S. cerevisiae* has numerous other elements, some of which are well characterized, others of which are not. Table 12.5 lists these elements together with what is known about their associated phenotypes and/or nucleic acid species.

**Table 12.5.** Nonkiller Cytoplasmic Determinants in Yeast

| Cytoplasmic determinant | Associated phenotype(s) | Nucleic acid | References |
|---|---|---|---|
| {rho} | Respiratory competence; drug resistance | Mit DNA | See reviews by Evans and by Wilkie, this book |
| {CIR} | Postulated oligomycin resistance—not substantiated | 2 μm DNA | Guerineau, 1979; Guerineau *et al.* (1971,1974) |
| {URE.3} | Affects nitrogen metabolism in {rho}$^-$ cells | ? | Aigle and Lacroute (1975); Lacroute (1971) |
| {psi} | Enhances activity of ochre, amber, and some frame-shift suppressors | ? | Cox (1965); Culbertson *et al.* (1977); Lund (1982) |
| {MAL} | Affects fermentation | ? | Schamhart *et al.* (1975) |
| | Contains copies of the rRNA genes | 3 μm DNA | Clark-Walker and Azad (1980); Larionov and Grishin (1980) |
| | Various uncorrelated, nonmitochondrial drug-resistance phenotypes | ? | Carignani *et al.* (1977); Guerineau *et al.* (1974); Hughes and Wilkie (1972); Kruszewskar and Szczesniak (1978); Kunz and Ball (1977); Rotman (1975) |
| {20S RNA} | Production of 20S RNA during growth under sporulation conditions | ? | Garvik and Haber (1978) |

In an attempt to elucidate any possible relationship between these nonnuclear determinants and the killer-associated dsRNA genomes, various experimental approaches have been made, including both formal genetic analyses and curing. A summary of such work, pertaining to each cytoplasmic element in turn, follows.

### {rho}

Mitochondrial genes and M segregate independently in crosses and are transferred independently during cytoduction. Furthermore, $\{rho\}^-$ cells, totally lacking in any mitDNA, may possess M and vice versa (Al-Aidroos *et al.* 1973; Fink and Styles 1972; Mitchell 1974). The only common factor is a nuclear gene *PET.18*, considered previously under "Viral Maintenance," which controls the maintenance of both {rho} and M. {rho} has also similarly been tested for any possible relationship to {HOK} (Wickner and Toh-e 1982) or to {EXL} or {NEX} (Wickner 1980), but no such relationship was found.

### {CIR}

$\{CIR\}^-$ strains may possess M and {CIR}+ strains may lack M. In addition, curing and cytoduction experiments have lent support to the hypothesis that {CIR} is a separate entity from M, {HOK}, {EXL}, or {NEX} (Livingston 1977; Wickner 1980; Wickner and Toh-e 1982). Furthermore, the 2-$\mu$m DNA responsible for {CIR} does not depend on the *PET.18* gene, whereas M, {HOK}, and {EXL} do (Leibowitz and Wickner 1978).

### {URE.3}

The nucleic acid responsible for the {URE.3} phenotype has not yet been identified. Nevertheless, this element does not require the *PET.18* gene for maintenance, whereas M, {EXL}, and {HOK} all do, and {URE.3} may therefore be considered to be a separate entity (Leibowitz and Wickner 1978).

### {psi}

$\{psi\}^+$ strains exist that lack both L and M, and curing can differentially result in the loss of {psi} and {HOK} (Lund 1982). *PET.18*, as stated previously, is required for M, {EXL} and {HOK} but not for {psi} (Leibowitz and Wickner 1978). {psi} may therefore be concluded not to have any relationship to L, M, {HOK}, or {EXL}.

No information is available as to any possible relationship between {MAL}, 3 μm DNA, or any of the "unknown" drug-resistance determinants and any of the killer genomes.

### {20S RNA}

There is considerable speculation as to the nature of the genetic element responsible for the production of a 20S RNA species in yeast. During conditions under which diploid cells are normally induced to undergo meiotic sporulation, the synthesis of a single-stranded RNA species, of $\sim 0.8 \times 10^6$d and 20S, is induced, accounting for some 15% of all new RNA synthesis (Garvik and Haber 1978; Kadowaki and Halvorson 1971a,b). This novel species of RNA has been shown not to be methylated (Sogin *et al.* 1972) and not to be related to ribosomal RNA (Wejksnora and Haber 1978).

Garvik and Haber (1978) have looked for a possible correlation of {20S RNA} with the killer system. They have demonstrated that 20S RNA production is under the control of a cytoplasmic determinant but that that determinant is neither L nor M. Furthermore, they showed that strains that produce 20S RNA may be killers or nonkillers but that all nonproducers of 20S RNA are also always nonkillers. On the basis of this somewhat circumstantial evidence, Garvik and Haber (1978) have suggested that this 20S RNA species may be necessary either for the expression of the killer phenotype or for the maintenance of the M genome. Further work is obviously justified to determine whether or not this suggestion is in fact correct.

{HOK} a cytoplasmic element that may be, as discussed previously, a mutant form of L is not correlated with the presence of this 20S RNA species. Wickner and Toh-e (1982) have analyzed two of Garvik and Haber's (1978) strains that lack 20S RNA and demonstrated that they still carry {HOK}.

## Conclusions

The killer system in the yeast *Saccharomyces cerevisiae* is perhaps the best understood of all the known yeast–dsRNA viral systems. The two constituent particles, designated ScV-L and ScV-M, possess all the attributes, save that of infectivity, required of a segmented virus. The L genome, of the helper ScV-L virus, encodes the major capsid protein for both particles and possibly also the capsid-associated polymerase activities. The M genome, of the defective ScV-M virus, encodes an exotoxin, the killer toxin, and it almost certainly also encodes the resistance determinants. The M genome is dependent on a multiplicity of host gene products for its maintenance and functional expression while one form of L at least is independent of these same gene products.

Although many mutant forms of M exist, no mutant forms of L have yet been described. However, it has recently been found that most killer strains possess at least two different forms of L and that one of these is dependent on at least one of the M maintenance genes, *MAK.10* (Mitchell and Sherbeni, manuscript in preparation).

Several important points remain to be elucidated, not least among which is the question as to what the maintenance and expression genes code for and what functions are provided for by these gene products. Furthermore, the demonstration that the "second" protein encoded by the M genome is in actuality the resistant determinant has not yet been made. More importantly perhaps, the question as to what genome encodes the RNA polymerase(s) necessary for transcription and replication has not been answered, although it seems likely that it is the L genome. Finally, the nature of the cytoplasmic determinants {HOK} and {NEX} has not been elucidated. Are they mutant L genomes?

The killer system as a whole provides an excellent model for the elucidation of host–viral and viral–viral interactions or indeed the replication and its control of the viruses themselves. The control of expression of the M genome—the mechanism of processing of the toxin precursors and the mechanism by which the exotoxin itself is secreted—is of particular interest, since it may provide insight into the general area of exported proteins.

## References

Adams, J. M., Cory, S. (1975). Modified nucleotides and bizarre 5′-termini in mouse myeloma mRNA. Nature New Biol. 255:28–33.

Adler, J., Mackenzie, D. W. R. (1972). Abst. Ann. Meet. Amer. Soc. Microbiol. p. 8.

Adler, J., Wood, H. A., Bozarth, R. F. (1976). Virus-like particles from killer, neutral, and sensitive strains of *Saccharomyces cerevisiae*. J. Virol. 17:472–476.

Aigle, M., Lacroute, F. (1975). Genetic aspects of {URE 3}, a non-mitochondrial cytoplasmically inherited mutation in yeast. Mol. Gen. Genet. 136:327–335.

Al-Aidroos, K., Bussey, H. (1978). Chromosomal mutants of *Saccharomyces cerevisiae* affecting the cell wall binding site for killer factor. Can. J. Micro biol. 24:228–237.

Al-Aidroos, K., Somers, J. M., Bussey, H. (1973). Retention of cytoplasmic killer determinants in yeast cells after removal of mitochondrial DNA by ethidium bromide. Mol. Gen. Genet. 122:323–330.

Alwine, J. C., Kemp, D. J., Stark, G. R. (1977). Method for detection of specific RNAs in agarose gels by transfer to diazobenyloxy-methyl-paper and hybridisation with DNA probes. Proc. Nat. Acad. Sci. USA 74:5350–5354.

Berry, E. A., Bevan, E. A. (1972). A new species of double-stranded RNA from yeast. Nature New Biol. 239:279–280.

Bevan, E. A., Herring, A. J. (1976). The killer character in yeast: preliminary studies of virus-like particle replication. In: Genetics, Biogenetics and Bioenergetics of Mitochondria, edited by R. J. Bandelow, D. Y. Schweyen, K. Thomas, K. Wolf, F. Kaudewitz, pp. 153–162. Berlin: De Gruyter.

Bevan, E. A., Makower, M. (1963). The physiological basis of the killer character in yeast. In: Genetics Today, XIth Int. Congr. Genet., Vol. 1, edited by S. J. Geerts, pp. 202–203. Oxford: Pergamon Press.

Bevan, E. A., Mitchell, D. J. (1979). The killer system in yeast. In: Viruses and Plasmids in Fungi, edited by P. A. Lemke, pp. 161–169. New York: Marcel Dekker.

Bevan, E. A., Somers, J. M. (1969). Somatic segregation of the killer (k) and neutral (n) cytoplasmic genetic determinants in yeast. Genet. Res. 14:71–77.

Bevan, E. A., Somers, J. M., Theivendirarajah, K. (1969). In: Proceedings XIth International Botanical Congress (Seattle), p. 14.

Bevan, E. A., Herring, A. J., Mitchell, D. J. (1973). Preliminary characterisation of two species of dsRNA in yeast and their relationship to the "killer" character. Nature 245:81–86.

Blobel, G., Dobberstein, B. (1975). Transfer of proteins across membranes. I. J. Cell Biol. 67:835–851.

Bostian, K. A., Hopper, J. E., Rogers, D. J., Tipper, D. J. (1980a). Translational analysis of the killer-associated virus-like particle dsRNA genome of *S. cerevisiae*. M dsRNA encodes toxin. Cell 19:403–414.

Bostian, K. A., Sturgeon, J. A., Tipper, D. J. (1980b). Encapsidation of yeast killer double-stranded ribonucleic acid: Dependence of M on L. J. Bacteriol. 143:463–470.

Bostian, K. A., Burn, V. E., Jayachandran, S., Tipper, D. J. (1982a). Yeast killer dsRNA plasmids are transcribed *in vivo* to produce full and partial-length plus-stranded RNAs: models for their synthesis and for the functional sequence of $M_1$-dsRNA. Cell, in press.

Bostian, K. A., Jayachandran, S., Tipper, D. J. (1982b). A glycosylated protoxin in killer yeast: models for its structure and maturation. Cell, in press.

Botcham, M., Topp, W., Sambrook, J. (1976). The arrangement of simian virus 40 sequences in the DNA of transformed cells. Cell 9:269–287.

Brennan, V. E., Bobek, L. A., Bruenn, J. A. (1981a). Yeast dsRNA viral transcriptase pulse products: identification of the transcript strand. Nucl. Acids Res. 10:5049–5060.

Brennan, V., Field, L., Cizdziel, P., Bruenn, J. A. (1981b). Sequence of the 3′ ends of yeast dsRNA: proposed transcriptase and replicase initiation sites. Nucl. Acids Res. 9:4007–4021.

Bruenn, J. A. (1980). Virus-like particles of yeast. Ann. Rey. Microbiol. 34:49–68.

Bruenn, J. A., Brennan, V. R. (1980). Yeast viral double-stranded RNAs have heterogeneous 3′ termini. Cell 19:923–933.

Bruenn, J., Kane, W. (1978). Relatedness of the double-stranded RNAs present in yeast virus-like particles. J. Virol. 26:762–772.

Bruenn, J., Keitz, B. (1976). The 5′ ends of yeast killer factor RNAs are pppGP. Nucl. Acids Res. 3:2427–2436.

Bruenn, J. A., Bobek, L., Brennan, K., Held, W. (1980). Yeast viral RNA polymerase is a transcriptase. Nucl. Acids Res. 8(13):2985–2998.

Buck, K. W. (1980). Viruses and killer factors of fungi. In: "The Eukaryotic Microbial Cell," edited by G. W. Gooday, D. Lloyd, and A. P. J. Trinci, Soc. Gen. Micro. Symp. 30:329–375. Cambridge University Press.

Buck, K. W., Lhoas, P., Street, B. K. (1973). Virus particles in yeast. Biochem. Soc. Trans. 1:1141–1142.

Bussey, H. (1972). Effects of yeast killer factor on sensitive cells. Nature New Biol. 235:73–75.

Bussey, H. (1974). Yeast killer factor-induced turbidity changes in cells and sphaeroplasts of a sensitive strain. J. Gen. Microbiol. 82:171–179.

Bussey, H. (1981). Physiology of killer factor in yeast. Advances in Microbial Physiology, 22, edited by A. H. Rose and G. Morris. New York: Academic Press, pp. 93–122.

Bussey, H., Sherman, D. (1973). Yeast killer factor: ATP leakage and coordinate inhibition of macromolecular synthesis in sensitive cells. Biochem. Biophys. Acta. 298:868–875.

Bussey, H., Skipper, N. (1975). Membrane-mediated killing of *Saccharomyces cerevisiae* by glycoproteins from *Torulopsis glabrata.* J. Bacteriol. 124:476–483.

Bussey, H., Sherman, D., Somers, J. M. (1973). Action of yeast killer factor: A resistant mutant with sensitive spheroplasts. J. Bacteriol. 113:1193–1197.

Bussey, H., Saville, D., Hutchins, K., Palfree, R. G. E. (1979). Binding of yeast killer toxin to a cell wall receptor on sensitive *Saccharomyces cerevisiae.* J. Bacteriol. 140:888–892.

Bussey, H., Sacks, W., Galley, D., Saville, D. (1982). Yeast killer plasmid mutations affecting toxin secretion and activity, and toxin immunity function. Mol. Cel. Biol. 2(4):346–354.

Carignani, G., Lancashire, W. E., Griffiths, D. E. (1977). Extra-chromosomal inheritance of Rhodamine 6G resistance in *Saccharomyces cerevisiae.* Mol. Gen. Genet. 151:49–56.

Chow, N., Shatkin, A. J. (1975). Blocked and unblocked 5′ termini in reovirus genome RNA. J. Virol. 15:1057–1064.

Clare, J. J., Oliver, S. G. (1979). The regulation of RNA synthesis in yeast IV. Synthesis of double-stranded RNA. Mol. Gen. Genet. 171:161–166.

Clark-Walker, G. D., Azad, A. A. (1980). Hybridizable sequences between cytoplasmic ribosomal RNAs and 3 micron circular DNAs of *Saccharomyces cerevisiae* and *Torulopsis glabrata.* Nucl. Acids Res. 8(5):1009–1022.

Cohn, M. S., Tabor, C. W., Tabor, H. (1978a). Isolation and characterisation of *Saccharomyces cerevisiae* mutants deficient in S-adenosylmethionine decarboxylase spermidine and spermidine. J. Bacteriol. 134:208–213.

Cohn, M. S., Tabor, C. W., Tabor, H., and Wickner, R. B. (1978b). Spermidine or spermine requirement for killer double-stranded RNA plasmid replication in yeast. J. Biol. Chem. 253:5225–5227.

Cox, B. S. (1965). $\psi$, a cytoplasmic suppressor of super-suppressor in yeast. Heredity 20:505–521.

Culbertson, M. R., Charmos, L., Johnson, M. T., Fink, G. R. (1977). Frameshifts and frameshift suppressors in *Saccharomyces cerevisiae.* Genetics 86:745–764.

El-Sherbeini, M., Bevan, E. A., Mitchell, D. J. (1983). Current Genet. 7: 63–68.

Fink, G. R., Styles, C. A. (1972). Curing of a killer factor in *Saccharomyces cerevisiae.* Proc. Natl. Acad. Sci. USA 69:2846–2849.

Fried, H. M., Fink, G. R. (1978). Electron microscopic heteroduplex analysis of "killer" double-stranded RNA species from yeast. Proc. Nat. Acad. Sci. USA 75:4224–4228.

Furuichi, Y., Morgan, M., Muthukrishnan, S., Shatkin, A. J. (1975a). Reovirus messenger RNA contains a methylated, blocked 5′-terminal structure, $m^7G(5')$-ppp(5′)GmpCp-. Proc. Nat. Acad. Sci. USA 72:362–366.

Furuichi, Y., Muthukrishnan, S., Shatkin, A. J. (1975b). 5′-terminal $m^7G(5')$-ppp(5′)Gmp *in vivo:* identification in reovirus genome. Proc. Nat. Acad. Sci. USA 72:742–745.

Garvik, B., Haber, J. E. (1978). New cytoplasmic genetic element that controls 20S RNA synthesis during sporulation in yeast. J. Bacteriol. 134:261–269.

Guerineau, M. (1979). Plasmid DNA in yeast. In: Viruses and Plasmids in Fungi. edited by P. A. Lemke. New York: Marcell Dekker, pp. 540–593.

Guerineau, M., Grandchamp, C., Paoletti, C., Slonimski, P. (1971). Characterisation of a new class of circular DNA molecules in yeast. Biochem. Biophys. Res. Commun. 42:550–557.

Guerineau, M., Slonimski, P. P., Avner, P. R. (1974). Yeast episome: Oligomycin resistance associated with a small covalently closed, non-mitochondrial circular DNA. Biochem. Biophys. Res. Commun. 61:462–469.

Guerry-Kopecko, P., Wickner, R. B. (1980). Isolation and characterisation of temperature-sensitive *mak* mutants of *Saccharomyces cerevisiae*. J. Bacteriol. 144(3):1113–1118.

Guild, G. M., Stollar, V. (1977). Defective interfering particles of Sendai virus. V. Sequence relationships between $SV_{STD}$ 42S RNA and intracellular defective Viral RNAs[1,2]. Virology 77:175–188.

Gunge, N., Tamaru, A., Ozawa, F., Sakaguchi, K. (1981). Isolation and characterisation of linear deoxyribonucleic acid plasmids from *Kluyveromyces lactis* and the plasmid-associated killer character. J. Bacteriol. 145(1):382–390.

Hastie, N., Brennan, V., Bruenn, J. (1978). No homology between double-stranded RNA and nuclear DNA of yeast. J. Virol. 28:1002–1005.

Haylock, R. W., Bevan, E. A. (1981). Characterisation of the L dsRNA encoded mRNA of yeast. Curr. Genet. 4:181–186.

Herring, A. J., Bevan, E. A. (1974). Virus-like particles associated with double-stranded RNA species found in killer and sensitive strains of the yeast *Saccharomyces cerevisiae*. J. Gen. Virol. 22:387–394.

Herring, A. J., Bevan, E. A. (1975). Double-stranded RNA containing particles from the yeast *Saccharomyces cerevisiae* and their relationship to the killer character. In: Molecular Biology of Nucleocytoplasmic Relationships, edited by S. Puiseux-Dao. Amsterdam: Elsevier, pp. 149–154.

Herring, A. J., Bevan, E. A. (1977). Yeast virus-like particles possess a capsid-associated single-stranded RNA polymerase. Nature 268:464–466.

Hill, M., Hillova, J., Dankchev, D., Mariage, R., Goubin, G. (1974). Infectious viral DNA in Raw sarcoma virus-transformed nonproducer and producer animal cells. Cold Spring Harbor Symp. Quant. Biol. 39:1015–1016.

Holm, C. A., Oliver, S. G., Newman, A. M., Holland, L. E., McLaughlin, C. S., Wagner, E. K., Warner, R. C. (1978). The molecular weight of yeast P1 double-stranded RNA. J. Biol. Chem. 253:8332–8336.

Hopper, J. E., Bostian, K. A., Rowe, L. B., Tipper, D. J. (1977). Translation of the L-species dsRNA genome of killer-associated virus-like particles of *Saccharomyces cerevisiae*. J. Biol. Chem. 252:9010–9017.

Huang, A. S., Baltimore, D. (1977). Defective interfering animal viruses. In: Comprehensive Virology, edited by H. Fraenkel-Conrat and R. R. Wagner, Vol. 10. New York: Plenum Press, pp. 73–116.

Hughes, A. R., Wilkie, D. (1972). Genetic analysis of mitochondrial resistance to tetracycline in *Saccharomyces cerevisiae*. Heredity 28:117–127.

Ishii, K., Hashumoto-Goboh, T., Matsubara, K. (1978). Random replication and random assortment model for plasmid incompatibility in bacteria. Plasmid 1:435–445.

Johnson, L. D., Lazzarini, R. A. (1977). The 5′ terminal nucleotide of RNA from vesicular stomatitis virus defective interfering particles. Virology 77:863–866.

Kadowaki, K., Halvorson, H. O. (1971a). Appearance of a new species of ribonucleic acid during sporulation in *Saccharomyces cerevisiae*. J. Bacteriol. 105:826–830.

Kadowaki, K., Halvorson, H. O. (1971b). Isolation and properties of a new species of ribonucleic acid synthesised in sporulating cells of *Saccharomyces cerevisiae*. J. Bacteriol. 105:831–836.

Kandel, J. S., Stern, T. A. (1979). Killer phenomenon in pathogenic yeasts. Antimicrob. Agents Chemother. 15:568–571.

Kane, W., Pietras, D., Bruenn, J. (1979). The evolution of defective-interfering dsRNAs of the yeast killer virus. J. Virol. 32:692–696.

Keene, J. D., Rosenberg, M., Lazzarini, R. A. (1977). Characterisation of the 3′-terminus of RNA isolated from vesicular stomatitis virus and from its defective interfering particles. Proc. Natl. Acad. Sci. USA 74:1353–1357.

Keene, J. D., Schubert, M., Lazzarini, R. A., Rosenberg, M. (1978). Nucleotide sequence homology at the 3′-termini of RNA from vesicular stomatitis virus and its defective interfering particles. Proc. Natl. Acad. Sci. USA 75:3225–3229.

Kennedy, S. I. T. (1976). Sequence and relationships between genome and the intracellular RNA species of standard and defective-interfering Semliki forest virus. J. Mol. Biol. 108:491–511.

Koper-Zwarthoff, F. C., Bol, J. F. (1980). Nucleotide sequence of the putative recognition site for coat protein in the RNAs of alfalfa mosaic virus and tobacco streak virus. Nucl. Acids Res. 8:3307–3318.

Kruszewska, A., Szczesniak, B. (1978). Janus green resistance in *Saccharomyces cerevisiae:* Interaction of nuclear and cytoplasmic factors. Mol. Gen. Genet. 160:171–181.

Kunz, B. A., Ball, A. J. S. (1977). Glucosamine resistance in yeast. II. Cytoplasmic determinants conferring resistance. Mol. Gen. Genet. 153:169–177.

Lacroute, F. (1971). Non-Mendelian mutation allowing ureidosuccinic acid uptake in yeast. J. Bacteriol. 196:519–522.

Lanonov, V. L., Grishin, A. V. (1980). 3 micron DNA: A new class of extrachromosomal DNA of the yeast. Dokl. Acad. Nauk. USSR Ser. Biokhim. 250(1-6):78–81.

Leibowitz, M. J., Wickner, R. B. (1978). *pet.18*: a chromosomal gene required for cell growth and maintenance of mitochondrial DNA and the killer plasmid of yeast. Mol. Gen. Genet. 165:115–121.

Leppert, M., Kort, L., Kolakofsky, D. (1977). Further characterisation of Sendai virus D1-RNAs: A model for their generation. Cell 12:539–552.

Lhoas, P. (1972). Electron micrographs of cells of *Saccharomyces cerevisiae* infected with double-stranded RNA virus from *Aspergillus niger* and *Penicillium stoloniferum*. Nature: New Biol. 236:86–87.

Livingstone, D. M. (1977). Inheritance of the 2 $\mu$m DNA plasmid from *Saccharomyces*. Genetics, 86:73–84.

Lund, P. M. (1982). Studies on killer and {psi} phenomena in *Saccharomyces cerevisiae*. D. Phil. Thesis, Oxford University.

Makower, M. (1964). Inheritance of a killer reaction in yeast. D. Phil. Thesis, Oxford University.

Maule, A. P., Thomas, P. D. (1973). Strains of yeast lethal to brewery yeasts. J. Inst. Brew. (London) 79:137–141.

McFadden, J. J. P. (1982). Detection and characterisation of viruses in conidial and ascospore isolates of *Graeumannomyces graminis var. tritici*. Ph.D. Thesis, University of London.

Mitchell, D. J. (1974). The correlation between nucleic acid differences and the killer phenotypes in yeast. Ph.D. Thesis, University of London.

Mitchell, D. J., Herring, A. J., Bevan, E. A. (1976a). The genetic control of dsRNA virus-like particles associated with *Saccharomyces cerevisiae* killer yeast. Heredity 37:129–134.

Mitchell, D. J., Herring, A. J., Bevan, E. A. (1976b). Virus uptake in Yeasts. In: Microbial and Plant Protoplasts, edited by J. F. Peberdy, A. H. Rose, H. J. Rogers, and E. C. Cocking. London: Academic Press, pp. 91–105.

Miura, K. I., Watanake, K., Sugura, M. (1974). 5′-terminal nucleotide sequences of the double-stranded RNA of silkworm cytoplasmic polyhedrosis virus. J. Mol. Biol. 86:31–48.

Mossop, D. W., Franki, R. I. B. (1978). Survival of a satellite RNA *in vivo* and its dependence on cucumber mosaic virus for its replication. Virology 86:562–566.

Naumov, G. I., Naumova, T. I. (1973). Comparative genetics of yeast. XIII. Comparative study of killer strains of *Saccharomyces* from different collections. Genetika 9:140–145.

Naumov, G. I., Tyurina, L. V., Bur'Yan, N. I., Naumova, T. I. (1973). Wine-making, an ecological niche of type $K_2$ killer *Saccharomyces*. Biol. Nauki 16:103–107.

Naumova, T. I., Naumov, G. I. (1973). Comparative genetics of yeast. XII. Study of antagonistic interrelations in *Saccharomyces* yeast. Genetika 9:85–90.

Nesterova, G. F. (1974). Infection of germinating spores of *Saccharomyces cerevisiae* by cytoplasmic factor K. Genetika 10:78.

Newman, A. M., Elliot, S. G., McLaughlin, C. S., Sutherland, P. A., Warner, R. C. (1981). Replication of double-stranded RNA of the virus-like particles in *Saccharomyces cerevisiae*. J. Virol. 38:263–271.

Novick, R. P., Hoppensteadt, F. C. (1978). On plasmid incompatibility. Plasmid 1:421–434.

Novick, P., Field, C., Schekman, R. (1980). Identification of 23 complementation groups required for post-translational events in the yeast secretory pathway. Cell 21:205,215.

Oliver, S. G., McCready, S. J., Holm, C., Sutherland, P. A., McLaughlin, C. S., Cox, B. S. (1977). Biochemical and physiological studies of the yeast virus-like particle. J. Bacteriol. 130:1303–1309.

Ouchi, K., Akiyama, H. (1976). Breeding of useful killer Sake yeasts by repeated back-crossing. J. Ferment. Technol. 54:615–623.

Palfree, R., Bussey, H. (1979). Yeast killer toxin: Purification and characterisation of the protein toxin from *Saccharomyces cerevisiae*. Eur. J. Biochem. 93:487–493.

Philliskirk, G., Young, T. W. (1975). The occurrence of killer character in yeasts of various genera. Antonie van Leeuwenhoek 41:147–151.

Pietras, D. F., Bruenn, J. A. (1976). The molecular biology of the yeast killer factor. Int. J. Biochem. 7:173–179.

Puhalla, J. E. (1968). Compatibility reactions on solid medium and interstrain inhibition in *Ustilago maydis*. Genetics 60:461–474.

Rogers, D. T. (1976). The genetic and phenotypic characterisation of killer strains of yeast isolated from different sources. Ph.D. Thesis, University of London.

Rogers, D., Bevan, E. A. (1978). Group classification of killer yeasts based on cross-reactions between strains of different species and origin. J. Gen. Microbiol. 105:199–202.

Rogers, D. T., Saville, D., Bussey, H. (1979). *Saccharomyces cerevisiae* killer expression mutant *kex 2* has altered secretory proteins and glycoproteins. Biochem. Biophys. Res. Commun. 90:187–193.

Rothstein, R. J., Sherman, F. (1980). Dependence on mating type for the overproduction of iso-2-cytochrome c in the yeast mutant CYC7-H2. Genetics 94:891–898.

Rotman, A. (1975). Genetics of a primaquin-resistant yeast. J. Gen. Microbiol. 89:1–10.

Saksena, K. N., Lemke, P. A. (1978). Viruses in Fungi. In: Comprehensive Virology, edited by H. Fraenkel-Conrat and R. R. Wagner, Vol. 12. New York: Plenum Publishing, pp. 102–143.

Schamhart, D. H. J., Ten Berge, A. M. A., Van de Poll, K. W. (1975). Isolation of a catabolite repression mutant of yeast as a revertant of a strain that is maltose negative in the respiratory deficient state. J. Bacteriol. 121:747–752.

Shalitin, C., Fischer, I. (1975). Abundant species of poly(A)-containing RNA from *Saccharomyces cerevisiae*. Biochim. Biophys. Acta. 414:263–272.

Skipper, N., Bussey, H. (1977). Mode of action of yeast toxins: Energy requirement for *Saccharomyces cerevisiae* killer toxin. J. Bacteriol. 129:668–677.

Sogin, S. J., Haber, J. E., Halvorson, H. O., (1972). Relationships between sporulation-specific 20S ribonucleic acid and ribosomal ribonucleic acid processing in *Saccharomyces cerevisiae*. J. Bacteriol. 112:806–814.

Somers, J. M. (1973). Isolation of suppressive mutants from killer and neutral strains of *Saccharomyces cerevisiae*. Genetics 74:571–579.

Somers, J. M., Bevan, E. A. (1969). The inheritance of the killer character in yeast. Genet. Res. 13:71–83.

Sommer, S. S., Wickner, R. B. (1982). Co-curing of plasmids affecting killer double-stranded RNAs of *Saccharomyces cerevisiae:* {HOK}, {NEX}, and the abundance of L are related and further evidence that $M_1$ requires L. J. Bacteriol. 150(2):545–551.

Southern, E. M. (1975). Detection of specific sequences among DNA fragments separated by gel electrophoresis. J. Mol. Biol. 98:503–517.

Sweeney, T. K., Tate, A., Fink, G. R. (1976). A study of the transmission and structure of double-stranded RNAs associated with the killer phenomenon in *Saccharomyces cerevisiae*. Genetics 84:27–42.

Theivendirarajah, K. (1969). The genetic linkage relationships of genes concerned in the determination of the killer character in yeast. Ph.D. Thesis, University of London.

Toh-e, A., Wickner, R. B. (1979). A mutant killer plasmid whose replication depends on a chromosomal "superkiller" mutation. Genetics 91:673–682.

Toh-e, A., Wickner, R. B. (1980). "Superkiller" mutations suppress chromosomal mutations affecting double-stranded RNA killer plasmid replication in *Saccharomyces cerevisiae*. Proc. Natl. Acad. Sci. USA 77(1):527–530.

Toh-e, A., Guerry, P., Wickner, R. B. (1978). Chromosomal superkiller mutants of *Saccharomyces cerevisiae*. J. Bacteriol. 136:1002–1007.

Tzen, J. C., Somers, J. M., Mitchell, D. J. (1974). A dsRNA analysis of suppressive sensitive mutants of "killer" *Saccharomyces cerevisiae*. Heredity 33:132.

Vodkin, M. (1977). Homology between double-stranded RNA and nuclear DNA of yeast. J. Virol. 21:516–521.

Vodkin, M., Fink, G. R. (1973). A nucleic acid associated with a killer strain of yeast. Proc. Natl. Acad. Sci. USA 70:1069–1072.

Vodkin, M., Katterman, F., Fink, G. R. (1974). Yeast killer mutants with altered double-stranded ribonucleic acid. J. Bacteriol. 117:681–686.

Weber, H., Linder, R. (1975). Virusartige partikeln in hef esprots plasten I: Elektrinmikroshopischer nachweis. Z. Allg. Mikrobiol. 15:631–638.

Wejksnora, P. J., Haber, J. E. (1978). Ribonucleoprotein particle appearing during sporulation in yeast. J. Bacteriol. 134:246–260.

Welsh, J. D., Leibowitz, M. J. (1980). Transcription of killer virion dsRNA *in vitro*. Nucl. Acids Res. 8:2365–2375.

Welsh, J. D., Leibowitz, M. J. (1982). Localisation of genes for the double-stranded RNA killer virus of yeast. Proc. Natl. Acad. Sci. USA 79:786–789.

Welsh, J. D., Leibowitz, M. J., Wickner, R. B. (1980). Virion DNA-dependent RNA polymerase from *Saccharomyces cerevisiae*. Nucl. Acid Res. 8:2349–2363.

Wickner, R. B. (1974a). Killer character of *Saccharomyces cerevisiae:* Curing by growth at elevated temperatures. J. Bacteriol. 117(3):1356–1357.

Wickner, R. B. (1974b). Chromosomal and non-chromosomal mutations affecting the "killer character" of *Saccharomyces cerevisiae*. Genetics 76:423–432.

Wickner, R. B. (1976b). Mutants of the killer plasmid of *Saccharomyces cerevisiae* dependent on chromosomal diploidy for expression and maintenance. Genetics 82:273–285.

Wickner, R. B. (1977). Deletion of mitochondrial DNA bypassing a chromosomal gene needed for maintenance on the killer plasmid of yeast. Genetics 87:441–452.

Wickner, R. B. (1978). Twenty-six chromosomal genes needed to maintain the killer double-stranded RNA plasmid of *Saccharomyces cerevisiae*. Genetics 88:419–425.

Wickner, R. B. (1979a). The killer double-stranded RNA plasmids of yeast. Plasmid 2:303–322.

Wickner, R. B. (1979b). Mapping chromosomal genes of *Saccharomyces cerevisiae* using an improved genetic mapping method. Genetics 92:803–821.

Wickner, R. B. (1980). Plasmids controlling exclusion of the $K_2$ killer double-stranded RNA plasmid of yeast. Cell 21:217–226.

Wickner, R. B. (1981). Killer systems in *Saccharomyces cerevisiae*. In: The Molecular Biology of the Yeast *Saccharomyces:* Life Cycle and Inheritance edited by J. N. Strathern, E. W. Jones, J. R. Broach, Cold Spring Harbor Laboratory, pp. 415–444.

Wickner, R. B., Leibowitz, M. J. (1976a). Two chromosomal genes required for killing expression in killer strains of *Saccharomyces cerevisiae*. Genetics 82:429–442.

Wickner, R. B., Leibowitz, M. J. (1976b). Chromosomal genes essential for replication of a double-stranded RNA plasmid of *Saccharomyces cerevisiae:* The killer character of yeast. J. Mol. Biol. 105:427–443.

Wickner, R. B., Leibowitz, M. J. (1977). Dominant chromosomal mutation bypassing chromosomal genes needed for killer RNA plasmid replication in yeast. Genetics 87:453–469.

Wickner, R. B., Leibowitz, M. J. (1979). *Mak* mutants of yeast: Mapping and characterisation. J. Bacteriol. 140:154–160.

Wickner, R. B., Toh-e, A. (1982). {HOK}, a new yeast non-Mendelian trait, enables a replication-defective killer plasmid to be maintained. Genetics, in press.

Woods, D. R. (1966). Studies of the nature of the "killer" factor in yeast. D. Phil. Thesis, Oxford University.

Young, T. W., Yagiu, M. (1978). A comparison of the killer character in different yeasts and its classification. Antonie van Leeuwenhoek J. Microbiol. Ser. 44:59–77.

Zakian, V. A., Wagner, D. W., Fangman, W. L. (1981). Yeast L-dsRNA is synthesised during the G1 phase but not the S phase of the cell cycle. Mol. Cel. Biol. 1:673–679.

# 13

# Approaches to the Genetic Analysis and Breeding of Brewer's Yeast

Morten C. Kielland-Brandt*,
Torsten Nilsson-Tillgren**, Jens G. Litske Petersen*, Steen Holmberg*, and Claes Gjermansen*

## Introduction[1]

Man has subjected many plants to breeding to suit his needs better. Some plant species have been bred for thousands of years. In modern times plant breeding has become increasingly intentional and systematic. *Saccharomyces* yeasts belong, according to their use, to man's old, cultivated plants, but with the exception of pure culture selection (Hansen 1888) they have undergone little intentional or systematic breeding.

Our ignorance of the detailed genetic constitution of industrial *Saccharomyces* strains is in extreme contrast to the vast and rapidly increasing amount of genetic and biochemical information that exists on the *Saccharomyces cerevisiae* strains that are being used in fundamental biological studies ("laboratory strains," "academic yeasts").

One reason for this difference is the way the strains behave in crosses and sporulation. Often during their derivation the laboratory strains are subjected

---

*Department of Physiology, Carlsberg Laboratory, Gamle Carlsberg Vej 10, DK-2500 Copenhagen Valby, Denmark.

**Institute of Genetics, University of Copenhagen, Øster Farimagsgade 2A, DK-1353 Copenhagen K, Denmark.

[1]*Gene symbols* (cf. Mortimer and Schild 1980): *ade2,* adenine requirement, red color of the yeast; *can1*, canavanine resistance; *cen3* and *cen5*, centromeres of chromosomes III and V; *cyh2,* cycloheximide resistance; *his1* and *his4,* histidine requirement; *HML* and *HMR*, silent mating-type information; *ilv1,* isoleucine requirement; *kar1,* karyogamy deficiency; *leu2,* leucine requirement; *MAT,* mating type; *rad3,* radiation sensitivity; rho$^-$, mitochondrially inherited respiratory deficiency; *thr4,* threonine requirement; *ura3,* uracil requirement. Uppercase and lowercase letters designate dominant and recessive alleles respectively, wild type being dominant for all genes mentioned here.

to selection for efficient sexual reproduction, but this criterion has probably seldom been considered in the selection and maintenance of industrial strains, including brewer's yeasts, which this chapter deals with. Because of the need for constant properties of each strain, there may even have been an unintentional selection against efficient sexual reproduction. In any case, most brewing strains, whether classified as *S. cerevisiae* or *S. carlsbergensis*, show no or poor mating ability, poor sporulation, and low spore viability (Winge 1944; Thorne 1951; Johnston 1965a; Fowell 1969; Anderson and Martin 1975).

As a consequence, it is generally hopeless to do tetrad analysis directly on brewing strains. The study by Thorne (1951) on the genetics of flocculence of *S. cerevisiae* is illustrative. He found that out of 20 top-fermenting English brewing strains, 12 sporulated adequately but that their spore viability only averaged 12%. Out of 372 different attempted crosses of the spores, only eight hybrids were obtained, none of which were amenable to tetrad analysis. In order to study the inheritance of the flocculation character of the brewing strains, Thorne found it necessary to cross some of the spores to nonbrewing strains. This approach is probably rather generally applicable. After one or more backcrosses to a suitable nonbrewing parent, spore viability may be adequate for tetrad analysis (Thorne 1951; Lewis *et al.* 1976). However, the approach gives only limited information on the genetic constitution of the brewing strain. Below we shall review another approach, single chromosome transfer, which is also based on the principle of analyzing limited parts of a brewing yeast genome in the genetic background of a nonbrewing yeast but which has proven more informative on the genetic structure of the chromosomes of the brewing yeast.

Difficulties in genetic analysis obviously may affect breeding programs rather directly. As will be discussed, spores of brewing strains are probably very often aneuploid. We believe that this phenomenon has been an important obstacle to breeding programs, including, for instance, regular backcrosses. However, as we shall also discuss, if the occurrence of aneuploidy is taken correctly into account, crossbreeding and mutation breeding both seem to hold promise. Furthermore, the recent developments in basic yeast genetics are suggesting new approaches not only to genetic analysis but also to breeding of brewer's yeast. Related reviews have been recently published by Tubb (1979), Wettstein (1979), Stewart and Russell (1979), and Kielland-Brandt (1981).

## Genetic Characterization of Brewing Strains

A basic question on the genetic organization of a yeast is, "What is the number of chromosome sets ('ploidy')?" On the basis of the segregation of mating and sporulation phenotypes, Emeis (1961) suggested that brewing yeasts generally are homothallic or polyploid. Uneven segregations consistent with triploidy were observed in many cases. However, the low spore viability, although being consistent with triploidy, made it difficult to exclude alternative interpretations.

Several more recent observations should be noted here. Bisexual behavior has been observed in some of the clones derived directly (Clayton *et al.* 1972; Anderson and Martin 1975; Gjermansen and Sigsgaard 1981) or indirectly (Spencer and Spencer 1977b) from single spores of industrial strains. Furthermore, C. Gjermansen (unpublished) has found that some of the spore-derived clones that are first clearly classified as nonsporulating are found, after further culturing, to have sporulation ability. Thus in brewing strains and their progeny there may be homothallism that sometimes only acts slowly, or mitotic homo- or hemizygotization of mating-type genes may be frequent, so that the meiotic segregations are obscured. Furthermore, as will be described, we have made observations that can explain irregularity of meiosis without invoking polyploidy.

Attempts have also been reported to estimate the number of chromosome sets by measuring the DNA content per cell (Stewart *et al.* 1975; Lewis *et al.* 1976; Russell and Stewart 1979), assuming that the amount of DNA per haploid genome is constant throughout the strains studied. Determinations corresponding to polyploidy were common. However, a difficulty with this approach is the fact that DNA makes up a rather small proportion of the yeast, about 1–2% of the nucleic acid (Cryer *et al.* 1975). Although the method of Burton (1956) for the determination of DNA is sensitive and very specific (Herbert *et al.* 1971), it still might be short of accuracy in the presence of possible interfering compounds derived from the yeast. Indeed, Stewart and Russell (1977) reported on "difficulties encountered in correlating the DNA content of cells with a possible ploidy determination of each strain." Regular multiplicities of the DNA content measured in a series of euploid laboratory strains (haploid, diploid, etc.) provide only circumstantial evidence for the specificity of a given method.

In summary, it is not unlikely that many brewing strains are polyploid, but we are still lacking satisfactory means to determine the overall genetic constitution of a given brewing strain.

Until recently, single genes were thus the only entities from the nuclear genomes of brewing strains that could be studied genetically in detail, namely, after crossing the genes into nonbrewing strains with high spore viability. The advent of the single-chromosome transfer technique has improved the situation. The basic idea is the same: to place the entity to be studied in a strain with regular genetic behavior, but now this entity is one chromosome.

## Single-Chromosome Transfer

The single-chromosome transfer technique makes use of the mutation *kar1-1* isolated by Conde and Fink (1976) as a mutation that impedes karyogamy after mating even if present in only one of the parents. Thus a genetic consequence of having *kar1* in a cross is a high yield of progeny with the nuclear genotype of either parent and cytoplasmic genetic elements of the other (Conde and Fink 1976). This phenomenon, which has been termed cytoduction (Zakharov and Yarovoy 1977), is observed also in normal crosses, but with a much lower frequency (Wright and Lederberg 1957). The use of the *kar1* gene for chromosome characterization rests on the fact that not only cytoplasmic elements are moved from one haploid strain to another in these crosses. The 2-$\mu$m DNA plasmid present in most laboratory (Cameron *et al.* 1977; Livingston 1977) and brewing (Tubb 1980) *Saccharomyces* strains is likely to have a nuclear location (Kielland-Brandt *et al.* 1980), but is transferred rather efficiently (in the order of half the times that functional mitochondrial DNA is transferred) in crosses involving *kar1,* as first shown by Livingston (1977) and confirmed by others (Kielland-Brandt *et al.* 1980; Brown *et al.* 1981). A composite 2-$\mu$m DNA-based yeast plasmid which is 2.5 times larger was also transferred, but with the much lower frequency of $10^{-3}$–$10^{-4}$ (Kielland-Brandt *et al.* 1979a). Importantly, even a whole chromosome can be transferred (Nilsson-Tillgren *et al.* 1979, 1980; Dutcher 1981). Nilsson-Tillgren *et al.* (1980) crossed two haploid *S. cerevisiae* laboratory strains of the genotypes *MATα his4 ade2 can1 kar1* [rho$^-$] and *MAT*a with one another and plated the mixture on minimal medium with adenine and canavanine in order to select for clones of the genotype of the first parent that had received the *HIS4* wild-type gene of the second parent. The first parent would not grow, since it requires histidine, and because of canavanine sensitivity the second parent would not grow, nor would normal diploid products of mating, the gene for canavanine resistance, *can1,* being recessive. Red colonies appeared at low frequency. They proved to be respiratory sufficient, histidine prototrophic, adenine requiring, canavanine resistant, nonsporulating, and nonmating. This indicates that they had the genotype of the first parent, except that they had received functional mitochondrial DNA and chromosome III with *MAT*a and *HIS4* (cf. Fig. 13.1) from the second parent, but few or no other chromosomes. At least chromosome V (with the *CAN1* wild-type gene) and chromosome XV (with *ADE2*) were not received. Because of a rare occurrence of mating cells, prototrophs could be obtained by mating these clones to appropriate haploid tester strains. Tetrad analysis of prototrophs from five different crosses confirmed the disomic nature of the two clones that were crossed, and the rare mitotic events that gave rise to mating cells were identified. Chromosome III, as identified by four markers distributed on both arms (Fig. 13.1), was transferred with an efficiency of $10^{-5}$–$10^{-6}$. In that study, cotransfer of five other chromosomes was

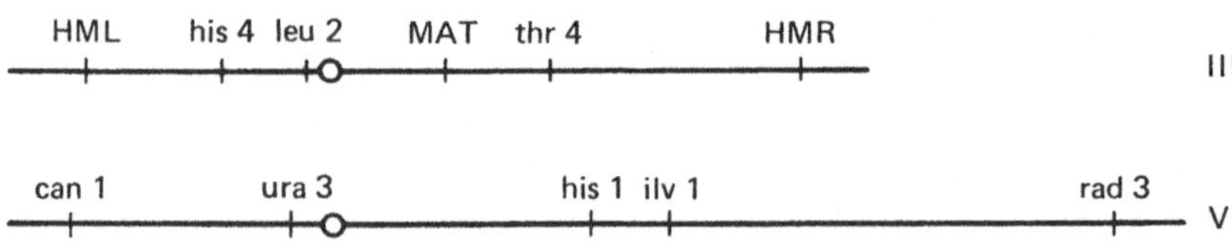


**Figure 13.1.** Maps of chromosomes III and V (only relevant markers are shown). (From Mortimer and Schild 1980.)

looked for but was not detected. In an independent study of crosses involving *kar1-1*, Dutcher (1981) has observed the transfer of several different chromosomes. The frequency of transfer was inversely correlated with the known map size; chromosome III was transferred 100–1000 times more frequently than the rate found by Nilsson-Tillgren *et al.* (1980), a fact probably due to strain differences. Dutcher (1981) found that transfer of more than one chromosome to the same cell was much rarer than the transfer of one chromosome but was more frequent than that calculated for independent events. In crosses where a *S. carlsbergensis* strain was a donor, Nilsson-Tillgren *et al.* (1981) observed some progeny that may be explained by events intermediary between complete nuclear fusion and the transfer of a single chromosome. Dutcher (1981) suggested that chromosome loss takes place in the transient heterokaryons with a frequency exceeding that of chromosome transfer. The mechanism of transfer, however, is unknown.

Single-chromosome transfer into genetically marked cells can be used to determine the gene content of chromosomes in genetically uncharacterized strains. Most brewing strains do not mate readily, probably because they are heterozygous for the mating type. On the other hand, even from those that sporulate poorly and produce spores of low viability, some maters can be selected upon germination of the spores. Such progeny can be subsequently mated to a haploid laboratory strain that carries the *kar1* mutation as well as appropriate genetic markers allowing the selection of progeny that, in addition to the chromosomes of the laboratory strain, have one specific chromosome from the brewing strain. In the following we shall review the principles of the genetic characterization of a lager yeast by this technique.

The yeast is a production strain of the Carlsberg Breweries and is designated *S. carlsbergensis*, strain 244. It shows the same low mating ability ("rare mating") as is exhibited by diploid laboratory strains that are heterozygous for the two alleles for mating type *MAT*a and *MAT*$\alpha$. It was classified as nonsporulating (Kielland-Brandt *et al.* 1979b) until incubation on sporulation medium was tried at temperatures lower than the standard (30°C) for laboratory strains. At 22°C and lower it is able to produce asci with one, two, or three spores. Asci with four spores have not been seen, and the spore viability is low (Gjermansen and Sigsgaard 1981). Some of the germinating spores yielded clones that mate. When one of these was mated to a haploid laboratory strain, the resulting strain again showed low spore viability (Gjermansen and Sigsgaard 1981). Thus the obstacles to a genetic characterization of this strain are similar to those reported for brewer's yeast in general.

## Chromosomes III and V of *S. carlsbergensis* Strain 244

Nilsson-Tillgren *et al.* (1981) crossed a mater derived from strain 244 to a strain carrying *kar1* and *his4*, thus defective in the biosynthesis of histidine. We selected progeny that could grow without histidine (i.e., carrying a chromosome from strain 244 with a wild type *HIS4* gene). With the help of other genes (*ade2* and *can1*) progeny was selected and identified that contained no or few other chromosomes from the brewing strain. The strains were nonmaters, indicating that a mating type gene had been imported from the brewing strain, together with the *HIS4* gene, so that the strain was now heterozygous for mating type. Rare spontaneous loss of either of the mating type alleles did, however, allow crosses to other genetically marked haploid laboratory strains. The products of these crosses showed high spore viability and could therefore be subjected to tetrad analysis, yielding information on the imported chromosome. This chromosome turned out to be *functionally* indistinguishable from chromosome III of *S. cerevisiae* laboratory strains. Not only did it carry wild type alleles of all the genes tested for (*HIS4, LEU2, MAT, THR4,* and probably *HML*—cf. Fig. 13.1), but it could provide the haploid progeny of these crosses with all the functions essential for viability that are provided by the *S. cerevisiae* chromosome III. Nevertheless, the *structure* of the transferred *S. carlsbergensis* chromosome was different. It did not recombine with *S. cerevisiae* chromosomes III in the *his4–MAT* region, whereas recombination was normal between *MAT* and *thr4*. The possibility was considered that the structural difference consisted in differences in nucleotide sequence throughout genes with identical functions. Identical proteins could be coded for by somewhat different nucleotide sequences because of the degeneracy of the genetic code, and outside the coding regions there might be even more room for sequence variations with little effect on essential functions. Such differences could explain the absence of recombination observed. Because of our previous isolation and molecular characterization of the *his4* region in *S. cerevisiae* (Kielland-Brandt *et al.* 1979a; Holmberg *et al.* 1979a, 1979b), it was a straightforward matter to look for nucleotide sequence differences in this region. The technique is based on sequence-specific enzymatic scissions of total DNA extracted from the yeast strain to be investigated. After electrophoretic separation by size of the many different sorts of DNA fragments formed, a separate preparation of radioactive DNA, with yeast nucleotide sequences of the *his4* region only, is hybridized to the DNA in the electropherogram (Southern 1975; Holmberg *et al.* 1979a, 1979b; Nilsson-Tillgren *et al.* 1981). The few fragments in the electropherogram that were formed from the *his4* region will bind molecules of the radioactive DNA and thereby reveal themselves, among the thousands of sizes of fragments, as black bands when the electropherogram is placed on a photographic film. The result is a pattern of bands that can be regarded as a fingerprint characteristic for a certain DNA sequence. Thus the wild type *HIS4* DNA sequence from *S. cerevisiae* yields the patterns designated I when the sequence-specific enzymes (restriction endonucleases) called EcoRI and HindIII are used (Fig. 13.2). The sizes of the fragments are given in thousands of nucleotide pairs (abbreviated kb, meaning kilobases). A small difference in nucleotide sequence will usually not be detected by the technique. Thus lanes a and f represent DNA from a strain carrying a point mutation in the *his4* gene. The pattern is indistinguishable from the wild-type pattern

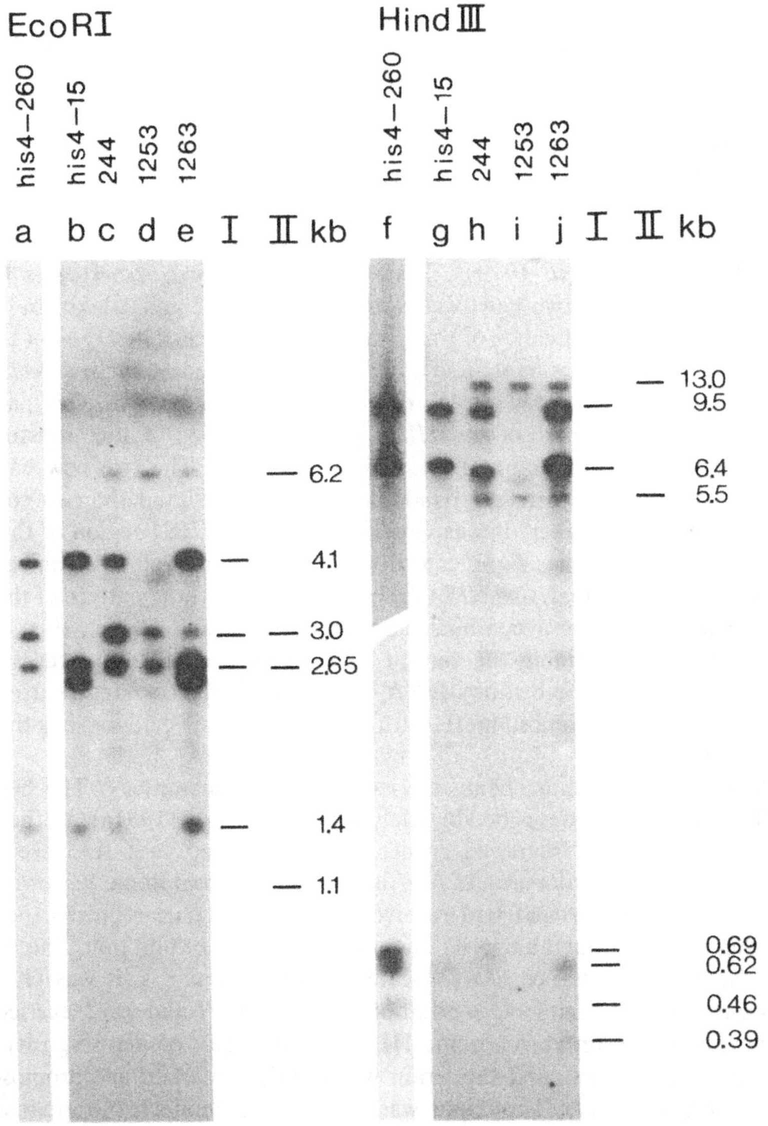


**Figure 13.2.** Detection in a *S. carlsbergensis* production strain of a wild-type *HIS4* region different in nucleotide sequence from that found in standard *S. cerevisiae* strains. DNA was cleaved with sequence-specific nuclease EcoRI or HindIII, subjected to electrophoresis and hybridized (Southern 1975) with radioactive *HIS4* DNA and the electropherogram was autoradiographed. Result of Nilsson-Tillgren *et al.* (1981). Lanes a and f show the *S. cerevisiae* wild-type pattern (I) and lanes b and g show a deletion mutant. The DNA in lanes d and i is from a strain with the *S. carlsbergensis* variant *HIS4* region (pattern II), whereas the production strain (lanes c and h) is heterozygous for the two wild-type regions.

(Holmberg *et al.* 1979a), and the change in sequence that causes the dysfunction of the gene is probably a change of a single base pair. In the following discussion we shall restrict ourselves to the patterns obtained with the sequence-specific enzyme EcoRI. The HindIII patterns are in accordance with the conclusions drawn. The pattern in lane b may be regarded as a fingerprint with a scar. Like the first strain, this strain also carries a mutation destroying the gene function, but this time the change is more radical; about 500 base pairs are missing, whereby the 3.0-kb band has moved down to the 2.5-kb position (Holmberg *et al.* 1979a). Lane c shows the *S. carlsbergensis* brewing strain and contains two new, weak bands of sizes 6.2 and 1.1 kb. In order to understand the significance of this, lane d is considered. The DNA of this lane is from a *S. cerevisiae* haploid in which the transferred *S. carlsbergensis* chromosome III has replaced the *S. cerevisiae* chromosome III. This pattern is therefore characteristic of the *HIS4* wild type region of the transferred *S. carlsbergensis* chromosome and is called pattern II. The two new bands are present; two of the pattern I bands have disappeared and the two conserved bands are much weaker. It was concluded that the *HIS4* region of the transferred chromosome has significant deviations from, as well as significant similarities to, the *S. cerevisiae HIS4* region. Interestingly, the pattern of the brewing yeast (lane c) is a combination of patterns I and II, i.e., this strain is heterozygous, containing the variant wild-type allele as well as the standard wild-type allele. Lane e shows DNA from a haploid *S. cerevisiae* strain that, in addition to its chromosome III with the 500 base pair deletion, has the transferred chromosome.

Analogous studies are being carried out on chromosome V. The beginning of these studies was described by Kielland-Brandt *et al.* (1981a). In the studies on chromosome III reviewed earlier, *can1* was used as a selective marker against diploids and donors. As *can1* is located on chromosome V, however, the selection of progeny that had received this chromosome required the use of another marker selectable in its hemizygous recessive condition. The gene for cycloheximide resistance, *cyh2,* located on chromosome VII, was chosen for this purpose. The recipient, in addition to having *kar1* and *cyh2,* carried *ade2,* as in the studies on chromosome III, as an indicator of hemizygosity of yet another chromosome, and the strain was multiply marked on chromosome V *(can1 ura3 his1 ilv1).* The donor was an $\alpha$ mater obtained (Gjermansen and Sigsgaard 1981) by sporulation of strain 244. Transfer of *URA3, HIS1,* and *ILV1* was selected, singly and in combination. These three wild-type genes were always transferred together, indicating that they are linked on one transferred *S. carlsbergensis* chromosome. This notion was confirmed by crossing the presumed disomic progeny to an appropriately marked haploid strain for subsequent sporulation and tetrad analysis. Data for two of the crosses were given by Kielland-Brandt *et al.* (1981a). Especially noteworthy was the fact that two different chromosomes V could be transferred from the $\alpha$ mater derived from strain 244. One did not recombine with a *S. cerevisiae* chromosome V in any of the intervals between *ura3*, *cen5*, *his1*, and *ilv1*, whereas the other showed normal recombination among these markers. Further analysis, to be published in detail elsewhere, has shown that the nonrecombining *S. carlsbergensis* chromosome V is *functionally* homologous to chromosome V of *S. cerevisiae:* It not only carries wild-type genes *CAN1* and RAD3, in addition to *URA3*, *HIS1*, and *ILV1*, but it can provide the cell with all the functions essential for viability that are provided by the *S. cerevisiae* chromosome V.

We speculate that the decreased recombination that we have now observed with a chromosome V as well as a chromosome III is due to general nucleotide sequence inhomology throughout regions with functional homology, as we have seen it in the *his4* region (Nilsson-Tillgren *et al.* 1981; cf. Fig. 13.2). This hypothesis is being tested by restriction endonuclease analysis of other regions of the two chromosomes. The results obtained so far on chromosome III (Holmberg 1982) are in full accordance with this view.

On the basis of our observations we propose that the original natural sources of the Carlsberg production strain were at least two rather diverged *Saccharomyces* genotypes. Thus our studies may have taxonomic significance. We find it likely that the strain contains several functionally homologous chromosome pairs that recombine poorly. Since recombination is a prerequisite for proper disjunction of chromosomes in meiosis, such a situation evidently leads to poor sporulation, low spore viability, and aneuploidy of the viable spores. Such phenomena can therefore not—at least not in the case of strain 244—be taken as evidence for polyploidy.

The restriction endonuclease analysis of specific genetic regions can be used in strain identification. This is shown already by the differences found in the *his4* region and as more regions are being analyzed, even more strains should be distinguishable by this technique. In the brewery it will be useful to be able to tell if an anomalously behaving yeast is a contaminant, a genetic variant of the production strain or a population of cells of the production strain that is different for a purely physiological reason.

To summarize, the technique of single chromosome transfer has proven to be an essential tool for the genetic characterization of whole chromosomes present in maters derived from a brewing strain. Such characterizations have yielded information important for our understanding of the genetic behavior of the strain. In particular, we tentatively conclude that aneuploidy is very frequent among the spores of strain 244. Restriction endonuclease analysis of the *HIS4* regions of several lager brewing strains shows that the sequence heterozygosity found in strain 244 is not limited to the strains used by Carlsberg or other Danish breweries (M. B. Pedersen, Carlsberg Laboratory, personal communication). It should be noted that the characterized chromosomes contain no genetic material from the nonbrewing strains used in the analysis. Such recombinational admixture would be an unavoidable possibility if tetrad analysis had been carried out after backcrosses to haploid laboratory strains instead of the single step of chromosome transfer. It may also be noted that the genetic technique of single-chromosome transfer has been efficiently complemented by the molecular techniques of restriction endonuclease analysis and DNA hybridization.

The possibility of extending the single-chromosome transfer technique by using somatic hybridization with protoplasts from the genetically marked strain carrying *kar1* and the brewing strain remains to be investigated.

## Breeding of Brewing Strains

We wish to discuss here mainly general approaches and principles in breeding and shall thus only shortly refer to some of the findings that are relevant for specific goals. The flocculation characteristics of a brewing strain are important in different ways, according to the technology used; too strong flocculence may lead to problems with attenuation or problems with removal of diacetyl during lagering, whereas too weak flocculence during lagering may lead to filtration problems if centrifugation is not employed. Several studies on the genetics of flocculence have been carried out and a chapter in this volume by Johnston is devoted to this subject; suffice it here to state that several genes are known that affect flocculence. Genes have been identified that enable certain *Saccharomyces* strains to ferment dextrin and starch (Tamaki 1978; Erratt and Stewart 1978) and thereby to superattenuate the wort, a trait that is desired for the production of low-carbohydrate types of beer. By blocking steps in the biosynthesis of isoleucine and valine, yeasts have been constructed that produce no diacetyl (Cabane *et al.* 1974; Ramos-Jeunehomme and Masschelein 1977), the amount of which normally has to be reduced by lagering. A gene responsible for "phenolic off-flavor" of the beer has been identified (Goodey and Tubb 1981). All these findings are important examples that the aims of the brewer, stated in terms of quality and economy, can be reformulated into more specific genetic terms by research in brewing chemistry, biochemistry, and yeast genetics.

### Crossbreeding

The first production of new yeast types based on recombination of existing variation between strains was described by Winge and Laustsen (1938, 1939). Since that time several attempts have been made to produce strains of increased brewing performance by sexual hybridization. Examples are the works by Johnston (1965b), Emeis (1971), and Clayton *et al.* (1972). But there are difficulties. The number of desired characteristics of the yeast and the lack of a clear definition of them (Winge and Laustsen 1939) suggest an approach in which a single character is imported from one parent into the range of characters of the other. Thus the brewer will often be in the situation that he has, among available yeasts, chosen the one that best suits his purposes but that among its many properties there are one or a few, often "technological," properties that he would like to change while keeping the others constant. In the crossbreeding of higher plants this can be done by repeated backcrosses with one parent. However, this is tedious in industrial yeasts with irregular sporulation and low spore viability.

The recent work by Gjermansen and Sigsgaard (1981), however, suggests that crossbreeding has important potential. Pairwise crosses between some of the maters derived by sporulation from strain 244 were performed and the resulting hybrids were screened, first for colony size and aerobic growth rate; then the best strains were tested for fermentation behavior in 2 liters of wort in EBC tubes. The hybrid that most resembled strain 244 was used in a 28-kg pilot beer production, yielding beer with very similar analytical values to that produced with strain 244. The two strains were subsequently compared in 600 hl production trials and the close similarity was confirmed (Kielland-Brandt *et al.* 1981a). The hybrid was slightly more flocculent and gave a beer with somewhat less diacetyl than the parent, trends that were observed already in the pilot experiments (Gjermansen and Sigsgaard 1981) and that are advantageous with the production techniques used at the Carlsberg Breweries. The beer from the hybrid was classified by the taste panel as being normal. It has thus been demonstrated that the nuclear genomes of brewer's yeast can be segregated into spores and clones competent in mating and that they can be recombined by mating into a new hybrid that produces a normal beer in production scale.

This work opens possibilities for strain improvement in two different ways. First, the results indicate that it is possible to obtain improvements by selection among newly produced hybrids. This may be regarded as a crossbreeding taking advantage of the natural variation inherent in the heterozygosity of strain 244. Second, it has important implications for mutation breeding (see below).

The techniques of protoplast fusion (Solingen and Plaat 1977; Gunge and Tamaru 1978; Christensen 1979; Russell and Stewart 1979) and respiratory deficient-auxotroph rare mating (Gunge and Nakatomi 1972; Spencer and Spencer 1977a) can be used in crossbreeding as a supplement to mating. They overcome the requirement for opposite mating types of strains to be crossed, thereby extending the possibilities for crosses to be done.

## Mutation Breeding

Mutation breeding could seem a rational approach in cases where a large number of performance parameters are to be kept constant while one is changed. But although mutations are probably induced with the same frequency in diploids or polyploids as they are in haploids, they do not usually reveal themselves because of the presence of nonmutated alleles. Only if the mutation is dominant is a phenotypic effect obtained without the need for additional alterations (Kielland-Brandt *et al.* 1979b). Secondary events that lead to the expression of a recessive mutant phenotype in a diploid or polyploid yeast include chromosome loss, mitotic recombination, mutation of the remaining corresponding wild type allele(s), or sporulation. Treatments that induce these events are, however, expected to cause further mutations or gross rearrangements of the genome, resulting in unwanted changes. Again, however, the work by Gjermansen and Sigsgaard (1981) suggests a frame in which mutation breeding may have an important role. Thus, the two mating parents of a well-behaving hybrid provide a good basis for the introduction of recessive mutations. These may have value either directly (e.g., for less diacetyl production) (Cabane *et al.* 1974; Ramos-Jeunehomme and Masschelein 1977) or for selection purposes, as outlined below. Recessive mutations are extremely difficult to introduce directly in strain 244 (Kielland-Brandt *et al.* 1979b), as is expected for a diploid or a polyploid yeast, whereas several have been introduced in the mating parents of the hybrid (unpublished).

## Genetic Transformation

Genetic transformation is the change of the genetic setup of an organism by the introduction of purified DNA. In *Saccharomyces,* transformation can be carried out with four different types of DNA. First, circular molecules can be used that have been constructed *in vitro* by combining a selectable yeast gene with a bacterial plasmid (Hinnen *et al.* 1978). The second type of DNA is like the first, except that the yeast DNA part includes a region that confers autonomous replication (growth) to the molecule in yeast (Beggs 1978; Struhl *et al.* 1979). In both cases the transformants will, in addition to the yeast gene(s) introduced, contain genes from foreign species. The third type of DNA also represents circular *in vitro* constructed plasmids, but only yeast DNA has been involved and the DNA has been through no foreign organism (Kielland-Brandt *et al.* 1979a; Holmberg *et al.* 1979b; Kielland-Brandt *et al.* 1981b). The fourth type is total native yeast DNA (Jansen *et al.* 1979). The last two types of DNA may be more appealing for use in breeding of an organism used for beverage production than the first two. It should be mentioned, however, that it is possible to remove all foreign DNA sequences from a yeast that has been transformed with the first type of DNA, leaving behind just the yeast gene that was introduced (Scherer and Davis 1979).

The only type of DNA that has until now been used with genes of specific brewing interest is native yeast DNA (Barney *et al.* 1980a, 1980b). Unfortunately, transformation with native yeast DNA is also the least characterized method for transforming yeast. The statistics of the numbers of transformants versus the controls given by Jansen *et al.* (1979) are convincing, but in contrast to transformation with other types of DNA (Hinnen *et al.* 1978; Beggs 1978; Struhl *et al.* 1979; Kielland-Brandt *et al.* 1979a, 1981b) it was not possible to distinguish transformants from revertants. The use of DNA from a strain with a distinctly recognized disorder in the gene selected for, for example, a temperature-sensitive mutation, would be helpful for this purpose. Furthermore, the transformation frequency was highly gene specific; some mutations seemed not to be transformable to wild type at all. A gene for flocculence, *FLO1* (Russell *et al.* 1980) has been transferred by this kind of transformation (Barney *et al.* 1980b) as well as a gene for dextrin utilization (Barney *et al.* 1980a).

## Limited Transfer of Genetic Elements in Crosses Involving *kar1*

This phenomenon and its use in genetic analysis have been reviewed here in detail. Obviously, it can also be used as a breeding technique, as illustrated by the *S. cerevisiae* strain whose chromosome III was replaced by the *S. carlsbergensis* chromosome III. The reverse (which for practical breeding may be more important), the transfer of a chromosome from a haploid laboratory strain to a *S. carlsbergensis* mater, should be equally possible. Furthermore, the transfer of genetic elements other than chromosomes can be important in breeding. Thus, as mentioned earlier, *in vitro* constructed plasmids with nuclear markers can be transferred in *kar1* crosses. A killer determinant has, with the help of *kar1*, been introduced into a brewing strain (Young 1981) to make it immune to killing by wild yeasts. Conde Zurita and Mascort Suárez (1981) have shown that replacing certain cytoplasmic determinants of a brewing strain with those from other strains can cause alterations of the beer flavor.

## Auxotrophic Mutations

As has been mentioned, auxotrophic mutations may have value directly in breeding, or they may be a tool to inform the breeder of biochemical interactions important in his program. But they have a further use, namely, in the selection of progeny in a variety of breeding techniques. Selection is essential in techniques like *kar1* crosses, protoplast fusion, rare mating, and transformation, where the desired progeny occurs with low frequency. For instance, a nonselectable gene can be introduced readily by a *kar1* cross or by transformation into a strain with an auxotrophic mutation if it is located on a chromosome or combined plasmid together with the wild type allele that will complement the auxotrophy.

## A General Remark

As the research in recent years has provided the brewing yeast geneticist with a series of new tools for the recombination of genes, knowledge of the biological system with which he is working has become an increasingly limiting factor for progress in breeding. This is a situation shared by most attempts to tailor living organisms genetically and should be recognized when breeding programs are initiated.

## References


Anderson, E., Martin, P. A. (1975). The sporulation and mating of brewing yeasts. J. Inst. Brew. 81:242–247.

Barney, M. C., Jansen, G. P., Helbert, J. R. (1980a). Use of spheroplast fusion and genetic transformation to introduce dextrin utilization into *Saccharomyces uvarum*. Am. Soc. Brew. Chem. J. 38:1–5.

Barney, M. C., Jansen, G. P., Helbert, J. R. (1980b). Use of genetic transformation for the introduction of flocculence into yeast. Am. Soc. Brew. Chem. J. 38:71–74.

Beggs, J. D. (1978). Transformation of yeast by a replicating hybrid plasmid. Nature 275:104–109.

Brown, A. J. P., Goodey, A. R., Tubb, R. S. (1981). Interstrain transfer of the 2$\mu$m DNA plasmid of *Saccharomyces* by cytoduction. J. Inst. Brew. 87:234–238.

Burton, K. (1956). A study of the conditions and mechanism of the diphenylamine reaction for the colorimetric estimation of deoxyribonucleic acid. Biochem. J. 62:315–323.

Cabane, B., Ramos-Jeunehomme, C., Lapage, N., Masschelein, C. A. (1974). Vicinal diketones—the problem and prospective solutions. Am. Soc. Brew. Chem. Proc. 1973:94–99.

Cameron, J. R., Philippsen, P., Davis, R. W. (1977). Analysis of chromosomal integration and deletions of yeast plasmids. Nucl. Acids Res. 4:1429–1448.

Christensen, B. E. (1979). Somatic hybridization in *Saccharomyces cerevisiae:* Analysis of products of protoplast fusion. Carlsberg Res. Commun. 44:225–233.

Clayton, E., Howard, G. A., Martin, P. A. (1973). Yeast hybridization. Am. Soc. Brew. Chem. Proc. 1972: 78–81.

Conde, J., Fink, G. R. (1976). A mutant of *Saccharomyces cerevisiae* defective for nuclear fusion. Proc. Natl. Acad. Sci. USA 73:3651–3655.

Conde Zurita, J., Mascort Suárez, J. L. (1981). Effect of the mitochondrial genome on the fermentation behaviour of brewing yeast. Eur. Brew. Conv. Congr., Copenhagen, 177–186.

Cryer, D. R., Eccleshall, R., Marmur, J. (1975). Isolation of yeast DNA. In: Methods in Cell Biology, edited by D. M. Prescott. New York: Academic Press, Vol. XII, pp. 39–44.

Dutcher, S. K. (1981). Internuclear transfer of genetic information in *kar1-1/KAR1* heterokaryons in *Saccharomyces cerevisiae.* Mol. Cell. Biol. 1:245–253.

Emeis, C. C. (1961). Polyploide Kulturhefen. Eur. Brew. Conv. Congr., Wien, 205–215.

Emeis, C. C. (1971). A new hybrid yeast for the fermentation of wort dextrins. Am. Soc. Brew. Chem. Proc. 58–62.

Erratt, J. A., Stewart, G. G. (1978). Genetic and biochemical studies on yeast strains able to utilize dextrins. Am. Soc. Brew. Chem. J. 36:151–161.

Fowell, R. R. (1969). Sporulation and hybridization of yeasts. In: The Yeasts, edited by A. H. Rose, J. S. Harrison. London and New York: Academic Press, Vol. 1, pp. 303–383.

Gjermansen, C., Sigsgaard, P. (1981). Construction of a hybrid brewing strain of *Saccharomyces carlsbergensis* by mating of meiotic segregants. Carlsberg Res. Commun. 46:1–11.

Goodey, A. R., Tubb, R. S. (1981). A yeast gene responsible for the production of 4-vinyl guaiacol from ferulic acid. Soc. Gen. Microbiol. Quart. 8:103 (abstr.).

Gunge, N., Nakatomi, Y. (1972). Genetic mechanisms of rare matings of the yeast *Saccharomyces cerevisiae* heterozygous for mating type. Genetics 70:41–58.

Gunge, N., Tamaru, A. (1978). Genetic analysis of products of protoplast fusion in *Saccharomyces cerevisiae*. Japan. J. Genet. 53:41–49.

Hansen, E. C. (1888). Undersøgelser fra gjæringsindustriens praxis. Meddelelser fra Carlsberg Laboratoriet 2:257–322.

Herbert, D., Phipps, P. J., Strange, R. E. (1971). Chemical analysis of microbial cells. In: Methods in Microbiology, edited by J. R. Norris, D. W. Ribbons. London and New York: Academic Press, Vol. 5B, pp. 209–344.

Hinnen, A., Hicks, J. B., Fink, G. R. (1978). Transformation of yeast. Proc. Natl. Acad. Sci. USA 75:1929–1933.

Holmberg, S. (1982). Genetic differences between *Saccharomyces carlsbergensis and S. cerevisiae*. II. Restriction endonuclease analysis of genes in chromosome III. Carlsberg Res. Commun. 47:233–244.

Holmberg, S., Kielland-Brandt, M. C., Nilsson-Tillgren, T., Petersen, J. G. L. (1979a). Molecular characterization of three *his4* deletion mutants in *Saccharomyces cerevisiae*. Carlsberg Res. Commun. 44:283–288.

Holmberg, S., Petersen, J. G. L., Nilsson-Tillgren, T., Kielland-Brandt, M. C. (1979b). Molecular characterization of a *Saccharomyces* plasmid containing the *HIS4* gene. Carlsberg Res. Commun. 44:269–282.

Jansen, G. P., Barney, M. C., Helbert, J. R., Esposito, M. S. (1979). Nonvectorassociated transformation in *Saccharomyces*. J. Appl. Biochem. 1:369–376.

Johnston, J. R. (1965a). Breeding yeasts for brewing, I. Isolation of breeding strains. J. Inst. Brew. 71:130–135.

Johnston, J. R. (1965b). Breeding yeasts for brewing, II. Production of hybrid strains. J. Inst. Brew. 71:135–137.

Kielland-Brandt, M. C. (1981). The genetics of brewer's yeast. Eur. Brew. Conv. Congr., Copenhagen, 263–276.

Kielland-Brandt, M. C., Nilsson-Tillgren, T., Holmberg, S., Petersen, J. G. L., Svenningsen, B. A. (1979a). Transformation of yeast without the use of foreign DNA. Carlsberg Res. Commun. 44:77–87.

Kielland-Brandt, M. C., Petersen, J. G. L., Mikkelsen, J. D. (1979b). Mutants in the biosynthesis of isoleucine in a non-mating, non-sporulating brewing strain of *Saccharomyces carlsbergensis*. Carlsberg Res. Commun. 44:27–36.

Kielland-Brandt, M. C., Wilken, B., Holmberg, S., Petersen, J. G. L., Nilsson-Tillgren, T. (1980). Genetic evidence for nuclear location of 2-micron DNA in yeast. Carlsberg Res. Commun. 45:119–124.

Kielland-Brandt, M. C., Gjermansen, C., Nilsson-Tillgren, T., Petersen, J. G. L., Holmberg, S., Sigsgaard, P. (1981a). Genetics of a lager production yeast. Tech. Quart. Master Brew. Assoc. Am. 18:185–187.

Kielland-Brandt, M. C., Nilsson-Tillgren, T., Petersen, J. G. L., Holmberg, S. (1981b). Transformation in yeast without the involvement of bacterial plasmids. In: Molecular Genetics in Yeast, edited by D. von Wettstein, J. Friis, M. Kielland-Brandt, A. Stenderup. Copenhagen: Munksgaard, pp. 369–382.

Lewis, C. W., Johnston, J. R., Martin, P. A. (1976). The genetics of yeast flocculation. J. Inst. Brew. 82:158–160.

Livingston, D. M. (1977). Inheritance of the 2μm DNA plasmid from *Saccharomyces*. Genetics 86:73–84.

Mortimer, R. K., Schild, D. (1980). Genetic map of *Saccharomyces cerevisiae*. Microbiol. Rev. 44:519–571.

Nilsson-Tillgren, T., Petersen, J. G. L., Holmberg, S., Kielland-Brandt, M. C. (1979). Stabilization by chromosomal integration of *HIS4* from an unstable yeast transformant. Meeting on: The Molecular Biology of Yeast. Cold Spring Harbor, N.Y.: Cold Spring Harbor Laboratory, abstract p. 125.

Nilsson-Tillgren, T., Petersen, J. G. L., Holmberg, S., Kielland-Brandt, M. C. (1980). Transfer of chromosome III during *kar* mediated cytoduction in yeast. Carlsberg Res. Commun. 45:113–117.

Nilsson-Tillgren, T., Gjermansen, C., Kielland-Brandt, M. C., Petersen, J. G. L., Holmberg, S. (1981). Genetic differences between *Saccharomyces carlsbergensis* and *S. cerevisiae*. Analysis by single chromosome transfer. Carlsberg Res. Commun. 46:65–76.

Ramos-Jeunehomme, C., Masschelein, C. A. (1977). Controle genetique de la formation des dicetones vicinales chez *Saccharomyces cerevisiae*. Eur. Brew. Conv. Congr., Amsterdam, 267–283.

Russell, I., Stewart, G. G. (1979). Spheroplast fusion of brewer's yeast strains. J. Inst. Brew. 85:95–98.

Russell, I., Stewart, G. G., Reader, H. P., Johnston, J. R., Martin, P. A. (1980). Revised nomenclature of genes that control yeast flocculation. J. Inst. Brew. 86:120–121.

Scherer, S., Davis, R. W. (1979). Replacement of chromosome segments with altered DNA sequences constructed *in vitro*. Proc. Natl. Acad. Sci. USA 76:4951–4955.

Solingen, P. van, Plaat, J. B. van der (1977). Fusion of yeast spheroplasts. J. Bacteriol. 130:946–947.

Southern, E. M. (1975). Detection of specific sequences among DNA fragments separated by gel electrophoresis. J. Mol. Biol. 98:503–517.

Spencer, J. F. T., Spencer, D. M. (1977a). Hybridization of non-sporulating and weakly sporulating strains of brewer's and distiller's yeasts. J. Inst. Brew. 83:287–289.

Spencer, J. F. T., Spencer, D. M. (1977b). Apparent bisexual behavior of yeast strains obtained from hybridization of industrial yeasts of the genus *Saccharomyces* with auxotrophic diploids. Antonie van Leeuwenhoek 43:245–254.

Stewart, G. G., Russell, I. (1977). The identification, characterization and mapping of a gene for flocculation in *Saccharomyces* sp. Can. J. Microbiol. 23:441–447.

Stewart, G. G., Russell, I. (1979). Current use of the "new" genetics in research and development of brewer's yeast strains. Eur. Brew. Conv. Congr., Berlin, 475–490.

Stewart, G. G., Russell, I., Goring, T. (1975). Nature–nurture anomalies—further studies in yeast flocculation. Am. Soc. Brew. Chem. Proc. 33:137–147.

Struhl, K., Stinchcomb, D. T., Scherer, S., Davis, R. W. (1979). High-frequency transformation of yeast: Autonomous replication of hybrid DNA molecules. Proc. Natl. Acad. Sci. USA 76:1035–1039.

Tamaki, H. (1978). Genetic studies of ability to ferment starch in *Saccharomyces*: Gene polymorphism. Mol. Gen. Genet. 164:205–209.

Thorne, R. S. W. (1951). The genetics of flocculence in *Saccharomyces cerevisiae*. Compt. Rend. Lab. Carlsberg, Sér. Physiol. 25:101–140.

Tubb, R. S. (1979). Applying yeast genetics. A current assessment. J. Inst. Brew. 85:286–289.

Tubb, R. S. (1980). 2$\mu$m DNA plasmid in brewery yeasts. J. Inst. Brew. 86:78–80.

Wettstein, D. von (1979). Biochemical and molecular genetics in the improvement of malting barley and brewer's yeast. Eur. Brew. Conv. Congr., Berlin, 587–629.

Winge, Ø. (1944). On segregation and mutation in yeast. Compt. Rend. Lab. Carlsberg, Sér. Physiol. 24:79–96.

Winge, Ø., Laustsen, O. (1938). Artificial species hybridization in yeast. Compt. Rend. Lab. Carlsberg, Sér. Physiol. 22:235–247.

Winge, Ø., Laustsen, O. (1939). On 14 new yeast types, produced by hybridization. Compt. Rend. Lab. Carlsberg, Sér. Physiol. 22:337–355.

Wright, R. E., Lederberg, J. (1957). Extranuclear transmission in yeast heterokaryons. Proc. Natl. Acad. Sci. USA 43:919–923.

Young, T. W. (1981). The genetic manipulation of killer character into brewing yeast. J. Inst. Brew. 87:292–295.

Zakharov, I. A., Yarovoy, B. Ph. (1977). Cytoduction as a new tool in studying the cytoplasmic heredity in yeast. Mol. Cell. Biochem. 14:15–18.

# 14

# Genetic Improvement of Wine Yeast

Richard Snow*

## Introduction

Until recently, most of the attention directed toward wine improvement has been concentrated upon improvement of grape varieties and their culture and on fermentation and wine-making practices. Relatively little has been paid to improvement of the other major organism involved in wine production, the wine yeast. This is now changing; papers reporting on genetic studies of wine yeasts are appearing with increasing frequency. For the wine maker it is extraordinarily fortunate that yeast has become one of the premier organisms for basic genetics research and is at the same time one of the leading industrial microorganisms. The wine maker can thus profit from both the impressive arsenal of genetic techniques and the extensive biochemical and molecular information on the yeast cell and also from decades of practical experience in handling it in large-scale industrial processes. A very comprehensive review of the genetics of all types of industrial yeasts has recently appeared (Johnston and Oberman 1979). Much of the information on brewing, baking, and distilling yeasts is of direct interest to the wine yeast geneticist. The ability to utilize both conventional mutagenesis, hybridization, and selection programs in addition to the newly developed techniques of molecular cloning and protoplast fusion give yeast an advantage shared by few other organisms and should make possible the design of strains that a few years ago would have been impossible.

*Department of Genetics, University of California, Davis, California 95616 U.S.A.

## Origin and Classification of Wine Yeasts

The original wine yeast strains were of course derived from the yeasts that occur naturally on the skins of grapes. No deliberate inoculation was made to start the fermentation. Under these conditions there appears to be a progression of yeast species involved in the fermentation, including species of *Saccharomyces, Kloeckera, Hanseniaspora,* and others (Kunkee and Amerine 1970). With the rise of the ethanol concentration to 3–4% most other species are inhibited and *Saccharomyces* then outgrows them and completes the fermentation. In some established vineyards of Europe, reliance is still had on the natural yeast flora present. However, many winemakers now use pure yeast cultures, following the lead of Hansen (1886, 1888), who introduced their use into brewery practice. As early as 1890 Müller-Thurgau sent out pure cultures for wine making from Geisenheim-am-Rhein, and many other cultures were later released from this and other laboratories, some of which are still preserved and in use today (Kunkee and Amerine 1970). In the past, extravagant claims have sometimes been made for the use of pure cultures, with the result that they have fallen into some disrepute when these claims could not be sustained (Rankine 1977).

Although one uses the term *pure culture,* it should be emphasized that this does not mean the culture is genetically uniform. The cultures are pure in the sense that they have been derived ultimately from a single cell. However, they have usually been propagated for years by mass transfer on agar slants and have had ample time to produce many mutations, a number of which may be expressed (even though they are recessive) through mitotic crossing-over or gene conversion. An example of the heterogeneity of a "pure" culture is given by Zimmermann (1978), who was able to isolate a strain with considerably improved characteristics from successive single-cell cultures of an Epernay yeast. No mutagens were employed, with selection simply being made of the best of a series of tested cultures. The genetic heterozygosity of pure cultures becomes very obvious when they are sporulated and one examines the ascosporal clones. The common experience of all who have done this is that sporulation is often poor, spore viability is generally quite poor, and there is considerable variation in growth rate between spore clones (Thornton and Eschenbruch 1976; Cummings and Fogel 1978; Snow 1979), although wine yeasts appear to be better in these respects than brewing yeasts (Johnston 1965; Anderson and Martin 1975). Some of this variation is undoubtedly caused by the segregation of aneuploid chromosome complements from an aneuploid or polyploid parental strain; the rest must reflect segregation of lethals or genes affecting growth in some manner.

Wine yeasts of the genus *Saccharomyces* have been referred taxonomically to at least 29 different species or varieties (Kunkee and Goswell 1977; Lodder 1970). Most of these species are separated primarily on the basis of their sugar fermentation and assimilation patterns. In many cases mutation in a single gene can cause a strain to lose the capacity to ferment or utilize a sugar. Few would consider that such a mutant strain had suddenly been transformed into a new species. A more fundamental means of classification is based on DNA base sequence homology. Vaughan and Martini (1980) have reported that *Sacch. cerevisiae, bayanus, chevalieri, italicus,* and *uvarum* have a DNA GC content ranging from 38.5 to 39.5% and hold more than 90% of their base sequences in common. They conclude that it is best to consider all these strains as belonging to one species, *Sacch. cerevisiae* having taxonomic priority.

## Life Cycle and Ploidy Level

Many strains of *Sacch. cerevisiae* are homothallic diploids. When sporulated, such strains produce ascospores, all of which grow into clones of cells that can themselves be made to sporulate without deliberate mating. This behavior is caused by the allele *HO* (Harashima, Nogi, and Oshima 1974; Hicks and Herskowitz 1976), such strains being genotypically *HO/HO*. The surveys that have been made of wine yeasts indicate that this situation is probably typical of most of them (Naumov *et al.* 1974; Thornton and Eschenbruch 1976; Cummings and Fogel 1978; Takahashi 1978; Snow 1979; Kusewicz and Johnston 1980). Because of poor spore viability it has in many cases not been possible to isolate complete tetrads, and even when complete tetrads can be obtained, some spore clones may not sporulate because of segregation for respiratory deficiency or for other genetic causes, so that it is impossible to say confidently whether the original strain is genotypically *HO/HO* or *HO/ho*. In some cases spore viability and growth were good enough to conclude that the strain was *HO/ho* (Snow 1979). Taken as a whole, however, the information we have indicates that the majority of wine yeasts are genotypically *HO/HO*.

The problem of determining the chromosome number of yeast strains is a difficult one, since direct chromosome counts are not possible. Where strains have been crossed to laboratory haploids, segregations of genetic markers can provide considerable evidence. Thus Cummings and Fogel (1978) were able to show that two wine yeasts were almost certainly normal diploids, since matings of ascospores with cells of laboratory strains gave regular 2:2 segregations for markers on 13 of the 16 known yeast chromosomes. Furthermore, this regularity of segregation shows that there are no gross chromosomal differences between the strains. One of the strains studied by Thornton and Eschenbruch (1976) was also probably diploid, as it gave mostly 2:2 segregations for markers on six different chromosomes. On the other hand, Takahashi (1978), in a study of a widely used commercial German wine yeast (Hefix 1000), concluded that it was about, if not exactly, tetraploid and had an $a/a/a/\alpha$ mating locus genotype.

A segregation analysis coupled with DNA determinations has been used by Sakai and Takahashi (1972) to estimate ploidies of brewing yeasts. They showed that estimates based on cell size, quite a popular method in the past, gave unreliable results and are probably of value only when used with artificially synthesized, closely related series of strains. Given the rather large number of chromosomes in the haploid genome of yeast, its tolerant attitude toward aneuploidy, and the normal variance encountered in DNA determinations, it seems that the most reliable way to determine the exact chromosome number would be by genetic analysis of segregation of centromere markers located on each of the chromosomes.

## Wine Yeast Improvement

There are four methods that may be used in yeast improvement programs: (1) clonal selection without mutagenesis; (2) clonal selection after mutagenesis; (3) hybridization between strains or species, use being made of the primary hybrid product or of later derivatives of it (included in this category would be protoplast fusion techniques); (4) transformation of small amounts of foreign DNA into a strain (included here are the recombinant DNA techniques). All these methods are now being used with wine yeasts; examples of them will be given subsequently.

### Clonal Selection of Variants

This method depends on the genetic variation normally present in all wine yeast strains. An example of its successful use has already been cited (Zimmermann 1978). It has also been used to select nonfoaming mutants (Ouchi and Akiyama 1971; Eschenbruch and Rassell 1975). It requires the testing of large numbers of clones derived from single cells of the parental strain, saving those that meet predetermined standards. Dramatic improvements for most characteristics would not be expected; nevertheless it should often be possible to make significant gains in many areas, especially if analysis of variance tests is used (Azevedo *et al.* 1978). This method might be of most use in cases where a strain had degenerated to some extent through the accumulation of undesirable mutations or somatic recombinants.

### Mutagenesis Followed by Clonal Selection

This method has been widely used in the development of improved strains of many industrial microorganisms (Elender and Chang 1979). There are a number of reports describing its use with wine yeasts (Alikhanyan and Nalbandyan 1971; Alikhanyan *et al.* 1971a,b; Kusewicz 1976; Kishkovskaya and Bur'yan 1980; Kusewicz and Oberman 1980; Rous 1981). It has to some extent been employed with brewing yeasts (Johnston and Oberman 1979, and references therein). The mutagens used have been radiations (ultraviolet light or x-rays) or chemicals (usually ethylmethane sulfonate, nitrosoguanidine, or diethyl sulfate), administered according to one of the standard procedures (for instance, Fink 1970).

The heavy mutagen treatment frequently used could often be expected to produce mutations in addition to the one of interest, a possible disadvantage. Since many of the desirable mutations are recessive, their expression in diploid or polyploid wine strains has resulted from homozygosity brought about by gene conversion or mitotic crossing-over.

## Hybridization

Hybridization has not yet been widely employed with wine yeasts. There are, however, some reports of its use (Mosiashvili and Shalutashvili 1971; Ouchi and Akiyama 1976; Thornton and Eschenbruch 1978; Ouchi *et al.* 1979; Hara *et al.* 1980). It has been quite widely used for a number of years in the brewing industry (references in Johnston and Oberman 1979). Recently, notable success has been obtained with brewing strains using spheroplast fusion techniques (Russell and Stewart 1979; Barney *et al.* 1980a; Skatrud *et al.* 1980). Fusion between spheroplasts of *Sacch, cerevisiae* and *Saccharomycopsis lipolytica* has also been accomplished (de van Broock *et al.* 1980).

Needless to say, hybridization programs are not helped by the homothallism of wine strains and by their frequently poor sporulation and spore viability. The respiratory-deficient–auxotrophic-selection technique of Gunge and Nakatomi (1972) should be very useful and has been used to overcome such problems with nonsporulating and weakly sporulating strains of brewer's and distiller's yeasts (Spencer and Spencer 1977).

## Transformation

The gene-cloning–recombinant-DNA methodologies offer exciting prospects for improving wine yeasts. The demonstration of yeast transformation by Hinnen *et al.* (1978) and the development of plasmids that will replicate in either yeast or *Escherichia coli* (Botstein *et al.* 1979) are exceedingly important advances in yeast recombinant DNA methodology and have made yeast the organism of choice for many purposes. Recently, two reports have appeared describing transformation using purified, native DNA, without a plasmid vector (Barney *et al.* 1980b; Russell and Stewart 1980). This technique will further extend the usefulnes of yeast.

Plasmids capable of replicating in yeast use sequences from the 2-$\mu$ plasmid widely found in yeast strains. This plasmid has been found in 10 brewing and two nonbrewing strains (Tubb 1980). As far as I am aware, it has not been reported from wine yeasts, but is to be expected.

Transformation has not yet been used in wine yeast betterment, but it has been successfully applied to brewing yeasts. In one case, transformation of

*Sacch. cerevisiae* with purified DNA from *Sacch. diastaticus* resulted in clones of the former species that could ferment dextrins (Barney *et al.* 1980a). In a second case, a haploid yeast strain was transformed with DNA from a ploy-ploid ale yeast and clones that could ferment maltotriose were produced (Russell and Stewart 1980).

There are two major advantages offered by transformation techniques as far as wine yeast improvement is concerned. First, it should be possible to introduce literally any desired gene into yeast if some means for signaling successful transformation can be devised and there is adaquate expression of the gene in its new host to make it of practical value. Second, wholesale disruption of the wine yeast genotype, such as would occur with selection of recombinant derivatives following hybridization, will be avoided. Hence, other desirable characteristics of the wine yeast strain are not likely to be lost on acquiring a new biochemical function after transformation.

## Genetics and Examples of Strain Improvement

Although wine yeasts have received genetic attention only relatively recently, the genetic basis for several characteristics of enological importance has been determined, and there are already several examples of successful attempts at improvement.

### Nonfoaming Mutants

The production of a froth-head during fermentation is an undesirable trait of some wine yeasts, as up to 5% of the capacity of the fermentation vessel may have to be reserved to prevent the froth from spilling out. In the case of sake fermentations, where the mash consists of steamed rice, koji (a culture of *Aspergillus oryzae* on steamed rice), and water, the situation can be much worse as the froth-head can amount to 1–1.5 times the height of the mash itself. In addition to a significant reduction in the effective fermentation volume, the foam level must be controlled by a mechanical foam breaker or by protective walls. Selection of nonfoaming mutants of the widely used Kyokai No. 7 strain of sake yeast was achieved by Ouchi and Akiyama (1971) in either of two ways: by a cell agglutination or a froth flotation method.

The cell agglutination method is based on the observation that a normal sake yeast was agglutinated with mixed in acid solution with certain species of lactobacilli, whereas a nonfoaming strain was not (Momose *et al.* 1968). Ouchi and Akiyama were able to enrich for nonfoaming mutants by a series of repeated selection cycles. Washed cells from a culture of normal Kyokai No. 7 were mixed with cells of *L. plantarum* in a dilute citric acid solution. After mixing and allowing the cell aggregates to settle, an aliquot of the upper part of the suspension was used as the inoculum for another round of yeast growth.

The bacteria in the inoculum were killed by incorporation of antibiotics in the yeast culture medium. After the ninth selection step most of the clones sampled were nonfoaming. When cells of the parental strain were exposed to UV light (0.03% survival) before starting the selection cycles, 100% of the sampled clones at the seventh cycle were nonfoamers.

Selection by the froth flotation method depended on work of Akiyama and colleagues (Akiyama *et al.* 1971), who found that cells from the normal foaming strain adhered to the $CO_2$ gas bubbles formed during fermentation, while cells of a nonfoaming strain did not. Nonfoaming mutants were progressively enriched in repeated selection cycles by bubbling air through the culture. This effectively removed a large part of the foaming cells in the froth. Aliquots of the culture, now enriched for nonfoaming mutants, were used as the inoculum for the next round. About 50% of the tested clones were nonfoamers after nine selection cycles; when the series was started with UV-irradiated cells, about 80% of the sampled clones were nonfoamers after seven cycles. All mutants retained certain distinguishing characteristics of the parental strain, showing that they were not contaminants. Large-scale trials with two of the mutants proved their worth under the usual conditions of sake fermentations.

Using the froth flotation technique, Eschenbruch and Rassell (1975) were also able to select nonfoaming mutants from two strains of New Zealand wine yeasts.

In studies of the nature of the cell walls, Ouchi *et al.* (1973) found no marked differences between the mutants and the parental strain in total carbohydrates, nitrogen, ash, glucan, mannan, glucosamine, phosphorus, and amino acid composition. However, the mutants do have fewer hydrophobic groups on their surface, probably because of masking by phosphomannan (Ouchi and Nunokawa 1973). This difference appears to be the basis for differences in agglutinability by *Lactobacillus* species.

The genetic basis for foaming has been investigated by Kasahara *et al.* (1974) and by Thornton (1978a,b). The former authors found that the nonfoaming sake mutations were recessive and that the foaming character was under the control of at least two genes, as indicated by the 4:0, 3:1, and 2:2 segregations they observed in tetrads. Using grape wine yeasts, Thornton also showed that two dominant genes control foaming, that they are allelic to the sake yeast genes, and that they are linked on chromosome VII, 21 cM from one another and near *ade3*.

## Sedimentation and Flocculation

Alikhanyan *et al.* (1971b) reported the selection of a strain of champagne yeast with improved sedimentation properties. It was obtained after screening 800 colonies that survived treatment with x-rays, diethyl sulfate, or both. The mutant strain formed a coarse-grained sediment that settled rapidly and completely and was easily transferred to the cork without clouding the wine during remuage. The parent strain formed a dusty, viscous, silty sediment that could only be removed from the bottom of the bottle by shaking, thus clouding the wine. No data on the inheritance of this characteristic were given.

In a survey of 70 strains of wine yeasts, Zambonelli *et al.* (1973) found variation in their growth in must or semisynthetic medium. Cells of some strains remained dispersed, whereas those of others were highly flocculent. Tetrad data were presented for one cross, which showed that the dispersed–nondispersed characteristic segregated 2:2 in agitated cultures but 1:3 in static cultures. Involvement of two genes is suggested.

There are four known flocculation genes in brewing yeasts (Russell *et al.* 1980; Lewis *et al.* 1976), the flocculation characteristic being semidominant. One or more of these may be involved in the work with wine yeasts cited earlier. It is also possible that the sedimenting characteristics could be due to clumpy growth (delayed separation of mother and daughter cells) rather than to true flocculence (attachment of cells to one another by ionic bridging, brought about by divalent ions, especially calcium). Genetic studies of clumpy growth with laboratory strains have been made by Gilmore and Murphy (1972) and Faris and Gilmore (1974), who described three recessive genes. Earlier studies on what is probably the same phenomenon were made by Pomper and Burkholder (1949), Roman *et al.* (1951), and Roman and Sands (1953).

## Genetic Control of "Flor" Formation

The formation of a yeast film on the surface of wines containing about 15% ethanol is a characteristic stage in the production of high-aldehyde, low-sugar "flor" sherries (Kunkee and Amerine 1970). The majority of cells in the flor are one or more species of *Saccharomyces,* but *Pichia* species may also be found. Santa Maria and Vidal (1973) and Santa Maria *et al.* (1973) found that for four species of flor yeasts, the film-forming characteristic segregated 2:2 in asci, indicating that flor formation is under the control of a single dominant Mendelian gene and that the gene segregates independently of the mating-type locus. An odd feature of the data was the finding that this group of *Saccharomyces* species showed 2:2 segregation for both the flor character and for homothallism-heterothallism, in contrast to most wine strains which produce only homothallic spores (and in that case would not be expected to segregate for the flor character). The authors point out that it would be easy to obtain diploids homozygous for the flor gene, which might have improved properties.

## Ethanol Tolerance

Alikhanyan *et al.* (1971a), using UV and diethyl sulfate as mutagens, were able to select several mutant strains of the film-forming yeast *Sacch. oviformis* that could grow well at an ethanol concentration of 17.5% from a parent strain that could tolerate no more than 14.4%. The mutants produced high amounts of aldehydes and acetals (important sherry flavor components) and caused no gross sensory degradation of the wine. No data were presented on the genetic basis for the increased ethanol tolerance.

In making sake, saccharification of steamed rice by a culture of *Aspergillus oryzae* proceeds simultaneously with fermentation by the yeast. Ethanol concentrations can reach 20–23%, considerably higher than the normal concentrations that can be attained by most wine yeasts. Hayashida *et al.* (1974) showed that the *Aspergillus* produces an unsaturated fatty-acid-containing proteolipid complex that gives the yeast this tolerance. Even top and bottom beer yeasts could produce ethanol concentrations greater than 20% when grown with it (Hayashida and Hongo 1976). Protoplasts prepared from cells grown in its presence are also alcohol endurable. The compound prevents leakage of proteases and other cell components in high ethanol concentrations. Ethanol-tolerant cells have a much higher amount of linoleic acid in their membranes than usual (Hayashida and Ohta 1978). This work suggests that mutants with altered amounts of fatty acids in the cell membrane could show differences in their ethanol tolerance. It might also be possible to clone the proteolipid genes from *Aspergillus* and introduce them into a wine yeast.

In a study of the effect of ethanol inhibition on yeast growth and fermentation, Brown *et al.* (1981) showed that the effect of this alcohol on growth has two components, a reduction in growth rate and cell killing. For a laboratory haploid the inhibition constant $(K_i)$ for killing (0.29 *M*) was greater than the $K_i$ for growth inhibition (0.44 *M*), and showed noncompetitive kinetics. These figures also apply to a highly ethanol-tolerant sake yeast. In the case of the sake yeast, however, the $K_i$ for ethanol inhibition of fermentation was 1.32 *M* but for the laboratory strain it was only 0.97 *M*. Thus the difference between the strains lies in the ability of the sake yeast to continue to ferment at ethanol concentrations that suppress fermentation of the laboratory strain. The authors point out that their findings have significant implications for efforts to select ethanol-tolerant mutants. Both strains were completely killed by concentrations of about 12%—an all-or-none effect. Mutants able to survive higher concentrations might be selected by simple plate tests. However, mutants with faster growth or fermentation rates are not likely to be obtained by such means, since these effects are graded. Chemostat selection should be the most efficient way to obtain this class of mutants and was successfully employed (Brown and Oliver 1980).

## Killer Factor

Killer strains of yeast were recognized by Bevan and Makower (1963). These strains, which harbor a piece of double-stranded RNA, kill sensitive strains without this RNA by excreting a protein toxin into the medium (Wood and Bevan 1968; Wickner 1979). Killer strains are widespread among yeasts used in wine and beer making (Naumov and Naumova 1973; Naumov 1974; Tyurina and Bur'yan 1975; Ouchi and Kawashima 1974; Kreil *et al.* 1975). In sake fermentations contamination by wild yeasts has been a serious problem because of a slow-down of fermentation and an inferior taste of the product. The situation is particularly bad if the contaminating wild yeast is a killer, since the most widely used sake yeast is sensitive. Ouchi and Akiyama (1976) and Hara *et al.* (1980) introduced the killer plasmid into sake and wine yeasts by crossing them with a wild killer strain isolated as a contaminant from a sake mash. Repeated backcrossing with selection at each generation for certain characteristics of the sake strain gave killer hybrids that produced sake or grape wine of comparable quality to the parent. The killer hybrids could exclude sensitive contaminants even if outnumbered 10:1. Furthermore, pseudo-film-forming spoilage yeasts were inhibited in wine made from grape must, while they were abundant in that made from the sensitive parental strain unless $SO_2$ was added after fermentation.

The backcrossing program of Ouchi and Akiyama was hampered by poor sporulation and spore viability. This was circumvented by Ouchi *et al.* (1979) through the use of the *kar1-1* mutation. Haploid strains carrying this mutation form zygotes normally, but nuclear fusion usually does not occur (Conde and Fink 1976). A haploid *kar 1-1* laboratory yeast with the killer plasmid was crossed with a haploid *KAR1-1* respiratory-deficient killer-sensitive derivative of the sake Kyokai No. 7 strain. From this cross a haploid could be selected that had the killer plasmid and mitochondria from the laboratory yeast but had the nucleus from the sake strain (hence was *KAR1-1*). The mitochondia were removed from this haploid by ethidium bromide treatment and the resulting killer *KAR1-1* haploid was mated to a *KAR1-1* respiratory-sufficient derivative of Kyokai No. 7 to produce a killer diploid that had the nuclear and mitochondrial genomes of Kyokai No. 7. This strain gave results in trail fermentations that were even better than the parental strain with regard to rapidity of fermentation, volatile flavor components, and acidity. These examples represent an ingenious use of phenomena heretofore restricted to the laboratory (killer, cytoduction) to produce a commercially valuable wine yeast strain.

## Sulfite and Sulfide Production

The formation of $SO_2$ and $H_2S$ by wine yeasts is of great practical importance. Sulfur dioxide is regularly added to disinfect fermentation equipment, to control organisms that would compete with the yeast during fermentation, and to prevent excessive oxidation of the wine. Health concerns have led to efforts to restrict its use as an additive; hence the production of $SO_2$ by the yeast itself has become a matter of some importance. While $SO_2$, when properly used, has beneficial effects, the same cannot be said of $H_2S$, which is one of the most undesirable of yeast metabolities affecting the smell and taste of wines.

The formation of these two substances during wine making has been well reviewed by Eschenbruch (1974). Both are complex processes not yet fully understood that are influenced by a great many factors. From our viewpoint of genetics, however, it is especially pertinent that wide variation has been found between yeast strains. Most *Sacch. cerevisiae* strains produce between 10 and 30 mg/liter of $SO_2$ when tested under comparable conditions, but some form as little as 10 mg while other form in excess of 100 mg/liter (Eschenbruch 1974; Eschenbruch and Bonish 1976b). Strain variation in $H_2S$ production has been revealed by surveys of Zambonelli (1964a,b), Rankine (1968), and Eschenbruch *et al.* (1978).

In addition to the yeast strain, sulfite formation is affected by the fermentation medium and by the concentration of its precursor, sulfate (Eschenbruch 1974), and sometimes by the initial pH (Eschenbruch and Bonish 1976a). The first enzyme in the conversion of intracellular sulfate to sulfite is ATP-sulfurylase (Dott *et al.* 1976). Heinzel and Trüper (1976) showed that in high- and low-sulfite-producing strains this enzyme had the same specific activity in crude extracts but differed in its substrate saturation concentration and regulation. It was saturated by an ATP concentration of 4 m*M* in the low strain but required 30 m*M* for saturation in the high strain—an unphysiological concentration. Furthermore, the enzyme from the low strain was strongly feedback inhibited by adenosine-5′-phosphosulfate (APS) and sulfide while the high strain showed only slight inhibition by APS and was not inhibited by sulfide at all. There are obviously two molecular forms of ATP-sulfurylase and it should be possible to cross the type from the low $SO_2$ strain into other wine yeasts. Enzymatic differences were found by Dott and Trüper (1976) in the kinetic properties of sulfite reductases in high- and low-sulfite-producing strains, and by Dott *et al.* (1977) in the active uptake of sulfate and its regulation by methionine, which also suggest that these differences could be taken advantage of genetically.

Sulfide can be formed from sulfate or sulfite, elemental sulfur applied to grapes as a fungicide or from cysteine (Eschenbruch 1974). In addition to the yeast strain, its formation can be indirectly influenced by the amount of yeast growth, pantothenate or pyridoxine deficiencies or excess levels of certain amino acids (which cause a methionine deficiency and consequently lead to higher levels of $H_2S$), metal ions (copper, manganese, zinc), and yeast cell autolysis. However, both Schütz and Kunkee (1977) and Eschenbruch *et al.* (1978) concluded that the important source of $H_2S$ in wine is from residual colloidal sulfur from fungal sprays. The reduction of elemental sulfur may occur both enzymatically or nonenzymatically by reaction with thiol groups. Given these various sources of sulfide and the many influences on its formation, one cannot expect to find a single gene mutation that would eliminate it. But since strains do vary in their capacity to produce it, there is considerable natural genetic heterogeneity that could be exploited. Deliberate introduction of mutations in certain enzymes of the sulfur, sulfur amino acids, and pantothenate and pyridoxine pathways might well allow stepwise progress to be made.

Most of the genetic work on $SO_2$ and $H_2S$ production has been performed by Zambonelli and his colleagues. In studies of color stabilization of white wines and of resistance to $SO_2$ (Zambonelli *et al.* 1972a,b), it was found that a high-$SO_2$-producing strain that was effective in preventing the browning of white wine typically produced 2:2 segregations when crossed with five low $SO_2$ strains. In each case the hybrid was a low producer, indicating that low production is dominant and is controlled by a single gene. There was an inverse correlation between $SO_2$ production in grape musts and production of $H_2S$ on the acid bismuth yeast (ABY) agar test medium of Nickerson (1953). In a later study (Zambonelli *et al.* 1976), monosporal descendants of the highest-$SO_2$-producing strains were examined. Tetrad data unfortunately could not be obtained in the earlier work with it, because of difficulties with spore viabilities of its hybrids. Segregation for $SO_2$ production was found in an examination of 45 spore clones. Twelve produced $SO_2$ only in synthetic medium but not in wine, 23 produced $SO_2$ in synthetic medium and in wine, while 10 clones produced $SO_2$ in all wines and synthetic media tested. However, although some descendant strains produced quite high levels of $SO_2$, this was not always sufficient to prevnet browning. The authors suggested that factors such as nitrogen metabolism may also be important.

Compared to other microorganisms present in fresh grape must, wine yeasts are quite highly resistant to $SO_2$. Zambonelli *et al.* (1972b) found great differences between strains for this characteristic. In a cross between a sensitive and a resistant strain, the hybrid was resistant, and resistance–sensitivity segregated 2:2 in the ascospore clones. A cross between two resistant strains gave only resistant progeny. A single dominant gene for resistance is obviously indicated.

In a series of five papers, Zambonelli (1964a,b; 1965a,b,c) has described his work on the variation and genetic control of $H_2S$ production. In screening 100 strains for $H_2S$ production from sulfate or sulfite he found four that produced none under all conditions tested; the rest formed varying amounts, from traces to over 200 $\mu g$/50 ml of his standard culture medium. When tested on the ABY medium mentioned earlier, the four $H_2S$ negative strains formed white streaks while the others gave streaks of varying shades of brown. When $H_2S$ positive strains were crossed with negative strains, the hybrid produced $H_2S$ and segregation in the spores was 2:2 as could easily be demonstrated by their color reaction on ABY agar. However, since the amount of $H_2S$ produced by the positive clones varied considerably, segregation of modifying genes is indicated. Modifiers are also strongly suggested by the results of crosses of strains from natural habitats with varying levels of sulfide production. These hybrids had quantitatively intermediate levels, but in one case that was examined in some detail the spore clones exhibited levels that were outside the range of the parents.

In two later papers Zambonelli *et al.* (1975) and Romano *et al.* (1976) reported genetic results with mutants of sulfite reductase. Prototrophic strains had been obtained without any *in vitro* activity for this enzyme. Since it is essential in the biosynthesis of the sulfur amino acids, the prototrophy of the reductase-negative strains was ascribed either to leakiness of the reductase mutation or to an alternate metabolic pathway. If leakiness were the case, then a metabolic block after the reducing step could result in methionine auxotrophy and at the same time lead to production of $H_2S$. Methionine auxotrophs were selected with phenylmercuric nitrate, which favors the survival of cells producing high levels of $H_2S$. The auxotrophs obtained proved to have a second mutation leading to a methionine requirement that was independent of the reductase mutation, and as a result of this new mutation the strains accumulated $H_2S$. The authors therefore concluded that the sulfite-reductase-negative strains carry a leaky mutation whose activity cannot be detected by *in vitro* assays but is nevertheless great enough to support growth of the cells in minimal medium. From a survey of several strains it was concluded that a number of factors influence the production of sulfite from sulfate: inhibition of sulfite reductase by endogenous factors, reduced function of the reductase caused by mutation, enzymatic blockage after the reductase step causing methionine auxotrophy, and the state of genetic heterozygosity of the cell.

In an abstract, Tauro and Rupela (1980) state that they have selected mutants which excrete less $H_2S$, and that these mutants produce wine with lowered amounts. No genetical data were given.

## Fusel Oil

The higher alcohols (fusel oil) are formed from intermediates in the branched chain amino acids pathway leading to leucine, isoleucine, and valine by transamination, decarboxylation, and reduction (Webb and Ingraham 1963). Although they have undesirable flavor and odor characteristics, they are usually present in wines below the flavor threshold, and may even in some cases contribute to wine quality (Kunkee and Amerine, 1970). Their reduction in wines that are to be distilled (say, for brandy production) could be of considerable importance, however, since they are concentrated by the distilling process. The use of auxotrophic mutants might offer a way to reduce the fusel oil level, and Ingraham and Guymon (1960) and Ingraham *et al.* (1961) were indeed able to produce unusually low levels of isobutanol and active and isoamyl alcohols in fermentations carried out with valine, isoleucine, and leucine mutants. These mutants would not be of much commercial value, however, as they grew and fermented much more slowly than good wine yeasts should.

Following up on this work, Rous *et al.* (1980) and Rous (1981) reported on studies with auxotrophs of the branched chain amino acid pathway obtained from a widely used wine yeast (Montrachet, UCD Enology 522). One leucine mutant reduced the isoamyl alcohol in the fermentation product by over 50%. The mutant fermented at the same rate as the nonmutant parent strain and could successfully compete with the natural yeast flora in unsterilized grape musts. Taste panel trials indicated no difference between wines produced with it and the parent strain. Enzymatic activity for isopropylmalate dehydratase is missing, as the mutant does not complement a *leu1* laboratory haploid (Snow, unpublished).

## Esters and Esterases

A wide variety of esters are produced in wine by the yeast and are important flavor components (Kunkee and Amerine 1970; Kunkee and Goswell 1977). Nykänen and Nykänen (1977) surveyed 52 strains of wine yeasts along with several strains of related species. As a whole, the wine yeasts produced significantly more isoamyl acetate and ethyl caproate than the other yeasts and very significiantly more ethyl caprylate and ethyl caprate. There was considerable variation in production of the various esters by the wine strains.

Wöhrmann and Lange (1980) identified four esterase loci in samples of wine yeasts from 40 localities in Europe. Two of the loci had both fast and slow alleles (as determined by their electrophoretic migration patterns). All strains had at least two active loci (*EST1* and *EST2*); some had all four. Beneficial flavor changes might be obtained by using yeast strains carrying esterase mutations that reduce, increase, or alter the balance of various esters. This seems like a promising line of future work—if it can be decided which esters ought to be altered!

## Malolactic Fermentation

The decarboxylation of malic to lactic acid by several species of bacteria is termed the malolactic fermentation. This process has three significant aspects for the winemaker: (1) reduction in acidity; (2) microbiological stability following growth of the bacteria; and (3) changes in wine flavor caused by products of the bacterial fermentation (Kunkee and Goswell 1977). The malolactic fermentation is desirable where the wine is too acid (as is often the case in cool grape-growing regions) or for purposes of microbiological stability when the acidity is in the proper range. The additional flavor complexity added by the process may be desirable in some cases but is not the primary reason for inducing the fermentation. The fermentation can be brought about by holding the wine under conditions that are favorable for the growth of the bacteria already present, or an inoculation with the appropriate bacterial species can be made (Kunkee 1974).

It would be much more convenient if yeast were able to carry out this process at the same time as the alcoholic fermentation. Toward this end, the gene coding for the malolactic enzyme from a *Lactobacillus* species has been successfully cloned and introduced into a haploid laboratory strain, where it is expressed (Williams 1982; Williams, Snow, and Kunkee, unpublished). Future work will be directed toward increasing the level of expression of the foreign DNA and assuring its genetic stability in the new host and of transferring the gene from the laboratory strain into a commercially useful wine strain.

## Strain Maintenance

As was pointed out earlier, a pure culture of yeast is really a collection of subclones with a varying and unknown number of genotypes. Many of these genotypes would have no discernible effect on the wine-making qualities of the yeast even if they could be identified and isolated in some way, whereas others would. In spite of this genetic heterogeneity, it is the common experience of wine makers (and brewers) that the fermentaiton properties of their strains remain quite constant. This is partly because many of the mutations that might strongly affect these properties are recessive and would not be expressed in the diploid or polyploid genome of the commercial yeast except through the relatively rare genetic events of mitotic crossing-over or gene conversion that would lead to their becoming homozygous. This constancy is probably also due in part to the fact that many mutations of an undesirable nature are under a selective disadvantage in the usual conditions of use so that their relative frequency in the culture as a whole remains low. As wine yeasts become better defined and improved genetically, improvements in storage methods will be necessary.

The standard storage method is on agar slants kept under cool temperatures or in the refrigerator, sometimes with mineral oil to retard drying. New slants are prepared by mass transfer every few months. This storage method does not prevent the accumulation of genetic heterogeneity in the strain. The ideal method would be one that allows no cell growth yet maintains cell viability at a high level for an indefinite time. For long-term storage it appears that the best method at present is liquid nitrogen (Russell and Stewart 1981). After two years, viability of several brewing and laboratory strains remained very high (over 90%), there was no change in several strain characteristics of importance to brewing, and no increase in mutation rate.

With genetically engineered strains, methods of monitoring purity will be very desirable. For strains carrying a plasmid, a marker could be used for assurance that it was still present in the strain. In other cases an innocuous mutation (such as a conditional lethal that would not be expressed under any normally occurring winemaking conditions) might be put into the strain during the breeding program. Restriction enzyme patterns could also be useful in this respect.

## Concluding Remarks

With increasingly sophisticated and controlled fermentation technologies wine makers will want strains more highly tailored for specific functions. Certainly it will not be possible to develop a single, best strain; rather the wine maker can look forward to having a number of strains with different characteristics to choose from, all of which will perform suitably under his specific conditions. The haphazard, trial-and-error methods of the past will be replaced increasingly by planned, scientific improvement programs.

To facilitate breeding programs it would be most desirable to produce yeast strains of known ploidy, preferably diploid, to convert them from the homothallic to the heterothallic mating system and to eliminate the assortment of lethals and other mutations that have a deleterious effect of spore viability and growth that seem to be common to all of them. Such strain reconstruction has been achieved with a brewing yeast (Gjermansen and Sigsgaard 1981).

In the design of yeast strains there are many improvements that are readily specified (lower $H_2S$ production or ability to ferment to higher ethanol levels, for instance). In the subtle areas of taste and aroma, however, it is often not yet obvious just what should be changed in the physiology of the yeast to produce better wine. These same problems confront the brewer also, as Stewart (1978) pointed out. He quotes Winge, the father of yeast genetics, who commented in 1938 that "undefined demands cannot be answered to order," and Lindegren, who commented in 1956 that "if the brewer can prepare the specifications, the geneticist can tell him what characters are hereditary and what kinds of yeast he can produce." "Preparing the specifications" is certainly going to be a major step in any wine yeast improvement program.

## References

Akiyama, H., Kumagai, C., Tanaka, Y., Ouchi, K. (1971). Studies on non-foaming yeast. VII. Selection of the practical non-foaming yeasts for sake brewing from various stock cultures. J. Soc. Brew. Japan 66:516–522.

Alikhanyan, S. I., Nalbandyan, G. M. (1971). Selection of wine yeasts with the use of mutagens. I. Selection of *Saccharomyces cerevisiae* strains used in the production of natural strong table wines. Sov. Genet. 7:1200–1205.

Alikhanyan, S. I., Nalbandyan, G. M., Avakyan, B. P. (1971a). Selection of wine yeasts using mutagens. II. Selection of highly active, alcohol resistant strains of *Saccharomyces oviformis* for the production of Kheres wines. Sov. Genet. 7:1287–1293.

Alikhanyan, S. I., Nalbandyan, G. M., Avakyan, B. P. (1971b). Selection of wine yeasts using mutagens. III. Selection of new strains of *Saccharomyces vini* used in the production of champagne wines. Sov. Genet. 7:1452–1457.

Anderson E., Martin, P. A. (1975). The sporulation and mating of brewing yeasts. J. Inst. Brew. 81:242–247.

Azevedo, J. L., Tavares, F. C. A., Cruz, M. R. M. (1978). Genetic improvement of yeasts. In: Biochemistry and Genetics of Yeasts: Pure and Applied Aspects, edited by M. Bacila, B. L. Horecker, O. M. Andres. New York: Academic Press, pp. 563–575.

Barney, M. C., Jansen, G. P., Helbert, J. R. (1980a). Use of spheroplast fusion and genetic transformation to introduce dextrin utilization into *Saccharomyces uvarum*. J. Amer. Soc. Brew. Chem. 38:1–5.

Barney, M. C., Jansen, G. P., Helbert, J. R. (1980b). Use of genetic transformation for the introduction of flocculence into yeast. J. Amer. Soc. Brew. Chem. 38:71–74.

Bevan, E. A., Makower, M. (1963). The physical basis of the killer character in yeast. Proc. XIth Int. Cong. Genet. 1:202–203.

Botstein, D., Falco, S. C., Stewart, S. E., Brennan, M., Scherer, S., Stinchcomb, D. T., Struhl, K., Davis, R. W. (1979). Sterile host yeasts (SHY): A eukaryotic system of biological containment for recombinant DNA experiments. Gene 8:17–24.

de van Broock, M. R., Sierra, M. F., de Figeuroa, L. I. (1980). Intergeneric fusion of yeast protoplasts. VI Int. Ferm. Symp./V Int. Symp. Yeasts, London (Canada).

Brown, S. W., Oliver, S. G. (1980). Isolation of ethanol-tolerant mutants of yeast by continuous selection. Tenth Int. Conf. Yeast Genetics and Molecular Biology, Louvain-la-Neuve.

Brown, S. W., Oliver, S. G., Harrison, D. E. F., Righelato, R. C. (1981). Ethanol inhibition of yeast growth and fermentation: differences in the magnitude and complexity of the effect. Eur. J. Appl. Microbiol. Biotechnol. 11:151–155.

Conde, J., Fink, G. R. (1976). A mutant of *Saccharomyces cerevisiae* defective for nuclear fusion. Proc. Natl. Acad. Sci. USA 73:3651–3655.

Cummings, J., Fogel, S. (1978). Genetic homology of wine yeasts with *Saccharomyces cerevisiae*. J. Inst. Brew. 84:267–270.

Dott, W., Heinzel, M., Trüper, H. G. (1976). Sulfite formation by wine yeasts. I. Relationships between growth, fermentation, and sulfite formation. Arch. Microbiol. 107:289–292.

Dott, W., Heinzel, M., Trüper, H. E. (1977). Sulfite formation by wine yeasts. IV. Active uptake of sulfate by "low" and "high" sulfite producing wine yeasts. Arch. Microbiol. 112:283–285.

Dott, W., Trüper, H. G. (1976). Sulfite formation by wine yeasts. III. Properties of sulfite reductase. Arch. Microbiol. 108:99–104.

Elender, R. P., Chang, L. T. (1979). Microbial culture selection. In: Microbial Technology, vol. II, edited by H. J. Peppler, D. Perlman. New York: Academic Press.

Eschenbruch, R. (1974). Sulfite and sulfide formation during wine making—a review. Amer. J. Enol. Vitic. 25:157–161.

Eschenbruch, R., Bonish, P. (1976a). The influence of pH on sulphite formation by yeasts. Arch. Microbiol. 107:229–231.

Eschenbruch, R., Bonish, P. (1976b). Production of sulphite and sulphide by low- and high-sulphite forming wine yeasts. Arch. Microbiol. 107:299–302.

Eschenbruch, R., Bonish, P., Fisher, B. M. (1978). The production of $H_2S$ by pure culture wine yeasts. Vitis 17:67–74.

Eschenbruch, R., Rassell, J. M. (1975). The development of non-foaming yeast strains for wine-making. Vitis 14:43–47.

Faris, J. S., Gilmore, R. A. (1974). Genetics of flocculence in *Saccharomyces cerevisiae*. Genetics 77:s21–22.

Fink, G. R. (1970). The biochemical genetics of yeast. Meth. Enzymol. 17A:59–78.

Gilmore, R. A., Murphy, J. (1972). Flocculence in *Saccharomyces cerevisiae*. Genetics 71:s19–20.

Gjermansen, C., Sigsgaard, P. (1981). Construction of a hybrid strain of *Saccharomyces carlsbergensis* by mating of meiotic segregants. Carlsberg Res. Commun. 46:1–11.

Gunge, N., Nakatomi, Y. (1972). Genetic mechanism of rare matings of the yeast *Saccharomyces cerevisiae* heterozygous for mating type. Genetics 70:41–58.

Hansen, E. C. (1886). Undersogelser over alkoholsjaersvanpenes fysiologi og morfologi. V. Methoder til fremstilling af renkulturer af saccharomyceter og ligende mikroorganismer. Meddr. Carlsberg Lab. 2:152–167.

Hansen, E. C. (1888). Undersogelser fra gjaeringsindustriens praxis. Meddr. Carlsberg Lab. 2:257–322.

Hara, S., Iimura, Y., Otsuka, K. (1980). Breeding useful killer wine yeasts. Amer. J. Enol. Vitic. 31:28–33.

Harashima, S., Nogi, Y., Oshima, Y. 1974. The genetic system controlling homothallism in *Saccharomyces* yeasts. Genetics 77:639–650.

Hayashida, S., Feng, D. D., Hongo, M. (1974). Function of the high concentration alcohol-producing factor. Agric. Biol. Chem. 38:2001–2006.

Hayashida, S., Hongo, M. (1976). The mechanism of formation of high concentration alcohol in sake brewing. Fifth Int. Ferm. Symp., Berlin.

Hayashida, S., Ohta, K. (1978). Cell structure of yeasts grown anaerobically in *Aspergillus oryzae*-proteolipid-supplemented media. Agric. Biol. Chem. 42:1139–1145.

Heinzel, M., Trüper, H. G. (1976). Sulfite formation by wine yeasts. II. Properties of ATP-sulfurylase. Arch. Microbiol. 107:293–297.

Hicks, J., Herskowitz, I. (1976). Interconversion of mating types in yeast. I. Direct observations of the action of the homothallism *(HO)* gene. Genetics 83:245–258.

Hinnen, A., Hicks, J. B., Fink, G. R. (1978). Transformation of yeast. Proc. Natl. Acad. Sci. USA 75:1929–1933.

Ingraham, J. L., Guymon, J. F. (1960). The formation of higher aliphatic alcohols by mutant strains of *Saccharomyces cerevisiae*. Arch. Biochem. Biophys. 88:157–166.

Ingraham, J. L., Guymon, J. F., Crowell, E. A. (1961). The pathway of formation of *n*-butyl and *n*-amyl alcohols by a mutant strain of *Saccharomyces cerevisiae*. Arch. Biochem. Biophys. 95:169–175.

Johnston, J. R. (1965). Breeding yeasts for brewing. I. Isolation of breeding strains. J. Inst. Brew. 71:130–135.

Johnston, J. R., Oberman, H. (1979). Yeast genetics in industry. Prog. Ind. Microbiol. 15:151–205.

Kasahara, H., Ouchi, K., Kurata, K., Ishido, T., Nunokawa, Y. (1974). Genetic characters of non-foaming mutants of sake yeast. J. Ferment. Technol. 52:348–351.

Kishkovskaya, S. A., Bur'yan, N. I. (1980). Thermo-tolerant variants of wine yeasts. Microbiology 49:132–135.

Kreil, H., Kleber, W., Teuber, M. (1975). Wirkung von Killerfaktoren auf Bierhefen. Proc. 15th Eur. Brew. Conf., pp. 323–331.

Kunkee, R. E. (1974). Chemistry of Winemaking, edited by A. D. Webb. Advances in Chemistry Series, No. 137. American Chemical Society, Washington.

Kunkee, R. E., Amerine, M. A. (1970). Yeasts in wine-making. The Yeasts, vol. III, edited by J. S. Harrison and A. H. Rose. New York: Academic Press, pp. 5–71.

Kunkee, R. E., Goswell R. W. (1977). Table wines. Economic Microbiology, edited by A. H. Rose. New York: Academic Press, pp. 315–386.

Kusewicz, D. (1976). The characteristics of cryophilic wine yeast. II. The production and characteristics of cryophilic UV irradiated wine yeast. Acta Aliment. Pol. 2:61–72.

Kusewicz, D., Johnston, J. (1980). Genetic analysis of cryophilic and mesophilic wine yeasts. J. Inst. Brew. 86:25–27.

Kusewicz, D. Oberman, H. (1980). Improvement of technological properties of cryophilic wine yeast. VI Int. Ferm. Symp./V Int. Symp. Yeast, London (Canada).

Lewis, C. W., Johnston, J. R., Martin, P. A. (1976). The genetics of yeast flocculation. J. Inst. Brew. 82:158–160.

Lodder, J. (1970). The Yeasts. Amsterdam: North-Holland.

Momose, H., Okazaki, N., Tonoike, R. (1968). Studies on the aggregation of yeasts caused by lactic acid bacteria. I. Aggregation of yeast in mixture of pure-cultured yeast cells and lactic acid bacterial cells. J. Soc. Brew. Japan. 63:686–688.

Mosiashvili, G. I., Shalutashvili, J. D. (1971). (Yeast hybrids for use in winemaking.) Vinodel. i Vinograd. 31:19–20. (In Russian.)

Naumov, G. I. (1974). Comparative genetics of yeast. 14. Analysis of Saccharomyces wine strains neutral to K-2 killer. Sov. Genet. 10:100–105.

Naumov, G. I., Kondrat'eva, V. I., Tolstorukov, I. I. (1974). 15. Determination of the character delayed self-diploidization in the yeast *Saccharomyces bayanus* strain M-180. Sov. Genet. 10:1167–1171.

Naumov, G. I., Naumova, T. I. (1973). Comparative genetics of yeast. 13. Comparative study of killer strains of *Saccharomyces* from different collections. Sov. Genet. 9:1459–1462.

Nickerson, W. J. (1953). Reduction of inorganic substances by yeast. I. Extracellular reduction of sulfite by species of *Candida*. J. Infectious Diseases 93:43–48.

Nykänen, L., Nykänen, I. (1977). Production of esters by different yeast strains in sugar fermentations. J. Inst. Brew. 83:30–31.

Ouchi, K., Akiyama, H. (1971). Nonfoaming mutants of sake yeasts. Selection by cell agglutination method and by froth flotation method. Agric. Biol. Chem. 35:1024–1032.

Ouchi, K., Akiyama, H. (1976). Breeding of useful killer sake yeasts by repeated back-crossing. J. Ferment. Technol. 54:615–623.

Ouchi, K., Kawashima, H. (1974). (Distribution of killer and neutral strains of sake yeast in the type cultures.) J. Soc. Brew. Japan 69:629–630. (In Japanese.)

Ouchi, K., Nunokawa, Y. (1973). Nonfoaming mutants of sake yeast. Their physicochemical characteristics. J. Ferment. Technol. 51:85–95.

Ouchi, K., Takahashi, K., Suzuki, S., Nunokawa, Y. (1973). Nonfoaming mutants of sake yeast. Comparison of the cell wall composition between the parent and mutants. J. Gen. Appl. Microbiol. 19:429–437.

Ouchi, K., Wickner, R. B., Toh-e, A., Akiyama, H. (1979). Breeding of killer yeasts for sake brewing by cytoduction. J. Ferment. Technol. 57:483–487.

Pomper, S., Burkholder, P. R. (1949). Studies on the biochemical genetics of yeast. Proc. Natl. Acad. Sci. USA 35:456–464.

Rankine, B. C. (1968). The importance of yeasts in determining the composition and quality of wine. Vitis 7:22–49.

Rankine, B. C., (1977). Modern developments in selection and use of pure yeast cultures for winemaking. Aust. Wine, Brewing, and Spirit Rev. 96(8):31–33; 96(9):32–34.

Roman, H., Hawthorne, D. C., Douglas, H. C. (1951). Polyploidy in yeast and its bearing on the occurrence of irregular genetic ratios. Proc. Natl. Acad. Sci. USA 37:79–84.

Roman, H., Sands, S. M. (1953). Heterogeneity of clones of *Saccharomyces* derived from haploid ascospores. Proc. Natl. Acad. Sci. USA 39:171–179.

Romano, P., Zambonelli, C., Soli, M. G. (1976). Biosynthesis of sulphur amino acids in *Saccharomyces cerevisiae*. II. Analysis of sulphur-dioxide-producing strains. Arch. Microbiol. 108:211–215.

Rous, C. V. (1981). Reduction of wine fusel oil by fermentation with a yeast leucine auxotroph. M.S. Thesis, University of California, Davis.

Rous, C. V., Kunkee, R. E., Snow, R. (1980). Wine fermentations with amino acid-less mutant strains of Montrachet resulting in lowered production of higher alcohols. Thirty-first Annual Meeting of the American Society of Enologists, Los Angeles.

Russell, I., Stewart, G. G. (1979). Spheroplast fusion of brewer's yeast strains. J. Inst. Brew. 85:95–98.

Russell, I., Stewart, G. G. (1980). Transformation of maltotriose uptake ability into a haploid strain of *Saccharomyces* spp. J. Inst. Brew. 86:55–59.

Russell, I., Stewart, G. G. (1981). Liquid nitrogen storage of yeast cultures compared to more traditional storage methods. J. Amer. Soc. Brew. Chem. 39:19–24.

Russell, I., Stewart, G. G., Reader, H. P., Johnston, J. R., Martin, P. A. (1980). Revised nomenclature of genes that control yeast flocculation. J. Inst. Brew. 86:120–121.

Sakai, K., Takahashi, T. (1972). Estimation of ploidies in brewery yeasts. Bull. Brew. Sci. 18:29–36.

Santa Maria, J., Vidal, D. (1973). Genetic control of flor formation by *Saccharomyces*. J. Bacteriol. 113:1078–1080.

Santa Maria, J., Vidal, D., Rodriguez-Marin, M. A. (1973). Análisis genetico de la formación de velo por especies del género *Saccharomyces*. An. Inst. Nac. Invest. Agrar. (Ser. Gen.) 2:11–20.

Schütz, M., Kunkee, R. E. (1977). Formation of hydrogen sulfide from elemental sulfur during fermentation by wine yeast. Amer. J. Enol. Vitic. 28:137–144.

Skatrud, P. L., Jaeck, D. M., Kot, E. J., Helbert, J. R. (1980). Fusion of *Saccharomyces uvarum* with *Saccharomyces cerevisiae:* genetic manipulation and reconstruction of a brewer's yeast. J. Amer. Soc. Brew. Chem. 38:49–53.

Snow, R. (1979). Toward genetic improvement of wine yeast. Amer. J. Enol. Vitic. 30:33–37.

Spencer, J. T. F., Spencer, D. M. (1977). Hybridization of non-sporulating and weakly sporulating strains of brewer's and distiller's yeasts. J. Inst. Brew. 83:287–289.

Stewart, G. G. (1978). Application of yeast genetics within the brewing industry. A review. J. Amer. Soc. Brew. Chem. 36:175–185.

Takahashi, T. (1978). Genetic analysis of a German wine yeast. Bull. Brew. Sci. 24:39–47.

Tauro, P., Rupela, O. P. (1980). Hydrogen sulfide excretion in yeast and the isolation of low $H_2S$ excreting wine yeasts. VI Int. Ferm. Symp./V Int. Symp. Yeast, London (Canada).

Thornton, R. J. (1978a). The mapping of two dominant genes for foaming in wine yeasts. FEMS Lett. 4:207–209.

Thornton, R. J. (1978b). Investigation on the genetics of foaming in wine yeasts. Eur. J. Appl. Microbiol. Biotechnol. 5:103–107.

Thornton, R. J., Eschenbruch, R. (1976). Homothallism in wine yeasts. Antonie van Leeuwenhoek 42:503–509.

Thornton, R. J., Eschenbruch, R. (1978). The improvement of pure-culture wine yeasts by hybridization. Sixth Int. Special. Symp. Yeasts, Montpellier.

Tubb, R. S. (1980). 2-micron DNA plasmid in brewery yeasts. J. Inst. Brew. 86:78–80.

Tyurina, L. V., Bur'yan, N. I. (1975). Phenotypes (killer, neutral, sensitive) of yeasts of the genus *Saccharomyces* in viticulture and methods of their determination. Microbiology 44:357–361.

Vaughan, A. E., Martini, A. (1980). DNA base composition and DNA-DNA reassociation of the "wine" species of the yeast genus *Saccharomyces* Hansen. VI Int. Ferm. Symp./V Int. Symp. Yeast, London (Canada).

Webb, A. D., Ingraham, J. L. (1963). Fusel oil: a review. Adv. Appl. Microbiol. 5:317–353.

Wickner, R. B. (1979). The killer double-stranded RNA plasmids of yeast. Plasmid 2:303–322.

Williams, S. A. (1982). Cloning and expression of the malolactic gene of *Lactobacillus delbrueckii* in *E. coli* and yeast. Ph.D. Thesis, University of California, Davis.

Wöhrmann, K., Lange, P. (1980). The polymorphism of esterases in yeast *(Saccharomyces cerevisiae)*. J. Inst. Brew. 86:174.

Wood, D. R., Bevan, E. A. (1968). Studies on the nature of the killer factor produced by *Saccharomyces cerevisiae*. J. Gen. Microbiol. 51:115–126.

Zambonelli, C. (1964a). Ricerche biometriche sulla produzione di idrogeno solforato da solfati e solfiti in *Saccharomyces cerevisiae* var. *ellipsoideus*. Ann. Microbiol. 14:129–141.

Zambonelli, C. (1964b.) Ricerche genetiche sulla produzione di idrogeno solforato in *Saccharomyces cerevisiae* var. *ellipsoideus*. Ann. Microbiol. 14:143–153.

Zambonelli, C. (1965a). Richerche genetiche sulla produzione di idrogeno solforato in *Saccharomyces cerevisiae* var. *ellipsoideus*. II. Ereditarietà del carattere dal punto di vista quantitativo. Ann. Microbiol. 15:89–97.

Zambonelli, C. (1965b). Ricerche genetiche sulla produzione di idrogeno solforato in *Saccharomyces cerevisiae* var. *ellipsoideus*. III. Ulteriori studi sulla ereditarietà del carattere quantitativo. Ann. Microbiol. 15:99–106.

Zambonelli, C. (1965c). Stabilità ed ereditarietà delle variazione nella produzione de $H_2S$ indotte dal nitrato fenil-mercurico in *Saccharomyces cerevisiae* var. *ellipsoideus*. Ann. Microbiol. 15:181–195.

Zambonelli, C., Guerzone, M. E., Nanni, M., Gianstefani, G. (1972a). Selezione genetica nei lieviti della fermentazione vinaria. 2. Il potere stabilizzante del colore. Revista Vitic. Enol. 25:111–129.

Zambonelli, C., Guerzoni, M. E., Nanni, M. (1972b). Selezione genetica nei lieviti della fermentazione vinaria. 3. La resistenza alla anidride solforosa. Revista Vitic. Enol. 25:170–179.

Zambonelli, C., Guerzoni, M. E., Nanni, M. (1973). Selezione genetica nei lieviti della fermentazione vinaria. 5. La modalita di sviluppo. Revista Vitic. Enol. 26:104–112.

Zambonelli, C., Mutinelli, P., Pachetti, G. (1975). Biosynthesis of sulphur amino acids in *Saccharomyces cerevisiae*. I. Genetic analysis of leaky mutants of sulphite reductase. Arch. Microbiol. 102:247–251.

Zambonelli, C., Soli, M. G., Romano, P., Grazia, L. (1976). Influenza del ceppo di lievito sulla stabilità del colore dei vini bianchi. II. Analisi del ceppo 633. Ann. Microbiol. 26:109–118.

Zimmermann, F. K. (1978). On the use of systematically selected yeasts in large wineries. Proc. 5th Int. Oenol. Symp., Aukland.

# 15

# Aspects of the Biochemistry and Genetics of Sugar and Carbohydrate Uptake by Yeasts

G. G. Stewart* and I. Russell*

## Introduction

Microorganisms that metabolize glucose or compounds that can give rise to glucose as a source of energy in the absence of oxygen use a variety of different pathways to regenerate $NAD^+$. All of these pathways lead to the formation of so-called fermentation products, which include such acids as lactic, acetic, and propionic and such compounds as acetone, butanol, isopropanol, and acetylmethylcarbinol. They also include ethanol. Although many microorganisms produce ethanol as an end product of their energy-generating metabolism, they are a minority of the total microbial population. However, pride of place in this minority group must be given to species of the genus *Saccharomyces*, which have, for several thousands of years, been used to produce ethanol, both in a potable and nonpotable form, from sugars. The main deficiencies of this ethanol-producing microorganism are its inability to ferment pentose sugars, which can be made available in large quantity by hydrolyzing wood, and its lack of depolymerizing enzymes, which might yield mono- and disaccharides from cellulose. *S. cerevisiae* is also unable to hydrolyze starch and dextrin, but, as will be discussed later, there is a closely related species, *S. diastaticus*, that can utilize starch by virtue of the production of glucoamylase (E.C, 3.2.1.3), which will hydrolyze the $\alpha$-1,4 linkages of starch and dextrin to glucose.

A definition of the class of organisms collectively known as yeasts is not easily postulated. They are unicellular, eukaryotic microorganisms that typically reproduce by budding. This is true for the majority of yeasts, but there are exceptions; there are species with a propensity for mycelium formation, others with a fission type of cell division, and still another genus that bears conidia or

*Labatt Brewing Company Ltd., London, Ontario, Canada N6A 4M3.

sterigmata. Indeed, Lodder (1970) is of the view that the term "yeasts" developed historically and really includes a heterogeneous group of microorganisms. Phaff *et al.* (1978) have reviewed the etymology of "yeast" and equivalent terms in other languages and point out the consistent relationship to fermentation and an ability to metabolize a wide range of sugars. The number of defined yeast species is now in excess of 400. In addition, there are a number of fungi with "yeastlike" cells as part of their life cycle. At the other extreme, some biochemists incorrectly use the term "yeast" as synonymous with *S. cerevisiae.* Although most of the genetic and biochemical studies have been conducted on this species, there are a rich variety of yeast genera and species that await biological investigation; indeed the more exotic species may offer advantages for experimental study. In this chapter the authors have drawn heavily on data obtained with *S. cerevisiae* and related species, but wherever possible reference has been made to non-*Saccharomyces* species.

No other group of microorganisms has been more intimately associated with the progress and well-being of the human race than the yeasts. Their contribution to man's progress has been based very largely on the capacity of certain yeasts to effect a rapid and efficient conversion of sugars into ethanol and carbon dioxide and to conduct an alcoholic fermentation of sugary liquids such as grape juice, grain extracts, and milk. It has often been said that yeasts are the oldest of cultivated plants. Yeasts are still economically the most important microorganism exploited on this globe. It is the opinion of some biotechnologists that in the future the industrial importance of this group of microorganisms will diminish because they produce few secondary metabolites of commercial interest (e.g., they do not produce any biologically active antibiotics). However, many species of yeast produce a very important primary metabolite, namely, ethanol, used both as a beverage and for industrial/fuel purposes. A second major contribution that yeasts have made to man's progress has been their use in the elucidation of the basic biochemical, metabolic, and genetic processes of living cells (Rose 1977). The third important "contribution" that yeasts make is as pathogens of both plants and animals. In man, *Candida albicans,* although often present on the skin and mucosal surfaces without causing disease, is an important pathogen. Many other yeasts cause disease in man or other animals as secondary pathogens or in hosts that lack normal immunity. The conditions under which yeasts induce disease are as varied as the species. They include diabetes, leukemia, various types of cancer, many chronic pulmonary diseases, obesity, pregnancy, maceration of skin surfaces, and various conditions associated with old age (Odds 1979).

At present, only *S. cerevisiae* and related species are of major industrial importance. Yet some of the many yeast species could certainly assume a considerable role in industry, as a source of protein, for producing ethanol, and for carrying out other chemical transformations, such as those involved in synthesizing precursors of important natural products. The following research program would further such ends: (1) The carbohydrate utilization capabilities of the more versatile yeasts should be characterized genetically and biochemically. (2) Artificial means of genetically manipulating yeasts should be developed further in order to make a wider range of yeasts available. It may no longer be essential to choose yeasts having an amenable, natural system of genetic recombination; and the useful physiological characteristics found in various strains that have been little studied could be transferred, by gene cloning or fusion, to a commonly used but less versatile strain. (3) To assist this endeavor more must be found out about genetic and physiological differences between yeast genera and species.

*S. cerevisiae* has the ability to take up and ferment a wide range of sugars; for example, sucrose, glucose, fructose, galactose, maltose, and maltotriose. In addition, the closely related species *S. disataticus* and *S. uvarum (carlsbergensis)* are able to utilize dextrins and melibiose, respectively. The first step in the utilization of any sugar by yeast is usually either:

1. Its passage intact across the cell membrane.
2. Initial hydrolysis outside the membrane followed by entry into the cell by some or all of the hydrolysis products.

Unlike some bacteria, which ferment a wide range of organic compounds, yeasts (with a few notable exceptions) only ferment those metabolized via the Embden-Meyerhof and Entner-Doudoroff pathways. Furthermore, cytochrome-deficient mutants apart, a yeast that uses sugar anaerobically also uses it aerobically (Barnett 1976, 1981).

*S. cerevisiae* is constitutive for the ability to metabolize glucose, mannose, and fructose. With the exception of certain mutant yeast strains (which will be discussed in a later section), all yeasts are able to utilize these sugars. In *S. cerevisiae* a number of transport systems for sugars have been described: a constitutive system common for glucose, fructose, and mannose; inducible systems for galactose, α-methyl-D-glucoside, maltose, and maltotriose, respectively. The transport of hexoses by the constitutive and by the inducible galactose system is equilibrative and does not require energy. On the other hand, the α-methyl-D-glucoside, maltose, and maltotriose systems have been characterized as active transport. For example, maltose transport, although independent of the intracellular ATP level, is coupled to the electrochemical gradient of protons (Serrano 1977). Sucrose is hydrolyzed outside the cell membrane by the extracellular enzyme invertase (β-D-fructofuranoside fructohydrolase, E.C. 3.2.1.26) to glucose and fructose. Six genes *SUC1, 2, 3, 4, 5,* and *SUC80* code for the production of invertase, and a number of them have been mapped (Mortimer and Schild 1980). The disaccharide maltose and the trisaccharide maltotriose have independent uptake mechanisms (maltose or maltotriose permease), which transport the two sugars across the cell membrane into the cell. Once inside the cell, both sugars are hydrolyzed to glucose units by the α-glucosidase system (Hautera and Lovgren 1975; Stewart *et al.* 1979).

As previously stated, there is a species closely related to *S. cerevisiae* that is able to utilize dextrin material, dextrins being polymers of glucose, greater than three units of the monosaccharide. In 1952 Andrews and Gilliland (1952) isolated a strain of yeast capable of utilizing starch and dextrin by virtue of its ability to produce the extracellular enzyme glucoamylase or amyloglucosidase

($\alpha$-1,4-glucan glucohydrolase, E.C. 3.2.1.3). Because of its dextrinase activity, it was defined as a new species of yeast—*S. diastaticus*. The species *S. uvarum (carlsbergensis)* is also differentiated from *S. cerevisiae* on the basis of sugar fermentation characteristics, *S. uvarum (carlsbergensis)* being able to ferment the disaccharide melibiose by virtue of the production of the extracellular enzyme melibiase ($\alpha$-galactosidase).

The disaccharide lactose consists of glucose and galactose units. Lactose is found in milk, and whey, the by-product of cheese manufacture, contains it in high concentrations. There are two species of yeast, *Kluyveromyces fragilis* and *Kluyveromyces lactis,* that can utilize lactose. The uptake mechanism is very similar to that of maltose. The lactose is transported across the cell membrane by means of a lactose permease system; once inside the cell it is hydrolyzed by $\beta$-galactosidase ($\beta$-D-galactoside galactohydrolase, E.C. 3.2.1.23) into its constituent monosaccharides glucose and galactose, which then enter the common glycolytic pathways of the cell.

A yeast developing on a carbohydrate source must either be able to ferment it or to utilize it by respiration. It is customary in taxonomic work (Lodder 1970) to refer to the first process as "fermentation" and to the second as "assimilation." Although these terms are firmly established, they are somewhat misleading, since actual assimilation of the carbohydrate (i.e., uptake by the cell) occurs during both the fermentative and the oxidative dissimilatory processes.

The classical concept of the alcoholic fermentation by yeasts has been that of a process that takes place under strictly anaerobic conditions. The ability or inability to ferment carbohydrates to ethanol and carbon dioxide is a most useful characteristic for differentiating species. Although it finds less application in the definition of genera, it is nevertheless occasionally of some use. For example, the genus *Saccharomyces* is characterized by the vigorous fermentation of a number of sugars, whereas other genera (e.g., *Lipomyces* and *Rhodotorula*) are strictly nonfermentative, and in other genera (e.g., *Hansenula*) an entire range from nonfermentative to strongly fermentative species may be found.

## D-Galactose

The critical steps in the utilization of D-galactose by extracts of galactose-adapted yeast are described in the following reactions (Cardini and Leloir 1953). The first three are often referred to as the Leloir pathway.

$$\text{Galactose} + \text{ATP} \rightarrow \text{Gal-1-P} + \text{ADP} + \text{H}^+$$
galactokinase (E.C. 2.7.1.6)

$$\text{Gal-1-P} + \text{UDPG} \rightarrow \text{Gluc-1-P} + \text{UDP-galactose}$$
galactose-1-P uridylyltransferase (E.C. 2.7.7.10)

$$\text{UDP} - \text{galactose} \rightarrow \text{UDP-glucose}$$
UDP-D-galactose 4-epimerase (E.C. 5.1.3.2)

$$\text{Gluc-1-P} \rightarrow \text{Gluc-6-P}$$
phosphoglucomutase (E.C. 2.7.5.1)

The genetic system that encodes the enzymes responsible for the conversion of exogenous galactose to endogenous glucose-1-phosphate in *Saccharomyces* has been well studied. After transport across the cell membrane by the *GAL2*-encoded permease (Douglas and Hawthorne 1966), galactose is converted to glucose-1-phosphate by reactions catalyzed by the products of the *GAL1*, *GAL7*, and *GAL10* genes (i.e., galactokinase, galactose-1-P uridylyltransferase, and UDP-D-galactose 4-epimerase, respectively). These three genes are located in a tightly linked cluster, six map units from the centromere of chromosome II (Douglas 1961). The genetic order of these genes has been determined by Bassell and Douglas (1968) as centromere-*GAL7*-*GAL10*-*GAL1*.

The enzymes encoded by the galactose gene cluster are coordinately induced by the addition of galactose to appropriately adapted cultures (Hartwell 1970). Broach (1979) has shown that the galactose enzymes are translated as separate polypeptides. These data are consistent with those of Hopper *et al.* (1978), which showed that the *GAL1* and *GAL7* genes specify distinct and separate mRNAs and are the result of *de novo* synthesis after induction.

Expression of the *GAL7*-*GAL10*-*GAL1* gene cluster is known to be controlled primarily by two unlinked regulatory loci, *GAL80* and *GAL4* (Kew and Douglas 1976). These loci also affect the expression of the galactose permease and the extracellular enzyme α-galactosidase (St. John and Davis 1979). The *GAL80* locus behaves genetically as a negative regulator. Both dominant and uninducible superrepressor mutations (Klar and Halvorson 1974) and recessive constitutive mutations (Matsumoto *et al.* 1978) have been isolated at this locus. The *GAL4* locus is required for induction and expression of the galactose enzymes. It has been shown (Perlman and Hopper 1979) that the wild-type *GAL4* function is required for the expression of the messenger RNAs, encoding at least two of the galactose gene cluster enzymes—the kinase *(GAL1)* and the transferase *(GAL7)*. Dominant *GAL80* mutations result in the constitutive expression of all the galactose enzymes (Nogi *et al.* 1977). These mutations have been shown to drop within the *GAL4* structural gene (Matsumoto *et al.* 1980).

## Maltose and Maltotriose

The disaccharide maltose and the trisaccharide maltotriose pass intact across the yeast plasma membrane. Both sugars are transported by inducible α-glucoside permeases and are hydrolyzed by an α-glucosidase (maltase α-D-glucoside glucohydrolase, E.C. 3.2.1.20) that is also inducible. The transport, hydrolysis, and fermentation of maltose are particularly important in brewing and baker's yeast strains, since maltose is the main sugar component of both brewing wort and wheat dough. There are (at least) five unlinked polymeric genes that control the ability of yeast to produce α-glucoside permease in response to maltose—*MAL1*, *MAL2*, *MAL3*, *MAL4*, and *MAL6*. Strains carrying an active allele at any one of these loci are inducible; strains carrying inactive alleles at all loci are uninducible but produce basal levels of α-glucosidase (Khan *et al.* 1973). The role of *MAL* genes is still not fully understood. It has been suggested that the *MAL* loci are either (1) structural genes for α-glucosidase, (2) regulatory genes controlling both α-glucosidase and α-glucoside permease, or (3) complex loci containing both regulatory and structural elements. The role of the *MAL* genes has been investigated genetically in two ways, either by analyzing mutations at a single *MAL* locus (ten Berge *et al.* 1973a) or by comparing the effects of the different *MAL* loci (Tauro and Halvorson 1966). As an example of the former approach, ten Berge and his associates have analyzed the effects of mutations in the *MAL6* locus. They have found that both noninducible and constitutive mutations map at the *MAL6* locus (ten Berge *et al.* 1973b) and that two strains that are temperature sensitive for α-glucosidase induction contain α-glucosidase indistinguishable from that of the wild type.

These and similar results with other *MAL* loci (Khan *et al.* 1973) are consistent with the proposal that the *MAL* genes are regulatory loci that code for a protein required for α-glucosidase induction (Zimmerman and Eaton 1974). These results do not rule out the possibility (discussed later) that the *MAL* genes are complex loci that also include structural genes (Naumov 1971) for the enzymes of maltose metabolism.

Halvorson and co-workers (Rudert and Halvorson 1963; Tauro and Halvorson 1966) have compared the effects of several different *MAL* loci on the synthesis of α-glucosidase and have found that strains carrying different active *MAL* loci produce different amounts of the enzyme. Mowshowitz (1979) has reexamined the role of the *MAL* loci and the influence of *MAL* gene dosage because (1) it is generally assumed that regulatory genes do not exhibit gene dosage effect because their products are present in excess, (2) structural genes for inducible and repressible enzymes do not necessarily show strictly additive gene dosage effects, and (3) it has not been demonstrated that the differences in α-glucosidase levels are actually due to differences in the *MAL* gene dosage. Inbred strains of *S. cerevisiae* carrying *MAL1*, *MAL2*, or *MAL6* in a common background were used to construct (1) homozygous or heterozygous diploids carrying one or two active alleles of a single *MAL* locus (*MAL1*, *MAL2*, or *MAL6*) and (2) triploids carrying one, two, or three active alleles of *MAL2*. The diploid and triploid strains were employed to investigate gene dosage effects on the different rates of α-glucosidase synthesis. All three *MAL* loci exhibited a gene dosage effect on the rate of α-glucosidase synthesis; *MAL2* also exhibited a gene dosage effect on the enzyme kinetics of induction. The dosage effects of the *MAL1* and *MAL6* loci were additive, but the effects of the *MAL2* locus were not of the same magnitudes, the gene dosage effect decreasing with increasing gene dosage. In a further study Mowshowitz (1981) has shown that the alleles *MAL1*, *MAL2*, and *MAL6* have different effects on the regulation of α-glucosidase accumulation, and it has been documented that the *MAL* loci themselves regulate α-glucosidase accumulation. However, it has not been demonstrated whether all *MAL1*, *MAL2*, or *MAL6* alleles are identical, so that it is possible that properties ascribed to each *MAL* locus are only attributable to a particular allele of that locus. Therefore it is possible that there is as much variation between the different alleles of any one *MAL* locus as that observed between the alleles of three different *MAL* loci.

Fermentation studies with the *MAL2* triploids (Stewart 1981) have been conducted in brewer's wort. In 12° Plato wort the strains containing the *MAL2/MAL2/mal* and *MAL2/MAL2/MAL2* genotype fermented at a much faster rate than a production ale yeast strain *(S. cerevisiae)*. In 16° Plato wort the gene dosage effect was more pronounced, and again a faster rate of fermentation with the multiple *MAL2* strains was seen, this time compared with a production lager brewing strain [*S. uvarum (carlsbergensis)*]. It now appears that the five *MAL* loci (i.e., *MAL1, MAL2, MAL3, MAL4,* and *MAL6*) contain regulatory genes controlling the induction of the maltose fermentation enzymes. It has also been proposed that the *MAL* loci include the structural genes for α-glucosidase and maltose permease. Naumov (1971, 1976) has analyzed the progeny of crosses between various naturally occurring maltose-negative and maltose-positive isolates of *Saccharomyces* and concluded that maltose fermentation requires the presence of two genes, termed *MALp* and *MALg,* and that each *MAL* locus is composed of these two linked genes. A maltose-negative strain could arise if it lacked either a functional *MALp* or a functional *MALg* gene. A strain carrying *MALg* and *MALp,* either linked or at unlinked loci, allows the cell to utilize maltose as a carbon source. Federoff *et al.* (1982) have constructed a clone bank from a donor yeast strain that carries the *MALp MALg* locus. The cloned yeast DNA was then amplified in *E. coli* K-12 strain RR101. The recombinant plasmid DNA isolated from pools of 500 *E. coli* (pYT) transformants was used to transform a *Mal*$^-$ yeast strain. *Mal*$^+$ colonies were identified as transformants. Most of the *Mal*$^+$ colonies were found to be highly unstable during mitotic growth on nonselective media as they segregated mostly *Mal*$^-$ clones. DNA prepared from several *Mal*$^+$ yeast transformants grown under selective conditions were transferred to *E. coli* and the plasmid pMAL26 identified by this approach. This plasmid was shown by positive hybridization translation, as well as Southern and Northern blot experiments, to carry a *MAL6* structural gene for α-glucosidase.

## Pentoses

Because of concern about the long-term cost and availability of petroleum and natural gas, interest is increasing in processes that make use of renewable resources, such as plant biomass, for the generation of liquid fuels and chemical feedstocks. The energy content of biomass can be utilized most simply by direct burning. It is also possible to convert biomass to higher-value gaseous and liquid fuels and chemicals by thermochemical and biological means. The major organic components of land-plant biomass are cellulose, hemicellulose, and lignin. Cellulose is a high-molecular-weight, insoluble, linear β-1,4 linked polymer of D-glucose. Lignin is a high-molecular-weight, insoluble, three-dimensional random polymer of sinapyl, coniferyl, and *p*-coumaryl alcohols. The hemicelluloses are alkali-soluble, linear, and branched heteropolymers of D-xylose, L-arabinose, D-mannose, D-glucose, D-galactose, and D-glucuronic acid. In addition, homopolymers of xylose, mannose, and galactose are known to occur (Rosenberg 1980). According to Sciamanna *et al.* (1977), pentoses comprise nearly one-third and hexoses nearly two-thirds of the reducing sugars obtained from lignocellulose hydrolysis. While a number of modern processes have been designed or proposed for the conversion of cellulose, starch, and hexose sugars to fuels and chemicals (Shuster 1978), less interest has been shown in bioconversion schemes involving pentose sugars, although they represent a third or more of the total carbohydrate content of biomass. Several researchers (Barnett 1976; Suomalainen and Oura 1971) have reported that yeasts are able to assimilate aldopentoses oxidatively but are unable to ferment them to ethanol.

It has previously been stated that species of the genus *Saccharomyces* are unable to utilize D-xylose. However, when strains of *S. cerevisiae* were tested on D-xylulose (D-threopentulose), an intermediate of D-xylose metabolism, surprisingly good growth was obtained. Catabolism of D-xylose, in yeasts that can utilize it, proceeds by intracellular conversion to xylitol, then to D-xylulose via mediation of the enzymes D-xylose reductase and xylitol dehydrogenase (Horecker 1962). D-Xylulose is subsequently phosphorylated by D-xylulose kinase to an intermediate of the pentose phosphate pathway. Although the pathway for this conversion has not been absolutely experimentally defined, it is believed (Mahler and Cordes 1974) that 6 moles of phosphorylated D-xylulose are converted to 5 moles of D-glucose-6-phosphate (G-6-P) and that each mole of G-6-P is in turn converted to 2 moles of $CO_2$ and 2 moles of ethanol via the Embden-Meyerhof-Parnas pathway. It has been shown that several yeasts (e.g., *Schizosaccharomyces pombe, S. cerevisiae,* and *Kluyveromyces lactis*) that do not grow aerobically on D-xylose will grow aerobically on D-xylulose. The effect is specific to D-xylulose in the sense that the other intermediate of D-xylose metabolism in yeast, xylitol, will not support growth of these strains. The data suggest that these strains contain at least one enzyme that allows entry of D-xylulose into a major metabolic pathway.

The nature of the enzymatic activity involved in D-xylulose utilization by species that do not assimilate D-xylose or xylitol has yet to be identified. Since sugar catabolism is thought to require an initial phosphorylation step, phosphorylation to produce D-xylulose-5-phosphate could be the critical reaction, with this ketose phosphate then entering the pentose phosphate pathway. One possibility is that D-xylulose is phosphorylated by a kinase whose normal substrate is some other sugar. The rapid commencement of growth on D-xylulose of glucose-grown cells is in accord with this view (Wang *et al.* 1980a,b). Alternate possibilities are that the strains contain a specific D-xylulose kinase that represents the remaining portion of a D-xylose-utilizing system they once possessed or that phosphorylation occurs during transport (van Steveninck 1972).

*Petite* mutants of *S. cerevisiae* are noteworthy because, in contrast to their isogenic parents, they do not grow on D-xylulose. This indicates that D-xylulose catabolism in this species involves mitochondria in some way, or respiration *per se*. The suggestion that the loss of the ability of petite mutants to utilize certain sugars is due to the involvement of respiring mitochondria in their transport (Wang and Schneider 1980) will be discussed in relation to maltose. Further, the need for oxygen in the transport of some sugars will be discussed (Wilkie and Evans 1982) in detail later in this chapter. However, it is unclear if such views are relevant to D-xylulose utilization by *S. cerevisiae* because it may not be transported by an active process.

Because of the inability of *S. cerevisiae* to metabolize D-xylose, methods have been sought to convert D-xylose to D-xylulose in a fermenter. To this end, a method has been developed (Schneider *et al.* 1981) that involves the inclusion of glucose isomerase in the medium, an enzyme that can isomerize D-xylose to D-xylulose. In this system Wang *et al.* (1980b) have employed 5% xylose, *Schiz. pombe,* and glucose isomerase, but the ethanol produced was only 10% of the theoretical. It is thought that this low ethanol yield may involve one or several of the following factors:

1. A drop in cell viability that occurred after 2–3 days.
2. A buildup of proteolytic activity in the medium, which inactivated the glucose isomerase despite the presence of excess soluble protein in the medium.
3. A buildup of Maillard reaction products, which could inhibit growth.
4. Denaturation of soluble glucose isomerase by the continual shaking of the culture.

The rate of ethanol production in 5% xylose plus glucose isomerase was only approximately 25% of that when 5% D-xylulose was employed. This difference was due to the relatively low concentration of D-xylulose present in D-xylose–glucose isomerase mixtures.

Chiang *et al.* (1981) have employed the technique of cell immobilization for ethanol production from pentoses. The results indicate a slow fermentation rate when pentose was used as a substrate. As expected, *S. cerevisiae* had a superior fermentation rate on D-xylulose and this system could be used to produce ethanol at high concentration from D-xylose by employing immobilized glucose isomerase and yeast cells in separate columns.

Several mycelial fungal systems that ferment D-xylose to ethanol (e.g., *Rhizopus* spp., *Fusarium* spp. and *Mucor* spp.) are not practical because of their slow rates of growth and conversion. Consequently, the use of a single organism such as yeast that could ferment both hexose and pentose to ethanol in high yields would be ideal. A mutant strain of *Candida* sp. XF217 was found to produce ethanol from D-xylose aerobically as well as anaerobically (Gong *et al.* 1981). The rate of ethanol production under aerobic conditions was greater, indicating an oxygen requirement for the uptake of D-xylose in strain XF217. Ethanol was also produced by XF217 when D-glucose, D-fructose, sucrose, or maltose were used as substrates. This D-xylose-fermenting yeast strain is a potential organism to employ for ethanol production from renewable biomass-derived hexoses and pentoses.

The yeast *Pachysolen tannophilus* has been found to produce ethanol rapidly and in good yield from D-xylose (Slininger *et al.* 1982; Schneider *et al.* 1981). This finding promises a potential solution to the problem of utilizing the pentoses obtained from lignocellulose hydrolysis and also disproves the belief that yeasts cannot convert aldopentoses to ethanol (Barnett 1976). If it is assumed that *Pa. tannophilus* can convert each mole of D-xylose to D-xylulose and that each mole of D-xylulose follows the mechanism hypothesized previously (Gong *et al.* 1981; Wang *et al.* 1980a), the overall conversion can be summarized as follows:

$$\underset{\text{D-xylose}}{C_5H_{10}O_5} \rightarrow \tfrac{5}{3}\, \underset{\text{ethanol}}{C_2H_5OH} + \tfrac{5}{3}\, CO_2$$

The theoretical ethanol yield based on this stoichiometry would be 0.51 g/g. However, Slininger *et al.* (1982) have reported that the ethanol yield was only 60% of this. Assuming that the yeast cell is composed of 50% carbon on a dry weight basis, 11% of the xylose was consumed to form biomass. This means that approximately 30% of the xylose was being used in cell maintenance and/or formation of metabolic products besides ethanol and carbon dioxide.

Slininger *et al.* (1982) have found that optimum ethanol production from D-xylose by *Pa. tannophilus* NRRLY-2460 was obtained when the incubation temperature was maintained at 32°C, when the medium pH was controlled at 2.5, when the substrate concentration was not higher than 50 g/liter and when the ethanol concentration did not exceed 20 g/liter. They have also reported on the ability of *Pa. tannophilus* to convert D-xylose to ethanol. They have found that allowing limited access of air to the medium enhances the rate of conversion, although this was not an absolute requirement.

Maleszka and Schneider (1982) have surveyed 15 yeast strains, which are classified as being able to ferment D-glucose and to assimilate D-xylose, for the ability to convert this pentose to ethanol. Included in this number were representatives of seven genera, different species within genera and variants within species. All grew aerobically on a defined medium consisting of yeast nitrogen base plus 4% D-xylose. However, some grew more rapidly on a rich medium consisting of yeast nitrogen base, casein amino acids, and 4% D-xylose. The data indicate that the conversion of D-xylose to ethanol by yeasts when air has access to the medium is probably a common phenomen, since it occurred in six of the seven genera tested. The data also indicate that D-xylose fermentability can vary among species within a genus and between variants of a single species. Thus there are many candidates from which to choose that would be optimal for a particular application. Demonstrations of the ability of two *Candida* species to ferment D-xylose aerobically have also been made by Jeffries (1981) and by Gong *et al.* (1981).

Detroy (1981) has studied the bioconversion of wheat straw cellulose and hemicellulose to ethanol by *S. uvarum* and *Pa. tannophilus*. Agricultural straw was subjected to thermal and alkali purifying prior to enzymatic saccharification. A sodium hydroxide extract yielded a 60% cellulosic pulp and a hemicellulose fraction available for fermentation to ethanol. The cellulosic pulp was subjected to cellulose hydrolysis at 55°C to provide hexoses for fermentation, and the hemicellulose was recovered from the liquor filtrates by acid–alcohol precipitation followed by acid hydrolysis to xylose for fermentation. Fermentation of these substrates was inhibited by substances introduced by thermal and alkali treatment of the straw.

As a result of a comparison of the relative fermentability of D-xylose, xylitol, and D-xylulose by a number of yeast strains, two relationships have become apparent. (1) The fermentabilities of D-xylose and D-xylulose appear to be inversely correlated; the strains that produced relatively more ethanol from D-xylose produced less from D-xylulose. (2) Xylitol is not converted to ethanol by any of the xylose-fermenting strains tested. Because of the nonfermentability of this pentitol, the problem of its formation as an unwanted by-product during D-xylose fermentation remains. It is only as a result of greater knowledge of the genetics and biochemistry of the catabolism of D-xylose, xylitol, and D-xylulose in *Pa. tannophilus* in particular and in yeasts in general that a solution to this problem (i.e., xylitol nonfermentability as well as rationalization of the relative fermentabilities of D-xylose and D-xylulose) will be achieved.

## Extracellular Enzymes

It has already been stated that the first step in the utilization of many sugars and carbohydrates by yeast is their "initial hydrolysis outside the membrane followed by entry into the cell by some or all of the hydrolytic products." This initial hydrolysis is facilitated by the presence of extracellular enzymes. These enzymes have been discussed in detail by Arnold (1981) and fall into two classes: first, those that cleave substrates that do not permeate the plasma membrane, and, second, those enzymes involved with the turnover of envelope polymers during growth. The first group includes the enzyme systems in question in this chapter. A consideration of the second group is beyond the scope of this chapter but includes both synthetic and degradative enzymes that have been sparingly studied; indeed, their existence in some cases is only being inferred from experiments with inhibitors.

Arnold (1981) has established certain criteria that can be assembled as hallmarks of this group of enzymes. Enzymes involved in the hydrolysis of sugars and carbohydrates in yeast that fulfill most if not all of these criteria include invertase ($\beta$-fructofuranosidase, E.C. 3.2.1.26), glucoamylase (E.C. 3.2.1.3), melibiase (E.C. 3.2.1.22), and $\alpha$-amylase (E.C. 3.2.1.1). Extracellular enzymes involved in the hydrolysis of noncarbohydrate material would include acid phosphatase (E.C. 3.1.3.2), catalase (E.C. 1.11.1.6), and phospholipase (E.C. 3.1.1.4).

## Sucrose

This disaccharide is first hydrolyzed by the extracellular enzyme invertase into constituent monosaccharides, which are then taken up into the yeast cell.

The specificity of the enzyme is absolute for the unsubstituted or unchanged β-fructofuranoside terminal:

$$\text{R-}\beta\text{-fructofuranoside} + H_2O \rightarrow \text{fructose} + \text{ROH}$$

Thus α-fructofuranosides, fructopyranosides, turanose, and melezitose are not acted upon, whereas methyl-β-D-fructofuranoside, sucrose, raffinose, stachyose, and verbascose are cleaved at the fructosidic linkage on the fructose side of the bridge oxygen. The bulk of the literature is based on *S. cerevisiae* as source microorganism with sucrose as the substrate, sucrose being the best substrate for the enzyme with the lowest $K_m$ and highest $V_{max}$.

In *S. cerevisiae,* the ability to hydrolyze sucrose is conferred by any one of six (or more) polymeric sucrose genes (denoted *SUC1* to *SUC6*) that reside at loci distributed throughout the genome. Strains unable to ferment sucrose can arise either by segregation during crosses of strains containing different *SUC* genes or by mutation of a known *SUC* gene. Segregational negatives are termed *suc0* to distinguish them from the mutational nonfermenters *suc1, suc2 . . . suc6* (Carlson *et al.* 1981).

The presence of any one of the *SUC* genes in the genome leads to the production of two forms of invertase: a large, secreted mannoprotein that is about 50% carbohydrate and has a molecular weight of 270,000 d and a smaller intracellular nonglycosylated form of 120,000 d molecular weight. The production of these forms is controlled by catabolite repression and can vary widely, with the level of the large enzyme varying over a much greater range than that of the small enzyme. Invertase biosynthesis in yeast is not inducible by its substrates (e.g., sucrose and raffinose), but is subject to repression by the glucose concentration in the culture medium. It has been shown (Rodriguez *et al.* 1981) that external invertase increases proportionately much more than does the internal invertase when cells are released from glucose repression.

The biosynthetic relationship between the two forms of invertase is not fully understood. Large invertase probably consists of two identical subunits and the small enzyme is also probably a dimer. There are antigenic and kinetic similarities between the large and small enzymes, but attempts to demonstrate definitively a precursor–product relationship have been inconclusive. Tunicamycin, which inhibits the synthesis of external glycoproteins (Kuo and Lampen 1974), was found to have no effect on the formation of the small invertase, whereas the large form of the enzyme was not produced in the presence of this antibiotic. Furthermore, there was no accumulation of invertase activity with the properties of the small internal invertase. Similar results were also obtained by Kuo and Lampen (1974) using spheroplasts.

Mutagenesis of the sucrose-fermenting *(SUC1) S. cerevisiae* strain 4059-358D yielded an invertase-negative mutant (D10) (Mizunaga *et al.* 1981). Although no mutant lacking only one form of invertase has been described, subsequent mutation tests of D10 gave a sucrose-fermenting invertase D10-ER1, which, under a variety of growth conditions, has a low level of small intracellular nonglycosylated invertase but a normal level of large invertase. It has been suggested that strain D10-ER1 carries a defect specific to a gene for small invertase. Limited genetic evidence indicated that these defects in D10 and D10-ER1 were allelic to the *SUC1* gene. The large invertases from D10-ER1 and 4059-358D have been purified and compared. The two enzymes have similar specific activity and $K_m$ for sucrose, cross-react immunologically, and show the same subunit molecular weight after removal of the carbohydrate with end-β-*N*-acetyl glucosaminidase H. However, the large invertase from strain D10-ER1 is rapidly inactivated at 55°C, whereas that from the parent strain is relatively stable at 65°C. The small invertase in extracts of D10-ER1 is also heat sensitive as compared to the small enzyme from the original parent strain. The low level of small invertase in mutant D10-ER1 may reflect increased intracellular degradation of this heat-labile form. In several crosses of D10-ER1 with strains carrying the *SUC1* or *SUC3* gene, the temperature sensitivity of the large and small invertases and the low cellular level of small invertase appeared to cosegregate. These findings are evidence that *SUC1* is a structural gene for invertase and that both large and small forms are encoded by a single gene.

The following model for the biosynthesis of invertase has been proposed by Rodriguez *et al.* (1981). The protein moiety for the external (large) enzyme is synthesized at the level of the rough endoplasmic reticulum. This protein carries a hydrophobic N-terminal peptide that permits its insertion into the lipid bilayer and its passage to the lumen of the rough endoplasmic reticulum. Glycosylation starts while the protein is still attached to the polysomes and continues in the smooth endoplasmic reticulum. The addition of sugar residues could be either sequential or due to the transfer of a relatively bulky oligosaccharide. This latter mechanism would explain the changes in molecular weight and the ratio of carbohydrate to protein between the membrane-bound and the external isoenzymes. As far as the cytoplasmic or small enzyme is concerned, its synthesis might occur in free polysomes that release the protein into the cytoplasm or into an organelle. The organelle could disrupt during preparation of a cell-free extract and would account for the cytoplasmic location of this invertase. In living cells, the enzyme would be degraded at a rate similar to that of its synthesis. The metabolic significance of this small invertase is obscure.

The invertase found in the osmotolerant yeast *S. rouxii* is "cryptic" in young cells but expressed in aged cells (Arnold 1974). The enzyme is still cryptic after the formation of spheroplasts and penetration of the cell membrane does not lead to the expression of the enzyme. Arnold and Garrison (1979) concluded that the cryptic invertase must be located either in the protoplasm or in some organelle such as the periplasmic body because membrane disruption was necessary to achieve enzyme expression.

## Melibiose

α-Galactosidase (α-D-galactoside galactohydrolase E.C. 3.2.1.22) or melibiase is an inducible enzyme, found in a number of *Saccharomyces* spp., which hydrolyzes the sugar melibiose to units of galactose and glucose. The enzyme is found mainly outside the cell membrane; however, an intracellular form of the enzyme also exists (Lazo *et al.* 1977). Unlike invertase, the internal α-galactosidase is the same molecular size as the external enzyme. Similar to most other extracellular enzymes in yeast, carbohydrate material has been found to be associated with the enzyme, indicating that α-galactosidase is a glycoprotein (Lazo *et al.* 1977) where the carbohydrate portion accounts for 57% of the molecule (Lazo *et al.* 1978). Furthermore, α-galactosidase is similar in molecular weight to the extracellular enzymes invertase and acid phosphatase, being approximately 300,000 d. When α-galactosidase was partly inactivated with 3 *M* urea, two proteins corresponding to the denatured enzyme resulted. The molecular weights of the two proteins were calculated to be 20,000 and 100,000 d (Lazo *et al.* 1978).

As previously discussed in this chapter, the existence of extracellular hydrolytic enzymes in yeast (or any other microorganism) can be attributed to the impermeability of the cell membrane to their corresponding substrates. One way to explain the induction of these enzymes is the appearance in the cytoplasm of hydrolysis products of the substrate resulting from the presence of constitutive levels of enzyme on the cell surface. In the case of α-galactosidase this possibility is in agreement with the fact that galactose (a part of the melibiose molecule) is able to induce its synthesis (Kew and Douglas 1976). Moreover, when α-galactosides were used as inducers, the level of enzyme correlated with the number of galactose residues in their molecule, thus supporting the notion that the galactose, not the glucose moiety, is responsible for the enzyme induction. Indeed, it is logical that galactose moiety is responsible for the induction, since the other half of the melibiose molecule, glucose, has been shown to be a good repressor of the enzyme. Furthermore, coordinate induction of α-galactosidase and galactose-1-phosphate uridylyltransferase (E.C. 2.7.7.10), an enzyme involved in the utilization of galactose, has been reported by Kew and Douglas (1976). Indeed, it is difficult to envisage how the induction mechanism can be triggered when melibiose is hydrolyzed at the cell surface, since an inducer, galactose, and a repressor, glucose, are the products of the reaction.

Lazo *et al.* (1981) have reported that constitutive levels of α-galactosidase can be detected in cells, but this activity is very clearly distinguishable from that of the cells growing in the presence of an inducer such as melibiose. The appearance of both enzymes (constitutive and induced) is efficiently repressed by glucose and fructose. A kinetic analysis of the repression has revealed both a transient and a permanent repression similar to that described for inducible enzymes in *Escherichia coli.*

The genetic control of α-galactosidase synthesis in *Saccharomyces* sp. has not been the subject of detailed study. One gene—a *MEL* gene—has been identified to code for the production of this enzyme, but the presence of other polymeric *MEL* genes cannot be precluded. Indeed, the taxonomic difference between the two brewing yeast species, *S. cerevisiae* (ale yeast) and *S. uvarum (carlsbergensis)* (lager yeast), is based on the fact that strains of *S. uvarum (carlsbergensis)* produce this extracellular enzyme, whereas strains of *S. cerevisiae* are not able to produce α-galactosidase (Lodder 1970).

## Dextrin and Starch

The enzymatic production of sugars from starch is one of the oldest processes practiced by mankind. The saccharification of starchy materials by malt, in Western countries, and by fungal koji, in the Orient, for beverage fermentations was practiced before there were written records. Fungal koji has been used for centuries in the Orient instead of malt for saccharifying starchy mashes in the production of the traditional alcoholic beverage saké.

In the traditional processes (e.g., brewing, potable distilled spirits, and saké) the production of nearly all of the fermentable substrate from starch occurs separately from the fermentation of the substrate. However, economic benefit could be achieved if all or a large part of the starch was directly utilized by the fermenting microorganism. Of the more than 400 yeast species that are currently recognized (Lodder 1970), approximately 100 contain strains capable of using starch as a carbon and energy source. As previously discussed, brewer's wort, in addition to the fermentable sugars sucrose, glucose, fructose, maltose, and maltotriose, contains unfermentable carbohydrates, collectively termed dextrins. As the wort is fermented by the brewing yeasts *S. cerevisiae* and *S. uvarum (carlsbergensis),* the wort sugars are utilized; as a result, the specific gravity falls. The attenuation limit, which is a fixed characteristic of the wort and the yeast strain, is the lowest specific gravity that can be attained. Occasionally, a beer with a specific gravity much lower than the usual attenuation limit is obtained; this phenomenon has been called *superattenuation.*

In search for the cause of superattenuation, Andrews and Gilliland (1952) isolated a yeast that could convert wort dextrins to fermentable sugars and also ferment the sugars produced. As a result of this "dextrinase" activity, it was defined as a new species of yeast—*S. diastaticus. S. diastaticus* metabolizes or assimilates glucose, fructose, galactose, maltose, sucrose, one-third of the raffinose molecule, dextrin, and starch. It is more heat resistant than most cultured or wild yeasts and survives 59°C temperatures for 10 min when suspended in water. This is about 8°C higher than the lethal temperature for ale yeasts (lager yeast strains are much less heat tolerant than ale strains) and is higher than the lethal temperatures for most wild yeasts tested under similar conditions.

During the last few years there has been considerable interest in the production of low-carbohydrate, or "light," beers. Although "light" (lite) beers have been popular only for the last several years, research into the techniques of low-carbohydrate beer brewing has been going on for many years. Various methods have been described—for example, Gluek (1948), Pauls and Quatz (1957), Gablinger (1968), Saletan (1968), Green (1969), Clayton (1970), and Sfat and Morton (1974). Emeis (1971) attempted to produce a palatable diabetic beer by hybridizing *S. diastaticus* with *S. uvarum (carlsbergensis)*. After repeated back crosses, a hybrid that produced a palatable product was achieved. However, the apparent degree of attenuation (i.e., the extent of fermentation) was not sufficient to meet the requirements of a diabetic beer, as defined in West Germany. Many of the previously mentioned processes involved the addition of fungal (non-*Saccharomyces* sp.) glucoamylase to the fermenter. However, this enzyme is not inactivated by pasteurization, which results in an increase in the concentration of glucose in the beer during storage.

Hopkins (1955) reported that the amylolytic enzyme of *S. diastaticus* was glucoamylase (syn. amyloglucosidase, α-1,4-glucan glucohydrolase) E.C. 3.2.1.3). Glucoamylase has been reported to have the capability of cleaving α-1,3, α-1,4, and α-1,6 glycosidic linkages and theoretically can completely hydrolyze starch to glucose (Marshall 1972). However, as will be described later, glucoamylase isolated and purified from strains of *S. diastaticus* has only been found to be capable of cleaving α-1,4 glycosidic linkages.

To date, six genes associated with glucoamylase activity have been identified: *DEX1*, *DEX2*, and *DEX3* (Erratt and Stewart 1981a) and *STA1*, *STA2*, and *STA3* (Tamaki 1978). Studies have been conducted in this laboratory to determine whether allelism exists between any of these genes. Both *DEX3* and STA3 were isolated from *S. diastaticus* strains that were brewing contaminants, and tetrad analysis has revealed that *DEX3* and *STA3* are allelic. Allelism studies with *STA1* have encountered serious difficulties. Although *STA1* haploid strains mated easily with *DEX1* and *DEX2* haploids and cell fusion was observed microscopically, the resulting diploids when dissected yielded four spores that on germination only possessed the characteristics and markers of the *STA1* strains. It is believed that although cell fusion occurred, nuclear fusion did not; therefore tetrad analysis could not be utilized to determine allelism. Indeed, spore stain studies (in collaboration with C. F. Robinow of the University of Western Ontario) have produced evidence that some of the spores in the asci are binucleate.

Studies on the *MAL5*-containing strain 645-1A (from the Yeast Genetic Stock Center of the University of California at Berkeley) have been limited. This strain exhibits very slow maltose fermentation ability. However, a haploid isolate containing *MAL5* (coded I-159) and a *MAL5/MAL5* diploid (coded I-

139) (both obtained from the Research Department of the Miller Brewing Company, Milwaukee, Wis.) were found to ferment maltose more rapidly than the parent strain (645-1A) from which they were both derived. It was observed that not only did these strains ferment maltose but that all three were capable of fermenting dextrin. This raised the possibility that either, rather than a *MAL* gene a *DEX* gene coding for the production of glucoamylase was present, or both a *MAL* gene and a *DEX* gene were present.

To ascertain which was correct, the *MAL5* strain I-159 was mated with a strain incapable of fermenting either maltose or dextrin. Tetrad analysis revealed a 2:2 segregation, indicating that only a single gene was present and that this gene was responsible for both maltose and dextrin fermentation. Allelism studies with *DEX1, DEX2,* and *DEX3* would indicate that *MAL5* and *DEX2* appear to be allelic. It is interesting to note that the original *MAL5*-containing strain was isolated by Gilliland (1954) from a *S. diastaticus* strain and was described at that time as governing slow or delayed fermentation of maltose.

During the course of this laboratory's investigation into dextrin-utilizing yeasts, two different types of dextrin material have been employed, a commercially obtained preparation (produced from the hydrolysis of cornstarch) and the dextrin remaining in fermented beer (freeze-dried beer). The commercial dextrin was almost totally composed of larger dextrins, whereas the freeze-dried beer contained significant amounts of G4 (maltotetrose), G5 (maltopentose), G6, G7, and larger dextrins.

Glucoamylase hydrolyzes successive glucose units from the nonreducing ends of starch molecules, and as a consequence glucose is the sole hydrolysis product. Glucoamylase from *S. diastaticus* is an extracellular enzyme and the enzyme located in the culture medium has been purified from a strain containing a single *DEX1* gene. Two mannoproteins have been isolated with nearly identical molecular weights ($\sim$ 312,000 d). However, their ratios of carbohydrate to protein differed, as did their rates of carbohydrate hydrolysis, comparing a number of substrates (Erratt and Stewart 1981a). Nevertheless, the relative order of hydrolysis rates of carbohydrate to carbohydrate remained constant, with the corn dextrin having the fastest rate, followed by freeze-dried beer, soluble starch, and maltose. Neither enzyme possessed significant activity toward the hydrolysis of the $\alpha$-1,6 bond of isomaltose or pullulan. The temperature optimum was 50°C and the pH optimum at 40°C was 5.2. When maltose was used as the substrate the $K_m$ was $1.0 \times 10^{-2}$ *M*.

Searle and Tubb (1981) have attempted to determine the factors that are important in regulating the synthesis and excretion of glucoamylase by *S. diastaticus* strains in brewer's wort. The wort dextrin was hydrolyzed most effectively when the glucoamylase was produced by the yeast early in the fermentation, but under these conditions the yeast failed to ferment the wort completely. This situation presents a problem if strains that can excrete glucoamylase are to be used for the production of low-carbohydrate beer. However, it has been found that lack of contact with oxygen or unsaturated lipid will induce early glucoamylase production (Searle and Kirsop 1979). Consequently, failure to ferment wort fully could be avoided by provision of air or oxygen to the wort late in the fermentation.

Attempts, in this and other laboratories, to produce palatable beer with strains of *S. diastaticus* have proved unsuccessful (Erratt and Stewart 1981b; Tubb *et al.* 1981; Freeman 1981). Whereas a trained taste panel would rate a beer produced by a brewer's yeast strain that possessed no dextrin fermentative ability as being "full-bodied, fresh, smooth and well-balanced," the beers produced from polyploid and haploid *S. diastaticus* strains varied in taste but were generally described as "medicinal and phenolic" (Erratt and Stewart 1981b). The source of this phenolic off-flavor has been found to be 4-vinyl guaicol (4-VG), which is also called 2-methoxy-4-vinyl phenol. Wackerbauer *et al.* (1977) found that brewing strains that produce 4-VG are rare, but 50% of all non-brewing strains produced beers with phenolic off-flavors due to the formation of 4-VG.

4-VG is produced as a result of the enzymatic decarboxylation of ferulic acid (a constituent of most brewer's worts derived from malt and hops). Tubb *et al.* (1981) have reported that a single dominant nuclear gene, designated *POF* for *p*henolic *o*ff-*f*lavor, has been found to code for the ferulic acid decarboxylating enzyme. In this laboratory a survey of yeast strains has found that brewing strains (dextrin negative) did not produce 4-VG, whereas all polyploid *S. diastaticus* and haploid and diploid *DEX1*-, *DEX2*-, and *DEX3*-containing strains tested produced 4-VG to varying degrees (Russell *et al.* 1983). It was therefore critical to produce yeast strains that contained *DEX* genes and that did not contain the gene for POF and to employ such strains as hybridization and/or fusion partners in order to produce brewing yeast strains that were able to ferment wort dextrin material. Thus, a haploid that was *DEX* positive and that carried the *POF* characteristic was mated with a *dex, pof* haploid. The resulting diploid fermented dextrin and decarboxylated ferulic acid. Upon tetrad dissection of asci from this diploid both *DEX* and *POF* exhibited a 2:2 segregation. In addition, *DEX* and *POF* segregated independently of each other, making it possible to select haploids that were *DEX* positive and that lacked *POF*. As a result of the studies discussed earlier the "construction" of a brewing yeast strain with the ability to produce glucoamylase and thus ferment wort to produce a palatable low-carbohydrate beer is now a reality.

As glucoamylase from *S. diastaticus* cannot hydrolyze the $\alpha$-1,6 glycosidic linkages of amylopectin, the enzyme cannot completely degrade starch. Therefore, an $\alpha$-amylase ($\alpha$-1,4-glucan glucohydrolase, E.C. 3.2.1.1) that hydrolyzes random $\alpha$-1,4 glycosidic linkages throughout the starch molecule, or a debranching enzyme such as pullulanase (amylopectin-1,6-glucohydrolase, E.C. 3.2.1.9) that hydrolyzes the $\alpha$-1,6 glycosidic linkages, could be essential for complete hydrolysis of starch.

A number of yeasts have been reported to produce $\alpha$-amylase. These include *Saccharomycopsis fibuligera* [formerly *Endomycopsis fibuligera* (Fukumoto *et al.* 1960)], *Saccharomycopsis capsularis* (Ebertova 1966), *Lipomyces starkeyi*

(Moulin and Galzy 1977), *Lipomyces kononenkoae* (Spencer-Martins and van Uden 1979), *Pichia burtonii* and *Schwanniomyces castellii* (Oteng-Gyang *et al.* 1980).

*Saccharomycopsis fibuligera* has been found to produce both α-amylase and glucoamylase (Hattori and Takeuchi 1961; Sills and Stewart 1982a); however, very little debranching activity has been detected. It would appear that glucose has a negative effect on α-amylase activity when the former is present at concentrations above $5.5 \times 10^{-4}$ *M*. On the other hand, glucoamylase activity does not appear to be repressed by glucose. Both α-amylase and glucoamylase have been isolated and purified from *Schw. alluvius* (Wilson and Ingledew 1982). α-Amylase had an optimum pH of 6.3 and was stable from pH 4.5 to 7.5. The enzyme had an optimum temperature of 40°C and a $K_m$ for soluble starch of 0.364 mg/ml. Glucoamylase had a pH optimum of 5.0 and was stable from pH 4.0 to 8.0 with an optimum temperature of 50°C and a $K_m$ value of 12.7 mg/ml for soluble starch and 0.72 m*M* for maltose. *Schw. occidentalis* also produced α-amylase and glucoamylase (Sills and Stewart 1982a), with the α-amylase only being produced when the glucose content of the medium was low. However, glucoamylase activity was detected early in the fermentation and did not appear to be subject to glucose repression. In addition, debranching activity was also present in the culture medium after 24 h of incubation with subsequent diminishing activity as the growth cycle progressed. With *Schw. castellii* similar patterns of glucoamylase and debranching activities of that of *Schw. occidentalis* have been found; however, α-amylase demonstrated three times the activity of *Schw. occidentalis*.

The effect of glucose and 2-deoxy-glucose (DOG) on α-amylase activity in resting cells of *Schw. castellii* has been investigated (Sills and Stewart 1982b). As DOG is an analogue of glucose that is not metabolized by yeast cells, it can imitate the glucose effect for longer periods than glucose, which eventuallly becomes metabolized (van Uden *et al.* 1980). In both glucose and DOG treated and untreated cultures, α-amylase activity could be detected at the time of inoculation as a result of carryover from the previous stage. Addition of glucose caused a drop in α-amylase activity; a more severe drop occurred with added DOG. The low activity of catabolite-sensitive enzymes in the presence of glucose (Mahler *et al.* 1981) has been correlated with a reduced content of cyclic AMP (cAMP) in the cells, and externally added cAMP is able to overcome repression of these enzymes. It has been found that addition of cAMP together with DOG diminished the repressive effect of DOG on α-amylase activity in *Schw. castellii*. The amylases produced by *Lipomyces starkeyi* and *Lipomyces kononenkoae* have been studied by van Uden and his colleages (Spencer-Martins and van Uden 1977, 1979; van Uden *et al.* 1980; Sá-Correia and van Uden 1981). *L. konoenkoae* produced three extracellular amylases: an α-amylase, a glucoamylase, and a debranching transferase.

Interest in α-amylase-producing yeasts has centered on their possible industrial use either for direct fermentation of starch into ethanol (both potable and fuel) or for the production of yeast single-cell protein (SCP) from starch. The preference for yeasts, rather than other fungi, bacteria, or algae for SCP, is partly due to a cultural bias and partly due to a consequence of the good nutritional quality of yeast proteins (Schrimshaw 1975; Cooney *et al.* 1980). Unfortunately, yeasts that produce appreciable quantities of α-amylase are usually near-obligate aerobes and consequently do not produce significant amounts of ethanol. In an attempt to produce a facultative anaerobic yeast with α-amylase activity, Wilson *et al.* (1982) have attempted to fuse a strain of *Schw. alluvius* with a strain of *S. uvarum.* Although nuclear fusion (karyogamy) appears to have occurred between the two organisms, the fusion product was unstable and spontaneously segregated to the original parent cultures. It is possible that the genetic diversity between these two genera (i.e., *Schwanniomyces* and *Saccharomyces*) may be too great to allow stable coexistence of the two genomes within a single nucleus. It would appear that the use of plasmid vectors to transfer α-amylase genes into a fermentative yeast may be a route to generate ethanol-producing yeast species with α-amylase activity.

## Mitochondrial Control of Sugar Utilization

Wilkie and co-workers (e.g., Evans and Wilkie 1976; Mahler and Wilkie 1978; Wilkie and Nudd 1981) have studied the relationship between the functional state of a yeast cell's mitochondria and the culture's ability to utilize one or more of the sugars galactose, maltose, and α-methyl-D-glucoside. They have found that when a number of wild-type strains of *S. cerevisiae* (all capable of utilizing the sugars galactose, maltose, and α-methyl-D-glucoside for growth) were converted by ethidium bromide mutagenesis to stable cytoplasmic *petite* ($rho^-$) mutants, the latter lost the ability to grow on one or more of these sugars. Initial interpretations of these data (Evans and Wilkie 1978) suggested that sugar utilization was at least partially under the control of the mitochondrial genome. A later study (Mahler and Wilkie 1978) was undertaken to test this hypothesis with two types of experiments: (1) a comparison of the patterns of sugar utilization by a number of *petite* mutants of the same strain that differed in the extent of their retention of the mitochondrial genome and (2) a comparison of these patterns with those of *mit* mutants, which differ in the location of the genetic lesion on their mitochondrial DNA but in all of which this lesion probably does not extend beyond the confines of a single gene.

The results from these experiments suggest that the functional state of the mitochondria can affect nuclear genes or their products. For example, small errors in the assembly of the inner mitochondrial membrane and the failure to produce or insert a single mitochondrially coded polypeptide into cytochrome oxidase are sufficient to generate the α-methyl-D-glucoside negative characteristic of many respiratory-deficient mutants of *Saccharomyces* species.

Genetic analysis of strain differences (Evans and Wilkie 1976) between *petite* and *grande* cultures have indicated that mitochondrial changes alter the activity or expression of nuclear genes involved in the uptake and use of sugars.

It has also been found that a *petite* mutant unable to use galactose cannot be induced to synthesize galactokinase, although the parent *grande* strain does produce the enzyme in the presence of the sugar. However, pretreatment of *petite* cells with dimethylsulfoxide (DMSO), followed by exposure to the sugar, restored inducibility. Since the DMSO makes membranes permeable, it has been concluded that a defect in permeation contributes to the *petite* cell's inability to use the sugar. It is not clear how mitochondria influence the nuclear genes, but the effect of the abberation in *petites* does not appear to be due to respiratory incapacity per se. All *petites* are equally unable to produce ATP by oxidative phosphorylation but can differ widely in their use of sugar. Also, the addition of small amounts of glucose to the culture medium does not restore ability to use other sugars. Wilkie and co-workers (Wilkie and Nudd 1981; Spencer *et al.* 1981) have shown that mitochondrial aberrations in yeast usually result in alterations in cell surface characteristics that can be detected in a number of ways other than sugar uptake. These include changes in (1) agglutinability of cells with concanavalin A, (2) cellular tolerance to drugs, (3) secretion of amylase, (4) degree of flocculence in flocculating strains, (5) cell morphology, and (6) movement of cells in a biphasic polymer system.

Wilkie and Evans (1982) are of the opinion that the cell surface changes that occur in *petite* yeast cultures are analogous to those observed in neoplastic transformation in mammalian cells. The results from the yeast experiments described above have been extrapolated to animal cells and a connection between mitochondrial abnormalities and cell surface change postulated. The striking similarity between yeast and mammalian mitochondria in many fundamental respects justify, in the view of Wilkie and Evans (1982), the construction of a model of chemical carcinogenesis based on mitochondrial mutagenesis as a primary step. Genetic lesions would result in alteration of the organelle's inner membrane, leading to modulation of the activity of certain nuclear genes involved in cell surface biogenesis. An important factor of the model is a requirement for cell division in order for segregation of mutant mitochondria to occur. This would correspond to the well-documented "promotion" phase of neoplastic transformation. Obviously there is more than one route to the neoplastic state and the mitochondrial model is open to further development and testing, not only in the amenable yeast cell but also through mammalian somatic cell genetics.

## References

Andrews, J., Gilliland, R. B. (1952). J. Inst. Brew. 58, 189.

Arnold, W. N. (1974). J. Bacteriol. 120, 886.

Arnold, W. N. (1981). Yeast Cell Envelopes: Biochemistry, Biophysics, and Ultrastudies. Boca Raton, Fl.: CRC Press, Inc.

Arnold, W. N., Garrison, R. G. (1979). J. Biol. Chem. 254:4919.

Barnett, J. A. (1976). Adv. Carbohydr. Chem. Biochem. 34:125.

Barnett, J. A. (1981). Adv. Carbohydr. Chem. Biochem. 39:347.
Bassel, J., Douglas, H. C. (1968). J. Bacteriol. 95:1103.
Broach, J. R. (1979). J. Mol. Biol. 131:41.
Cardini, C. E., Leloir, L. F. (1953). Arch. Biochem. Biophys. 45:55.
Carlson, M., Osmond, B. C., Botstein, D. (1981). Genetics 98:1.
Chiang, L-C., Hsiao, H.-Y., Ueng, P. R., Chen, L. -F., Tsao, G. T. (1981). Biotechnol. Bioeng. Symp. No. 11, 263.
Clayton, D. H. (1970). U.K. Pat. 1,183,772.
Cooney, C. L., Chokyun, Rha, Tannenbaum, S. R. (1980). Advan. Food Res 26:1.
Detroy, R. W. (1981). In: Trends in the Biology of Fermentations for Fuels and Chemicals, edited by A. Hollaender, New York: Plenum Press.
Douglas, H. C. (1961). Biochim. Biophys. Acta 52:209.
Douglas, H. C., Hawthorne, D. C. (1966). Genetics 54:911.
Ebertova, H. (1966). Folia Microbiol. (Prague) 11:14.
Emeis, C. C. (1971). Amer. Soc. Brew. Chem. Proc. 58.
Erratt, J. A., Stewart, G. G. (1981a). In: Current Developments in Yeast Research, edited by G. G. Stewart, I. Russell. Toronto: Pergamon Press.
Erratt, J. A., Stewart, G. G. (1981b). Dev. Ind. Microbiology 22:577.
Evans, I. H., Wilkie, D. (1976). In: The Genetic Function of Mitochondrial DNA, edited by C. Saccone, A. M. Kroon. Amsterdam: Elsevier North-Holland.
Evans, I. H., Wilkie, D. (1978). Genet. Res. 32:171.
Federoff, H. J., Cohen, J. D., Eccleshall, T. R., Needleman, R. B., Buchferer, B. A., Giacalone, J., Marmur, J. (1982). J. Bacteriol. 149:1064.
Freeman, R. F. (1981). Proc. 18th Congress Eur. Brew. Conv. London: IRL Press Ltd.
Fukomoto, J., Tsujisaka, Y., Araki, M. (1960). Kagaku To Kogyo 34, 423 (Chem. Abs. 55:24917 (1961)).
Gablinger, H. (1968). U.S. Pat. 3,379,534.
Gilliland, R. B. (1954). Nature 173:409.
Gluek, A. C. (1948). U.S. Pat. 2,442,806.
Gong, C.-S., Ladisch, M. R., Tsao, G. T. (1981). Biotechnol. Lett. 3:657.
Green, J. W. (1969). U.K. Pat. 1,174,618.
Hartwell, L. H. (1970). Ann. Rev. Genet. 4:373.
Hattori, Y., Takeuchi, J. (1961). Rika Gaku Kenkyusho Hokoku 37:37.
Hautera, P., Lovgren, I. (1975). J. Inst. Brew. 81:309.
Hopkins, R. H. (1955). Proc. 5th Congress Eur. Brew. Conv. Elsevier, Amsterdam.
Hopper, J. E., Broach, J. R., Rowe, L. B. (1978). Proc. Nat. Acad. Sci., U.S.A., 75:2878.
Horecker, B. L. (1962). Pentose Metabolism in Bacteria. New York: Wiley.
Jeffries, T. W. (1981). Biotechnol. Bioeng. Symp. No. 11:315.
Kew, O. M., Douglas, H. C. (1976). J. Bacteriol. 125:33.
Khan, N. A., Zimmerman, F. K., Eaton, N. R. (1973). Molec. Gen. Genet. 124:365.
Klar, A. J., Halvorson, H. O. (1974). Molec. Gen. Genet. 135:203.
Kuo, S. C., Lampen, J. O. (1974). Biochem. Biophys. Res. Commun. 58:287.
Lazo, P. S., Ochoa, A. G., Gascon, S. (1977). Eur. J. Biochem. 77:375.
Lazo, P. S., Ochoa, A. G., Gascon, S. (1978). Arch. Biochem. Biophys. 191:316.
Lazo, P. S., Florez, I. G., Ochoa, A. G., Gascon, S. (1981). Cell. Mol. Biol. 27:615.
Lodder, J. (1970). The Yeasts: A Taxonomic Study. Amsterdam: North-Holland.
Mahler, H. R., Cordes, E. H. (1974). Biological Chemistry. New York: Harper and Row.

Mahler, H. R., Wilkie, D. (1978). Plasmid 1:125.
Mahler, H. R., Jaynes, P. K., McDonough, J. P., Hanson, D. K. (1981). Curr. Top. Cell. Reg. 18:455.
Maleszka, R., Schneider, H. (1982). Can. J. Microbiol. 28:360.
Marshall, J. J. (1972). Wall. Lab. Commun. 35:49.
Matsumoto, K., Toh-e, A., Oshima, T. (1978). J. Bacteriol. 134:446.
Matsumoto, K., Adachi, T., Toh-e, A., Oshima, T. (1980). J. Bact. 141:508.
Mizunaga, T., Tkacz, J. S., Rodrigues, L., Hackel, R. A., Lampen, J. O. (1981). Mol. Cell. Biol. 1:460.
Mortimer, R. K., Schild, D. (1980). Microbiol. Rev. 44:519.
Moulin, G., Galzy, P. (1977). Agric. Biol. Chem. 43:1165.
Mowshowitz, D. B. (1979). J. Bacteriol. 137:1200.
Mowshowitz, D. B. (1981). Genetics 98:713.
Naumov, G. I. (1971). Genetika 7:141.
Naumov, G. I. (1976). Genetika 12:87.
Nogi, Y., Matsumoto, K., Toh-e, A., Oshima, Y. (1977). Molec. Gen. Genet 152:137.
Odds, F. (1979). Candida and Candidosis. Leicester: Leicester University Press.
Oteng-Gyang, K., Moulin, G., Galzy, P. (1980). Acta. Microbiol. Acad. Sci. Hung. 21:155.
Pauls, I. F. F. E., Quatz, R. R. A. (1957). U.S. Pat. 2,783,147.
Perlman, D., Hopper, J. E. (1979). Cell 16:89.
Phaff, H. J., Miller, M. W., Mrak, E. M. (1978). The Life of Yeasts. Cambridge, Mass.: Harvard University Press.
Rodriquez, L., Lampen, J. O., MacKay, V. L. (1981). Mol. Cell. Biol. 1:469.
Rose, A. H. (1977). In: Alcohol Industry and Research, edited by O. Forsander, K. Eriksson, E. Oura, P. Jounela-Eriksson. Helsinki: Alkon Keskuslaboratorio.
Rosenberg, S. L. (1980). Enzyme Microb. Technol. 2:185.
Rudert, F., Halvorson, H. O. (1963). Bull. Res. Council of Israel 11A4:377.
Russell, I., Hancock, I., Stewart, G. G. (1983). J. Amer. Soc. Brew. Chem., in press.
Sá-Correia, I., van Uden, N. (1981). Eur. J. Appl. Microbiol. Biotechnol. 13:24.
Saletan, L. T. (1968). Wall Lab. Commun. 31:33.
Schneider, H., Wang, P. T., Johnson, B. F., Shopsis, C. (1981), In: Current Developments in Yeast Research, edited by G. G. Stewart, I. Russell. Toronto: Pergamon Press.
Sciamanna, A. F., Freitas, R. P., Wilkie, C. R. (1977). Composition and Utilization of Cellulose for Chemicals from Agricultural Residues, Lawrence Berkeley Publication 5966, Berkeley, Calif.
Schrimshaw, N. S.: In: Single-Cell Protein II, edited by S. R. Tannenbaum, D. I. C. Wang. Cambridge, Mass.: MIT Press.
Searle, B. A., Tubb, R. S. (1981). J. Inst. Brew. 87:244.
Searle, B. A., Kirsop, B. H. (1979). J. Inst. Brew. 85:342.
Serrano, R. (1977). Eur. J. Biochem. 80:97.
Sfat, M. R., Morton, B. J. (1974). U.K. Pat. 1,355,698.
Shuster, W. W. (ed.) (1978). Proc. Second Annual Fuels from Biomass Symp.
Sills, A. M., Stewart, G. G. (1982a). J. Inst. Brew. 88:325.
Sills, A. M., Stewart, G. G. (1982b). Dev. Ind. Microbiol. 24, in press.
Slininger, P. J., Bothast, R. J., van Cauwenberge, J. E., Kurtzman, C. P. (1982). Biotechnol. Bioeng. 24, 371.
Spencer-Martins, I., van Uden, N. (1977). Eur. J. Appl. Microbiol. 4:29.
Spencer-Martins, I., van Uden, N. (1979). Eur. J. Appl. Microbiol. Biotechnol. 6:241.

Spencer, J. F. T., Miller, R., Spencer, D. M., Wilkie, D. (1981). In: Current Developments in Yeast Research, edited by G. G. Stewart, I. Russell, Toronto: Pergamon Press.

St. John, T. P., and Davis, R. W. (1979). Cell 16:443.

Stewart, G. G., Erratt, J., Garrison, I., Goring, T., Hancock, I. (1979). MBAA Tech. Quart. 16:1.

Stewart, G. G. (1981). Can. J. Microbiol. 27:973.

Suomalainen, H., Oura, E. (1971). In: The Yeasts, edited by A. H. Rose, J. S. Harrison. New York: Academic Press.

Tamaki, H. (1978). Molec. Gen. Genet. 164:205.

Tauro, P., Halvorson, H. O. (1966). J. Bacteriol. 92:652.

ten Berge, A. M. A., Zoutewelle, G., van de Poll, K. W. (1973a). Molec. Gen. Genet. 123:233.

ten Berge, A. M. A., Zoutewelle, G., van de Poll, K. W. (1973b). Molec. Gen. Genet. 125:139.

Tubb, R. S., Searle, B. A., Goodey, A. R., Brown, A. J. P. (1981). Proc. 18th Congress Eur. Brew. Conv. London: IRL Press Ltd.

van Steveninck, J. (1972). Biochim. Biophys. Acta. 274:575.

van Uden, N., Cabeca-Silva, C., Madeira-Lopes, A., Spencer-Martins, I. (1980). Biotechnol. Bioeng. 22:651.

Wackerbauer, K., Kossa, T., Tressl, R., (1977): Proc. 16th Congress Eur. Brew. Conv. Amsterdam: Elsevier.

Wang, P. Y., Schneider, H. (1980). Can. J. Microbiol. 26:1165.

Wang, P. Y., Shopsis, C., Schneider, H. (1980a). Biochem. Biophys. Research Commun. 9:248.

Wang, P. Y., Johnson, B. F., Schneider, H. (1980b). Biotechnol. Lett. 2:273.

Wilkie, D., Evans, I. (1982). Trends Biochem. Sci. 7:147.

Wilkie, D., Nudd, R. C. (1981). In: Current Developments in Yeast Research, edited by G. G. Stewart, I. Russell, Toronto: Pergamon Press.

Wilson, J. J., Ingledew, W. M. (1982). Appl. Envir. Microbiol. 44:172.

Wilson, J. J., Khachatourians, G. G., Ingledew, W. M. (1982). Molec. Gen. Genet., in press.

Zimmerman, F. K., Eaton, N. R. (1974). Molec. Gen. Genet. 134:261.

# Author Index

Numbers in italics indicate the page on which the complete reference is listed.

# Subject Index

# Species Index

## Yeasts

## Filamentous Fungi



## Bacteria

## Plants



## Birds



## Other Animals



## Algae



## Protozoa



## Slime Molds

Made in the USA
Las Vegas, NV
08 November 2024

11381209R00308